Bob S. ♡

Boone, IA

Y0-BBD-798

Best regards &
best wishes.
Gary Harman

Trichoderma and *Gliocladium*

Volume 2

Trichoderma and *Gliocladium* Volume 2

Enzymes, biological control and commercial applications

Edited by

GARY E. HARMAN
Cornell University, Geneva, NY, USA

and

CHRISTIAN P. KUBICEK
University of Technology, Vienna, Austria

with the capable editorial assistance of Kristen L. Ondik

TAYLOR & FRANCIS
ALERE FLAMMAM
1798 – 1998

UK Taylor & Francis Ltd, 1 Gunpowder Square, London, EC4A 3DE
USA Taylor & Francis Inc., 1900 Frost Road, Suite 101, Bristol, PA 19007

British Library Cataloguing in Publication Data

A catalogue record for this book is available from the British Library

ISBN 0-7484-0572-0 Vol. 1
ISBN 0-7484-0805-3 Vol. 2
ISBN 0-7484-0806-1 Two volume set

Library of Congress Cataloging in Publication Data are available

Cover design by Youngs Design in Production
Typeset in Times 10/12 pt by Santype International Ltd, Salisbury, UK
Printed by T. J. International Ltd, Padstow, UK

Contents

Preface

In the course of an investigation of damping off of Citrus seedlings, a strain of *Trichoderma lignorum* has been found to parasitize a number of pathogenic soil fungi in culture, i.e., *Rhizoctonia solani*, *Phytophthora parasitica*, *Pythium* spp., *Rhizopus* spp., and *Sclerotium rolfsii*.

The inhibition and death of the host hyphae were brought about in two ways: (a) in aerial hyphae by close contact or by coiling of *Trichoderma* around them; (b) in submerged hyphae by action at a little distance more frequently than by contact.

Weindling, R. 1932. *Trichoderma lignorum* as a parasite of other soil fungi.
Phytopathology 22: 837–845.

The above quotation is from the summary of one of Richard Weindling's landmark papers on the potential of *Trichoderma* spp. to control plant pathogenic fungi. This series of papers was among the first indications that these fungi have a wide-ranging commercial potential that is only now being realized.

These fungi are among the most widely distributed and common fungi in nature. They exist in climates ranging from the tundra to the tropics and colonize a wide range of habitats. They very commonly occur in soil and frequently are among the most prevalent fungi there, often occurring at levels of 100 to 10 000 propagules per gram of soil. They also exist in other environments and are frequently found in decaying wood or other cellulosic substrates such as decaying cotton fabrics in the tropics. Further, they may be found on aerial, as well as subterranean, plant parts.

They have a number of mechanisms for survival and proliferation, including physical attack of other fungi and degradation and utilization of complex carbohydrates. For the most part they are beneficial to man's economic interests, especially since they no doubt reduce plant diseases in nearly all agricultural environments. However, there are exceptions: for example, as noted in the Preface to Volume 1, some strains degrade cotton fabrics while other specific strains are pathogenic on commercial mushrooms.

The mechanisms which these remarkable fungi have evolved can be exploited for commercial applications, and many of these are documented in this volume. First and foremost, they are extremely efficient producers of a number of enzymes. Such enzymes include cellulases and hemicellulases that can be used for many purposes, including fabric treatments, enhancing nutritional composition of animal feeds, and improving properties and enhancing yields of plant oils and fruit juices. They also produce quantities of other enzymes, especially chitinases and glucanases, that are strongly fungitoxic. There is substantial potential for use of these enzymes as antifungal theraputants in numerous applications. Not only are these useful enzymes produced in large quantities, but also there are multiple forms of each class of enzyme. These multiple forms are nearly all synergistic and together form extremely potent enzyme mixtures for various commercial applications.

These fungi not only produce potent antifungal enzymes but also have evolved complex systems of parasitism of other fungi and produce several antibiotic compounds. It was this combination of properties that Weindling began to study in the 1920s and 1930s. Finally, sixty years later, the abilities of these fungi are beginning to be exploited for plant disease control in commercial agriculture. The first successful products are now on the market and should provide potent tools for integrated plant disease management. Not only do strains of *Trichoderma* control disease, but they also appear to enhance plant growth by mechanisms that only now are being explored and exploited.

Finally, there has been very substantial progress in the development of molecular techniques involving these fungi. The fungi themselves have been modified to very efficiently produce industrial enzymes. Many genes themselves have been cloned and are being used for a variety of purposes. They have been placed under the control of highly effective promoters, also obtained from *Trichoderma*, and used for high-efficiency production of various enzymes. They also have been placed in plants where they are efficiently expressed and result in plants with a high level of resistance to several plant pathogenic fungi.

Clearly, therefore, these fungi have substantial potential for a variety of commercial applications. As we further understand their genetic and physiological abilities, we will be able to use these fungi, and their genes and gene products, in a variety of useful ways. It has taken sixty years to progress from Weindling's initial studies to the present early commercial use of these fungi. The next decade should see very dramatic progress in understanding these fungi, as well as their utilization for a wide variety of applications useful to mankind.

The Editors

List of Contributors

EDUARDO AGOSIN
Department of Chemical and Bioprocess Engineering
School of Engineering
Pontificia Universidad Católica de Chile
Casilla 306-22
Santiago
Chile

JOSÉ MIGUEL AGUILERA
Department of Chemical and Bioprocess Engineering
School of Engineering
Pontificia Universidad Católica de Chile
Casilla 306-22
Santiago
Chile

B. A. BAILEY
Biocontrol of Plant Diseases Laboratory
USDA
ARS
Beltsville
MD 20705
USA

NICOLE BENHAMOU
Recherche en Sciences de la vie et de la santé
Pavillon C. E. Marchand
Université Laval
Sainte-Foy
Québec
Canada

TAHÍA BENÍTEZ
Departamento de Genética
Facultad de Biología
Universidad de Sevilla
E-41080 Sevilla
Spain

PETER BIELY
Institute of Chemistry
Slovak Academy of Sciences
842 38 Bratislava
Slovakia

T. BJÖRKMAN
Department of Plant Pathology
Cornell University
Geneva
NY 14456
USA

J. BUCHERT
VTT Biotechnology and Food Research
PO Box 1500
FIN-02044
Finland

ILAN CHET
Otto Warburg Center for Agricultural Biotechnology
The Hebrew University of Jerusalem
Faculty of Agriculture
Rehovot 76-100
Israel

ALBERTO DE CONTI
Laboratory of Biotechnology
Central R&D
Lamberti s.p.a.
Albizzate (VA)
Italy

JESÚS DELGADO-JARANA
Departamento de Genética
Facultad de Biología
Universidad de Sevilla
E-41080 Sevilla
Spain

YVES M. GALANTE
Laboratory of Biotechnology
Central R&D
Lamberti s.p.a.
Albizzate (VA)
Italy

SHOSHAN HARAN
Otto Warburg Center for Agricultural Biotechnology
The Hebrew University of Jerusalem
Faculty of Agriculture
Rehovot 76-100
Israel

G. E. HARMAN
Departments of Horticultural Sciences and Plant Pathology
Cornell University
Geneva
NY 14456
USA

LINDA HJELJORD
Agricultural University of Norway
Department of Biotechnological Sciences
N-1432
Aas
Norway

C. R. HOWELL
USDA-ARS
Cotton Pathology Research Unit
College Station
TX 77845
USA

A. KOIVULA
VTT Biotechnology and Food Research
PO Box 1500
FIN-02044
Finland

CHRISTIAN P. KUBICEK
Institut für Biochemische Technologie und Mikrobiologie
TU Wien
Getreidemarkt 9/172.5
A-1060 Wien
Austria

M. LINDER
VTT Biotechnology and Food Research
PO Box 1500
FIN-02044
Finland

M. LORITO
Instituto di Patologia Vegetale
Università degli Studi di Napoli
"Federico II" and Centro di Studio
CNR sulle Tecniche di Lotta
Biologica (CETELOBI)
Via Università 100
80055 Portici (NA)
Italy

CARMEN LIMÓN
Departamento de Genética
Facultad de Biología
Universidad de Sevilla
E-41080 Sevilla
Spain

R. D. LUMSDEN
Biocontrol of Plant Diseases
Laboratory
USDA
ARS
Beltsville
MD 20705
USA

ARJA MÄNTYLÄ
Röhm Enzyme Finland Oy
PO Box 26
05200 Rajamäki
Finland

RICCARDO MONTEVERDI
Laboratory of Biotechnology
Central R&D
Lamberti s.p.a.
Albizzate (VA)
Italy

T. OKSANEN
VTT Biotechnology and Food
Research
PO Box 1500
FIN-02044
Finland

MARJA PALOHEIMO
Roal Oy
PO Box 57
05200 Rajamäki
Finland

MERJA E. PENTTILÄ
VTT Biotechnology and Food
Research
PO Box 1500
FIN-02044
Finland

J. PERE
VTT Biotechnology and Food
Research
PO Box 1500
FIN-02044
Finland

MANUEL REY
Departamento de Genética
Facultad de Biología
Universidad de Sevilla
E-41080 Sevilla
Spain

DAVID SEABY
Applied Plant Science Division
Department of Agriculture for
Northern Ireland
Newforge Lane
Belfast 9
UK

M. SIIKA-AHO
VTT Biotechnology and Food
Research
PO Box 1500
FIN-02044
Finland

PIRKKO SUOMINEN
Roal Oy
PO Box 57
05200 Rajamäki
Finland

A. SUURNÄKKI
VTT Biotechnology and Food Research
PO Box 1500
FIN-02044
Finland

T. T. TEERI
VTT Biotechnology and Food Research
PO Box 1500
FIN-02044
Finland

MAIJA TENKANEN
VTT Biotechnology and Food Research
PO Box 1500
FIN-02044
Finland

ARNE TRONSMO
Agricultural University of Norway
Department of Biotechnological Sciences
N-1432
Aas
Norway

L. VIIKARI
VTT Biotechnology and Food Research
PO Box 1500
FIN-02044
Finland

PART ONE

Degradation of Polysaccharides and Related Macromolecules

1

Structure–function relationships in *Trichoderma* cellulolytic enzymes

A. KOIVULA, M. LINDER and T. T. TEERI

VTT Biotechnology and Food Research, Espoo, Finland

1.1 Introduction

Lignocellulose represents a considerable challenge to enzymatic hydrolysis on account of its heterogeneous composition and physical structure evolved to resist degradation. Its main component, cellulose, is composed of long, unbranched glucose polymers packed onto each other to form highly insoluble crystals. To meet the challenge of crystalline cellulose degradation, potent cellulolytic organisms, including *Trichoderma*, produce complex mixtures of enzymes all required for efficient solubilization of the substrate. Studies of the cellulolytic enzyme systems of *Trichoderma* species have a long history (for comprehensive coverage see Béguin and Aubert, 1994; Knowles *et al.*, 1987; Montenecourt, 1983; Nevalainen and Penttilä, 1995; Reese *et al.*, 1950; Teeri, 1997; Wood and McCrae, 1979; Wood and Garcia-Campayo, 1990), and today *T. reesei* is probably the most extensively studied cellulolytic organism. Its many different cellulolytic enzymes are efficiently secreted into the culture medium and they act synergistically to bring about complete solubilization of the highly crystalline native cellulose (Fägerstam and Pettersson, 1980; Henrissat *et al.*, 1985; Irwin *et al.*, 1993; Medve *et al.*, 1994; Nidetzky *et al.*, 1993, 1994a; Wood and McCrae, 1972, 1979).

All cellulases have identical chemical specificities towards the β-1,4-glycosidic bonds, but they differ in terms of the site of attack on the solid substrates. Traditionally, this macroscopic difference in their modes of action has been defined as exoglucanase and endoglucanase activities (Wood and McCrae, 1972, 1979). The exoglucanases, or cellobiohydrolases (CBH) (1,4-β-D-glucan cellobiohydrolase, EC 3.2.1.91), cleave cellobiose units from the ends of the polysaccharide chains and typically exhibit relatively high activities on crystalline cellulose. *T. reesei* produces two different cellobiohydrolases, with CBHI presumed to attack the cellulose chains from their reducing ends and CBHII from their non-reducing ends (Barr *et al.*, 1996; Divne *et al.*, 1994; Teleman *et al.*, 1995; Vrsanská and Biely, 1992). The endoglucanases (EG) (1,4-β-D-glucan-4-glucanohydrolase, EC 3.2.1.4) make more random cuts in the middle of the long chains, thereby producing new chain ends for the cellobiohydrolases to act upon. The endoglucanases seem to prefer the amorphous regions of cellulose and, in contrast to the cellobiohydrolases, can also hydrolyse substituted celluloses, such as carboxymethylcellulose (CMC) and hydroxyethylcellulose (HEC). Finally, β-glucosidases (EC 3.2.1.21) cleave cellobiose and other soluble oligosaccharides to glucose, which is an important step since cellobiose is an end product inhibitor of many cellulases.

The combination of genetic engineering and structural biology has in recent years led to significant advances in our understanding of the basic mechanisms of cellulase action. Since the cellulases from the filamentous fungus *Gliocladium* have not yet been studied at the structural level, this chapter will give an overview of the structure–function studies of *T. reesei* cellulases.

1.2 Domain structures of cellulolytic enzymes

Most *T. reesei* cellulases exhibit a two-domain organization containing a large catalytic domain and a small, distinct cellulose-binding domain (CBD) joined by a linker peptide (Teeri *et al.*, 1987; Tomme *et al.*, 1988; van Tilbeurgh *et al.*, 1986) (Figure 1.1). Small-angle X-ray scattering studies of *T. reesei* cellobiohydrolases suggest an elongated, tadpole-like shape in which the catalytic domain forms the ellipsoidal "head" and the linker adopts an extended conformation between the catalytic domain and the CBD (Abuja *et al.*, 1988a,b). So far, crystallization of intact fungal cellulases has not been successful, probably due to the properties of the glycosylated and extended interdomain linker peptide. However, structures of the isolated catalytic or cellulose-binding domains of many fungal and bacterial cellulases have been determined successfully.

1.2.1 Families of cellulase catalytic domains

Using extensive sequence comparisons by hydrophobic cluster analysis (HCA), the catalytic domains of all known glycosyl hydrolases, currently comprising about 950 sequences, have been classified into 58 families (Henrissat, 1991; Henrissat and Bairoch, 1993, 1996; also on the ExPASy server at http://expasy.hcuge.ch). Different cellulolytic enzymes have been allocated to 11 families (families 5–10, 12, 26, 44, 45, 48), and among these, *Trichoderma* cellulases are found in five (Henrissat and Bairoch, 1996; Henrissat *et al.*, 1989; Tomme *et al.*, 1995a). Determination of a total of nine crystal structures of cellulase catalytic domains from six different families has confirmed that the members of each different family have similar three-dimensional folds and probably share similar reaction mechanisms (Davies and Henrissat, 1995; Tomme *et al.*, 1995a) (see also section 1.3.2). However, the current classification does not follow the traditional enzyme nomenclature as established by the International Union of Biochemistry and mainly based on substrate specificity. Thus, a given cellulase family often contains both exo- and endoglucanases, and in addition to cellulases some families contain other enzymes, e.g., xylanases, mannanases and dextrinases (Henrissat, 1994; Henrissat and Bairoch, 1996).

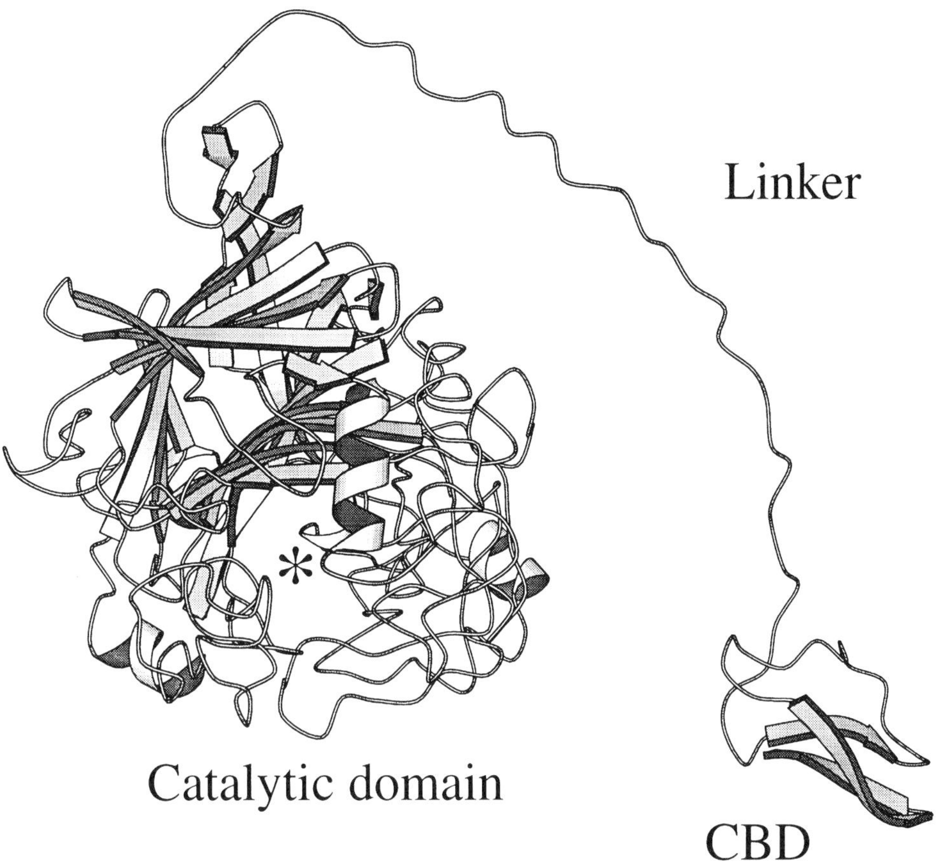

Figure 1.1 Schematic representation of the intact domain structure of *T. reesei* CBHI drawn using the Molscript program of Kraulis (1991) (CBD, cellulose-binding domain). The location of the active site tunnel is indicated by an asterisk. There is currently no structure available for intact CBHI, and the linker peptide has been model-built here to illustrate the physical linkage of the two functional domains. Coordinates of the model-built structure were kindly provided by Dr. Christina Divne (Uppsala University, Sweden).

1.2.2 Crystal structures of the catalytic domains

So far, crystal structures have been determined for the catalytic domains of the *T. reesei* CBHI, CBHII and EGI (Divne *et al.*, 1994; Kleywegt *et al.*, 1997; Rouvinen *et al.*, 1990), and the crystallization of *T. reesei* EGIII has been reported (Ward *et al.*, 1993). The catalytic domain structure of *T. reesei* CBHII, belonging to the family 6, was the first cellulase crystal structure solved at the atomic level (Rouvinen *et al.*, 1990). This polypeptide folds into an α/β barrel structure similar to triose phosphate isomerase (TIM) but contains seven instead of eight β-strands (Figure 1.2A). The active site of CBHII has been localized in a tunnel formed by two surface loops stabilized by disulphide bridges. The tunnel is 20 Å long, spans through the whole catalytic domain and is apparently able to accommodate a

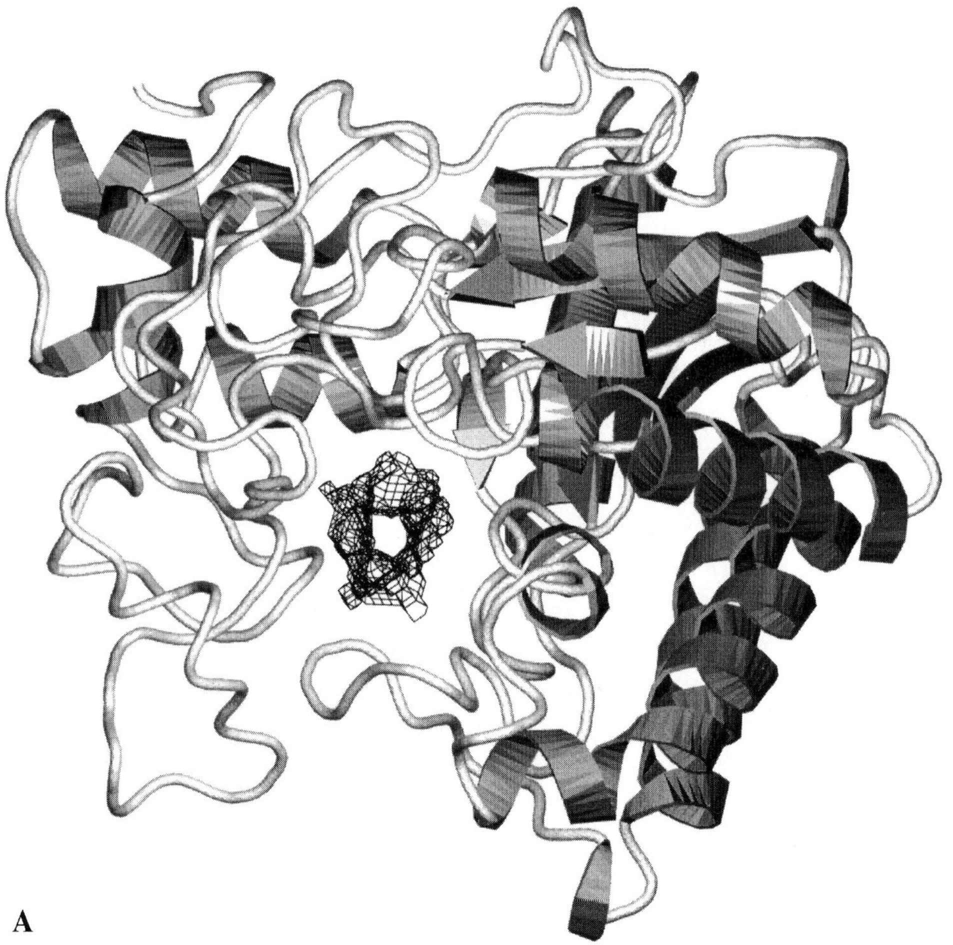

A

Figure 1.2 Schematic drawings showing the overall folds and the active site tunnels of the catalytic domains of *T. reesei* (A) (Rouvinen *et al.*, 1990) and (B) (Divne *et al.*, 1994). The secondary structure elements are colored dark grey with β-strands represented by arrows and α-helices by spirals. For both proteins the view is down their active site tunnels, which have been indicated by showing a cross-section of their water-accessible surfaces. The surfaces were calculated using the programs VOIDOO (Kleywegt and Jones, 1994a) and MAMA (Kleywegt and Jones, 1994b). The figure was generated using O and OPLOT (Jones *et al.*, 1991) and was kindly provided by Dr. Christina Divne.

single cellulose glycan chain entering from one end of the tunnel (Rouvinen *et al.*, 1990). Sequence and structural comparison with endoglucanases in family 6 revealed that the loops enclosing the active site tunnel in CBHII are missing or have adopted different conformations in the structures of related endoglucanases (Rouvinen *et al.*, 1990; Spezio *et al.*, 1993). This immediately provided a structural explanation for the different modes of action of the endo- and exoglucanases: in spite of their similar overall structures, endoglucanases have more open active sites allowing random binding and hydrolytic activity in the middle of the cellulose glucan chains, while the exoglucanases are confined to the chain ends by their tunnel-shaped active sites.

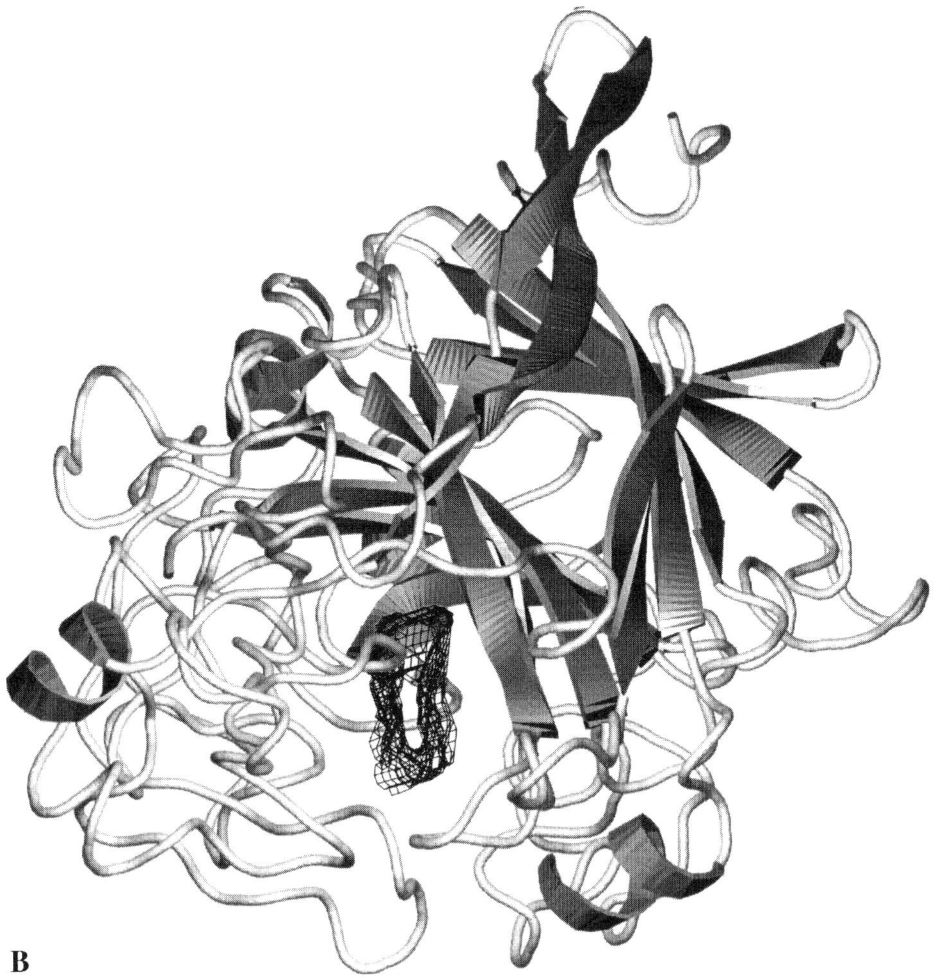

B

Figure 1.2 (*cont*)

The crystal structure of the catalytic domain of *T. reesei* CBHI (family 7) reveals an even longer active site tunnel extending from a central structural motif that consists of two large anti-parallel β-sheets stacked face-to-face to form a β-sandwich (Divne *et al.*, 1994) (Figure 1.2B). Again, a related endoglucanase, EGI, from *T. reesei* (Divne *et al.*, 1994; Kleywegt *et al.*, 1997; Penttilä *et al.*, 1986) lacks these loops and has a more open active site supporting the general difference between the exo- and endoglucanases.

Davies and Henrissat (1995) have compared all available three-dimensional structures of glycosyl hydrolases and found that their active site topologies can be divided into three main classes regardless of the catalytic mechanism or the overall fold of the enzyme. Pocket- or crater-like active sites are encountered in enzymes such as β-glucosidases, glucoamylases and β-amylases that hydrolyse monosaccharide units from the chain ends of polymeric carbohydrates. This kind of active site is apparently not optimal for a substrate like native crystalline cellulose, which contains few accessible chain ends. Instead, the optimal topology evolved for this purpose seems to be an extended active site providing binding sites for a number of sugar units. In the first variation of this theme, the active site is buried in a tunnel as shown above for the *T. reesei* CBHI and CBHII and also suggested for some bacterial cellobiohydrolases (Meinke *et al.*, 1995; Zhang *et al.*, 1995). In the other variant of this theme, the loops forming the active site are shorter or turned away, resulting in a more easily accessible cleft or groove on the enzyme surface. Such an open active site structure allows binding randomly along the entire length of the polysaccharide chain and is commonly found in endo-acting enzymes, such as α-amylases, endocellulases, xylanases and lysozymes (Davies and Henrissat, 1995).

1.2.3 *The active site tunnels of* T. reesei *cellobiohydrolases*

Complex structures of *T. reesei* CBHII with several different ligands have revealed four glucosyl binding sites (−2, −1, +1, +2) within its active site tunnel (Rouvinen *et al.*, 1990; Rouvinen, 1990). Although each subsite is somewhat flattened, they have different cross-sections. In the subsites −2, +1 and +2, tryptophan side chains W135, W367 and W269, respectively, make significant contributions to the formation of the sugar binding sites (see Figure 1.3). In site −2, in particular, this results in a narrower tunnel cross-section. In addition, the ligand structures have revealed a 20° twist in the chain between subsites +1 and +2. Modelling studies with longer oligosaccharides suggest favourable van der Waals interactions also between W272 and the sixth glucosyl in the putative binding site +4 at the very entrance of the tunnel (Rouvinen, 1990). This, however, requires a further 110° twist in the cellulose chain between the fourth and sixth glucosyl units, and no obvious sugar–protein interactions can be seen at the putative subsite +3 (Rouvinen, 1990). Besides the tryptophan residues, the active site tunnel contains many residues that

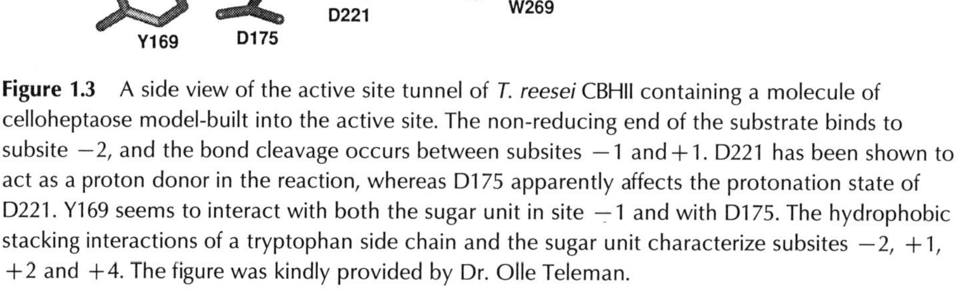

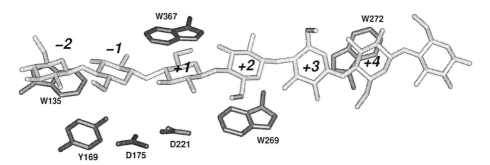

Figure 1.3 A side view of the active site tunnel of *T. reesei* CBHII containing a molecule of celloheptaose model-built into the active site. The non-reducing end of the substrate binds to subsite −2, and the bond cleavage occurs between subsites −1 and +1. D221 has been shown to act as a proton donor in the reaction, whereas D175 apparently affects the protonation state of D221. Y169 seems to interact with both the sugar unit in site −1 and with D175. The hydrophobic stacking interactions of a tryptophan side chain and the sugar unit characterize subsites −2, +1, +2 and +4. The figure was kindly provided by Dr. Olle Teleman.

form hydrogen bonds to the substrate (Rouvinen *et al.*, 1990). Most of these interactions involve charged residues, which are conserved within the glycosyl hydrolase family 6.

The active site tunnel of CBHI is about 50 Å long, formed by four loops partly stabilized by disulphide bridges (Divne *et al.*, 1994; Ståhlberg *et al.*, 1996). Similar to CBHII, the active site tunnel of CBHI is flattened to accommodate at least seven glucose units of a single glucan chain and has a narrow point in the centre which forces the chain to adopt a twisted conformation (Divne *et al.*, 1994; T. A. Jones, personal communication). Binding of the glucose units occurs via hydrogen bonding and, in four subsites, stacking onto the tryptophan residues W376, W367, W38 and W40.

1.3 Catalytic reaction mechanisms of *T. reesei* cellobiohydrolases

Cellobiohydrolases, like all glycosidases, catalyse the hydrolysis of the glycosidic bonds by general acid catalysis. The reaction involves two essential catalytic carboxylates, the proton donor and the nucleophile, which are generally located on opposite sides of the scissile bond. Depending on the spatial arrangement of these two residues, the reaction proceeds either by retention or by inversion of the anomeric configuration at the C_1 carbon (Sinnott, 1990) (Figure 1.4). *T. reesei* CBHI has been shown to be a retaining enzyme, similar to the hen egg-white lysozyme, whereas CBHII is an inverting glycosylase together with certain sialidases and glucoamylases (Claeyssens *et al.*, 1990; Knowles *et al.*, 1988).

Residues potentially involved in substrate binding and catalysis can be identified from the three-dimensional structure of an enzyme and verified experimentally by chemical modification and/or site-directed mutagenesis. In the case of retaining enzymes, the active site carboxylic acids can also be identified by the use of affinity labels or mechanism-based inactivators to specifically derivatize the key amino acids. 2-Deoxy-2-fluoro-derivatives of glucose and cellobiose have successfully been used as mechanism-based inhibitors which covalently bind the nucleophile in the retaining enzymes (McCarter and Withers, 1994; Withers and Aebersold, 1995). Structure determination of *T. reesei* and other cellulases has paved the way for detailed studies of their reaction mechanisms.

1.3.1 Catalytic residues of CBHII

Since CBHII is an inverting enzyme, the hydrolysis is postulated to proceed through a single-displacement reaction involving a general acid to donate a proton and a base to assist the nucleophilic attack of water (Koshland, 1953) (see Figure 1.4). The active sites of inverting glycosidases are usually constructed such that a water molecule can be accommodated between the base and the anomeric carbon. The nucleophilic water thus approaches the anomeric carbon (C_1) from the direction opposite the proton donor (McCarter and Withers, 1994). Two aspartic acids, D221 and

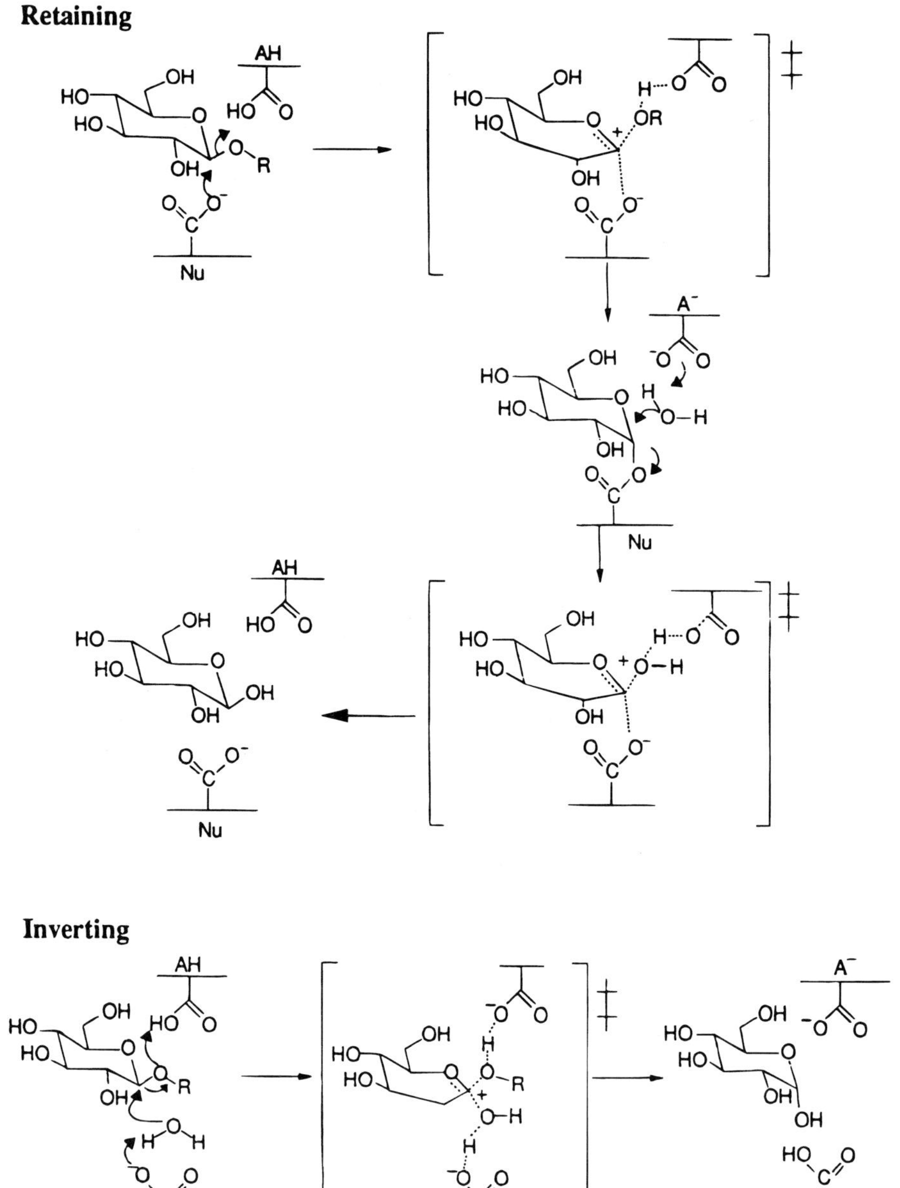

Figure 1.4 The two different catalytic mechanisms proposed for glycoside hydrolases (Davies and Henrissat, 1995; Koshland, 1953; McCarter and Withers, 1994; Sinnott, 1990). Retaining enzymes use a double-displacement mechanism. In the first step an acid catalyst (AH) donates a proton to the glycosidic oxygen while a nucleophile (Nu) displaces the glycosidic O4 oxygen of the leaving group. The anomeric carbon of the resulting glycosyl–enzyme intermediate now has an inverted configuration. In the second displacement step, water hydrolyses the glycosyl–enzyme intermediate restoring the original configuration of the anomeric carbon. Inverting enzymes use a single-displacement mechanism which involves an acid (AH) donating a proton to the leaving glycosidic oxygen, and a base (B^-) assisting the nucleophilic attack of water. The proposed transition states are shown in square brackets denoted ‡.

D175, have been identified at the centre of the active site tunnel of CBHII between the subsites −1 and +1 and very near the O-glycosidic linkage (Figure 1.3). However, the two residues are located on the same side of the scissile bond within hydrogen bonding distance of each other (Rouvinen *et al.*, 1990). D221 is in an environment in which it should be protonated, whereas D175 is more likely to be charged. The mutation D221A abolishes practically all of the catalytic activity of CBHII on small soluble oligosaccharides (Table 1.1) but does not affect the binding properties on small ligands. Mutation D175A severely impairs the catalytic activity of CBHII but also slightly alters its binding behaviour (Ruohonen *et al.*, 1993). In addition, the D175A mutant was shown to have about 2% residual activity on a soluble polymeric substrate, whereas D221A was completely inactive (L. Ruohonen, personal communication). It thus seems that D221 acts as the proton donor while the role of D175 may be to ensure the protonation of D221, stabilize reaction intermediates, or do both.

The active site of CBHII also contains a third carboxylate, D401, which is correctly oriented to act as the catalytic base but is salt-linked to two nearby residues, R353 and K395. Therefore, if D401 participates in the catalysis, it probably functions as a relatively weak base. However, some kinetic studies have recently cast doubt on the classical single-displacement mechanism for CBHII, suggesting that a base may not be absolutely required (Konstantinidis *et al.*, 1993; Sinnott, 1990). In particular, the hydrolysis of α-cellobiosyl fluoride by CBHII does not seem to proceed with the Hehre-type resynthesis–hydrolysis mechanism diagnostic of a classical single-displacement mechanism (Kasumi *et al.*, 1987). This is in striking contrast to an endoglucanase A from *Cellulomonas fimi*, also of family 6, which has been

Table 1.1 Catalytic constants for the CBHII wild-type and mutant enzymes

	k_{cat} (min^{-1})				
Protein	Glc_3	Glc_4	Glc_5	Glc_6	Ref.
CBHII wt	3.7	220	60	840	1, 2
D221A	0.01	≤0.2	nd	nd	3
D175A	0.03	≤0.2	nd	nd	3
Y169F	0.9	57	nd	nd	1
W135F	0.1	75	nd	nd	3, 4
W135L	0.02	1	nd	nd	3

Hydrolysis experiments were performed in 10 mM sodium acetate buffer, pH 5.0 at 27°C. Samples were taken at different time points and analysed with HPLC as described earlier (Ruohonen *et al.*, 1993; Teleman *et al.*, 1995). Kinetic constants were calculated by a non-linear regression analysis (Enzfitter) or by analysing whole progress-curves as described by Teleman *et al.* (1995). The error is estimated to be 10% on the basis of four repeated experiments.
wt = wild type; nd = not determined.
References: 1, Koivula *et al.*, 1996; 2, Harjunpää *et al.*, 1996; 3, Ruohonen *et al.*, 1993; 4, Koivula, 1996.

shown to use a typical inverting mechanism involving both the acid and base catalyst (Damude *et al.*, 1996). This implies that the endoglucanases and exoglucanases in family 6 may follow different mechanisms.

1.3.2 ***Ring distortion as an element of the catalytic mechanism of CBHII***

Both inverting and retaining enzymes are thought to operate through transition states with substantial oxocarbonium ion character, and in some hydrolases, distortion of the sugar ring towards the transition state seems to occur already upon substrate binding (Davies *et al.*, 1995; McCarter and Withers, 1994; Taylor *et al.*, 1995). The involvement of the oxocarbonium ion-like transition states in the reaction catalysed by CBHII gains support from kinetic studies (Konstantinidis *et al.*, 1993) and the high association constant observed for a putative transition state analogue, cellobionolactonoxime (van Tilbeurgh *et al.*, 1989). In the active site of CBHII, the ring distortion towards the transition state is anticipated in the subsite −1 preceding the scissile bond (Figure 1.3). This subsite is structurally quite different from the other binding sites since it lacks the sugar-binding tryptophan and contains a protrusion which may allow alternative sugar conformations (Koivula *et al.*, 1996). In most complex structures of CBHII with different ligands, the sugar density in this subsite is poor and has, in one case, been interpreted as a distorted sugar conformation (Koivula *et al.*, 1996; Rouvinen, 1990).

A tyrosine residue, Y169, is strictly conserved in family 6 and forms part of the subsite −1 of CBHII. The OH-group of this residue is estimated to be at hydrogen bond distance both from the methyl hydroxyl of the glucose ring and the carboxylate, D175 (Rouvinen *et al.*, 1990). The role of this residue has been studied by introducing a mutation, Y169F, which removes the interacting OH-group (Koivula *et al.*, 1996). Surprisingly, the mutant exhibited somewhat decreased catalytic rate but significantly increased association constants on small soluble ligands. In the case of methylumbelliferyl-cellobioside (MeUmb$(Glc)_2$), the Y169F mutant exhibited over 50-fold higher affinity than the wild-type CBHII (Koivula *et al.*, 1996) (Figure 1.5). A very similar increase in the affinity has been previously observed on binding of another ligand, MeUmbGlcXyl, to the wild-type CBHII (van Tilbeurgh *et al.*, 1989). Assuming that this ligand binds to the CBHII active site similar to the MeUmb$(Glc)_2$, the site −1 is occupied by the xylose residue lacking the methylhydroxyl group present in the corresponding glucose residue. It has been contemplated that these changes indicate conformational distortion of the sugar ring which can be relaxed by removing the interacting group either from the ligand, as in xylose, or from the enzyme, as in the mutant Y169F.

1.3.3 ***Role of tryptophan residues in the active site of CBHII***

The hydrophobic stacking interactions of aromatic residues are considered important for providing high affinity towards the polysaccharide substrate both in the catalytic and in the substrate binding domains of various carbohydrates (Linder *et al.*, 1995a; Toone, 1994; Vyas, 1991). In the active site of CBHII, tryptophan residues provide such stacking interactions in the subsites −2, +1 and +2 and in the putative binding site +4 (Figure 1.3). Earlier binding studies have revealed that the

Binding mode (-2, -1, +1)	**CBHII wildtype** K_{ass} at 25°C (M^{-1})	**Y169F** K_{ass} at 16°C (M^{-1})
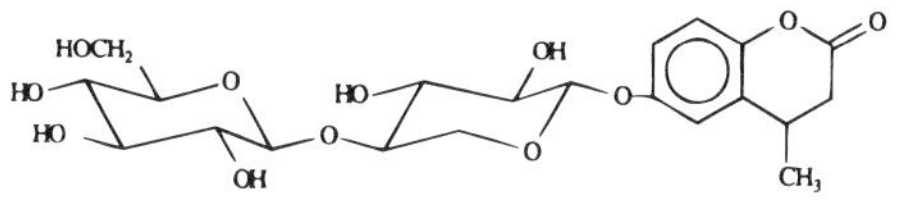 MethylUmbelliferyl(GlcXyl)	$4.9 \cdot 10^7$	nd
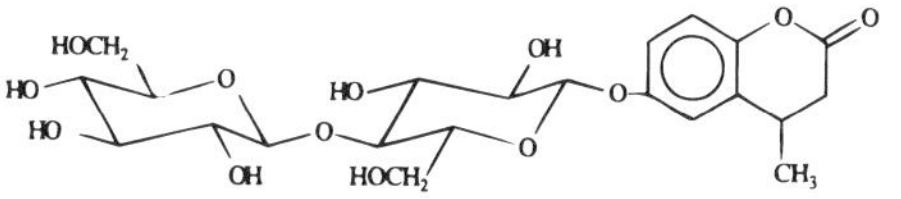 MethylUmbelliferyl(Glc)$_2$	$2.0 \cdot 10^5$ ($4 \cdot 10^5$ at 16°C)	$2 \cdot 10^7$

Figure 1.5 The association constants of the MeUmbGlcXyl (top) and MeUmb(Glc)$_2$ (bottom) ligands bound in the active sites of wild type CBHII or the Y169F mutant (respectively). The location of the MeUmb(Glc)$_2$ has been identified in the crystal structure of CBHII (Rouvinen *et al.*, 1990), and it is assumed that the binding of the umbelliferyl group over site +1, and partially over site +2, will pull the MeUmbGlcXyl ligand in an analogous position in the mutant Y169F. Reproduced with permission from Koivula *et al.*, Structure–function studies of two polysaccharide degrading enzymes, VTT publications 277, Technical Research Centre of Finland, 1996. ND means not determined.

subsite −2 is very specific for an intact D-glucopyranose configuration (van Tilbeurgh *et al.*, 1989), and complex structures of CBHII with different ligands confirm that it is the tightest binding site (Koivula *et al.*, 1996; Rouvinen, 1990; Rouvinen *et al.*, 1990). Furthermore, NMR studies failing to detect α-glucose among the hydrolysis products of CBHII demonstrate that this subsite must be occupied for hydrolysis to occur (Ruohonen *et al.*, 1993; Teleman *et al.*, 1995). When the indole ring of the tryptophan residue W135 at subsite −2 was substituted by a phenolic ring in the mutant W135F, decreased binding on small ligands and subsequent reduction in the catalytic efficiency of CBHII were observed (Table 1.1). Complete removal of the aromatic ring structure by the mutation W135L led to even more pronounced effects on catalysis and binding (Table 1.1) (Koivula, 1996). Since the unliganded mutant enzyme structures of W135F or W135L do not reveal any detectable changes (T. A. Jones, personal communication), the tight binding of the glucose at site −2 by W135 seems to be an important determinant of the catalytic efficiency of CBHII.

There is so far no published data on the roles of the other tryptophan residues in the activity of CBHII. However, contributions of the other binding sites have been evaluated by analysing the catalytic rates for several different oligosaccharides ranging from three to six glucose units (Harjunpää *et al.*, 1996; Ruohonen *et al.*, 1993; Teleman *et al.*, 1995). The specificity constant (k_{cat}/K_M) for cellotriose hydrolysis is significantly lower than that for cellotetraose, indicating that the sugar binding at the subsite + 2 contributes to the transition state stabilization. The rate of cellopentaose hydrolysis is lower than that of cellotetraose but increases again as cellohexaose is used as substrate (Harjunpää *et al.*, 1996; Ruohonen *et al.*, 1993). Preliminary results obtained by mutating the tryptophan residue W272 at the putative binding site +4 suggest that these phenomena arise from non-productive binding of the shorter oligosaccharides through this site (Koivula, 1996).

1.3.4 The reaction mechanism of CBHI

T. reesei CBHI has been shown to belong to retaining enzymes that use a double-displacement mechanism for catalysis (Sinnott, 1990) (Figure 1.4). The active site of *T. reesei* CBHI contains three carboxylates, E212, D214 and E217, that are positioned around the scissile bond between subsites −1 and +1 (Divne *et al.*, 1994). The spatial arrangement of the carboxylates observed in a complex structure of CBHI with an *o*-iodobenzyl-1-thio-*β*-D-glucoside ligand hinted that E217 acts as the proton donor and E212 as the nucleophile in the double-displacement reaction. According to this scheme, the role of D214 would be to position and/or control the protonation state of E212 (Divne *et al.*, 1994). An analysis of the behaviour of the mutants E212Q, D214N and E217Q towards chloro-nitrophenyl-lactoside (CNP-Lac) shows that all the three residues are needed for efficient catalysis (Ståhlberg *et al.*, 1996). Mutation of the putative nucleophile E212 produced the largest decrease in k_{cat}, suggesting that deglycosylation is the limiting step on this substrate. Mutagenesis of the residue E217 also resulted in a dramatic loss of activity, supporting its role as the proton donor. The mutant D214N had the highest residual activity, again consistent with its proposed supporting role in the catalytic reaction of CBHI.

1.4 Role and function of the cellulose-binding domains (CBDs)

A characteristic feature of cellulases and many other enzymes degrading insoluble carbohydrates is the occurrence of distinct substrate-binding domains connected to their catalytic domains. A CBD was first identified in the CBHI of *T. reesei* (Bhikhabhai *et al.*, 1984; van Tilbeurgh *et al.*, 1986). Since then, homologous CBDs have been found either at the N- or the C-terminal end of the catalytic domains of all but one of the cellulases so far identified in *T. reesei* (Nevalainen and Penttilä, 1995; Teeri *et al.*, 1987). The CBDs have been classified into 10 families sharing similar primary and tertiary structures but exhibiting a more conserved species specificity than the catalytic domains (Tomme *et al.*, 1995b).

The effect of the CBD on the enzyme function is typically seen in experiments where the CBD is removed by proteolysis or genetic engineering. The truncated forms, consisting of only the catalytic domain, have an unaffected activity on soluble substrates, but their binding and activities on insoluble cellulose are clearly reduced (Tomme *et al.*, 1988; van Tilbeurgh *et al.*, 1986; reviewed in Linder and Teeri, 1997).

1.4.1 The structure and properties of the CBHI CBD

All the *T. reesei* and other fungal CBDs belong to the family I (Tomme *et al.*, 1995b). The family I CBDs are small molecules consisting of only about 36 amino acid residues (Hoffrèn *et al.*, 1995; Teeri *et al.*, 1987; Tomme *et al.*, 1995b). The structure of a synthetic peptide corresponding to the CBHI CBD has been determined by NMR and shown to consist of an irregular triple stranded β-sheet stabilized by two or, in some cases, three disulfide bridges (Hoffrèn *et al.*, 1995; Kraulis *et al.*, 1989). A similar arrangement of disulfides, sometimes called cystine knots, has been identified in several different neurotoxins and protease inhibitors which typically exhibit high stability and structural rigidity (Pallaghy *et al.*, 1994).

The overall shape of the CBHI CBD resembles a wedge exposing two distinctly different faces (Figure 1.6). The flat face of the wedge contains a row of three tyrosines that are either strictly conserved or replaced by a tryptophan or phenylalanine in other fungal CBDs (Hoffrèn *et al.*, 1995). Amino acid replacement studies on the CBHI CBD have demonstrated that the three tyrosines are principally responsible for its interaction with crystalline cellulose although some other hydrogen bond forming residues also contribute (Linder *et al.*, 1995a,b; Reinikainen *et al.*, 1995).

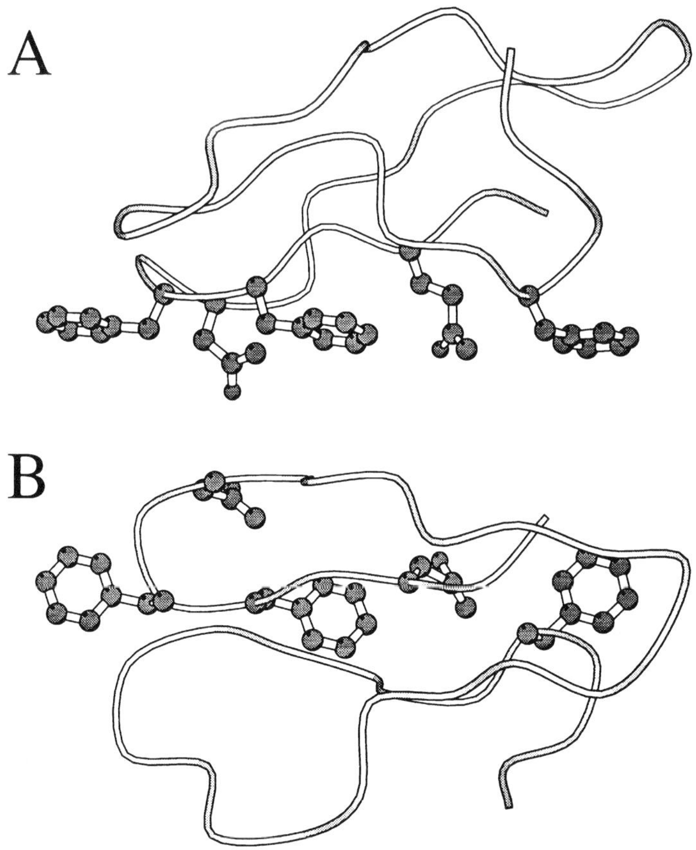

Figure 1.6 The backbone structure of the *T. reesei* CBHI CBD showing the side chains known to mediate its binding to cellulose (Kraulis *et al.*, 1989; Linder *et al.*, 1995a,b). In (A) the CBD is viewed from the side to illustrate the two faces of the wedge. In (B) the angle is tilted 90° so that the view is down onto the cellulose surface. The spacing between the key tyrosine residues equals the distance between every second glucose ring in an extended cellulose chain. The figure was drawn using the Molscript program of Kraulis (1991).

Participation of aromatic residues in protein–carbohydrate interactions is apparently a general phenomenon (Toone, 1994; Vyas, 1991). The other face of the wedge is rougher, less conserved and, based on substitution studies, unlikely to mediate direct interaction with the cellulose surface (Linder *et al.*, 1995a,b; Reinikainen *et al.*, 1995).

1.4.2 *CBDs and the action of intact cellulases*

A clear relationship between the binding and the hydrolytic activity of cellulases on solid substrates has been established by experiments with truncated enzymes lacking their CBDs (see above) and by comparing enzymes of different origin (Klyosov, 1990). However, the details of how the CBD interacts with the cellulose on the one hand and with the catalytic domain on the other hand are still in part unclear. It has been clearly demonstrated that, although the CBDs facilitate the binding of intact cellulases, most catalytic domains also have affinity for crystalline cellulose. Characterization of a recombinant double CBD, mimicking the domain structure of an intact cellulase, shows that the covalent linkage between two binding domains results in a synergistic, high affinity binding (Linder *et al.*, 1996). On the other hand, experiments with a labelled CBHI CBD have shown that the binding of an individual CBD is readily reversible (Linder and Teeri, 1996). It is thus possible to imagine a process in which the initial tight binding of the enzyme occurs through cooperative binding of both domains. Once the hydrolysis has begun, the movement of the catalytic domain along the substrate surface is allowed by the relatively high exchange rate of the CBD (Linder and Teeri, 1996). Since the CBD properties can be easily affected in nature by domain exchange or point mutations, the affinity of each different CBD has probably been fine tuned by evolution to work optimally with its corresponding catalytic domain in a selected environment.

Why then is the CBD so important in crystalline cellulose degradation? In the simplest case, the CBD could merely increase the effective enzyme concentration near the surface due to the partitioning of the enzyme between the liquid and the solid phases. However, the situation seems to be more complicated than that. The affinity of the *T. reesei* EGI CBD has been shown to be markedly higher than that of the CBHI CBD (Linder *et al.*, 1995a), but domain-swapping experiments demonstrate that improved CBD affinity does not increase or decrease the activity of CBHI on crystalline cellulose (Srisodsuk *et al.*, 1997). Furthermore, linker deletion studies of CBHI have shown that the linker length influences both the binding and activity of the enzyme indicating some kind of an interplay between the two domains (Srisodsuk *et al.*, 1993) (see below).

1.5 Aspects of crystalline cellulose degradation

It has been stated that the distinguishing feature of cellulase mixtures effective on crystalline cellulose is the presence of one or more cellobiohydrolases (Wood and Garcia-Campayo, 1990). This statement certainly seems valid for the *T. reesei* system in which the two cellobiohydrolases represent more than 80% of the total cellulolytic protein (Grizali and Brown, 1979). The characteristic structural feature of the cellobiohydrolases is their tunnel shaped active site designed to accommodate a single glucan chain. Another emerging feature of the cellobiohydrolases, and probably of some selected endoglucanases, seems to be their processive mode of action, leading to many consecutive bond cleavages without the dissociation of the enzyme–substrate complex (Davies and Henrissat, 1995; Divne *et al.*, 1994; Rouvinen *et al.*, 1990; Ståhlberg *et al.*, 1996; Teeri, 1997). In the cellobiohydrolases the processive mode of action is apparently dictated by the shape and design of their active site tunnels that seem to favour the entrance of the polymeric glucose chain from one end of the tunnel, followed by its tight binding over a number of subsites and subsequent hydrolysis and release of the product cellobiose from the far end of the tunnel (Divne *et al.*, 1994; Rouvinen *et al.*, 1990) (Figure 1.7). It is theoretically possible that the active site loops open to allow random binding as the first step of crystalline cellulose hydrolysis, and a significant conformational change in a loop embracing a glycan chain has indeed been observed in the active site of a fungal endoglucanase (Davies *et al.*, 1995). However, similar conformational changes have not so far been observed in CBHI and CBHII. Furthermore, both produce only cellobiose and glucose from crystalline cellulose and do not significantly decrease the average degree of polymerization of crystalline substrates even in prolonged hydrolysis experiments (Kleman-Leyer *et al.*, 1996; Srisodsuk, M., Kleman-Leyer, K., Keränen, S., Kirk, T. K. and Teeri, T. T., manuscript in preparation). Finally, it has been shown that during cellohexaose hydrolysis by CBHII, no cellotetraose is formed from the non-reducing end of the oligosaccharide, demonstrating that, if at all possible, loop opening must be a very rare event (Harjunpää *et al.*, 1996; Nidetzky *et al.*, 1994b).

In order to understand how a processive enzyme with a tunnel shaped active site can efficiently operate on a crystalline surface, one must look for a mechanism capable of releasing and feeding in single glucose chains. The two-domain structure and tight binding of cellulases have been shown to be important determinants of their ability to attack crystalline cellulose (see above). Therefore, the secret of crystalline cellulose degradation can be at least partially unravelled by striving to understand the domain interplay in the cellobiohydrolases. A first clue for the existence of such an interplay is provided by recent mutagenesis studies of the tryptophan residue, W272, which forms the proposed sixth subsite at the "mouth" or entrance of the CBHII active site tunnel (see Figure 1.3). This part of the protein including the tryptophan is altogether lacking in the corresponding endoglucanases, and interactions through this site seem to worsen, not facilitate, the hydrolysis of small soluble oligosaccharides (Koivula, 1996). However, our recent results of W272-directed mutants of CBHII suggest that this tryptophan is a very important determinant of efficient degradation of crystalline substrates (A. Koivula, unpublished data). It has been suggested that one of the functions of the CBDs is to help solubilize single glucan chain ends off the cellulose crystals (Knowles *et al.*, 1987). It is therefore easy to imagine that, due to the close physical linkage of the catalytic domain and the CBD, such ends could be caught by the tryptophan residue at the entrance of the tunnel, leading to further transport into the tunnel for catalysis. Although the details of such mechanisms are currently far from clear, these preliminary findings provide the basis for further experimentation.

In conclusion, the availability of several high resolution structures of different fungal and bacterial cellulases now provides a firm basis for detailed studies of their chemical reaction mechanisms by a combination of site-directed mutagenesis, chemical modification and structural biology. Comparison of the structures of selected cellulases with a very large number of sequences from other cellulases has led to

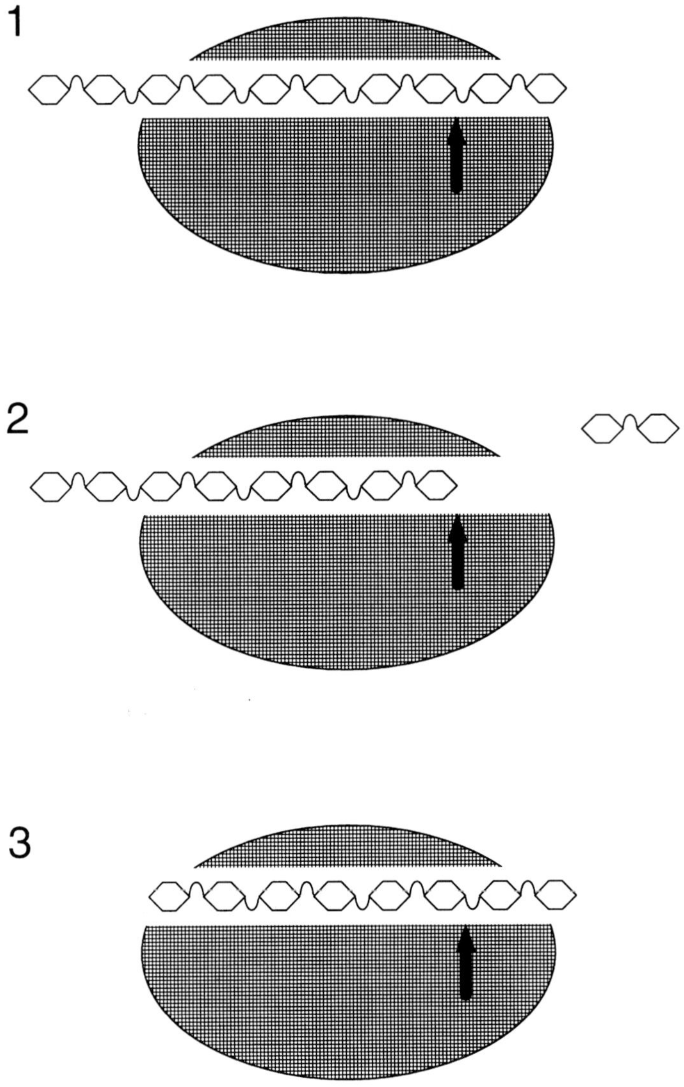

Figure 1.7 A schematic illustration of the processive action of the cellobiohydrolases. Upon hydrolysis of the first β-1,4-glycosidic linkage, cellobiose (or cellotriose) is released from the end of the cellulose glucan chain (step 1). The disaccharide product is liberated while the remaining polysaccharide remains bound to the enzyme through interactions at several subsites (step 2). In the next step, the chain end is translocated in place for another hydrolytic event by a currently unknown mechanism (step 3). In this way, the degradation can proceed through several catalytic cycles producing cellobiose. Figure adapted from Davies and Henrissat (1995).

new insight into their previously poorly understood endoglucanase/exoglucanase activities. Furthermore, studies of truncated and genetically modified enzymes are beginning to reveal how the catalytic and cellulose-binding domains interact with and degrade crystalline cellulose. These insights will help to engineer cellulase performance and properties and, as combined with novel recombinant production strains to produce tailored enzyme mixtures (see Chapter 13), will facilitate the development of new products and processes based on cellulose.

References

Abuja, P. M., Pilz, I., Claeyssens, M., and Tomme, P. 1988a. Domain structure of cellobiohydrolase II as studied by small X-ray scattering, close resemblance to cellobiohydrolase I. *Biochem. Biophys. Res. Comm.* **156**: 180–185.

Abuja, P. M., Schmuck, M., Pilz, I., Tomme, P., Claeyssens, M., and Esterbauer, H. 1988b. Structural and functional domains of cellobiohydrolase I from *Trichoderma reesei*. *Eur. Biophys. J.* **15**: 339–342.

Barr, B. K., Hsieh, Y.-L., Ganem, B., and Wilson, D. B. 1996. Identification of two functionally different classes of exocellulases. *Biochemistry* **35**: 589–592.

Béguin, P. and Aubert, J.-P. 1994. The biological degradation of cellulose. *FEMS Microbiol. Rev.* **13**: 25–58.

Bhikhabhai, R., Johansson, G., and Pettersson, G. 1984. Cellobiohydrolase from *Trichoderma reesei*: internal homology and prediction of secondary structure. *Int. J. Peptide Protein Res.* **25**: 368–374.

Claeyssens, M., Tomme, P., Brewer, C. F., and Hehre, E. J. 1990. Stereochemical course of hydrolysis and hydration reactions catalyzed by cellobiohydrolases I and II from *Trichoderma reesei*. *FEBS Lett.* **263**: 89–92.

Damude, H. G., Ferro, V., Withers, S. G., and Warren, R. A. J. 1996. Substrate specificity of endoglucanase A from *Cellulomonas fimi*: fundamental difference between endoglucanases and exoglucanases from family 6. *Biochem. J.* **315**: 467–472.

Davies, G. and Henrissat, B. 1995. Structures and mechanisms of glycosyl hydrolases. *Structure* **3**: 853–859.

Davies, G. J., Tolley, S. P., Henrissat, B., Hjort, C., and Schülein, M. 1995. Structure of oligosaccharide-bound forms of the endoglucanase V from *Humicola insolens* at 1.9 Å resolution. *Biochemistry* **34**: 16210–16220.

Divne, C., Ståhlberg, J., Reinikainen, T., Ruohonen, L., Pettersson, G., Knowles, J. K. C., Teeri, T. T., and Jones, T. A. 1994. The three-dimensional structure of cellobiohydrolase I from *Trichoderma reesei*. *Science* **265**: 524–528.

Fägerstam, L. and Pettersson, G. 1980. The β-1,4-glucan cellobiohydrolases of *Trichoderma reesei* QM9414: a new type of synergism. *FEBS Lett.* **119**: 97–101.

Grizali, M. and Brown, R. D. Jr. 1979. The cellulase system of *Trichoderma*: the relationship between purified enzymes from induced or cellulose-grown cells. *Adv. Chem. Ser.* **181**: 237–260.

Harjunpää, V., Teleman, A., Koivula, A., Ruohonen, L., Teeri, T. T., Teleman, O., and Drakenberg, T. 1996. Cello-oligosaccharide hydrolysis by cellobiohydrolase II from *Trichoderma reesei*; association and rate constants derived from an analysis of progress curves. *Eur. J. Biochem.* **240**: 584–591.

Henrissat, B. 1991. A classification of glycosyl hydrolases based on amino acid sequence similarities. *Biochem. J.* **280**: 309–316.

Henrissat, B. 1994. Cellulases and their interaction with cellulose. *Cellulose* **1**: 169–196.

Henrissat, B. and Bairoch, A. 1993. New families in the classification of glycosyl hydrolases based on amino acid sequence similarities. *Biochem. J.* **293**: 781–788.

Henrissat, B. and Bairoch, A. 1996. Updating the sequence-based classification of glycosyl hydrolases. *Biochem. J.* **316**: 695–696.

Henrissat, B., Driguez, H., Viet, C., and Schülein, M. 1985. Synergism of cellulases from *Trichoderma reesei* in the degradation of cellulose. *Biotechnology* **3**: 722–726.

Henrissat, B., Claeyssens, M., Tomme, P., Lemesle, L., and Mornon, J.-P. 1989. Cellulase families revealed by hydrophobic cluster analysis. *Gene* **81**: 83–95.

Hoffrèn, A.-M., Teeri, T. T., and Teleman, O. 1995. Molecular dynamic simulation of fungal cellulose-binding domains: differences in molecular rigidity but a preserved cellulose binding surface. *Protein Engin.* **8**: 443–450.

Irwin, D., Spezio, M., Walker, L. P., and Wilson, D. B. 1993. Activity studies of eight purified cellulases: specificity, synergism and binding domain effects. *Biotechnol. Bioeng.* **42**: 1002–1013.

Jones, T. A., Zou, J.-Y., Cowan, S. W., and Kjeldgaard, M. 1991. Improved methods for building protein models in electron density maps and the location of errors in these models. *Acta Crystallograph.* **A47**: 110–119.

Kasumi, T., Tsumuraya, Y., Brewer, C. F., Kersters-Hilderson, H., Claeyssens, M., and Hehre, E. J. 1987. Catalytic versatility of *Bacillus pumilus* β-xylosidase: glycosyl transfer and hydrolysis promoted with α- and β-D-xylosyl fluoride. *Biochemistry* **26**: 3010–3016.

Kleman-Leyer, K. M., Siika-aho, M., Teeri, T. T., and Kirk, T. K. 1996. The cellulases Endoglucanase I and Cellobiohydrolase II act synergistically to solubilize native cotton cellulose but not to decrease its molecular size. *Appl. Environ. Microbiol.* **62**: 2883–2887.

Kleywegt, G. and Jones, T. A. 1994a. Detection, delineation, measurement and display of cavities in macromolecular structures. *Acta Crystallograph.* **D50**: 178–185.

Kleywegt, G. and Jones, T. A. 1994b. Halloween . . . masks and bones. In: S. Bailey, R. Hubbard, and D. A. Walker (eds), *From First Map to Final Model.* SERC Daresbury Laboratory, Daresbury, pp. 59–66.

Kleywegt, G. J., Zou, Y.-J., Divne, C., Davies, G. J., Sinning, I., Ståhlberg, J., Reinikainen, T., Srisodsuk, M., Teeri, T. T., and Jones, T. A. 1997. The crystal structure of the catalytic core domain of endoglucanase I from *Trichoderma reesei* at 3.6 Å resolution, and a comparison with related enzymes. *J. Mol. Biol.* **272**: 383–397.

Klyosov, A. 1990. Trends in biochemistry and enzymology of cellulose degradation. *Biochemistry* **29**: 10577–10585.

Knowles, J. K. C., Lehtovaara, P., and Teeri, T. T. 1987. Cellulase families and their genes. *TIBTECH* **5**: 255–261.

Knowles, J. K. C., Lehtovaara, P., Murray, M., and Sinnott, M. L. 1988. Stereochemical course of the action of cellobioside hydrolases I and II of *Trichoderma reesei. J. Chem. Soc. Chem. Commun.* **1988**: 1401–1402.

Koivula, A. 1996. Structure–function relationships of two polysaccharide-degrading enzymes. Ph.D. Thesis, *VTT Publications* **277**, VTT Offsetpaino, Espoo, 97 pp. + app 45 pp.

Koivula, A., Reinikainen, T., Ruohonen, L., Valkeajärvi, A., Claeyssens, M., Teleman, O., Kleywegt, G., Szardenings, M., Rouvinen, J., Jones, T. A., and Teeri, T. T. 1996. The active site of *Trichoderma reesei* cellobiohydrolase II: the role of tyrosine-169. *Protein Engin.* **9**: 691–699.

Konstantinidis, A., Marsden, I., and Sinnott, M. L. 1993. Hydrolysis of α- and β-cellobiosyl fluorides by cellobiohydrolases of *Trichoderma reesei. Biochem. J.* **291**: 883–888.

Koshland, D. E. Jr. 1953. Stereochemistry and the mechanism of enzymatic reactions. *Biol. Rev.* **28**: 416–436.

Kraulis, P. 1991. MOLSCRIPT: a program to produce both detailed and schematic plots of protein structures. *J. Appl. Crystallograph.* **24**: 946–950.

Kraulis, P. J., Clore, G. M., Nilges, M., Jones, T. A., Pettersson, G., Knowles, J., and Gronenborn, A. M. 1989. Determination of the three-dimensional solution structure of the C-terminal domain of cellobiohydrolase I from *Trichoderma reesei*: a study using nuclear magnetic resonance and hybrid distance geometry-dynamical simulated annealing. *Biochemistry* **28**: 7241–7257.

LINDER, M. and TEERI, T. T. 1996. The cellulose-binding domain of the major cellobiohydrolase of *Trichoderma reesei* exhibits true reversibility and a high exchange rate on crystalline cellulose. *Proc. Natl. Acad. Sci. USA* **93**: 12251–12255.

LINDER, M. and TEERI, T. T. 1997. The roles and function of cellulose-binding domains. *J. Biotechnol.* **57**: 15–28.

LINDER, M., SALOVUORI, I., RUOHONEN, L., and TEERI, T. T. 1996. Characterization of a double cellulose-binding domain: synergistic high-affinity binding to cellulose. *J. Biol. Chem.* **271**: 21268–21272.

LINDER, M., LINDEBERG, G., REINIKAINEN, T., TEERI, T. T., and PETTERSSON, G. 1995a. The difference in affinity between two fungal cellulose-binding domains is dominated by a single amino acid substitution. *FEBS Lett.* **372**: 96–98.

LINDER, M., MATTINEN, M. L., KONTTELI, M., LINDEBERG, G., STÅHLBERG, J., DRAKENBERG, T., REINIKAINEN, T., PETTERSSON, G., and ANNILA, A. 1995b. Identification of functionally important amino acids in the cellulose-binding domain of *Trichoderma reesei* cellobiohydrolase I. *Protein Sci.* **4**: 1056–1064.

MCCARTER, J. D. and WITHERS, S. G. 1994. Mechanism of enzymatic glycoside hydrolysis. *Curr. Opin. Struct. Biol.* **4**: 885–892.

MEDVE, J., STÅHLBERG, J., and TJERNELD, F. 1994. Adsorption and synergism of cellobiohydrolase I and II from *Trichoderma reesei* during hydrolysis of microcrystalline cellulose. *Biotechnol. Bioeng.* **44**: 1064–1073.

MEINKE, A., DAMUDE, H. G., TOMME, P., KWAN, E., KILBURN, D. G., MILLER, R. C., WARREN, R. A. J., and GILKES, N. R. 1995. Enhancement of the endo-β-1,4-glucanase activity of an exocellulase by deletion of a surface loop. *J. Biol. Chem.* **270**: 4383–4386.

MONTENECOURT, B. 1983. *Trichoderma reesei* cellulases. *TIBTECH* **1**: 156–161.

NEVALAINEN, H. and PENTTILÄ, M. 1995. Molecular biology of cellulolytic fungi. The Mycota II. In U. Kück, (ed.), *Genetics and Biotechnology*. Springer-Verlag, Berlin, pp. 303–319.

NIDETZKY, B., HAYN, M., MACARRON, R., and STEINER, W. 1993. Synergism of *Trichoderma reesei* cellulases while degrading different celluloses. *Biotechnol. Lett.* **15**: 71–76.

NIDETZKY, B., STEINER, W., HAYN, M., and CLAEYSSENS, M. 1994a. Cellulose hydrolysis by the cellulases from *Trichoderma reesei*: a new model for synergistic interaction. *Biochem. J.* **298**: 705–710.

NIDETZKY, B., ZACHARIAE, W., GERCKEN, G., HAYN, M., and STEINER, W. 1994b. Hydrolysis of cellooligosaccharides by *Trichoderma reesei* cellobiohydrolases: experimental data and kinetic modeling. *Enzyme Microb. Technol.* **16**: 43–52.

PALLAGHY, P. K., NIELSEN, K. J., CRAIK, D. J., and NORTON, R. S. 1994. A common structural motif incorporating a cystine knot and a triple-stranded β-sheet in toxic and inhibitory polypeptides. *Prot. Sci.* **3**: 1833–1839.

PENTTILÄ, M. E., LEHTOVAARA, P., NEVALAINEN, H., BHIKHABHAI, R., and KNOWLES, J. K. C. 1986. Homology between cellulase genes of *Trichoderma reesei*: complete nucleotide sequence of the endoglucanase I gene. *Gene* **45**: 253–263.

REESE, E. T., SIU, R. G., and LEVINSON, H. S. 1950. Biological degradation of soluble cellulose derivatives. *J. Bacteriol.* **9**: 485–497.

REINIKAINEN, T., TELEMAN, O., and TEERI, T. T. 1995. Effects of pH and high ionic strength on the adsorption and activity of native and mutated cellobiohydrolase I from *Trichoderma reesei*. *Proteins* **22**: 392–403.

ROUVINEN, J. 1990. *Three-dimensional structure and function of cellobiohydrolase II*. Ph.D. Thesis. University of Joensuu, Joensuu, Finland, 38 pp.

ROUVINEN, J., BERGFORS, T., TEERI, T. T., KNOWLES, J. K. C., and JONES, T. A. 1990. Three-dimensional structure of cellobiohydrolase II from *Trichoderma reesei*. *Science* **249**: 380–386.

RUOHONEN, L., KOIVULA, A., REINIKAINEN, T., VALKEAJÄRVI, A., TELEMAN, A., CLAEYSSENS, N., SZARDENINGS, M., JONES, T. A., and TEERI, T. T. 1993. Active site of *T. reesei* cellobiohydrolase II. In P. Suominen and T. Reinikainen (eds), Trichoderma Reesei *Cellulases and Other Hydrolases.* Foundation for Biotechnical and Industrial Fermentation Research, Helsinki, Vol. **8**: 87–96.

SINNOTT, M. L. 1990. Catalytic mechanisms of enzymatic glycosyl transfer. *Chem. Rev.* **90**: 1171–1202.

SPEZIO, M., WILSON, D. B., and KARPLUS, A. 1993. Crystal structure of the catalytic domain of a thermophilic endocellulase. *Biochemistry* **32**: 9906–9916.

SRISODSUK, M., REINIKAINEN, T., PENTTILÄ, M., and TEERI, T. T. 1993. Role of the interdomain linker peptide of *Trichoderma reesei* cellobiohydrolase I in its interaction with crystalline cellulose. *J. Biol. Chem.* **268**: 20756–20761.

SRISODSUK, M., LEHTIÖ, J., LINDER, M., MARGOLLES-CLARK, E., REINIKAINEN, T., and TEERI, T. T. 1997. *Trichoderma reesei* cellobiohydrolase I with an endoglucanase cellulose-binding domain: action on bacterial microcrystalline cellulose. *J. Biotechnol.* **57**: 49–57.

STÅHLBERG, J., DIVNE, C., KOIVULA, A., PIENS, K., CLAEYSSENS, M., TEERI, T. T., and JONES, T. A. 1996. Activity studies and crystal structures of catalytically deficient mutants of cellobiohydrolase I from *Trichoderma reesei. J. Mol. Biol.* **264**: 337–349.

TAYLOR, J. S., TEO, B., WILSON, D. B., and BRADY, J. W. 1995. Conformational modeling of substrate binding to endoglucanase E2 from *Thermomonospora fusca. Protein Engin.* **8**: 1145–1152.

TEERI, T. T. 1997. Crystalline cellulose degradation: new insight into function of cellobiohydrolases. *TIBTECH.* **15**: 160–167.

TEERI, T. T., LEHTOVAARA, P., KAUPPINEN, S., SALOVUORI, I., and KNOWLES, J. K. C. 1987. Homologous domains in *Trichoderma reesei* cellulolytic enzymes: gene sequence and expression of cellobiohydrolase II. *Gene* **51**: 42–52.

TELEMAN, A., KOIVULA, A., REINIKAINEN, T., VALKEAJÄRVI, A., TEERI, T. T., DRAKENBERG, T., and TELEMAN, O. 1995. Progress-curve analysis shows that the glucose inhibits the cellotriose hydrolysis catalysed by cellobiohydrolase II from *Trichoderma reesei. Eur. J. Biochem.* **231**: 250–258.

VAN TILBEURGH, H., LOONTIENS, F., ENGELBORGS, Y., and CLAEYSSENS, M. 1989. Studies of the cellulolytic system of *Trichoderma reesei* QM 9414: binding of small ligands to the 1,4-β-D-glucan cellobiohydrolase II and influence of glucose on their affinity. *Eur. J. Biochem.* **184**: 553–559.

VAN TILBEURGH, H., TOMME, P., CLAEYSSENS, M., BHIKHABHAI, R., and PETTERSSON, G. 1986. Limited proteolysis of the cellobiohydrolase I from *Trichoderma reesei. FEBS Lett.* **204**: 223–227.

TOMME, P., WARREN, R. A. J., and GILKES, N. R. 1995a. Cellulose hydrolysis by bacteria and fungi. *Adv. Microb. Physiol.* **37**: 1–81.

TOMME, P., WARREN, R. A. J., MILLER, R. C. JR., KILBURN, D. G., and GILKES, N. R. 1995b. Cellulose-binding domains: classification and properties. *ACS Symp. Ser.* **618**: 142–163.

TOMME, P., VAN TILBEURGH, H., PETTERSSON, G., VAN DAMME, J., VANDEKERCKHOVE, J., KNOWLES, J. K. C., TEERI, T. T., and CLAEYSSENS, M. 1988. Studies on the cellulolytic system of *Trichoderma reesei* QM 9414. *Eur. J. Biochem.* **170**: 575–581.

TOONE, E. J. 1994. Structure and energetics of protein–carbohydrate complexes. *Curr. Opin. Struct. Biol.* **4**: 719–728.

VRSANSKÁ, M. and BIELY, P. 1992. The cellobiohydrolase I from *Trichoderma reesei* QM 9414: action on cello-oligosaccharides. *Carbohydrate Res.* **227**: 19–27.

VYAS, N. 1991. Atomic features of protein–carbohydrate interactions. *Curr. Opin. Struct. Biol.* **1**: 737–740.

WARD, M., WU, S., DAUBERMAN, J., WEISS, G., LARENAS, E., BOWER, B., REY, M., CLARKSON, K., and BOTT, R. 1993. Cloning, sequence and preliminary structural analysis of a small, high pI endoglucanase (EGIII) from *Trichoderma reesei*. In P. Suominen and T. Reinikainen (eds), Trichoderma Reesei *Cellulases and Other Hydrolases*. Foundation for Biotechnical and Industrial Fermentation Research, Helsinki, Vol. **8**: pp. 153–158.

WITHERS, S. G. and AEBERSOLD, R. 1995. Approaches to labelling and identification of active site residues in glycosidases. *Protein Sci.* **4**: 361–372.

WOOD, T. M. and GARCIA-CAMPAYO, V. 1990. Enzymology of cellulose degradation. *Biodegradation* **1**: 147–161.

WOOD, T. M. and MCCRAE, S. 1972. The purification and properties of the C1 component of *Trichoderma koningii* cellulase. *Biochem. J.* **128**: 1183–1192.

WOOD, T. M. and MCCRAE, S. 1979. Synergism between enzymes involved in the solubilization of native cellulose. *Adv. Chem. Ser.* **181**: 181–209.

ZHANG, S., LAO, G., and WILSON, D. B. 1995. Characterization of a *Thermomonospora fusca* exocellulase. *Biochemistry* **34**: 3386–3395.

2

Enzymology of hemicellulose degradation

P. BIELY* and M. TENKANEN†

* *Institute of Chemistry, Slovak Academy of Sciences, Bratislava, Slovakia, and* † *VTT Biotechnology and Food Research, Espoo, Finland*

2.1 Introduction

Strains of the genus *Trichoderma* are not classified solely as a group containing the best cellulase producers. They are also very efficient producers of other types of polysaccharide hydrolases, which include a whole array of hemicellulolytic enzymes. These enzymes hydrolyze the non-cellulosic plant polysaccharides which have been named "hemicelluloses" and which are found in plant cell walls in close association with cellulose (Eriksson *et al*., 1990). Hemicelluloses are heteropolysaccharides composed of two or more monosaccharides such as D-xylose, L-arabinose, D-mannose, D-glucose, D-galactose and 4-O-methyl-D-glucuronic acid. Some of them are partially esterified with acetic, ferulic and p-coumaric acids. The primary structure of hemicelluloses depends on the source and can vary even between different tissues of a single plant. Often two or three different hemicelluloses occur in the same plant species, but in different proportions.

A major hemicellulose component of hardwood is O-acetyl-4-O-methyl-D-glucurono-D-xylan (referred to here as glucuronoxylan) and a minor component is glucomannan (Timell, 1967; Wilkie, 1983). A major softwood hemicellulose is O-acetyl-D-galacto-D-gluco-D-mannan (referred to here as galactoglucomannan). The next most abundant softwood hemicellulose is L-arabino-D-glucurono-D-xylan (called arabinoglucuronoxylan here), which, in contrast to hardwood xylan, is not acetylated. Minor hemicelluloses found in wood are various types of galactans (e.g., in compression wood) and arabinogalactans (in larchwood). The most abundant hemicelluloses of cereals and grasses are arabinoxylans (Voragen *et al*., 1992; Wilkie, 1979).

In this chapter we shall concentrate on enzyme systems that are produced by *Trichoderma* species to hydrolyze a variety of xylans and mannans occurring in nature. These enzyme systems have been most extensively investigated, because of their great industrial potential. The largest number of papers has been published on *T. reesei* QM 9414 and the hypercellulolytic mutant, *T. reesei* Rut C-30 (Montenecourt and Eveleigh, 1979). Enzyme properties deduced from the gene data and three-dimensional structures are also addressed. A special chapter of this book is devoted to the regulation of the genes encoding the enzymes (Chapter 3).

2.2 Xylan structure and enzymes required for its hydrolysis

A hypothetical fragment of a plant xylan which shows the major structural features found in this group of hemicelluloses is depicted in Figure 2.1. Every xylan backbone is built of β-1,4-linked D-xylopyranosyl residues. The xylopyranosyl residues are randomly substituted with α-1,2-linked 4-O-methyl-D-glucuronic acid and α-1,3-linked L-arabinofuranose. In cereal xylans, one xylopyranosyl residue may carry two α-L-arabinofuranosyl substituents at positions 2 and 3. Hardwood glucuronoxylans are partially acetylated. Due to easy migration of the acetyl groups between the hydroxyl groups, it is difficult to determine their actual distribution between positions 2 and 3 in native polysaccharides. In cereals, arabinoglucuronoxylans composed of up to 10% of L-arabinofuranosyl residues are esterified at position 5 by ferulic acid or p-coumaric acid (Mueller-Harvey *et al.*, 1986).

The principal xylan depolymerizing enzyme is endo-β-1,4-xylanase (called xylanase here). This enzyme cleaves mainly in unsubstituted parts of the main chain. Other enzymes work with xylanase synergistically. They attack either the polymeric xylan, creating new sites for xylanase, or they attack the linear and substituted oligosaccharides liberated by xylanase. Strains of the *Trichoderma* genus are known to produce practically all of the xylanolytic enzymes shown in Figure 2.1. Numerous enzymes have been sufficiently purified and characterized, and their genes isolated and sequenced. Three *Trichoderma* xylanases have been crystallized and their tertiary structures established. Xylanolytic systems produced by strains of *Trichoderma*, mainly by *T. reesei*, are the best-characterized enzymes within this group of plant cell wall hydrolyzing enzymes (Table 2.1). There are only a few reports on the occurrence and production of hemicellulolytic enzymes from the *Gliocladium* genus (Da Silva Carvalho *et al.*, 1992; Dean *et al.*, 1989; Gomez *et al.*, 1989; Raghukumar *et al.*, 1994; Todorovic *et al.*, 1988; van Tilburg and Thomas, 1993). However, because the *Gliocladium* enzymes have not been isolated and characterized, this

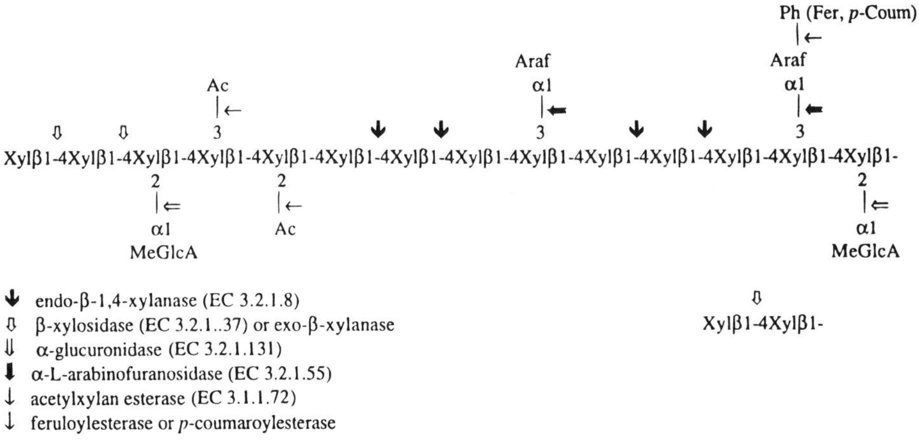

Figure 2.1 Hypothetical plant xylan and the enzymes required for its complete hydrolysis.

Table 2.1 Properties of *Trichoderma reesei* Rut C-30 hemicellulases

Enzyme activity	Enzyme designation	Gene	M_r[a] (kDa)	pI	pH-optimum	Activity against[b] oligo-	Activity against[b] polysacch.	Remarks[c]	Ref.[f]
Endo-β-1,4-xylanase	XYNI	xyn1	19	5.5	4.0–4.5			HCA 11 3-D structure known	1–4
	XYNII	xyn2	20	9.0	5.0–5.5			HCA 11 3-D structure known	1, 2, 5, 6
Endo-β-1,4-mannanase	MANI	man1	51	3.6–5.4[d]	3.5–5.0			HCA 5 contains a CBD	7–9
β-Xylosidase	BXLI	βxl1	100	4.7	4.0	+++	+	HCA 3	10–12
β-Glucosidase	BGLI	βgl1	71	8.7	4.6	+++	–	HCA 3	13, 14
	BGLII		114	4.8	4.0	+++	–		14
α-Arabino-furanosidase	ABFI	abf1	53	7.5	4.0	+++	+++	HCA 54	11, 15
α-Galactosidase	AGLI	agl1	50	5.2	4.0	+++	+	HCA 27	16–18
	AGLII	agl2	80[e]	nd	3.5–4.5	++	(+)	HCA 36	18
	AGLIII	agl3	66[e]	nd	3.5–4.5	+	(+)	HCA 27	18
α-Glucuronidase	GLRI	glr1	91	5.0–6.2	4.5–6.0	+++	(+)	HCA new	19, 20
Acetyl esterase	AE		45	6.0, 6.8[d]	5.5	++	–	Active on xylo- and mannooligos.	21–23
Acetylxylan esterase	AXEI	axe1	34	6.8, 7.0[d]	5.0–6.0	+++	+++	contains a CBD	22, 24, 25

[a] Determined by SDS-PAGE.
[b] Only the activity of accessory enzymes on oligosaccharides and polysaccharides is indicated.
[c] HCA and the number, glycosyl hydrolase family determined by hydrophobic cluster analysis; CBD, cellulose binding domain.
[d] Contains several enzyme forms.
[e] Calculated from the deduced amino acid sequence.
[f] References: 1, Tenkanen *et al.*, 1992; 2, Törrönen *et al.*, 1992; 3, Suominen *et al.*, 1992; 4, Törrönen and Rouvinen, 1995; 5, Saarelainen *et al.*, 1993; 6, Törrönen *et al.*, 1994; 7, Stålbrand *et al.*, 1993; 8, Arisan-Atac *et al.*, 1993; 9, Stålbrand *et al.*, 1995; 10, Poutanen and Puls, 1988; 11, Margolles-Clark *et al.*, 1996c; 12, Herrmann *et al.*, 1997; 13, Barnett *et al.*, 1991; 14, Chen *et al.*, 1992; 15, Poutanen, 1988; 16, Zeilinger *et al.*, 1993; 17, Kachurin *et al.*, 1995; 18, Margolles-Clark *et al.*, 1996b; 19, Siika-aho *et al.*, 1994; 20, Margolles-Clark *et al.*, 1996a; 21, Poutanen and Sundberg, 1988; 22, Poutanen *et al.*, 1990; 23, Tenkanen *et al.*, 1993; 24, Sundberg and Poutanen, 1991; 25, Margolles-Clark *et al.*, 1996d.

chapter is devoted only to hemicellulolytic enzymes of fungi in the genus *Trichoderma*.

2.2.1 Endo-β-1,4-xylanases

Extensive literature available on the main xylanases of *Trichoderma* species has been reviewed by Wong and Saddler (1992). A generalization that emerged from this review was that *Trichoderma* species (*T. harzianum, T. koningii, T. lignorum, T. longibrachiatum, T. pseudokoningii, T. reesei* and *T. viride*) produce multiple xylanases that can be divided into three groups (Table 2.2). Several strains produce a pair of low molecular mass xylanases having a molecular mass of between 18 and 23 kDa. Some of the strains produce an additional xylanase which is larger (29–33 kDa). Both these types of xylanases belong to the category of specific xylanases, i.e., they do not hydrolyze cellulose. There seems to be a difference in the catalytic properties of the two types of xylanase. A medium size xylanase of *T. harzianum* was found to hydrolyze xylan to a much higher degree than three other low MW xylanases from *Trichoderma* (Maringer *et al.*, 1995). Several *Trichoderma* strains were also found to produce a xylan-depolymerizing enzyme of 53–57 kDa. These enzymes have been identified as non-specific endo-β-1,4-glucanases, which are components of the cellulolytic rather than the xylanolytic system (Biely, 1985). The best-known non-specific endoglucanase is the EGI of *T. reesei* (Bailey *et al.*, 1993; Biely *et al.*, 1991; Claeyssens *et al.*, 1990; Zurbriggen *et al.*, 1991). While xylanases are produced during growth of the fungus on both cellulose and xylan or induced by xylobiose, EGI is produced only during growth on cellulose or may be induced by sophorose (Hrmová *et al.*, 1986).

Grouping of *Trichoderma* xylanases on the basis of molecular mass and pI values (Wong *et al.*, 1988) is in good agreement with the classification of glycosyl hydrolases on the basis of hydrophobic cluster analysis and sequence similarities (Davies and Henrissat, 1995; Gilkes *et al.*, 1991; Henrissat and Bairoch, 1993). The low molecular mass xylanases of *Trichoderma* belong to the glycanase family 11 (formerly family G). Complete amino acid sequences have been deduced from the gene sequence of four *Trichoderma* enzymes: two xylanases (XYNI and XYNII) from *T. reesei* (Törrönen *et al.*, 1992), one from *T. viride* (Yaguchi *et al.*, 1992a) and one from *T. harzianum* (Yaguchi *et al.*, 1992b). The xylanase of *T. harzianum* (Campbell *et al.*, 1993) and two xylanases of *T. reesei* (Törrönen *et al.*, 1994; Törrönen and Rouvinen, 1995) have been crystallized and their tertiary structures deter-

Table 2.2 Three different types of endo-β-1,4-xylanases known to be produced by strains of the genus *Trichoderma*. A summary based on the grouping of enzymes by Lappalainen (1988) and Wong and Saddler (1992)

Enzyme group	Molecular mass (kDa)	pI value	Hydrolysis of cellulose	Enzyme family
1	18–23	mostly 8–9.5	no	11
2	29–33	mostly 7–9.5	no	unknown, probably 10
3	53–57	4.5–5.3	yes	7 (non-specific glucanase)

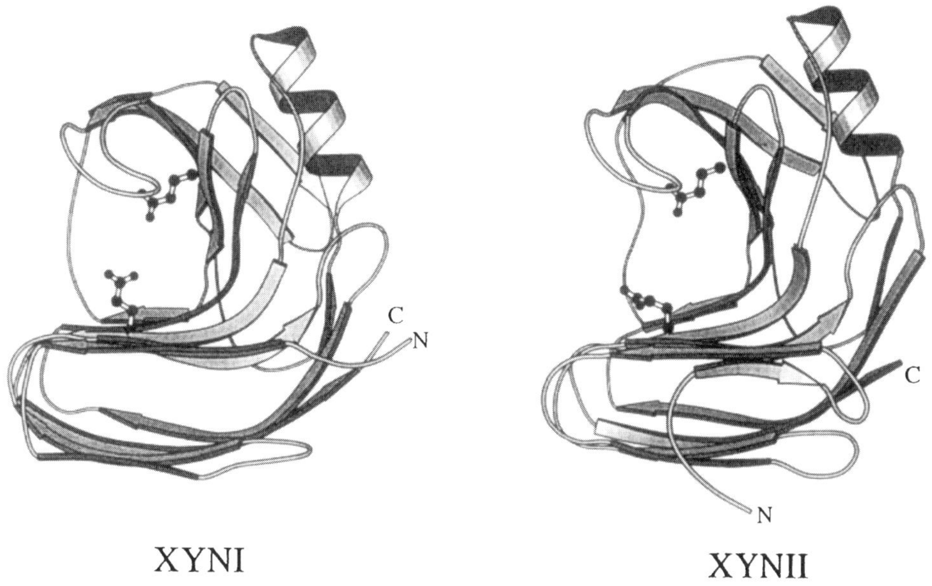

Figure 2.2 Ribbon representation of the main fold of the two specific xylanases of *Trichoderma reesei*. The carboxyl groups in the clefts are the carboxyl groups of two catalytic glutamic acid residues and they are indicated by balls and sticks. Despite the great similarity in tertiary structure, the enzymes show differences in catalytic properties, pointing to differences in the substrate binding sites. Figure is courtesy of Dr J. Rouvinen.

mined (Figure 2.2). These enzymes appear as small, well-packed molecules formed mainly of β-sheets. The β-sheet structure is twisted, forming a cleft where the catalytic groups are located. Two glutamic acid residues in the cleft function as catalytic groups. One residue serves as a proton donor and the other as a nucleophile/base. Both residues have been labelled in XYNII from *T. reesei* by epoxyalkyl-β-D-xylosides (Havukainen *et al.*, 1996). The loop hanging over the cleft (Figure 2.2) undergoes a conformational change on substrate binding (Havukainen *et al.*, 1996; Törrönen *et al.*, 1994).

The larger xylanases of *Trichoderma* species (Table 2.2) could belong to the glycanases of family 10 (formerly family F), but the experimental evidence for this is still lacking. None of the genes encoding larger xylanases has been sequenced or subjected to hydrophobic cluster analysis.

The two low molecular mass specific xylanases from *T. reesei* Rut C-30, one acidic (pI 5.5, 19 kDa) assigned as XYNI and one alkaline (pI 9.0, 20 kDa) assigned as XYNII (Tenkanen *et al.*, 1992; Törrönen *et al.*, 1992), are the most extensively studied enzymes of this group, and some of their catalytic properties might correspond to those of other *Trichoderma* xylanases. This will be particularly true for XYNII, which shows more than 90% homology with low molecular mass xylanases of *T. harzianum* (xylanase A) and *T. viride*. XYNI of *T. reesei* shows only 50% homology with the above-mentioned enzymes (Törrönen *et al.*, 1992). Despite the low homology, the tertiary structures of both enzymes appear similar (Figure 2.2).

Similarly, as has been demonstrated for other representative xylanases of family 11 (Gebler *et al.*, 1992), both xylanases of *T. reesei* utilize the mechanism of hydrolysis that is associated with the retention of the configuration of the glycosidic

Table 2.3 Structure of the shortest and shortest plus one residue heterooligosaccharides generated from the corresponding polysaccharides by xylanases and mannanase of *T. reesei* after extensive hydrolysis

Substrate	Enzyme	Shortest products	Shortest plus one products
Glucuronoxylan	XYNI	Xylβ1-4Xylβ1-4Xylβ1-4Xyl 2 | α1 MeGlcA	Xylβ1-4Xylβ1-4Xylβ1-4Xylβ1-4Xyl 2 | α1 MeGlcA
	XYNII	Xylβ1-4Xylβ1-4Xylβ1-4Xyl 2 | α1 MeGlcA	Xylβ1-4Xylβ1-4Xylβ1-4Xylβ1-4Xyl 2 | α1 MeGlcA [a]Xylβ1-4Xylβ1-4Xylβ1-4Xylβ1-4Xyl 2 | α1 MeGlcA
Arabinoxylan	XYNI, XYNII	Xylβ1-4Xylβ1-4Xyl 3 | α1 L-Ara*f*	Xylβ1-4Xylβ1-4Xylβ1-4Xyl 3 | α1 L-Ara*f* Xylβ1-4Xylβ1-4Xylβ1-4Xyl 3 | α1 L-Ara*f*
Galactoglucomannan	MANI	Glcβ1-4Man	Glcβ1-4Glcβ1-4Man [a]Glcβ1-4Manβ1-4Man Man1-4Man 6 | α1 Gal [a]Glcβ1-4Man 6 | α1 Gal

[a] Minor products.

linkage (Biely *et al.*, 1994) which is in accordance with the fact that both enzymes catalyze glycosyl transfer reactions at high substrate concentrations (Biely *et al.*, 1993).

The two xylanases of *T. reesei* do not differ significantly in their action on 4-O-methyl-D-glucuronoxylan. They hydrolyze the polysaccharide to about the same degree, but the ratio of xylose to xylobiose produced was higher with XYNI than with XYNII. The major, and the shortest, acidic fragment released by both enzymes is aldopentauronic acid of structure Xylβ1-4[MeGlcAα1-2]Xylβ1-4Xylβ1-4Xyl (Biely *et al.*, 1993) (Table 2.3). Differences are found in the structure of larger aldouronic acids found among the reaction products. Both endoxylanases produce aldohexauronic acid Xylβ1-4Xylβ1-4[MeGlcAα1-2]Xylβ1-4Xylβ1-4Xyl, however only XYNII produces the isomeric product, Xylβ1-4[MeGlcAα1-2]Xylβ1-4Xylβ1-4Xylβ1-4Xyl (M. Tenkanen, unpublished data). Differences have also been observed in the structure of larger fragments liberated by the enzymes from arabinoxylan. However, the shortest arabinose-containing fragment is in both cases identical, Xylβ1-4[Araα1-3]Xylβ1-4Xyl (M. Tenkanen, unpublished data) (Table 2.3). The enzymes also act differently on other polysaccharides. XYNI hydrolyzes to a much higher degree O-acetylglucuronoxylan and rhodymenan, a β-1,3-β-1,4-xylan, producing in both cases shorter oligomers as end products (Biely *et al.*, 1993).

There are also clear differences in the bond cleavage frequencies of xylooligosaccharides by XYNI and XYNII (Figure 2.3) (Biely *et al.*, 1993), which were interpreted in view of the differences in the arrangement of their substrate binding sites (Törrönen and Rouvinen, 1995). Other catalytic properties of XYNI, distinct from those of XYNII, were appreciable hydrolysis of xylobiose, degradation of aryl β-D-xylopyranosides at high substrate concentration accompanied by the synthesis of oligosaccharides and a β-1,3-xylosyl transfer as a side reaction to the main β-1,4-xylosyl transfer (Biely *et al.*, 1993). The greater tolerance to some xylan substituents and the greater catalytic versatility of XYNI remains a characteristic of xylanases of family 10 (Biely *et al.*, 1997). Therefore, the acidic XYNI of *T. reesei* seems to fulfill the role of a high molecular mass, low pI value xylanase. Although clearly a member of family 11, XYNI shows catalytic properties intermediate between the xylanases of family 10 and 11. The comparison of the tertiary structures of XYNI and XYNII

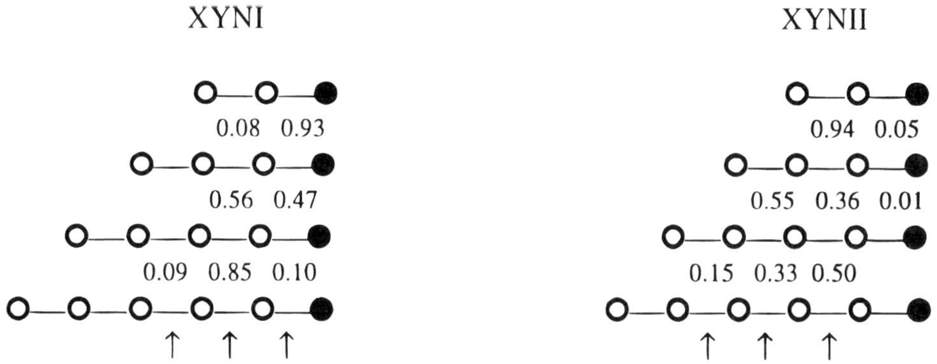

Figure 2.3 Initial bond cleavage frequencies of [1-^{3}H]-xylooligosaccharides and sites of cleavage of xylohexaose (arrows) by XYNI and XYNII of *T. reesei*. Concentration of substrates was 0.25 mM. Full circles mark the reducing-end xylopyranosyl residues. From Biely *et al.* (1993).

(Figure 2.2) strongly supports the view that they have evolved from the same ancestral gene (Törrönen *et al.*, 1993). Different ancestral genes may be assumed for the evolution of xylanases of family 10 (Harris *et al.*, 1994). Surprisingly, we still do not understand the reasons for the multiplicity of xylanases produced by microorganisms. The low molecular mass xylanases of some fungi, including XYNII of *T. reesei*, were shown to induce ethylene production in biosynthesis in a variety of plant tissues as an early response to invasion by pathogenic microorganisms (Dean *et al.*, 1989; Dean and Anderson, 1991). It is unknown whether the phenomenon is related to the generation of xylan fragments, because a similar effect of family 10 xylanases has not yet been demonstrated.

2.2.2 *β-Xylosidases*

There is much less information on this important saccharifying hydrolase than on xylanases. The main reason could be the fact that the strains of *Trichoderma* do not belong to the best producers of this enzyme (Reese *et al.*, 1973) and that it was unclear for a long time what kind of *β*-xylosidase is actually produced, whether it is a xylobiase type hydrolase or an exo-*β*-xylanase, or in other words whether the enzyme prefers dimers or long chain oligosaccharides/polymers as substrates.

The first *Trichoderma β*-xylosidase was studied by Dekker (1983) in *T. reesei* QM 94141. He pointed out the importance of the enzyme in overall hydrolysis yields of xylan and reported the first data on enzyme properties. *β*-Xylosidases that were purified later from *T. viride* and *T. reesei* exhibited very similar properties (Lappalainen, 1986; Matsuo and Yasui, 1984; Poutanen and Puls, 1988). The enzymes hydrolyzed xylooligosaccharides, 4-nitrophenyl *β*-D-xylopyranoside and showed a low activity on 4-nitrophenyl α-L-arabinofuranoside. Their aryl *β*-xylosidase activity is competitively inhibited by D-xylose.

Recent findings that the formation of *β*-xylosidase in *T. reesei* can be considerably specifically enhanced by D-xylose at low pH (Kristufek *et al.*, 1995) enabled the isolation of larger amounts of this enzyme and made it possible to investigate its catalytic properties in more detail (Herrmann *et al.*, 1997). It has been demonstrated that *β*-xylosidase hydrolyzed *β*-1,4-xylooligosaccharides with a degree of polymerization 2 to 7, at a fixed concentration, with the rate rising with increasing chain length. The enzyme also attacked to a certain extent a debranched beechwood xylan (Lenzing) and a glucuronoxylan, forming D-xylose as the only product (Herrmann *et al.*, 1997; Margolles-Clark *et al.*, 1996c). These results suggested that the enzyme could represent an exo-*β*-xylanase, specifically, a *β*-D-xylan xylohydrolase. The enzyme showed clear preference for *β*-1,4-xylosidic linkages which were hydrolyzed approximately three times faster than *β*-1,3-xylosidic linkages and more than 10 times faster than *β*-1,2-xylosidic linkages (Herrmann *et al.*, 1997). The enzyme is apparently a retaining hydrolase, because it readily catalyzed glycosyl transfer reactions at high concentrations of xylooligosaccharides and 4-nitrophenyl *β*-D-xylopyranoside.

A series of substrates, either chemically synthesized or obtained from enzymic hydrolysate of xylans, was used to define the *β*-xylosidase requirements for the environment of the non-reducing terminal xylopyranosyl residue (Herrmann *et al.*, 1997; Tenkanen *et al.*, 1996a). The enzyme liberates at a similar rate a xylopyranosyl residue which is *β*-1,4-linked to an unsubstituted xylopyranosyl residue (e.g., linear xylooligosaccharides) or a residue which is substituted at position 2, e.g., by 4-O-methylglucuronic acid or D-xylose. The enzyme does not liberate a terminal xylopyranosyl residue which is linked to a 3-substituted xylopyranosyl residue, e.g., by L-arabinofuranose or D-xylose. In such compounds the substitutent is apparently too close to the glycosidic oxygen, so the formation of the enzyme–substrate complex is sterically impaired. A summary of the structural requirements for the substrate of *T. reesei* β-xylosidase is shown in Figure 2.4.

The *Trichoderma* β-xylosidases were reported to exhibit the α-L-arabinofuranosidase activity when tested on the corresponding 4-nitrophenyl glycoside (Herrmann *et al.*, 1997; Matsuo and Yasui, 1984; Poutanen and Puls, 1988). This fact has been confirmed by cloning the *T. reesei* β-xylosidase gene in yeast (Margolles-Clark *et al.*, 1996c). The product expressed in yeast displayed α-L-arabinofuranosidase activity similar to the corresponding enzyme purified from the *T. reesei* gene. However, this activity does not seem to be relevant to the liberation of L-arabinose from arabinoxylan or arabinose-substituted xylooligosaccharides. The enzyme liberated no D-xylose or L-arabinose from the compound β-D-Xyl*p*-(1 → 4)-[α-L-Ara*f*-(1 → 3)]-β-D-Xyl*p*-O-Me (Herrmann *et al.*, 1997), and no L-arabinose from arabinoxylans (Margolles-Clark *et al.*, 1996c).

The isolation of the gene encoding β-xylosidase (*bxl1*) enabled the comparison of the enzyme with other hydrolases. The deduced amino acid sequence of the *T. reesei* β-xylosidase did not show any similarity with other β-xylosidases (glycosyl hydrolase families 39, 43 and 52) described so far, but it showed a significant homology with a conserved group of β-glucosidases grouped into the glycosyl hydrolases of family 3 (Margolles-Clark *et al.*, 1996c).

2.2.3 Enzymes acting on side groups in xylans

The role of α-glucuronidase, α-arabinofuranosidase, and acetyl xylan esterase is to liberate side groups from different xylans (Figure 2.1) or from substituted oligosaccharides formed from the polysaccharide by xylanase. The removal of side groups makes the xylopyranosyl residues of either the xylan main chain or xylooligosaccharides more accessible to degradation by xylanase or β-xylosidase. *T. reesei* is a good source of different side-group cleaving enzymes (Poutanen *et al.*, 1987). The properties and action of these enzymes from *T. reesei* Rut C-30 have been extensively studied, but there are only a few reports from other *Trichoderma* species.

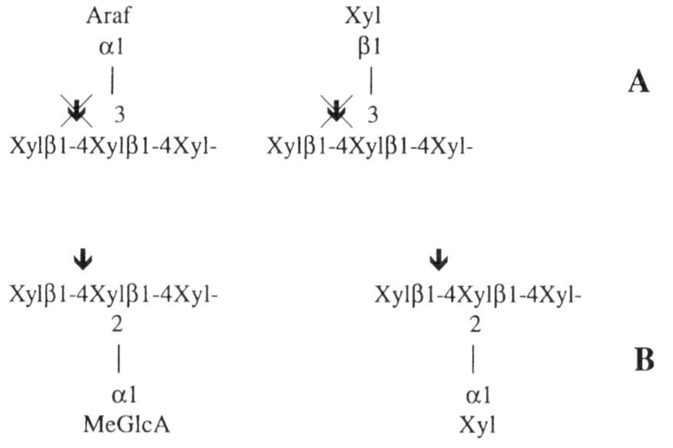

Figure 2.4 Structural arrangement with non-reducing terminal D-xylopyranosyl residues resistant (part A) and accessible (part B) to hydrolysis by *T. reesei* β-xylosidase/exo-β-xylanase.

α-Glucuronidases

The presence of α-glucuronidase in the hemicellulolytic system of *T. reesei* was first demonstrated by Dekker (1983). *T. reesei* produces one main α-glucuronidase, GLRI, which has been purified and characterized and the corresponding gene (*glr1*) has recently been isolated (Margolles-Clark *et al.*, 1996a; Siika-aho *et al.*, 1994). α-Glucuronidase is a rather large protein with a molecular mass of 91 kDa. The *glr1* gene was the first microbial α-glucuronidase gene cloned and characterized. The deduced amino acid sequence of α-glucuronidase does not show any significant similarity with any protein sequence available in the data bases. Probably the α-glucuronidases represent a completely new group of glycosyl hydrolases.

α-Glucuronidase has a very strict substrate requirement. Its activity on the polymeric substrate, a glucuronoxylan, is negligible. It acts almost exclusively on xylooligosaccharides which carry the 4-O-methylglucuronic acid linked to the terminal xylopyranosyl unit at the non-reducing end of the oligosaccharide (Siika-aho *et al.*, 1994). Thus α-glucuronidase of *T. reesei* works efficiently only in concert with xylanase and *β*-xylosidase (Figure 2.5). The α-glucuronidase studied from *T. viride* shows properties similar to the *T. reesei* α-glucuronidase, and the enzymes seem to be closely related (Ishihara *et al.*, 1990).

α-Arabinofuranosidases

T. reesei has been found to produce significant amounts of α-arabinofuranosidase (Poutanen *et al.*, 1987). The main enzyme form, ABFI, has been purified and characterized (Poutanen, 1988). Also, the gene encoding this enzyme (*abf1*) has been isolated (Margolles-Clark *et al.*, 1996c). The deduced amino acid sequence of α-arabinofuranosidase did not show similarity with any of the characterized bacterial α-arabinofuranosidases but it is highly similar throughout the complete sequence to one of the α-arabinofuranosidases of *Aspergillus niger*, ABF B, and with an enzyme encoded by the *xyl1* gene of *T. koningii*. These two enzymes have recently been classified to form family 54 of glycosyl hydrolases to which the α-arabinofuranosidase of *T. reesei* also belongs. *A. niger* and *T. reesei* α-arabinofuranosidases also show similar substrate specificity (Margolles-Clark *et al.*, 1996c; Poutanen, 1988; Rombouts *et al.*, 1988). The catalytic properties of the *T. koningii* enzyme encoded by the *xyl1* gene have not been sufficiently characterized, but given the high amino acid similarity to *T. reesei* α-arabinofuranosidase, the enzyme most probably is an α-arabinofuranosidase and not a *β*-xylosidase (Margolles-Clark *et al.*, 1996c).

The α-arabinofuranosidase of *T. reesei* has a much wider substrate specificity than the α-glucuronidase. The α-arabinofuranosidase is able to remove α-1,3-linked arabinofuranoside side groups from single substituted xylopyranosyl units both

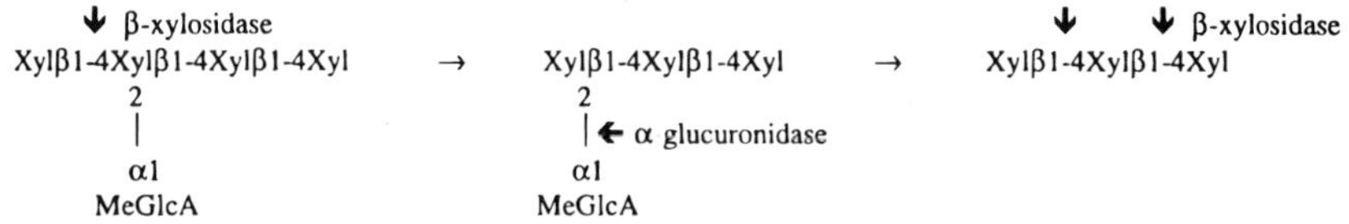

Figure 2.5 Sequence of enzymic reactions of the degradation of aldopentauronic acid liberated from glucuronoxylan by xylanases of *T. reesei*. Before the cleavage of 4-O-methylglucuronosyl residue takes place, the non-reducing xylopyranosyl residue has to be liberated by *β*-xylosidase.

from xylooligosaccharides and polymeric xylan (Margolles-Clark *et al.*, 1996c; Poutanen, 1988). Up to 70% of the arabinofuranosyl side groups in arabinoglucuronoxylan from pine kraft pulp could be removed by α-arabinofuranosidase alone (Margolles-Clark *et al.*, 1996c). The liberation of α-1,3-linked or α-1,2-linked arabinofuranoside groups in double substituted xylopyranosyl units has not been studied. A comparison of the amount of arabinose liberated from wheat and rye arabinoxylans and from a pine arabinoglucuronoxylan which contains only single α-1,3-linked arabinofuranosyl substituents, in relation to the total arabinose content in the polysaccharides, suggests that α-arabinofuranosidase is not able to attack the arabinofuranosyl residues in double substituted xylopyranosyl units (Margolles-Clark *et al.*, 1996c).

α-Arabinofuranosidase is somewhat synergistic with xylanase. The liberation of arabinose by 500 nkat of *T. reesei* α-arabinosidase per gram of arabinoxylan was clearly increased in the presence of xylanase (Poutanen, 1988). Thus, α-arabinosidase seems to show a preference for arabinose-substituted oligosaccharides above the polymeric substrate. However, the short arabinose-containing xylooligosaccharides with three or four xylose residues are not hydrolyzed as efficiently by the α-arabinofuranosidase of *T. reesei* as oligosaccharides with five or more xylose residues (E. Luonteri, unpublished data). *T. reesei* α-arabinosidase has also been reported to have some β-xylosidase activity on p-nitrophenyl-β-D-xylopyranoside (Margolles-Clark *et al.*, 1996c; Poutanen, 1988). However, because the enzyme does not hydrolyze xylobiose, it does not seem to have real β-xylosidase activity (Tenkanen *et al.*, 1996b). α-Arabinofuranosidase of *T. reesei* is also active on arabinofuranosyl side groups in arabinogalactans (E. Luonteri, unpublished data).

Esterases acting on xylans

T. reesei and *T. viride* were among the first group of microorganisms in which the acetylxylan esterase (AXE) was demonstrated for the first time (Biely *et al.*, 1985). The strains *T. reesei* QM 9414 and Rut C-30 produce the highest levels of the enzyme when grown on xylan, cellulose or a mixture of these two polysaccharides (Biely *et al.*, 1988). First fractionation of the enzymic systems produced on cellulose indicated the occurrence of multiple esterases in *T. reesei* (Biely *et al.*, 1987). Since then two different esterases acting on acetylated xylooligosaccharides and xylans have been purified from *T. reesei* (Poutanen and Sundberg, 1988; Sundberg and Poutanen, 1991). The esterase active on xylooligosaccharides was named acetyl esterase (AE) and the other, which was also active on polymeric xylan, was named acetylxylan esterase. Both enzymes were found to have several isoforms. The recently cloned *axe1* gene of *T. reesei* encodes at least two slightly different forms of acetylxylan esterase (Margolles-Clark *et al.*, 1996d).

The deduced amino acid sequence of acetylxylan esterase shows no similarity with the published acetylxylan esterase amino acid sequences. Interestingly, it shows some similarity with fungal cutinases, which are serine esterases that hydrolyze cutin. Although the overall similarity of acetylxylan esterase to the cutinases is low (about 10%), the protein alignment shows that the catalytically important residues and cysteins which form the necessary disulfide bridges in cutinases appear well conserved in acetylxylan esterase. Inhibition studies with PMSF confirmed that acetylxylan esterase is most probably a serine esterase (Margolles-Clark *et al.*, 1996d).

Another interesting feature is that the C-terminus of acetylxylan esterase contains a fungal type cellulose binding domain (CBD), as found in most *T. reesei* cellulases, separated from the catalytic core by a typical linker region. Acetylxylan esterase also binds specifically to cellulose, an interaction which was shown to be mediated by the CBD. However, the CBD was not found to be important for the catalytic activity of acetylxylan esterase. Removal of the CBD did not affect its action towards soluble or fibre bound xylan (Margolles-Clark *et al.*, 1996d). Acetylxylan esterase is highly active on polymeric xylan and can liberate up to 90% of the acetyl substituents (Poutanen *et al.*, 1990). Only very modest synergism of acetylxylan esterase with xylanase and β-xylosidase has been observed. The addition of α-glucuronidase also failed to significantly enhance the action of acetylxylan esterase (Tenkanen *et al.*, 1996b).

Investigations of the mechanism of acetylxylan esterase from *T. reesei* (P. Biely *et al.*, unpublished data) on penta- and di-O-acetyl derivatives of methyl β-D-xylopyranoside have shown that the primary targets of the enzyme are the acetyl groups at positions 2 and 3. Like the acetylxylan esterase of *Streptomyces lividans* (Biely *et al.*, 1996a), the enzyme catalyzes deacetylation of positions 2 and 3 about 20 times faster when the other position is not acetylated. In other words, the 2,3,4-tri-O-acetyl and 2,3-di-O-acetyl methyl β-D-xylopyranosides were deacetylated 20 times slower than 2,4- and 3,4-di-O-acetates. This suggested that the deacetylation at positions 2 or 3 involves an enzyme-catalyzed formation of a five-membered intermediate transition state (Biely *et al.*, 1996a). Consequently, the enzyme may be expected to deacetylate a natural polymeric substrate much faster if the xylopyranosyl residues are singly acetylated rather than doubly acetylated.

The other esterase of *T. reesei*, acetyl esterase (AE), has clear activity against short oligomeric and monomeric acetates (Poutanen and Sundberg, 1988; Poutanen *et al.*, 1990). It shows a somewhat similar substrate specificity to that of α-glucuronidase of *T. reesei*, having high synergy with β-xylosidase (Tenkanen and Poutanen, 1992). AE has also been shown to have high activity towards xylobiose acetylated on the non-reducing xylopyranosyl residue (Poutanen *et al.*, 1990). AE is needed, together with xylanase, β-xylosidase and α-glucuronidase, for the removal of the last 10% of acetyl substituents in acetylglucuronoxylan (Tenkanen *et al.*, 1996b). The acetyl groups not accessible to acetylxylan esterase but accessible to acetyl esterase seem to be located close to the 4-O-methylglucuronic acid side groups (Puls, 1992; Tenkanen *et al.*, 1996b).

Neither the acetyl esterase nor the acetylxylan esterase of *T. reesei* are active towards feruloyl or p-coumaroyl groups in arabinoxylans (Tenkanen, 1995). No activity against these side groups has been detected in *T. reesei* culture filtrate using wheat straw xylan fragments as a substrate (Puls and Poutanen, 1989). Smith *et al.* (1991), however, detected both esterase activities in *T. reesei* culture filtrate using methyl ferulate and methyl p-coumarate as substrates.

2.3 β-Mannan structures and enzymes required for their hydrolysis

The principal hemicellulose in softwoods, galactoglucomannan, occurs in two forms that differ in the galactose content (Timell, 1967). In the low galactose form the ratio of galactose:glucose:mannose is about 0.1:1:4, while in the high galactose form the ratio is 1:1:3. The backbone of both types of galactoglucomannan is built of β-1,4-linked D-mannopyranosyl and D-glucopyranosyl residues distributed randomly. The α-D-galactopyranosyl residues are linked to the D-mannopyranosyl residues in the backbone by α-1,6-linkages. Approximately every fourth hexopyranosyl residue in the backbone is partially acetylated at positions 2 or 3. Migration of the acetyl group is similar to the case of xylans, and thus it is difficult to determine their actual distribution in natural polymers. Hardwoods contain only some glucomannan (Timell, 1967). Glucomannans and galactomannans with several structural variations also occur in several annual plants, especially in seeds, tubers and bulbs (Aspinall, 1959; McCleary, 1979).

The main structural features of a plant galactoglucomannan and the enzymes required for its complete hydrolysis are shown in Figure 2.6. The enzyme hydrolyzing the backbone of mannans is endo-β-1,4-mannanase. The terminal residues of oligosaccharides can be liberated by β-mannosidase and β-glucosidase. The side groups linked to the backbone are hydrolyzed by α-galactosidase and acetylesterase, which is often acetylglucomannan esterase.

2.3.1 *Endo-β-1,4-mannanases*

The endo-β-mannanases are enzymes that hydrolyze the β-1,4-glycosidic linkages in mannans, galactomannans, glucomannans and galactoglucomannans. As glucomannans and galactoglucomannans contain, in addition to mannose units, glucose units in the backbone of the polymers, they might also be hydrolyzed by specific endoglycanases (endo-β-glucomannanases) which do not act on polymers with a mannan backbone and therefore cannot be called mannanases.

Mannanases of different *Trichoderma* species have been studied much less than the xylanases, and the only reports are on mannanases of *T. harzianum* E58 and *T. reesei* Rut C-30. Both species produce several forms of mannanase (Torrie *et al.*, 1990; Stålbrand *et al.*, 1993). Mannanase was produced when either cellulose or different mannans were used as a carbon source. *T. harzianum* produced almost the

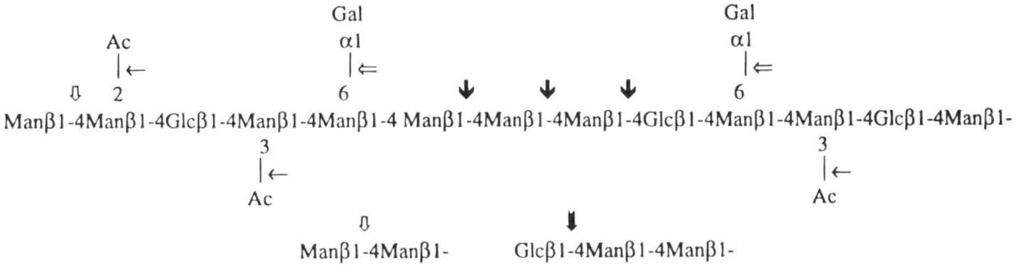

- ⬇ endo-β-1,4-mannanase (EC 3.2.1.78)
- ⇩ β-mannosidase or exo-β-mannanase (EC 3.2.1.25)
- ⇓ α-galactosidase (EC 3.2.1.22)
- ⬇ β-glucosidase (EC 3.2.1.21)
- ↓ acetylgalactoglucomannan esterase

Figure 2.6 Hypothetical plant galactoglucomannan and the enzymes required for its complete hydrolysis. The β-1,4-glucopyranosyl linkages in the main chain are hydrolyzed by endo-β-1,4-glucanases (Siika-aho, unpublished data).

same amount of mannanase activity on galactomannan as on cellulose, but the specific mannanase activity was clearly higher on galactomannan (Torrie *et al.*, 1990). *T. reesei*, on the other hand, was found to produce higher activity in the presence of cellulose than galactomannan (Arisan-Atac *et al.*, 1993; Rättö and Poutanen, 1988). However, the fungus did not grow well on galactomannan, and the amount of mannanases produced per biomass unit was similar on cellulose and galactomannan (Arisan-Atac *et al.*, 1993). No induction of mannanase synthesis by mannobiose or mannose was observed (Arisan-Atac *et al.*, 1993; Torrie *et al.*, 1990).

The only *Trichoderma* mannanase purified and studied in detail is from *T. reesei* Rut C-30 (Arisan-Atec *et al.*, 1993; Stålbrand *et al.*, 1993). At least five enzyme forms with mannanase activity could be detected in the culture filtrate (Stålbrand *et al.*, 1993). The two main enzyme forms with isoelectric points of 4.6 and 5.4 were purified and characterized. The enzymes had slightly different molecular masses at 51 kDa and 53 kDa, respectively. Otherwise the two mannanases had very similar protein properties (Stålbrand *et al.*, 1993). Also the amino acid compositions and N-terminal amino acid sequences seemed to be identical (Stålbrand, 1995; Stålbrand *et al.*, 1995). The mannanase purified by Arisan-Atac *et al.* (1993) had a pI value of 5.2 and a molecular mass of 46 kDa and is most probably the same as the pI 5.4 enzyme isolated by Stålbrand *et al.* (1993).

The gene *man1* encoding the mannanase of *T. reesei* has recently been isolated and expressed in *Saccharomyces cerevisiae* (Stålbrand *et al.*, 1995). It was found that the two purified forms of the *T. reesei* mannanases with pI 4.6 and 5.4, along with other forms of mannanases, are probably encoded by the *man1* gene (Stålbrand *et al.*, 1995). In accordance with this, the deletion of the *man1* gene in the *T. reesei* genome reduced the mannanase activity to below 10% of the parental strain (Primalco Biotec., Finland, unpublished results). The reason for the generation of different mannanase forms seemed at least partly to be due to post-translational modifications such as deamination (Stålbrand, 1995).

On the basis of gene sequence, the mannanase of *T. reesei* was found to have a multidomain structure similar to several cellulases: a catalytic domain and CBD separated by a linker peptide (Stålbrand *et al.*, 1995). The enzyme also strongly binds to cellulose but no specific binding to mannan has been detected (Tenkanen *et al.*, 1994). However, the enzyme does not have any cellulolytic activity. On the basis of hydrophobic cluster analysis, mannanase (MANI) belongs to family 5 of glycosyl hydrolases. This family contains several cellulases and three other mannanases that are from *Streptomyces lividans*, *Caldocellulosiruptor saccharolyticus* and *Aspergillus aculeatus* (Christgau *et al.*, 1994; Davies and Henrissat, 1995). *A. aculeatus* mannanase has a high amino acid sequence homology (70%) with the *T. reesei* mannanase but it does not contain a CBD (Stålbrand, 1995; Stålbrand *et al.*, 1995). The two bacterial mannanases appear to be more closely related to each other than to the fungal mannanases, thereby suggesting possible differences in the activity between the bacterial and fungal enzymes. Two glutamic acid residues suggested to be the catalytic residues of family 5 glycanases were also found in the *T. reesei* mannanase sequence (Stålbrand *et al.*, 1995).

Enzymes in family 5 employ the retaining type of hydrolytic mechanism (Gebler *et al.*, 1992) which has also been confirmed with the *T. reesei* mannanase (Harjunpää *et al.*, 1995). Accordingly, the *T. reesei* mannanase has a strong transglycosylating activity. Formation of mannohexaose at a high concentration of mannotriose was found to be the major transfer or condensation reaction catalyzed by mannanase (Harjunpää *et al.*, 1995). The hydrolysis experiments with mannooligosaccharides suggested that the substrate binding site of mannanase contains at least four subsites.

T. reesei mannanase degrades both soluble and insoluble mannans with different structures. The main end products from insoluble ivory nut mannan are mannobiose and mannotriose (Stålbrand *et al.*, 1993). The soluble galactomannan locust bean gum with a mannose to galactose ratio 10:3 is also efficiently hydrolyzed to short oligosaccharides. The smallest substituted oligosaccharide formed is Manβ1-4[Galα1-6]Man, thus the enzyme is able to cleave the linkage immediately after a mannosyl residue substituted with a galactosyl residue (Tenkanen *et al.*, 1997). The action of *T. reesei* mannanase on galactoglucomannan in pine kraft pulp (Man:Glc:Gal = 3.5:1:0.3) has also recently been studied. *T. reesei* mannanase is able to cleave not only linkages between two mannopyranosyl residues but also the mannopyranosyl linkage to glucose. However, it was unable to cleave the β-glucopyranosyl linkages to mannose (Tenkanen *et al.*, 1997). The enzyme first acts on the regions with successive unsubstituted mannopyranosyl residues, producing different mixed oligosaccharides containing mannose, glucose and galactose, which are later degraded further (Tenkanen *et al.*, 1997). Oligosaccharides that have been isolated after extensive hydrolysis of galactoglucomannan with mannanase are presented in Table 2.3. The specific mannanase is not the only enzyme of *T. reesei* that is able to degrade glucomannans. Recent results also show that all endoglucanases of *T. reesei* are able to hydrolyze the linkage between a glucopyranosyl and a mannopyranosyl residue in glucomannans, even though they were not active against polymers having a backbone containing only mannose residues (Siika-aho, unpublished data). The latter statement is in contrast to an observation that *T. reesei* EGII (formerly EGIII) exhibits endo-β-1,4-mannanase activity. Reversible and irreversible inhibitors affected both activities of EGII in the same manner, so that both activities were attributed to the same active site (Macarrón *et al.*, 1996).

When compared with several other mannanases, the mannanase of *T. reesei* was found to be superior in the hydrolysis of fibre-bound galactoglucomannan in softwood pulps (Rättö *et al.*, 1993; Suurnäkki *et al.*, 1996). The enzyme was also clearly better in the bleach-boosting of softwood kraft pulps than the other mannanases tested (Suurnäkki *et al.*, 1996). The reason for this is not clear, but part of it might be due to the cellulose binding domain of the enzyme. The removal of the CBD clearly decreased the action of MANI on softwood pulp (Primalco Biotec, Finland, unpublished results).

2.3.2 *Accessory enzymes in the hydrolysis of mannans*

β-Mannosidases and β-glucosidases

T. reesei has been reported to produce only very low amounts of extracellular β-mannosidase (Arisan-Atac *et al.*, 1993; Rättö and Poutanen, 1988) and it has been speculated that this may be one reason for the slow growth of the fungus on mannans (Arisan-Atac *et al.*, 1993). No β-mannosidase of *Trichoderma* has been purified or studied further. It might even be that the fungus does not produce any specific β-mannosidase and that the activity determined using p-nitrophenyl-Dmannopyranoside as a substrate is due to an artificial activity of some other exo-glycanases similar to the β-xylosidase activity of α-arabinofuranosidase. β-Mannosidase could also occur in the fungus but localized intracellularly. In this case the mannan fragments could be transported into the cells by a plasmalemma-bound transport system analogous to the *T. reesei* diglucoside permease, accompanying the cellulase biosynthesis (Kubicek *et al.*, 1993).

T. reesei is known to produce at least two β-glucosidases (Barnett *et al.*, 1991; Chen *et al.*, 1992). The activity of these enzymes towards glucomannooligosaccharides has not been studied, but β-glucosidases from other organisms used in hydrolysis experiments have been able to remove glucopyranosyl residues from the non-reducing termini of these oligosaccharides (Rättö *et al.*, 1993; Takahashi *et al.*, 1984). It has not been established whether the β-glucosidases involved in cellulose degradation participate in galactoglucomannan degradation.

α-Galactosidases

A screening of a *T. reesei* expression library established in yeast has demonstrated that *T. reesei* produces at least three different α-galactosidases (Margolles-Clark *et al.*, 1996b). The genes for all three enzymes (*agl1*, *agl2* and *agl3*) have been isolated and characterized. The corresponding enzymes (AGLI, AGLII and AGLIII) were produced in yeast in order to study their catalytic properties (Margolles-Clark *et al.*, 1996b).

AGLI was the most active of the three enzymes towards the polymeric galactomannan. However, the degree of hydrolysis by AGLI was rather low, and its action was clearly enhanced by the presence of mannanase, whilst the further addition of β-mannosidase did not have much effect. The α-galactosidases of *T. reesei* produced in transformed *Saccharomyces cerevisiae* have not been purified. There are, however, two reports on the purification of *T. reesei* α-galactosidase directly from the culture fluid of the fungus (Kachurin *et al.*, 1995; Zeilinger *et al.*, 1993). The molecular masses and isoelectric points for these α-galactosidases were reported to be very similar – 50 ± 3 and 54 kDa and 5.2 and 5.25, respectively – suggesting that the data define the same enzyme. According to molecular mass and hydrolytic properties the purified enzyme is identical to AGLI (Margolles-Clark *et al.*, 1996b; Zeilinger *et al.*, 1993). The p-nitrophenyl-α-D-galactoside activity of AGLI is competitively inhibited by galactose (Kachurin *et al.*, 1995; Zeilinger *et al.*, 1993).

The other α-galactosidase, AGLII, was almost inactive towards a polymeric substrate, but showed synergy with mannanase, and the degree of hydrolysis was increased even more with the addition of β-mannosidase. The action of AGLIII was also enhanced when mannanase and β-mannosidase were added to the reaction. However, the degree of hydrolysis by AGLIII was much lower than that obtained by AGLII (Margolles-Clark *et al.*, 1996b). The clear synergy of AGLII and AGLIII with β-mannosidase indicates that they prefer as substrates small oligosaccharides that carry the galactose substituent in the mannose unit at the non-reducing end of the oligosaccharide. AGLIII showed lower activity against p-nitrophenyl-α-D-galactopyranoside than against melibiose (Galα1-6Glc). This could explain why the enzyme has not been found in *T. reesei* Rut C-30 culture filtrates when p-nitrophenyl-α-D-galactopyranoside has been used as a substrate, even though the *agl3* gene seems to be well expressed (Margolles-Clark *et al.*, 1996b). On the other hand, the *agl2* gene seemed to be poorly expressed, which explains why AGLII has not been previously isolated. Zeilinger *et al.* (1993) reported only the presence of a minor α-galactosidase in *T. reesei* culture filtrate in addition to the main α-galactoside, AGLI.

According to the deduced amino acid sequences of AGLI and AGLIII, they belong to the glycosyl hydrolase family 27 which contains α-galactosidases of plant, animal, yeast and filamentous fungal origin. On the other hand, AGLII is similar to the bacterial α-galactosidases of family 36 and is thus the first reported eukaryotic α-galactosidase to show similarity with the corresponding prokaryotic enzymes. AGLIII carries an extra 230 amino acids at the N-terminus that are not found in other enzymes in family 27. This extra region seems to be unimportant for the catalytic activity and thus might be another functional domain which would differ from any polysaccharide binding domains described so far (Margolles-Clark *et al.*, 1996b). The active site of *T. reesei* AGLI was recently reported to contain a catalytically important methionine (Kachurin *et al.*, 1995).

Acetylglucomannan esterases

The deacetylation of galactoglucomannans by *T. reesei* culture filtrate is not as efficient as the deacetylation of glucuronoxylans (Tenkanen *et al.*, 1993), and no specific acetylglucomannan esterase has been isolated from any *Trichoderma*. However, no extensive screening of the specific acetylgalactomannan esterase of this fungus has been conducted. The acetyl esterase (AE) of *T. reesei*, which is able to liberate acetyl side groups from xylooligosaccharides, is also able to act on acetylated mannooligosaccharides (Tenkanen *et al.*, 1993). A similar property has been reported for the acetylxylan esterase of *Schizophyllum commune* (Biely *et al.*, 1996b). The AE was also found to act in synergy with acetylglucomannan esterase of *Aspergillus oryzae* in the hydrolysis of acetylated galactoglucomannan, similar to the action of acetylxylan esterase in the deacetylation of glucuronoxylans (Tenkanen, 1995). The synergism observed is believed to be at least partly due to the removal of the acetyl groups located close to the galactose side groups, which might be accessible to acetyl esterase but not to acetyl glucomannan esterase.

2.4 Conclusions

The knowledge of the hemicellulolytic enzymes of the genus *Trichoderma* is vast and is still growing. The hemicellulolytic enzyme system of the strain *T. reesei* Rut C-30 is the best characterized at both enzymic and gene levels, and with respect to its industrial potential, it remains competitive with similar systems of a number of other production microorganisms. The main information relevant to the enzymes and genes of this particular strain is summarized in Table 2.1. Future studies of hemicellulolytic systems of *Trichoderma* will lead to the discovery of additional hydrolases with new substrate specificities and novel biotechnological applications. New findings will not only contribute to our better understanding of the carbon cycle in nature but they will also add important information about the fine structure of plant cell walls and the interaction of plants with phytopathogenic microorganisms. Because of this the *Trichoderma* enzymes will remain a focus of basic and applied future research.

References

ARISAN-ATAC, I., HODITS, R., KRISTUFEK, D., and KUBICEK, C. P. 1993. Purification and characterization of a β-mannanase of *Trichoderma reesei* C-30. *Appl. Microbiol. Biotechnol.* **39**: 58–62.

ASPINALL, G. O. 1959. Structural chemistry of the hemicelluloses. *Adv. Carbohydr. Chem.* **14**: 429–468.

BAILEY, M. J., SIIKA-AHO, M., VALKEAJÄRVI, A., and PENTTILÄ, M. 1993. Hydrolytic properties of two cellulases of *Trichoderma reesei* expressed in yeast. *Biotechnol. Appl. Biochem.* **17**: 65–76.

BARNETT, C. C., BERKA, R. M., and FLOWLER, T. 1991. Cloning and amplification of the gene encoding an extracellular β-glucosidase from *Trichoderma reesei*: evidence for improved rates of saccharification of cellulosic substrates. *Bio/Technology* **9**: 562–567.

BIELY, P. 1985. Microbial xylanolytic systems. *Trends Biotechnol.* **3**: 286–290.

BIELY, P., MACKENZIE, C. R., and SCHNEIDER, H. 1988. Production of acetyl xylan esterases by *Trichoderma reesei* and *Schizophyllum commune. Can. J. Microbiol.* **34**: 767–772.

BIELY, P., PULS, J., and SCHNEIDER, H. 1985. Acetyl xylan esterases in fungal cellulolytic systems. *FEBS Lett.* **186**: 80–84.

BIELY, P., VRŠANSKÁ, M., and CLAEYSSENS, M. 1991. The endo-β-1,4-glucanase from *Trichoderma reesei*: Action on β-1,4-oligomers and polymers derived from D-glucose and D-xylose. *Eur. J. Biochem.* **200**: 157–163.

BIELY, P., KREMNICKÝ, L., ALFÖLDI, J., and TENKANEN, M. 1994. Stereochemistry of hydrolysis of glycosidic linkage by endo-β-1,4-xylanases of *Trichoderma reesei. FEBS Lett.* **356**: 137–140.

BIELY, P., MACKENZIE, C. R., PULS, J., and SCHNEIDER, H. 1987. The role of acetyl xylan esterases in the degradation of acetyl xylan by fungal xylanases. In J. F. Kennedy, G. O. Phillips, and P. A. Williams (eds), *Wood and Cellulosics.* Ellis Horwood, Chichester, pp. 283–290.

BIELY, P., VRŠANSKÁ, M., TENKANEN, M., and KLUEPFEL, D. 1997. Endo-β-1,4-xylanase families: Differences in catalytic properties. *J. Biotechnol.*, **57**: 151–166.

BIELY, P., CÔTÉ, G. L., KREMNICKÝ, L., WEISLEDER, D., and GREENE, R. V. 1996b. Substrate specificity of acetylxylan esterase from *Schizophyllum commune*: Mode of action on acetylated carbohydrates. *Biochim. Biophys. Acta* **1304**: 209–222.

BIELY, P., CÔTÉ, G. L., KREMNICKÝ, L., GREENE, R. V., DUPONT, C., and KLUEPFEL, D. 1996a. Substrate specificity and mode of action of acetylxylan esterase from *Streptomyces lividans. FEBS Lett.* **396**: 257–260.

BIELY, P., VRŠANSKÁ, M., KREMNICKÝ, L., TENKANEN, M., POUTANEN, K., and HAYN, M. 1993. Catalytic properties of endo-β-1,4-xylanases of *Trichoderma reesei.* In P. Suominen and T. Reinikainen (eds), Proceedings of the Second TRICEL *Symposium on* Trichoderma *Cellulases and Other Hydrolases.* Foundation for Biotechnical and Industrial Fermentation Research, Helsinki, pp. 125–135.

CAMPBELL, R. L., ROSE, D. R., WAKARCHUK, W. W., TO, R., SUNG, W., and YAGUCHI, M. 1993. A comparison of the structure of the 20 kd xylanases from *Trichoderma harzianum* and *Bacillus circulans.* In P. Suominen and T. Rainikainen (eds), Trichoderma *Cellulases and Other Hydrolases.* Foundation for Biotechnical and Industrial Fermentation Research, Helsinki, pp. 63–72.

Chen, H., Hayn, M., and Esterbauer, H. 1992. Purification and characterization of two extracellular β-glucosidases from *Trichoderma reesei*. *Biochim. Biophys. Acta* **1121**: 54–60.

Christgau, S., Kauppinen, S., Vind, J., Kofod, L. V., and Dalboke, H. 1994. Expression, cloning, purification and characterization of a β-1,4-mannanase from *Aspergillus aculeatus*. *Biochem. Mol. Biol. Int.* **33**: 917–925.

Claeyssens, M., van Tilbeurgh, H., Kamerling, J. P., Vrsanska, M., and Biely, P. 1990. Studies of the cellulolytic system of the filamentous fungus *Trichoderma reesei* QM 9414: Substrate specificity and transfer activity of endoglucanase I. *Biochem. J.* **270**: 251–256.

Da Silva Carvalho, S. M., Teixeira, M. F. S., Esposito, E., Machuca, A., Ferraz, A., and Duran, N. 1992. Amazonian lignocellulosic materials, I: Fungal screening from decayed laurel and cedar trees. *Appl. Biochem. Biotechnol.* **37**: 33–41.

Davies, G. and Henrissat, B. 1995. Structure and mechanism of glycosyl hydrolases. *Structure* **3**: 853–859.

Dean, J. F. D. and Anderson, J. D. 1991. Ethylene biosynthesis-inducing xylanase II: Purification and characterization of the enzyme produced by *Trichoderma viride*. *Plant Physiol.* **95**: 316–323.

Dean, J. F. D., Gamble, H. R., and Anderson, J. D. 1989. The ethylene biosynthesis-inducing xylanase: Its induction in *Trichoderma viride* and certain plant pathogens. *Phytopathology* **79**: 1071–1078.

Dekker, R. F. H. 1983. Bioconversion of hemicellulose: Aspects of hemicellulase production by *Trichoderma reesei* QM 9414 and enzymic saccharification of hemicellulose. *Biotechnol. Bioeng.* **25**: 1127–1146.

Eriksson, K.-E. L., Blanchette, R. A., and Aander, P. 1990. *Microbial and Enzymatic Degradation of Wood and Wood Components*. Springer-Verlag, Berlin, pp. 181–184.

Gebler, J., Gilkes, N. R., Claeyssens, M., Wilson, D. B., Beguin, P., Wakarchuk, W. W., Kilburn, D. G., Miller, R. C. Jr., Warren, R. A., and Withers, S. G. 1992. Stereoselective hydrolysis catalyzed by related β-1,4-glucanases and β-1,4-xylanases. *J. Biol. Chem.* **267**: 12559–12561.

Gilkes, N. R., Henrissat, B., Kilburn, D. G., Miller, R. C. Jr., and Warren, R. A. J. 1991. Domains of microbial β-1,4-glycanases: sequence conservation, function, and enzyme families. *Microbiol. Rev.* **55**: 303–315.

Gomez, J., Gomez, I., Esterbauer, H., Kreiner, W., and Steiner, W. 1989. Production of cellulase by a wild strain of *Gliocladium virens* – optimization of the fermentation medium and partial characterization of the enzymes. *Appl. Microbiol. Biotechnol.* **31**: 601–608.

Harjunpää, V., Teleman, A., Siika-aho, M., and Drakenberg, T. 1995. Kinetic and stereochemical studies of manno-oligosaccharide hydrolysis catalysed by β-mannanases from *Trichoderma reesei*. *Eur. J. Biochem.* **234**: 278–283.

Harris, G. W., Jenkins, J. A., Connerton, I., Cummings, H. N., Lo Leggio, L., Scott, M., Hazlewood, G. P., Laurie, J. L., Gilbert, H. J., and Pickersgill, R. W. 1994. Structure of the catalytic core of the family F xylanase from *Pseudomonas fluorescens* and identification of the xylopentaose-binding sites. *Structure* **2**: 1107–1116.

Havukainen, R., Törrönen, A., Laitinen, T., and Rouvinen, J. 1996. Covalent binding of three epoxyalkyl xylosides to the active site of endo-1,4-xylanase II from *Trichoderma reesei*. *Biochemistry* **35**: 9617–9624.

Henrissat, B. and Bairoch, A. 1993. New families in the classification of glycosyl hydrolases based on amino acid sequence similarities. *Biochem. J.* **293**: 781–788.

Herrmann, M. C., Vršanská, M., Jurickovà, M., Hirsch, J., Biely, P., and Kubicek, C. P. 1997. The β-xylosidase of *Trichoderma reesei* is a multifunctional β-D-xylan xylohydrolase. *Biochem. J.*, **321**: 375–381.

HRMOVÁ, M., BIELY, P., and VRŠANSKÁ, M. 1986. Specificity of cellulase and β-xylanase induction in *Trichoderma reesei* QM 9414. *Arch. Microbiol.* **144**: 307–311.

ISHIHARA, M., INAGAKI, S., HAYASHI, N., and SHIMIZU, K. 1990. 4-O-Methyl-D-glucuronic acid liberating enzyme in the enzymatic hydrolysis of hardwood xylan. *Bull. Forest Prod. Res. Inst.* **359**: 141–157.

KACHURIN, A. M., GOLUBEV, A. M., GEISOW, M. M., VESELKINA, O. S., ISAEVA-IVANOVA, L. S., and NEUSTROEV, K. N. 1995. Role of methionine in the active site of α-galactosidase from *Trichoderma reesei*. *Biochem. J.* **308**: 955–964.

KRISTUFEK, D., ZEILINGER, S., and KUBICEK, C. P. 1995. Regulation of β-xylosidase formation by xylose in *Trichoderma reesei*. *Appl. Microbiol. Biotechnol.* **42**: 713–717.

KUBICEK, C. P., MESSNER, R., GRUBER, F., MANDELS, M., KUBICEK-PRANZ, E. M. 1993. Triggering of cellulase biosynthesis in *Trichoderma reesei* – involvement of a constitutive, sophorose-inducible, glucose-inhibited β-diglucoside permease. *J. Biol. Chem.* **268**: 19364–19368.

LAPPALAINEN, A. 1986. Purification and characterization of xylanolytic enzymes from *Trichoderma reesei*. *Biotechnol. Appl. Biochem.* **8**: 437–448.

LAPPALAINEN, A. 1988. Cellulolytic and xylanolytic enzymes of *Trichoderma reesei*. VTT, Espoo.

MACARRÓN, R., ACEBAL, C., CASTILLÓN, M. P., and CLAEYSSENS, M. 1996. Mannanase activity of endoglucanase III from *Trichoderma reesei* QM 9414. *Biotechnol. Lett.* **18**: 599–602.

MARGOLLES-CLARK, E., SALOHEIMO, M., SIIKA-AHO, M., and PENTTILÄ, M. 1996a. The α-glucuronidase encoding gene of *Trichoderma reesei*. *Gene* **172**: 171–172.

MARGOLLES-CLARK, E., TENKANEN, M., LUONTERI, E., and PENTTILÄ, M. 1996b. Three α-galactosidase genes of *Trichoderma reesei* cloned by expression in yeast. *Eur. J. Biochem.* **240**: 104–111.

MARGOLLES-CLARK, E., TENKANEN, M., NAKARI-SETÄLÄ, T., and PENTTILÄ, M. 1996c. Cloning of genes encoding α-L-arabinofuranosidase and β-xylosidase from *Trichoderma reesei* by expression in *Saccharomyces cerevisiae*. *Appl. Environ. Microbiol.* **62**: 3840–3846.

MARGOLLES-CLARK, E., TENKANEN, M., SÖDERLUND, H., and PENTTILÄ, M. 1996d. Acetyl xylan esterase from *Trichoderma reesei* contains an active site serine and a cellulose binding domain. *Eur. J. Biochem.* **237**: 553–560.

MARINGER, U., WANG, K. K. Y., SADDLER, J. N., and KUBICEK, C. P. 1995. A functional comparison of two pairs of β-1,4-xylanases from *Trichoderma harzianum* E58 and *Trichoderma reesei* Rut C-30. *Biotechnol. Appl. Biochem.* **21**: 49–65.

MATSUO, M. and YASUI, T. 1984. Purification and some properties of β-xylosidase from *Trichoderma viride*. *Agric. Biol. Chem.* **48**: 1853–1860.

MCCLEARY, B. V. 1979. Enzymic hydrolysis, fine structure, and gelling interaction of legume-seed D-galacto-D-mannans. *Carbohydr. Res.* **71**: 205–230.

MONTENECOURT, B. S. and EVELEIGH, D. E. 1979. Selective screening for the isolation of high yielding mutants of *T. reesei*. *Adv. Chem. Ser.* **181**: 289–301.

MUELLER-HARVEY, L., HARTLEY, R. D., HARRIS, P. J., and CURZON, E. H. 1986. Linkage of p-coumaroyl and feruloyl groups to cell wall polysaccharides of barley straw. *Carbohydr. Res.* **148**: 71–85.

POUTANEN, K. 1988. An α-L-arabinofuranosidase of *Trichoderma reesei*. *J. Biotechnol.* **7**: 271–282.

POUTANEN, K. and PULS, J. 1988. Characteristics of *Trichoderma reesei* β-xylosidase and its use in the hydrolysis of solubilized xylans. *Appl. Microbiol. Biotechnol.* **28**: 425–432.

POUTANEN, K. and SUNDBERG, M. 1988. An acetyl esterase of *Trichoderma reesei* and its role in the hydrolysis of acetyl xylan. *Appl. Microbiol. Biotechnol.* **28**: 419–424.

POUTANEN, K., RÄTTÖ, M., PULS, J., and VIIKARI, L. 1987. Evaluation of different microbial xylanolytic systems. *J. Biotechnol.* **6**: 49–60.

POUTANEN, K., SUNDBERG, M., KORTE, H., and PULS, J. 1990. Deacetylation of xylans by acetylesterases of *Trichoderma reesei*. *Appl. Microbiol. Biotechnol.* **33**: 506–510.

PULS, J. 1992. α-Glucuronidases in the hydrolysis of wood xylans. In J. Visser, G. Beldman, M. A. Kuster-van Someren, and A. G. J. Voragen (eds), *Xylans and Xylanases*. Elsevier, Amsterdam, pp. 213–224.

PULS, J. and POUTANEN, K. 1989. Mechanism of enzymatic hydrolysis of hemicelluloses (xylans) and procedures for determination of the enzyme activities involved. In M. Coughlan and G. P. Hazlewood (eds), *Enzyme Systems for Lignocellulose Degradation*. Elsevier Applied Science, London, pp. 151–165.

RAGHUKUMAR, C., RAGHUKUMAR, S., CHINNARAJ, A., CHANDRAMOHAN, D., D'SOUZA, T. M., and REDDY, C. A. 1994. Laccase and other lignocellulose modifying enzymes of marine fungi isolated from the coast of India. *Bot. Mar.* **37**: 515–523.

RÄTTÖ, M. and POUTANEN, K. 1988. Production of mannan-degrading enzymes. *Biotechnol. Lett.* **10**: 661–664.

RÄTTÖ, M., SIIKA-AHO, M., BUCHERT, J., VALKEAJÄRVI, A., and VIIKARI, L. 1993. Enzymatic hydrolysis of isolated and fibre-bound galactoglucomannans from pine-wood and pine kraft pulp. *Appl. Microbiol. Biotechnol.* **40**: 449–454.

REESE, E. T., MAGUIRE, A., and PARRISH, F. W. 1973. Production of β-xylopyranosidases by fungi. *Can. J. Microbiol.* **19**: 1065–1074.

ROMBOUTS, F. M., VORAGEN, A. G. J., SEARLE-VAN LEEUVEN, M. F., GERADES, C. C. J. M., SCHOOLS, H. A., and PILNIK, W. 1988. The arabinases of *Aspergillus niger* – purification and characterization of two α-L-arabinofuranosidases and an endo-1,5-α-L-arabinase. *Carbohydr. Polym.* **9**: 25–47.

SAARELAINEN, R., PALOHEIMO, M., FAGERSTRÖM, R., SUOMINEN, P. L., and NEVALAINEN, H. 1993. Cloning, sequencing and enhanced expression of the *Trichoderma reesei* endoxylanase II (pI 9) gene, *xln2*. *Mol. Gen. Genet* **241**: 497–503.

SIIKA-AHO, M., TENKANEN, M., BUCHERT, J., PULS, J., and VIIKARI, L. 1994. An α-glucuronidase from *Trichoderma reesei* RUT C-30. *Enzyme Microb. Technol.* **16**: 813–819.

SMITH, D. C., BHAT, K. M., and WOOD, T. M. 1991. Xylan-hydrolyzing enzymes from thermophilic and mesophilic fungi. *World J. Microbiol. Biotechnol.* **7**: 475–484.

STÅLBRAND, H. 1995. *Hemicellulose-degrading enzymes from fungi*. PhD Thesis, Lund University, Sweden.

STÅLBRAND, H., SIIKA-AHO, M., TENKANEN, M., and VIIKARI, L. 1993 Purification and characterization of two β-mannanases from *Trichoderma reesei*. *J. Biotechnol.* **29**: 229–242.

STÅLBRAND, H., SALOHEIMO, A., VEHMAANPERÄ, J., HENRISSAT, B., and PENTTILÄ, M. 1995. Cloning and expression in *Saccharomyces cerevisiae* of a *Trichoderma reesei* β-mannanase gene containing a cellulose binding domain. *Appl. Environ. Microbiol.* **61**: 1090–1097.

SUNDBERG, M. and POUTANEN, K. 1991. Purification and properties of two acetylxylan esterases of *Trichoderma reesei*. *Biotechnol. Appl. Biochem.* **13**: 1–11.

SUOMINEN, P., MÄNTYLÄ, A., SAARELAINEN, R., PALOHEIMO, M., FAGERSTRÖM, R., PARKKINEN, E., and NEVALAINEN, H. 1992. Genetic engineering of *Trichoderma reesei* to produce suitable enzyme combinations for applications in the pulp and paper industry. In M. Kuwahara and M. Shimida (eds), *Biotechnology in Pulp and Paper Industry*. Uni Publishers, Tokyo, pp. 439–445.

SUURNÄKKI, A., CLARK, T., ALLISON, R., BUCHERT, J., and VIIKARI, L. 1996. Mannanase aided bleaching of softwood kraft pulp. In K. Messner and E. Srebotnik (eds), *Biotechnology in Pulp and Paper Industry – Advances in Applied and Fundamental Research*. WUA Universitätsverlag, Vienna, pp. 69–74.

TAKAHASHI, R., KUSAKABE, I., KUSAMA, S., SAKURAI, Y., MURAKAMI, K., MAEKAWA, A., and SUZUKI, T. 1984. Structures of glucomanno-oligosaccharides from the hydrolytic products of konjac glucomannan produced by a β-mannanase from *Streptomyces* sp. *Agric. Biol. Chem.* **48**: 2943–2950.

TENKANEN, M. 1995. *Characterization of esterases acting on hemicelluloses.* PhD Thesis, VTT, Finland.

TENKANEN, M. and POUTANEN, K. 1992. Significance of esterases in the degradation of xylans. In J. Visser, G. Beldman, M. A. Kuster-van Someren, and A. G. J. Voragen (eds), *Xylans and Xylanases.* Elsevier, Amsterdam, pp. 203–212.

TENKANEN, M., BUCHERT, J., and VIIKARI, L. 1994. Binding of hemicellulases on isolated polysaccharide substrates. *Enzyme Microb. Technol.* **17**: 499–505.

TENKANEN, M., LUONTERI, E., and TELEMAN, A. 1996a. Effect of side groups on the action of β-xylosidase from *Trichoderma reesei* against substituted xylooligosaccharides. *FEBS Lett.* **39**: 303–306.

TENKANEN, M., PULS, J., and POUTANEN, K. 1992. Two major xylanases of *Trichoderma reesei. Enzyme Microb. Technol.* **14**: 566–574.

TENKANEN, M., PULS, J., RÄTTÖ, M., and VIIKARI, L. 1993. Enzymatic deacetylation of galactoglucomannans. *Appl. Microbiol. Biotechnol.* **39**: 159–165.

TENKANEN, M., MAKKONEN, M., PERTTULA, M., VIIKARI, L., and TELEMAN, A. 1997. Action of *Trichoderma reesei* mannanase on galactoglucomannan in pine kraft pulp. *J. Biotechnol.*, **57**: 191–204.

TENKANEN, M., SIIKA-AHO, M., HAUSALO, T., PULS, J., and VIIKARI, L. 1996b. Synergism between xylanolytic enzymes of *Trichoderma reesei* in the degradation of acetyl-4-O-methylglucuronoxylan. In K. Messner and E. Srebotnik (eds), *Biotechnology in Pulp and Paper Industry – Advances in Applied and Fundamental Research.* WUA Universitätsverlag, Vienna, pp. 503–508.

TIMELL, T. E. 1967. Recent progress in the chemistry of wood hemicelluloses. *Wood Sci. Technol.* **1**: 45–70.

TODOROVIC, R., GRUJIC, S., PETROVIC, J., and MATAVULJ, M. 1988. Some properties of cellulolytic enzymes and xylanase from *Gliocladium virens* C_2R_1. *Mikrobiologija* (Serbo-Croatian) **25**: 143–151.

TORRIE, J. P., SENIOR, D. J., and SADDLER, J. N. 1990. Production of β-mannanases by *Trichoderma harzianum* E58. *Appl. Microbiol. Biotechnol.* **34**: 303–307.

TÖRRÖNEN, A. and ROUVINEN, J. 1995. Structural comparison of two major *endo*-1,4-xylanases from *Trichoderma reesei. Biochemistry* **34**: 847–856.

TÖRRÖNEN, A., HARKKI, A., and ROUVINEN, J. 1994. Three-dimensional structure of endo-1,4-β-xylanase II from *Trichoderma reesei*: two conformational states in the active site. *EMBO J.* **13**: 2493–2501.

TÖRRÖNEN, A., KUBICEK, C. P., and HENRISSAT, B. 1993. Amino acid sequence similarities between low molecular weight endo-1,4-β-xylanases and family H cellulases revealed by clustering analysis. *FEBS Lett.* **321**: 135–139.

TÖRRÖNEN, A., MACH, R. L., MESSNER, R., GONZALES, R., KALKKINEN, N., HARKKI, A., and KUBICEK, C. P. 1992. The two major xylanases from *Trichoderma reesei*: Characterization of both enzymes and genes. *Bio/Technology* **10**: 1461–1465.

VAN TILBURG, A.-U. B. and THOMAS, M. D. 1993. Production of extracellular proteins by the biocontrol fungus *Gliocladium virens. Appl. Environ. Microbiol.* **59**: 236–242.

VORAGEN, A. G. J., GRUPPEN, H., VERBRUGGEN, M. A., and VIETOR, R. J. 1992. Characterization of cereal arabinoxylans. In J. Visser, G. Beldman, M. A. Kuster-van Someren, and A. G. J. Voragen (eds), *Xylans and Xylanases.* Elsevier, Amsterdam, pp. 51–67.

WILKIE, K. C. B. 1979. Hemicelluloses of grasses and cereals. *Adv. Carbohydr. Chem. Biochem.* **36**: 215–264.

WILKIE, K. C. B. 1983. Hemicellulose. *Chem. Tech.* **13**: 306–319.

WONG, K. K. Y. and SADDLER, J. N. 1992. *Trichoderma* xylanases, their properties, and applications. *Crit. Rev. Biotechnol.* **12**: 413–435.

WONG, K. K. Y., TAN, L. U. L., and SADDLER, J. N. 1988. Multiplicity of β-1,4-xylanases in microorganisms: Functions and applications. *Microbiol. Rev.* **52**: 305–317.
YAGUCHI, M., ROY, C., UJIIE, M., WATSON, D. C., and WAKARCHUK, W. 1992a. Amino acid sequence of the low-molecular weight xylanase from *Trichoderma viride*. In J. Visser, G. Beldman, M. A. Kuster-van Someren, and A. G. J. Voragen (eds), *Xylans and Xylanases*. Elsevier, Amsterdam, pp. 149–154.
YAGUCHI, M., ROY, C., WATSON, D. C., ROLLIN, F., TAN, L. U. L., SENIOR, D. J., and SADDLER, J. N. 1992b. The amino acid sequence of the 20kD xylanase from *Trichoderma harzianum* E58. In J. Visser, G. Beldman, M. A. Kuster-van Someren, and A. G. J. Voragen (eds), *Xylans and Xylanases*. Elsevier, Amsterdam, pp. 435–438.
ZEILINGER, S., KRISTUFEK, D., ARISAN-ATAC, I., HODITS, R., and KUBICEK, C. P. 1993. Conditions of formation, purification, and characterization of an α-galactosidase of *Trichoderma reesei* RUT C-30. *Appl. Environ. Microbiol.* **59**: 1347–1353.
ZURBRIGGEN, B. D., PENTTILÄ, M., VIIKARI, L., and BAILEY, M. J. 1991. Pilot scale production of a *Trichoderma reesei* endo-β-glucanase by brewer's yeast. *J. Biotechnol.* **17**: 133–146.

3

Regulation of production of plant polysaccharide degrading enzymes by *Trichoderma*

C. P. KUBICEK* and M. E. PENTTILÄ†

* *Institut für Biochemische Technologie und Mikrobiologie, TU Wien, Wien, Austria,* and
† *VTT Biotechnology and Food Research, Espoo, Finland*

3.1 Introduction

Most species of *Trichoderma* and *Gliocladium* are saprophytes, and their respective teleomorphs are lignicolous ascomycetes (see Volume 1, Chapter 5). Hence, they encounter in nature a wide variety of polymeric substrates among which celluloses and hemicelluloses predominate. Cellulose chains are β-1,4-glucosidically-linked homopolymers of about 8000–12 000 glucose units, which are held together by hydrogen bonding to form essentially insoluble crystalline cellulose. Hemicellulose is a general term summarizing various heteropolysaccharides which are based on backbone polymers formed of xylose (xylans) or mannose (and glucose) (mannans, glucomannans) and which have additional side-chain substituents such as arabinose and galactose, and acetic and glucuronic acids (see Chapter 2). Pectins are even more complicated with various branched structures. In contrast to degradation of cellulose, which leads to glucose and glucooligomers, degradation of hemicelluloses leads to the accumulation of various mono- and disaccharides in various ratios depending on the hemicellulose type. Other polymers such as β-glucans, starch and protein account only for a smaller part of available carbon, although they may predominate in certain habitats (Chapter 6).

Complete degradation of cellulose and hemicellulose requires a large number of extracellular enzymes, which have received broad industrial interest. This has in turn led to investigations of their biochemical properties, three-dimensional protein structures and isolation of genes encoding them. A summary of the different cellulases from *Trichoderma* spp., characterized at the biochemical or genetic level, is given in Table 3.1. For a similar summary on hemicellulases, see Chapter 2.

Apart from studies on the properties of cellulases, there has been a continuing interest in understanding how the synthesis of these enzymes is regulated. Regulation studies are warranted to bring about understanding of the physiology of the organism and the role of fungi in the carbon turnover in nature. They will also help to improve enzyme production through nutritional strategies and bioprocess design,

Table 3.1 Cellulase genes characterized from *T. reesei* and other *Trichoderma* spp. (to date)

Cellulase	Strain	Gene	Ref.[a]
Cellobiohydrolase I	*T reesei*	*cbh1*	1, 2
	T. viride	*cbh1*	3
	T. koningii	*cbh1*	4
Cellobiohydrolase II	*T. reesei*	*cbh2*	5, 6
Endo-β-1,4-Glucanase I	*T. reesei*	*egl1*	7, 8
	T. longibrachiatum	*egl1*	9
Endo-β-1,4-Glucanase II	*T. reesei*	*egl2*	10
Endo-β-1,4-Glucanase III	*T. reesei*	*egl3*	11
Endo-β-1,4-Glucanase IV	*T. reesei*	*egl4*	12
Endo-β-1,4-Glucanase V	*T. reesei*	*egl5*	13
β-Glucosidase I	*T. reesei*	*bgl1*	14

[a] References: 1, Teeri *et al.* (1983); 2, Shoemaker *et al.* (1983); 3, Cheng *et al.* (1990); 4, Wey *et al.* (1994); 5, Chen *et al.* (1987); 6, Teeri *et al.* (1987); 7, Penttilä *et al.* (1986); 8, Van Arsdell *et al.* (1987); 9, Gonzalez *et al.* (1994); 10, Saloheimo *et al.* (1988); 11, Ward *et al.* (1993); 12, Saloheimo *et al.* (1997); 13, Saloheimo *et al.* (1994); 14, Barnett *et al.* (1991).

as well as to support efforts toward recombinant overproduction of cellulases and hemicellulases or of heterologous proteins under the control of cellulase or hemicellulase promoters (see Chapters 13 and 17).

Data on regulation of cellulase and hemicellulase production is almost nonexistent for *Gliocladium*, but a considerable amount of data has been accumulated for *Trichoderma*, and the present chapter will attempt to summarize the current knowledge in this area. The existing data are partly controversial, and sometimes their significance is difficult to judge. Problems encountered are, for example, that specific substrates for one particular enzyme rarely exist and some enzymes have broad specificities (e.g., the cellulase EGI also has xylanase activity, β-xylosidase also has α-arabinofuranosidase activity). Also, experiments based on enzyme activity measurements fail to distinguish possible differential expression patterns of even different types of enzymes, not to mention different enzymes of the same type (e.g., xylanases, see below). Furthermore, the number of different enzymes is still unknown and proteolysis in the medium complicates the picture. Part of the confusion in the literature is also due to the use of mutant strains or improper cultural conditions. For instance, several studies have been carried out with the *T. reesei* strain RutC-30, which was recently shown to be a carbon catabolite derepressed mutant (Ilmén *et al.*, 1996a; see later in this chapter). Also some studies that claimed to have been carried out in the presence of glucose were apparently analysed only after glucose depletion, or cultivations have been carried out on media containing organic nitrogen which *Trichoderma* can use also as a carbon source. Hence the following description focuses more on reports that are not impaired by the abovementioned drawbacks and in which the synthesis of individual enzymes has been established at least at the protein level and preferably at the mRNA level as well.

3.2 Regulation of cellulase expression

3.2.1 *Inducing conditions and inducer paradigms*

Many early studies (reviewed by Bisaria and Mishra, 1989; Kubicek *et al.*, 1993a) established culture conditions for production of cellulolytic activity. Abundant cellulase production occurs when the fungus is cultivated on media containing cellulose or mixtures of plant polymers. Also *β*-glucan and different xylans provoke expression (e.g., see Table 3.2). In addition to these polymers, pure oligosaccharides such as cellobiose (two *β*-1,4-linked glucose units), *δ*-cellobiono-1,5-lactone, lactose and sophorose (two *β*-1,2-linked glucose units) have been reported to cause cellulase production. Since the very potent natural inducing compound polymeric cellulose cannot enter the fungal cell, it is generally believed that the oligosaccharides or their derivatives, released from the polymer, serve as the actual compounds triggering the high level of induction of cellulase expression. Glucose, glycerol and fructose have been known to result in low or undetectable cellulase production.

Unlike glucose (see later), glycerol and sorbitol neither repress nor induce/derepress cellulase formation and can thus be considered as "neutral" carbon sources with respect to cellulase gene expression (Ilmén *et al.*, 1996a,b, 1997; slight repressing effect of sorbitol is possible, see Ilmén et al., 1996b; Table 3.2). While no cellulase transcription is observed on these carbon sources as such, the fact that addition of 2–4 mM sophorose, or xylobiose (Margolles-Clark *et al.*, 1997; Table 3.2), into these cultures induces cellulase expression shows that true induction mechanisms are operating in cellulase expression.

As a result of the concerted action of cellulases, the major soluble end-product formed from cellulose is cellobiose. Its appearance close to or in the cell could be considered as a signal for the presence of extracellular cellulose and would therefore be a logical candidate for the natural inducer of cellulase biosynthesis. However, cellobiose seems to be a tricky carbon source in this respect since growth on cellobiose or addition of cellobiose to resting mycelia sometimes leads to the formation of cellulases under some conditions (Fritscher *et al.*, 1990; Ilmén *et al.*, 1997; Vaheri *et al.*, 1979a), and adding cellobiose to *T. reesei* cultures growing on cellulose seems to inhibit rather than stimulate cellulase formation (Fritscher *et al.*, 1990). This is probably the result of cellobiose hydrolysis to glucose, which represses cellulase formation (Ilmén *et al.*, 1996b, 1997), and the effect of cellobiose therefore could depend on the activity of *β*-glucosidase. Consistent with this assumption, slow feeding of cellobiose or inhibition of extracellular cellobiose hydrolysis by the addition of *β*-glucosidase inhibitors leads to cellulase formation in amounts comparable to those observed on cellulose (Fritscher *et al.*, 1990; Vaheri *et al.*, 1979a). This suggests that the metabolism of cellobiose is a critical point determining whether cellobiose can act as an inducer, which it may do so only when its hydrolysis to glucose is slow. The findings that poor substrates of *β*-glucosidase, such as *δ*-cellobiono-1,5-lactone (Iyayi *et al.*, 1989; Szakmary *et al.*, 1991), provoke cellulase induction support this view.

Sophorose, which causes high levels of cellulase formation when added to cultures in small amounts and which is a poor substrate for *β*-glucosidase, has been considered for years as the natural inducer of cellulase formation (Mandels *et al.*, 1962; Mandels and Reese, 1960; Sternberg and Mandels, 1979). This assumption is supported by findings that its formation from cellobiose by transglycosylation activ-

Table 3.2 Expression levels of hydrolase genes in *T. reesei* QM 9414 cultivated for 3 days on different carbon sources[a] and in *T. reesei* RutC-30 cultivated on glucose

Carbon source[a]	Growth[b]	Gene probe *cbh1*	*bgl1*	*xyn1*	*xyn2*	*bxl1*	*abf1*	*glr1*	*axe1*	*man1*	*agl1*	*agl2*	*agl3*
QM 9414													
Cellulose	+++	++++[c]	+[c]	++[c]	++[c]	++[c]	−	+[c]	+++[c]	++[c]	+[c]	+[c]	+[c]
Arabinose	++	−	−	−	−	+	+	−	−	−	−	−	−
Arabitol	++	++	−	++	++	+++	++++	+	+	−	++	+++	−
Xylose	+++	−	−	−	−	−	−	−	−	−	−	−	−
Xylitol	+++	(+)	−	−	−	−	+	−	−	−	+	+	−
L. Xylan	+++	(+)	−	++	−	++	−	+	−	−	−	−	−
MeGlc-Xylan	+++	+	−	+	++	+(+)	−	(+)	++	−	−	−	−
O.S. Xylan	+++	+	−	+++	+	++	+++	+	(+)	−	+	+	−
β-Glucan	++	+++	+	−	−	+	+	−	(+)	+	+	+	−
Mannose	+++	−	−	−	−	−	−	−	−	−	+	−	−
Galactose	++	−	−	−	−	−	−	−	++	−	+++	++++	−
Lactose	+	++	−	−	−	+	−	−	−	−	−	−	−
Sorb	+	−	−	−	(+)	−	−	−	−	−	−	+	−
Sorb/Soph	+	++++	+	+	+(+)	++	+	−	+	−	+	+(+)	+
Sorb/$(Man)_2$	+	−	−	−	−	−	(+)	−	−	−	+	+	−
Sorb/$(Xyl)_2$	+	−	−	−	(+)	++	+	−	−	−	+	+	−
Sorb/$(Glc)_2$	+	+++	−	−	(+)	+	+	−	−	−	++	+++	−

Table 3.2 (*Cont*)

Carbon source[a]	Growth[b]	Gene probe											
		cbh1	*bgl1*	*xyn1*	*xyn2*	*bxl1*	*abf1*	*glr1*	*axe1*	*man1*	*agl1*	*agl2*	*agl3*
$(Glc)_2$	+	+++	–	–	(+)	+	+	–	–	–	+	++	–
Glys/$(Man)_2$	+	(+)	–	–	–	–	(+)	–	–	–	+	+	–
Glys/$(Xyl)_2$	+	+	–	++	+++	++++	–	++	+	–	+	+(+)	–
Glys	+	–	–	–	–	–	–	–	–	–	+	–	–
Glc	+++	–	–	–	–	–	–	–	–	–	–	–	–
Glc/Soph	+++	–	–	–	–	–	–	–	–	–	–	–	–
Glc/$(Man)_2$	+++	–	–	–	–	–	–	–	–	–	–	–	–
Glc/$(Xyl)_2$	+++	–	–	–	–	–	–	–	–	–	–	–	–
Glc depl.[d]	–	++++	++	–	++	++	+++	+	++	+	+++	+++	–
RutC-30													
Glc	+++	++++	–	+	–	+++	+	+	+	–	+++	+	+

The data are summarized from visual judgement of signal intensities from repeated hybridizations and various exposures of the Northern films. Signal intensities from strong to non-detectable: ++++, +++, ++, +(+), +, (+), –. Data from Margolles-Clark *et al.* (1997).

[a] Glc, glucose; Sorb, sorbitol; Glys, glycerol; Soph, sophorose; $(Man)_2$, mannosyl-β-1,4-mannose; $(Xyl)_2$, xylobiose; $(Glc)_2$, cellobiose; MeGlc-Xylan, 4-O-Methylglucuronoxylan; L. Xylan, Lenzing xylan; O.S. xylan, oat spent xylan.

[b] Growth was estimated visually (+++ best growth; + poor growth; – no growth after glucose depletion).

[c] Signal intensity was estimated from 0.5 μg RNA; all other samples contained 5 μg RNA.

[d] Glucose depleted fermenter cultivation as described by Ilmén *et al.* (1997).

ity of β-glucosidase was demonstrated *in vitro* (Vaheri *et al.*, 1979b). Genetic support for this was obtained with a recombinant strain, in which the β-glucosidase (*bgl1*) gene had been disrupted and consequently showed less efficient growth and cellulase production on cellulose (Fowler and Brown, 1992). Consistent results have also been reported using β-glucosidase inhibitors (Kubicek, 1987) or β-glucosidase-mutant strains (Mishra *et al.*, 1989; Strauss and Kubicek, 1990). Addition of sophorose to the medium restored cellulase induction and growth on cellulose in all of these cases, indicating that β-glucosidase is important for cellulase induction from cellulose and suggests that this may be related to its ability to form transglycosylation products.

It should be noted in this context, however, that the natural occurrence of δ-cellobiono-1,5-lactone and of other oxidized cellulose degradation products has also been demonstrated (Szakmari *et al.*, 1992; Vaheri, 1982a,b), and their additional action as inducers *in vivo* can therefore not be ruled out.

Transport of cellobiose, sophorose, gentiobiose and laminaribiose into *T. reesei* mycelium occurs by a constitutively formed β-linked disaccharide permease (Kubicek *et al.*, 1993b), with much higher affinity for this substrate than β-glucosidase but with much lower activity. This means that at low disaccharide concentrations, uptake is favoured over hydrolysis, and the situation becomes reversed upon further increase of cellobiose.

Based on the findings presented above, one might suggest that the relative rate of hydrolysis of cellulose-derived disaccharides by β-glucosidase determines cellulase induction. However, the role of β-glucosidase in the formation of the cellulase inducer is apparently more complicated. *T. reesei* strains carrying multiple copies of *bgl1* and displaying enhanced β-glucosidase activity showed an increased efficacy of cellulase induction not only by cellulose but also by sophorose (Mach *et al.*, 1995); sophorose induction of cellulase formation in a β-glucosidase I disruptant strain was still inhibited by the β-glucosidase inhibitor nojirimycin. This suggested the presence of at least one more β-glucosidase that is actively involved in the inductive action of sophorose. The location, constitutive nature and induction by methyl-β-D-glucoside suggest that this β-glucosidase II is identical to the enzyme described by Umile and Kubicek (1986) and Chen *et al.* (1992). However, as the gene encoding this second enzyme has not yet been cloned, its involvement in cellulase induction is at the moment merely hypothetical.

3.2.2 *Induction of cellulases at transcriptional level*

The first *T. reesei* cellulase genes were cloned by differential hybridization (Penttilä *et al.*, 1986; Saloheimo *et al.*, 1988; Shoemaker *et al.*, 1983; Teeri *et al.*, 1983, 1987) which showed that regulation of cellulase formation occurs at transcriptional level, and several studies based on Northern analyses have confirmed this (Abrahao-Neto *et al.*, 1995; El-Gogary *et al.*, 1989; Fowler and Brown, 1992; Gonzalez *et al.*, 1994; Ilmén *et al.*, 1996a,b, 1997; Messner and Kubicek, 1991; Morawetz *et al.*, 1992; Penttilä *et al.*, 1993). That transcriptional regulation is also the main mechanism of regulation is clearly suggested by the fact that no data yet exists that would contradict the idea that levels of individual enzymes (measured by ELISA, for example Seiboth *et al.*, 1997) closely correlate with the steady-state levels of the respective transcripts.

In experiments carried out thus far, expression of the different cellulase genes has been reported to be coordinate, that is, their relative expression levels are approximately the same in various inducing conditions (Abrahao-Neto *et al.*, 1995; Fowler and Brown, 1992; Ilmén *et al.*, 1997; Penttilä *et al.*, 1993; Seiboth *et al.*, 1997; Torigoi *et al.*, 1996). *Cbh1* is the most highly expressed gene, followed by *cbh2* and *egl1*. Expression levels of *egl5* are suprisingly high (Ilmén *et al.*, 1997) considering that this small cellulase had not been observed at the protein level before the cloning of the corresponding gene (Saloheimo *et al.*, 1994). The *bgl1* gene encoding *β*-glucosidase I is expressed at a much lower level than other cellulases, compared with *cbh1* in Table 3.2, but seems to have the highest expression in the same culture conditions as the other cellulases.

Coordinate expression was also observed when a set of isogenic *T. reesei* strains harbouring deletions in cellulase genes (strains Δ*cbh1*, Δ*cbh2*, Δ*egl1* and Δ*egl1*/Δ*egl2*) were studied by Fowler *et al.*, (1993), who reported that these strains produced all the other cellulase transcripts with the same kinetics as the parent strain during growth on lactose. On the other hand, Suominen *et al.* (1993), using isogenic strains in which either *cbh1*, *cbh2*, *egl1* or *egl2* had been replaced by the *A. nidulans amdS* marker gene and also using lactose as a carbon source, showed that deletion of *cbh1* resulted in a two-fold increase in formation of CBH II protein, whereas no such effect was observed with strains deleted for *egl1* or *egl2*. In subsequent experiments, Seiboth *et al.* (1997) reported that this up-regulation of *cbh2* occurred at the transcriptional level both on lactose and on cellulose as sole carbon sources. This increase in *cbh2* expression was due to the replacement of the *cbh1* promoter in the deletion strain, since the increase in *cbh2* transcription was not observed when *cbh1* was replaced by a $cbh1_p$:*xyn2* fusion construct. It is thus possible that the *cbh1* promoter withdraws transcription factors from *cbh2*, although this possibility has to be verified by more detailed analysis.

Some earlier mutant data suggest that separate regulation could exist for *β*-glucosidases, cellobiohydrolases and endoglucanases (Durand *et al.*, 1988; Montenecourt and Eveleigh, 1979; Nevalainen and Palva, 1978; Sheir-Neiss and Montenecourt, 1984; Shoemaker *et al.*, 1981). In addition, the cellulase negative *T. reesei* mutants (QM 9977, QM 9978 and QM 9979), which are unable to grow on cellulose (Mandels *et al.*, 1962) and are not inducible by sophorose or cellulose, produce cellulases on lactose (S. Zeilinger and C. P. Kubicek, unpublished data). This phenotype is probably due to a defect in a regulatory gene since neither the cellulase genes (Torigoi *et al.*, 1996) nor the uptake system for cellobiose or sophorose (Kubicek *et al.*, 1993b) are defective in these mutants. The apparent coordinate regulation generally seen does not, however, contradict the possibility of multiple separate regulation mechanisms. These could all contribute to keeping the relative levels of the different cellulases the same in standard conditions, but the separate mechanisms might be revealed in mutant backgrounds or in culture conditions not yet studied.

Ilmén *et al.* (1997) studied expression levels of the major cellulase genes when the fungus was growing on unlimited amounts of the carbon source, or on the neutral carbon sources sorbitol or glycerol supplemented with mM concentrations of sophorose, or some other disaccharides. Expression levels in a cellulose-based fermenter cultivation and in shake flasks 15 h after sophorose addition were comparable and very high. The mRNAs can be seen even 0.5–1 h after sophorose addition in Northern analysis (M. Ilmén, unpublished), but significant steady-state mRNA levels are observable somewhat later and continue to increase until at least 48 h after sophorose addition (Ilmén *et al.*, 1997). Lactose and cellobiose provoked expression levels in shake-flask cultures that were about fifty-fold lower than those in sophorose-containing cultures. It is interesting that lactose but not galactose provokes expression of cellulases (Margolles-Clark *et al.*, 1997; Morikawa *et al.*, 1995) and that this is reversed for α-galactosidases (see Table 3.2). Detectable, albeit lower, expression of cellulases is provoked by various polymeric xylans and xylobiose (Margolles-Clark *et al.*, 1997; Table 3.2). Interestingly, growth on arabitol results in expression comparable to lactose cultivations (see later).

In addition to carbon source regulation, other factors such as pH or nitrogen concentration might be regulating cellulase expression, but these have not yet been studied. On the other hand, Abrahao-Neto *et al.* (1995) studied the role of mitochondrial functions and suggest that mitochondrial activity is needed for cellulase expression and that decreasing the dissolved oxygen tension to 0.15 mg/l results in decreased cellulase transcription.

3.2.3 *Does* Trichoderma reesei *synthesize low levels of cellulases constitutively?*

In order to explain how the synthesis of cellulases can be turned on in the presence of polymeric cellulose, earlier reports suggested that the fungus produces low, constitutive cellulase levels in all culture conditions including glucose (Gritzali and Brown, 1979; Mandels *et al.*, 1962). This would enable the initial attack on cellulose and release of oligosaccharides which may act as inducers for more abundant cellulase biosynthesis. This hypothesis has more recently been evaluated using sensitive tools such as antibodies (El-Gogary *et al.*, 1989) and expression analysis (Henrique-Silva *et al.*, 1996). Results from all of these studies were in favour of the presence of very low levels of cellulases previously synthesized during growth on repressing carbon sources such as glucose.

In contrast to these results, Ilmén *et al.* (1997) could not demonstrate cellulase *cbh1*, *cbh2*, *egl1* and *egl2* transcript levels in glucose-grown cultures of *T. reesei* QM 9414 in Northern analyses nor in slot blot analyses overloaded with RNA. Similarly, no expression was observed on sorbitol or glycerol media. A comparison of the mRNA levels of the major cellulases showed that if any mRNA were present on glucose-based medium, the levels would be over one thousand-fold lower, and in the case of *cbh1* about five thousand-fold, than in cellulose- or sophorose-based cultures. Since the binding of a possible repressor of cellulase gene transcription to its target sequence in the promoter, even if very strong, is essentially in equilibrium, even strongly repressed genes can give rise to very low levels of "constitutive" expression which, however, may either escape detection or only be detectable by PCR. Whether this is the case with cellulases and whether it has any significant role for the fungus remains to be seen. The possible involvement of low cellulase activities in inducer formation is rendered less likely by the fact that glucose inhibits oligosaccharide uptake (Kubicek *et al.*, 1993b). In any case it seems clear that glucose has a strong repressing effect on cellulase expression.

One should bear in mind though that new *T. reesei* cellulases are still being discovered, *egl4* being the most recent one (Saloheimo *et al.*, 1997). Although expression of *egl3* (Ward *et al.*, 1993), *egl4* (Saloheimo *et al.*, 1997) and *egl5*

(Saloheimo *et al.*, 1994; Ilmén *et al.*, 1997) seems to be regulated coordinately with the other cellulases, these have not been extensively studied. In particular, some results suggest that *egl5* would be weakly expressed in glucose cultures under certain conditions (Ilmén *et al.*, 1997; Saloheimo, A., unpublished). It is possible that the different cellulase genes respond to glucose repression differently, and this might depend on extracellular glucose concentration. Furthermore, it is possible that glucose reduces mRNA half-lives, which might affect detection by Northern analysis. However, Abrahao-Neto *et al.* (1995) reported that cellulase mRNA half-lives would not be affected by glucose in *T. reesei.*

3.2.4 Triggering of cellulase biosynthesis by glucose depletion

Very significant expression of all cellulases can be observed after glucose has been depleted from the culture without an exogenous addition of any inducer (Ilmén *et al.*, 1997, unpublished data). For the major cellulases, this relief from glucose repression results in about 10% of the mRNA levels observed in mycelia cultivated on cellulose. The different cellulases seem to be expressed in the same relative levels as observed on cellulose or other inducing conditions. This triggering of cellulase synthesis by glucose depletion has also been observed earlier (Gong *et al.*, 1979). However, it seems that C- or N-starvation *per se* does not lead to significant expression (Ilmén *et al.*, 1997), and the results also indicate that growth in the presence of the carbon source is needed for expression to occur. Release of inducing compounds from fungal cell walls or inducer formation through transglycosylation of the glucose previously present in culture are yet unverified suggestions to explain these results. This strong derepression after glucose depletion occurs for all the cellulase genes analysed, including the β-glucosidase I-encoding *bgl1* (Margolles-Clark *et al.*, 1997; Table 3.2), and would be adequate to explain the release of inducing molecules from the polymeric substrates in nature.

The repressing/derepressing effects described above for glucose also occur with several other monosaccharides which are rapidly metabolized, e.g., galactose, fructose or mannose (Messner and Kubicek, 1991). Expression after carbon source depletion has also been observed in the *T. reesei* mutants (QM 9977, QM 9978 and QM 9979) described above whose cellulase system is not inducible by sophorose or cellulose but formed during growth on lactose (S. Zeilinger and C. P. Kubicek, unpublished data).

3.2.5 Role of conidial bound cellulases in cellulase induction

T. reesei conidia contain a whole set of different enzymes capable of hydrolysing a wide range of polysaccharides and related compounds (Kubicek *et al.*, 1988; Messner *et al.*, 1991). The conidial cellulases may therefore be an alternative for degradation of cellulose under conditions of nutritional stress. They are located on the conidial surface and formed irrespectively of the nutrient conditions used to induce sporulation (Kubicek *et al.*, 1988). Immunological analysis demonstrated the presence of CBHI and CBHII but not of EG I on the conidial surface (Messner *et al.*, 1991). The conidial cellulase system was able to hydrolyse crystalline cellulose

(Kubicek *et al*., 1988). Based on the immunological analysis, a relatively high proportion of CBHII was observed. Whereas normally only about 15–20% of total secreted protein is CBHII, and up to 60% is CBHI (Hayn and Esterbauer, 1985), the amount of CBHII on the conidial surface was approximately twice that of CBHI (Kolbe and Kubicek, unpublished; see Kubicek *et al*., 1993a). It is possible that elevated levels of CBHII may be beneficial for the initial attack on crystalline cellulose: the optimal ratio of purified CBHI : CBHII for cellulose hydrolysis *in vitro* has been reported to be 1 : 2 (Henrissat *et al*., 1985).

The relative importance of CBHII and some other cellulases in initiation of growth of *T. reesei* on cellulose has been investigated by Seiboth *et al*. (1992, 1997). Using the isogenic strains in which the major cellulase genes (*cbh1*, *cbh2*, *egl1* and/or *egl2*) had been replaced by the *A. nidulans amdS* gene (Suominen *et al*., 1993), they observed striking differences in the ability of the strains to grow on crystalline cellulose as the only carbon source. Strains in which *cbh2* and *egl2* had been deleted grew most poorly, whereas a strain containing the *egl1* deletion grew almost normally. Conidia from a strain in which both *cbh1* and *cbh2* had been replaced were almost unable to initiate growth on cellulose. The growth on cellulose of these strains roughly correlated with the respective levels of expression of the remaining cellulase genes. Addition of 2 mM sophorose to cultures of the strain in which both *cbh1* and *cbh2* had been deleted induced the transcription of *egl1* and *egl2* and restored the ability to attack cellulose. These results are in accordance with a mechanism by which conidial-bound cellulases can carry out an initial degradation of cellulose molecules, thereby generating the inducers of cellulase biosynthesis, and hence lead to growth on cellulose.

3.2.6 *Glucose repression*

The fact that distinct mechanisms for glucose repression occur in cellulase expression seems well established. Expression of the cellulases seems to be negligible on glucose (Table 3.2), but this does not necessarily indicate that true glucose repression mechanisms are operating, since this could just reflect the lack of an inducing compound. However, in the presence of high levels of glucose, sophorose fails to induce expression of the cellulase genes normally induced in cultures based on the "neutral" carbon sources glycerol and sorbitol (Ilmén *et al*., 1997; Margolles-Clark *et al*., 1997; Table 3.2). Glucose addition to cultures already induced leads to the disappearance of the transcripts of *cbh1*, *cbh2*, *egl1*, *egl2* and *egl5* cellulose (El-Gogary *et al*., 1989; Ilmén *et al*., 1997). As discussed earlier (Kubicek *et al*., 1993a; Ilmén *et al*., 1997), this may well reflect impaired inducer uptake. Nevertheless this strongly suggests that glucose repression mechanisms somehow affect cellulase expression.

Evidence for glucose directly repressing cellulase expression comes from cellulase promoter analysis (see below) and from the cloning of the gene encoding the general carbon catabolite repressor protein CREI. The *T. reesei cre1* gene was cloned by Strauss *et al*. (1995), Ilmén *et al*. (1996a) and Takashima *et al*. (1996a). Also the *T. harzianum cre1* equivalent has been characterized (Ilmén *et al*., 1996a). The two *Trichoderma* genes encode proteins showing 93% identity at the amino acid level. The *cre1* genes are the homologues of the *Aspergillus creA*, a well-characterized regulator of general carbon catabolite repression as studied by genetic (Arst and

Bailey, 1977; Bailey and Arst, 1975) and molecular means (Arst *et al.*, 1990; Cubero and Scazzochio, 1994; Dowzer and Kelly, 1989, 1991; Kulmburg *et al.*, 1993; Mathieu and Felenbok, 1994). Comparison of the *Trichoderma* and *Aspergillus* CREA/CREI proteins shows only 46% overall amino acid identity, and this allows one to search for conserved regions of putative functional importance. Considerable similarity is revealed in certain parts of the proteins, in the region containing two C2H2 type zinc fingers (aa_{54-120}), a region rich in serines and prolines ($aa_{258-296}$), and in another region rich in basic amino acids and proline ($aa_{360-377}$). The zinc fingers are similar to the ones found in the glucose repressor proteins Mig1 (Nehlin and Ronne, 1990) and Mig2 (Lutfiyya and Johnston, 1996) of *S. cerevisiae* and in some growth-related proteins of yeast and mammalian cells. Experiments done with *Aspergillus* and *Trichoderma* (see below) have shown that CREA and CREI also bind to similar nucleotide sequences as the yeast and mammalian proteins. The serine/proline-rich region was reported (Dowzer and Kelly, 1991; Ilmén *et al.*, 1996a) to have similarity to the yeast Rgr1 protein (Sakai *et al.*, 1990), which affects glucose repression, but highly similar serine/proline-rich regions are found in many eukaryotic transcription factors. It was thus hypothesized that this region could act as a general transactivating domain (Strauss *et al.*, 1995). That the role of CREI would not only be to act as a glucose repressor is indirectly suggested by the *a priori* surprising observations that its expression in *T. reesei* seems to be lower on glucose medium than, for instance, on cellulose-based medium. It seems to be partially repressible by glucose, which is consistent with the fact that binding sites for CREI itself can be found in the *cre1* promoters of *T. reesei* and *T. harzianum* (Ilmén *et al.*, 1996a). The possible autoregulation and alternative regulatory roles still need further experimental verification.

Ilmén *et al.* (1996a) showed that the commonly used *T. reesei* strain RutC-30 is mutated in the *cre1* gene. The strain has only 20% of the coding region remaining, and the truncated protein formed lacks all the C-terminal amino acids after the first zinc finger. Unlike an *Aspergillus* strain lacking regions after the two zinc fingers (Arst *et al.*, 1990), RutC-30 is viable. This could reflect differences between the two fungal species or be explained by some other suppressing mutations in RutC-30. RutC-30 was originally isolated as a hypercellulolytic strain by screening for growth on cellobiose in the presence of 2-deoxyglucose (Montenecourt and Eveleigh, 1979). Northern analysis confirmed that this strain produces cellulases in the presence of glucose (Ilmén *et al.*, 1996a). That this derepression is mediated by the *cre1-1* mutation was shown by transforming the wild type *cre1* gene into RutC-30 and by demonstrating that this abolished cellulase expression on glucose medium, thus confirming that the glucose repressor CREI is involved in cellulase regulation. It is not known, however, whether it is the only regulator mediating glucose repression.

3.2.7 ***Analysis of functional regions in the cellulase gene promoters***

A comparison of the 5′-upstream sequences of *cbh1*, *cbh2*, *egl1* and *egl2* from *T. reesei* and of cellulase genes from other *Trichoderma* spp. reveals a number of nucleotide consensus sequences that are shared between at least two of these genes and are known to bind transcription regulating proteins in other eukaryotes. These are for example the carbon catabolite repressor protein CREA (5′-SYGGRG-3′; Cubero and Scazzochio, 1994; Kulmburg *et al.*, 1993), the CBE-protein (5′-CCAAT-

3′; Johnson and McKnight, 1989), a cyclic-AMP response element (Chen *et al.*, 1987), and a stress response element (5′-CCCCT-3′; Marchler *et al.*, 1993). A summary of their occurrence in *Trichoderma* cellulase promoters is given in Figure 3.1. Only some of these, however, have been studied and functionally verified. It should be noted that the identification of conserved regulatory regions by sequence comparisons only is doubtful, and even when some putative regions are found, mutagenesis of these does not necessarily lead to changes in regulation (e.g., see Ilmén *et al.*, 1996b).

Cellobiohydrolase 1 promoter

In the 2.2 kb region of the *cbh1* promoter, 11 consensus binding sites (5′-SYGGRG-3′) for binding of the glucose repressor protein CREA/Mig1 are present, and some of them with the consensus 5′-GTGGGG have actually been shown to bind CREI. The zinc finger region of the *T. reesei* CREI has been produced in *E. coli* as a fusion to the maltose binding protein (Takashima *et al.*, 1996b) and glutathione S transferase (Straus *et al.*, 1995; Ilmén, 1997). Using these fusion proteins in *in vitro* gel shift assays, the consensus binding sites situated at −1500, −1150 and around −700 from the ATG were shown to bind CREI, whereas the site at −1000 did not

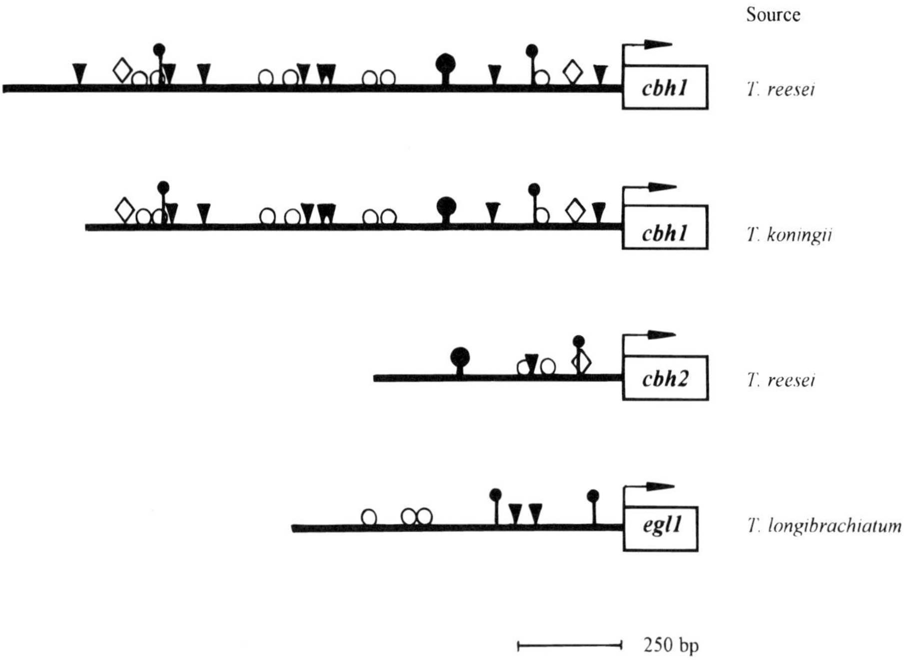

Figure 3.1 Occurrence of nt-motifs of potential transcriptional importance in the 5′-upstream regions of some cellulase genes from *Trichoderma* spp. Symbols: diamond, TATA box; open circles, CAAT or CCAAT box; small circle-capped long bar, stress response element, CCCCT; triangles, consensus for binding of Cre1 (SYGGRG); big circle-capped fat bar, cylic AMP responsive element. Data were taken from *T. reesei cbh1* (Ilmén *et al.*, 1996b; Takashima *et al.*, 1996b), *T. koningii cbh1* (Wey *et al.*, 1994), *T. reesei cbh2* (Stangl *et al.*, 1993), and *T. longibrachiatum egl1* (Gonzalez *et al.*, 1994).

bind efficiently (Takashima *et al.*, 1996b; Ilmén, 1997). In addition, Takashima *et al.* (1996) showed that the three individual putative CREI binding sites at the −700 region bind CREI fingers in *in vitro* footprint experiments. These results are consistent with the data available for *Aspergillus* that sites matching the consensus sequence 5′-(C/G)TGGGG mediate proper binding.

In vitro data does not necessarily reflect the functionality of the promoter regions *in vivo*. Ilmén *et al.* (1996b) mutated the consensus sites at around 700, 1000 and 1500 bp upstream of the protein coding region by replacing the native nt-sequences by the same number of nts of mutant sites in different combinations. These mutant promoters were fused to the *E. coli lacZ* gene as reporter and were transformed into *T. reesei* QM9414 so that they integrated as a single copy into the endogenous *cbh1* locus. The results indicated that the mutation of the site at −1000 may slightly affect glucose repression but it was evident that mutation of the region at around −700 resulted in clear derepression of *lacZ*-mRNA on glucose. The extent to which the three CREI binding sites present in this area each contribute to regulation is not clear at the moment. This result is consistent with the deletion series (see below; Ilmén *et al.*, 1996b) that showed that removal of the regions upstream of −500 resulted in glucose derepression in the same way as the specific mutation at −700.

The expression level from the *cbh1* promoter mutated in the three CREI binding sites situated at around −700 was about ten-fold lower on glucose-containing medium than in sophorose–sorbitol cultures (Ilmén *et al.*, 1996b). This indicates that CREI is not the only regulator mediating repression on glucose-containing medium and/or that full expression requires activating factors. Sophorose induction from the mutant promoters on sorbitol medium was comparable to that obtained from the wild type promoter, which suggests that at least not all the regions in the promoter needed for activation are overlapping with the CREI binding sites. Consistent with these results is the fact that the *T. reesei cre1-1* mutant strain RutC-30 does not produce full levels of cellulases on glucose medium.

To identify regions in the *T. reesei cbh1* promoter involved in cellobiose- and sophorose-mediated induction, Ilmén *et al.* (1996b) generated deletion series starting from the 5′ end of the promoter which was linked to the *E. coli lacZ* gene as reporter. A number of transformants of each deletion variant were analysed by microtitre plate assays and verified by Northern analyses. Interestingly, strong induction was still occurring even in promoter variants retaining only 30 bp upstream of the TATA box. Further deletions gave variable induction results, and removal of the TATA box abolished expression completely. Different types of experiments are needed to identify more precisely the regions responsible for induction. It is interesting, however, to envisage how induction would be mediated through this small promoter variant. If the action of a putative positive regulator is mediated through some of the regions near the TATA box, this "nearby" action could explain why the *cbh1* promoter is one of the strongest known promoters.

At first glance, these results are in contrast to the findings of Henrique-Silva *et al.* (1996) who, using the *E. coli* hygromycin B phosphotransferase as a reporter gene, reported that the *cbh1* promoter regions responsible for induction by cellulose are located between −241 and −72 bp upstream of TATA. They also reported that the 72 bp region adjacent to the TATA box is responsible for constitutive expression on glycerol-based medium. Solving the possible discrepancies between the results of these two reports (Henrique-Silva *et al.*, 1996; Ilmén *et al.*, 1996b) warrants further work. It must be taken into account, however, that in the experiments by Henrique-Silva *et al.* (1996) they did not analyse multiple transformants or the locus of integration, and also the copy number of the transformed reporter construct was not carefully characterized. It is therefore possible that the "basal" transcription observed was a result of the presence of multiple copies of the construct titrating out regulatory factors, or an improper place of integration, which both might lead to (partial) escape from carbon catabolite repression.

Cellobiohydrolase 2 promoter

Stangl *et al.* (1993) have studied the *cbh2* promoter of *T. reesei.* They used protein extracts from induced and non-induced mycelia in electrophoretic mobility shift assays (EMSA) with a 615 bp fragment of the *cbh2* promoter and detected a specific protein–DNA complex present only in sophorose-induced mycelium. Using various overlapping fragments and competitive oligonucleotides, the DNA target motif for this protein complex was identified as a double-inverted CCAAT box (Zeilinger *et al.*, unpublished). These findings bear some similarity to the taka-amylase A promoter of *A. oryzae* in which a similar CCAAT motif is recognized by proteins from starch-induced cell-free extracts of *A. nidulans* (Nagata *et al.*, 1993). CCAAT boxes have been observed in the regulatory regions of several fungal genes, and their function has been proven in *A. nidulans amdR* (van Heeswijck and Hynes, 1991) and *yA* (Aramayo and Timberlake, 1993). The *A. nidulans* gene *hapC* that encodes a homologue of *HAP3*, which is one of the three CCAAT-box-binding HAP proteins identified in *S. cerevisiae*, has recently been cloned and we thus assume its presence also in *T. reesei* by analogy.

It is intriguing in this context that no CCAAT box seems to exist in the *cbh1* promoter within the 30 bp region upstream of TATA which is still inducible by sophorose (Ilmén *et al.*, 1996b). A CCAAT box is present downstream of the *cbh1* TATA region and transcript start site, but the significance of this might be doubtful since functional regulatory regions have very rarely been reported to exist downstream of TATA in any fungal promoters, including *S. cerevisiae.*

In accordance with glucose repression of *cbh2*, Takashima *et al.* (1996b) have shown that *cbh2* promoter fragments bind the *E. coli* produced CREI fingers *in vitro.*

3.2.8 *Signalling pathways*

Only a few preliminary sets of data are currently available on the signalling mechanisms by which the inducer signal is transduced to the transcriptional machinery. Wang and Nuss (1995) provided evidence for the involvement of a G_α-protein in the induction of *cbh1* by cellulose in the fungus *Cryphonectria parasitica*, as a strain in which the respective gene had been inactivated by squelling was unable to induce *cbh1* transcription. However, in *T. reesei* the addition of mastoparam (an activator of G_α) was unable to replace sophorose in induction experiments (S. Gyamphi, S. Zeilinger and C. P. Kubicek, unpublished data), suggesting that sophorose induction might not act via G_α-protein activation in this fungus. Sestak and Farkaš (1993) reported that the efficacy of sophorose induction could be doubled by the addition of permeable cyclic-AMP derivatives or by the inhibitors of cyclic-AMP-phosphodiesterase, hence suggesting that the signalling by the inducer involves a cyclic-AMP dependent protein kinase. In accordance with this, unpublished results of D. Kristufek and C. P. Kubicek indicated that induction of cellulases as well as hemicellulases by sophorose or xylose was strongly inhibited by the addition of the serine/threonine protein kinase inhibitor H7. However, since RNA polymerase II is also dependent on phosphorylation during transcriptional initiation, these results are essentially preliminary and are only described here to provide hints for further research.

3.3 Regulation of hemicellulase expression

3.3.1 Regulation of production of xylan backbone degrading endoxylanases

T. reesei produces two xylanases encoded by the *xyn1* and *xyn2* genes, which degrade the β-1,4-D-xylan backbone of hemicelluloses (see Chapter 2). Hrmová *et al.* (1986) showed using isoelectric focusing gels coupled with staining based on xylanase and cellulase activity that *T. reesei* forms two specific xylanases and a single unspecific one (endoglucanase I) when cultured on cellulose or xylan. Upon induction by sophorose, only one of both xylanases was formed, whereas induction by xylobiose led to the formation of the two xylanases but not of endoglucanase I.

Zeilinger *et al.* (1996) investigated the regulation of xylanase expression at the level of transcription. Their studies confirmed the data of Hrmová *et al.* (1986) but in addition also provided evidence for a different behaviour of the two xylanases with respect to glucose. Transcription from the *xyn1* promoter was triggered by the presence of xylan and xylose and was not apparent in the presence of glucose. In contrast, the *xyn2* promoter enabled a low level of basal transcription also on glucose, which was enhanced in the presence of xylan and xylobiose (Zeilinger *et al.*, 1996). *Xyn2* transcription was also induced by sophorose and cellobiose, whereas transcription of *xyn1* was not.

Intriguingly, the addition of glucose to cultures growing on xylan reduced the expression of *xyn2* to a basal level (R. L. Mach, J. Strauss and C. P. Kubicek, unpublished data) but transcription of *xyn1* was not affected (Mach *et al.*, 1996). It should be noted that the uptake of xylose by *T. reesei* is not inhibited by glucose (C. P. Kubicek, unpublished data). Hence, in contrast to the cellulase system, there seems to exist evidence for a low constitutive level of *T. reesei* xylanase activity. The role of *xyn2* in the formation of the inducer from xylan therefore can be suggested but has yet to be proven. In contrast to the similar experiments carried out to study cellulase formation (Kubicek *et al.*, 1988), Herzog *et al.* (1992) reported that treatment of *T. reesei* conidia and mycelia with 0.1 M HCl or Tween 80 did not affect the ability of the fungus to form xylanases on xylan medium, although the treatment removed and inactivated the surface-bound xylanase activity. This would be consistent with a role of the secreted constitutive levels of XYNII in xylanase induction.

The results of Margolles-Clark *et al.* (1997; Table 3.2) seem to be partly different from those mentioned above. Sophorose induced expression of both *xyn1* and *xyn2*, as did xylobiose when added to glycerol medium. That the expression level of *xyn1* was somewhat lower than that of *xyn2* might explain the discrepancy between the different reports. No expression of *xyn1* or *xyn2* was observed when the fungus was grown on abundant amounts of glucose, but glucose depletion resulted in significant expression of *xyn2*. It is perhaps noteworthy that *xyn1* was exceptional amongst the many (hemi)cellulase genes studied thus far in not being derepressed after glucose depletion. On the other hand, *xyn2* was exceptional in not being expressed on glucose in the *cre1-1* mutant strain RutC-30, suggesting that it would not be under the control of *cre1*, at least not in a way similar to other (hemi)cellulase genes. No expression of *xyn1* or *xyn2* was observed when *T. reesei* was grown on abundant amounts of xylose. Although these results seem to contradict those presented above, it is possible that the differences merely reflect the amount of carbon source or inducing sugar present in the medium.

It is evident, however, that unlike the cellulase genes, *xyn1* and *xyn2* are not expressed in a coordinate way, and differential expression has been seen in many studies. This holds true also when various natural xylan polymers originating from different sources are used as carbon sources (Margolles-Clark *et al.*, 1997; Royer and Nakas, 1990; Senior *et al.*, 1989; see Table 3.2). The *bxl1* gene encoding *β*-xylosidase I is expressed in all conditions where either *xyn1* or *xyn2* is expressed (Table 3.2).

Analysis of xylanase promoters

Using promoter deletion analysis and the *E. coli* hygromycin B phosphotransferase as a reporter gene, the regions within the 5′-upstream non-coding sequences of *xyn1* and *xyn2* conferring regulation were localized within a 221 bp and a 55 bp fragment, respectively. These findings were corroborated by electrophoretic mobility shift assays with cell-free extracts from mycelia incubated with different inducers, which showed that these DNA fragments bind proteins from induced mycelia. With *xyn2*, a single DNA–protein complex of high mobility was observed under basal, non-induced conditions on glucose. This complex was replaced by a slow-migrating complex with cell-free extracts from xylan- or sophorose-induced mycelia. Oligonucleotide competition experiments revealed that both complexes were due to binding to a CCAAT box, thus possessing similarity with the promoter of the cellobiohydrolase II encoding gene *cbh2*. Interestingly, similar experiments with the promoter of *xyn1* also revealed this CCAAT box as responsible for binding of proteins from induced mycelia, whereas no binding was observed in the presence of the carbon catabolite repressor CREI, which binds to a nearby consensus motif organized as an inverted repeat (Mach *et al.*, 1996). In addition, *T. reesei* QM 9414 strains bearing a $xyn1_p$:*hph* reporter construct, in which four nucleotides from the middle of the inverted consensus for CREI binding had been removed, expressed *hph* on glucose at a level comparable to that observed during growth on carbon catabolite derepressing carbon sources such as lactose. Furthermore, Northern analysis of *xyn1* expression in the *cre1-1* mutant strain *T. reesei* RutC-30 grown on glucose also showed basal transcription of *xyn1*. Both in *T. reesei* RutC-30 as well as in the strain bearing the mutated $xyn1_p$:*hph* construct, a yet higher level of the *xyn1* transcript or reporter gene transcript, respectively, was observed upon induction by xylose or xylan in glucose-based cultures, suggesting that de-repression alone is not sufficient for obtaining full induction. These data, and the lack of repression by glucose in the presence of xylan, suggest that in *T. reesei xyn1*, CREI binding to the promoter interferes primarily with basal transcription.

Similar studies with the *xyn2* promoter indicated that *xyn2* gene expression might be via the CCAAT element only (Zeilinger *et al*., 1996). Whereas protein interaction with this motif was demonstrated under all conditions, induction by xylan (or respective oligomers) showed the formation of a larger protein–DNA complex. In extracts from mycelia cultivated in the simultaneous presence of xylan and glucose, only a smaller DNA–protein complex was observed, and it is tempting to speculate that here glucose interferes with the formation of an activating factor. We have as yet no evidence whether this involves CREI. Hence the simplest model to explain regulation of *xyn2* would be to postulate the binding of a *Trichoderma* equivalent of Hap2/Hap3 (and eventually Hap5) under basal conditions, which then associates with additional components upon induction by xylan or cellulose.

The same model may basically also be applicable to the regulation of *xyn1*, although the involvement of additional factors still has to be taken into account. Since the CCAAT box and the CREI binding consensus are separated by only 31 bp and the CREI binding site lies downstream of the CCAAT box, repression may act either by competition for binding sites or by inhibition of the contact with the RNA–polymerase II initiation complex.

3.3.2 Regulation of formation of other hemicellulases

Very few data have been available on the regulation of synthesis of other hemicellulases. Recently, however, the genes encoding three α-D-galactosidases (*agl1*, *agl2*, and *agl3*) (Margolles-Clark *et al*., 1996a), which release galactose side groups from hemicelluloses or the derived oligosaccharides, and an α-L-arabinofuranosidase (*abf1*) which similarly releases arabinose, have been isolated (Margolles-Clark *et al*., 1996b). Other genes include the one encoding a β-D-xylosidase (*bxl1*) (Margolles-Clark *et al*., 1996b), which mainly hydrolyses xylooligosaccharides and xylobiose, and those encoding an acetyl xylan esterase (*axe1*) (Margolles-Clark *et al*., 1996c) and an α-D-glucuronidase (*glr1*) (Margolles-Clark *et al*., 1996d), which release acetic and glucuronic acids from hemicelluloses, respectively. This has enabled analysis of the transcriptional regulation of these genes at a molecular level (Margolles-Clark *et al*., 1997).

Table 3.2 summarizes the results obtained by Margolles-Clark *et al*. (1997) from studies analysing the relative steady-state mRNA levels of the hemicellulase genes when *T. reesei* QM 9414 was growing on unlimited amounts of various carbon sources. Under these conditions the relative expression levels vary so that sometimes a gene or a set of genes is more highly expressed and at other times another gene(s) is(are) expressed more. Thus, it seems likely that different specific regulatory mechanisms are involved that modulate the expression levels. The genes *agl1* and *agl2* are induced by galactose but not by lactose in particular, and *axe1* and the xylanases including *bxl1* are induced by various xylans; *abf1* is especially strongly expressed on the arabinose-substituted xylan from oat spelt. It is noteworthy that *man1* and *agl3* are rather poorly expressed except on cellulose only, and interestingly it seems that *agl3* expression is otherwise revealed only in the *cre1-1* mutant strain RutC-30.

The two other *agl* genes, the xylanases *xyn1*, *xyn2* and *bxl1*, and the cellulase *cbh1* are the ones expressed in many conditions. Expression of the *β*-xylosidase-encoding gene *bxl1* is significantly higher in these conditions than the expression of the *β*-glucosidase-encoding gene *bgl1*. Whether the corresponding enzyme activities of any of these genes expressed in many conditions would contribute to induction of the other hemicellulase or cellulase genes in the presence of polymeric substrates would be of interest to study. Expression of the α-galactosidase genes does not seem to play this role since the galactose released provokes expression of only few of the genes. Xylobiose and sophorose (as well as the polymeric xylans and cellulose), on the other hand, seem to be rather general inducers, triggering expression of many of the genes. Similarly to the cellulases, specific induction mechanisms (and not just release from glucose repression) seem to be operating since most of the genes are not expressed on sorbitol or glycerol media, but are expressed after the addition of inducing compounds. As a conclusion, it seems that the fungus has mechanisms to ensure production of a vast number of the enzymes involved in hydrolysis of cellulose and various hemicelluloses, albeit at varying levels, when exposed to only one type of polymer. Also various heterooligosaccharides consisting of xylose and glucose have been reported to provoke enzyme production (Hrmová *et al.*, 1991).

Arabitol, the intracellular intermediate of arabinose (xylose) metabolism provokes high expression levels of many of the genes studied including *xyn1* and *xyn2*. This is of interest since it has been shown that *Aspergillus* mutants with high intracellular arabitol levels produce elevated amounts of extracellular arabinases (van der Veen *et al.*, 1993, 1994; Witteveen *et al.*, 1989).

As is the case with cellulases, expression of the hemicellulase genes on media containing high glucose amounts is very low or non-detectable in Northern analysis (Margolles-Clark *et al.*, 1997; Table 3.2), but significant expression is observed (except for *xyn1* and *agl3*) once glucose has been depleted from the culture. Studies with the *cre1-1* mutant RutC-30 and RutC-30 complemented with the wild type *cre1* gene revealed that at least *xyn1*, *bxl1*, *agl1*, *agl2*, *agl3* and *axe1* are apparently under *cre1*-mediated carbon catabolite repression (Margolles-Clark *et al.*, 1997).

3.4 Conclusions

Due to the development of efficient transformation systems (Volume 1, Chapter 10) and the availability of a number of the genes encoding various cellulases (Table 3.1) and hemicellulases (Chapter 2), *T. reesei* has now entered a stage where the regulation of degradation of plant polysaccharides can be studied at the molecular level. Research carried out with *T. reesei* clearly has the leading role with respect to regulation of production of cellulolytic enzymes in fungi, and *Trichoderma* contributes together with *Aspergillus* to our understanding of hemicellulase regulation as well. While the present insight is clearly only a beginning, it is already evident that many different regulatory mechanisms have evolved to control this special feature of fungi, extracellular enzyme production, which has a significant role in the nutrition of these organisms. Recent achievements have led to the identification of the first specific transcriptional activators for cellulase and hemicellulase genes (van Peij *et al.*, 1998; Saloheimo *et al.*, in press). Their availability will be a basis for further work endeavouring to reveal the intracellular pathways that mediate the signal from extracellular plant polysaccharides to the transcriptional level of cellulase and hemicellulase genes.

References

Abrahao-Neto, J., Rossini, C. H., El-Gogary, S., Henrique-silva, F., Crivellaro, O., and El-Dorry, H. 1995. Mitochondrial functions mediate cellulase gene expression in *Trichoderma reesei*. *Biochemistry* **34**: 10456–10462.

Aramayo, R. and Timberlake, W. E. 1993. The *Aspergillus nidulans yA* gene is regulated by *abaA*. *EMBO J.* **12**: 2039–2048.

van Arsdell, J. N., Kwok, S., Schweickart, V. L., Ladner, M. B., Gelfand, D. H., and Innis, M. A. 1987. Cloning, characterization, and expression in *Saccharomyces cerevisiae* of endoglucanase I from *Trichoderma reesei*. *Bio/Technology* **5**: 60–64.

Arst, H. N. and Bailey, C. R. 1977. The regulation of carbon metabolism in *Aspergillus nidulans*. In J. E. Smith and J. A. Pateman (eds), *Genetics and Physiology of* Aspergillus nidulans. Academic Press, London, pp. 131–146.

Arst, H. N., Tollervey, D., Dowzer, C. E. A., and Kelly, J. M. 1990. An inversion truncating the *creA* gene of *Aspergillus nidulans* results in carbon catabolite derepression. *Mol. Microbiol.* **4**: 851–854.

Bailey, C. R. and Arst, H. N. 1975. Carbon catabolite repression in *Aspergillus nidulans*. *Eur. J. Biochem.* **51**: 573–577.

Barnett, C. C., Berka, R. M., and Fowler, T. 1991. Cloning and amplification of the gene encoding extracellular *β*-glucosidase from *Trichoderma reesei*. *Bio/Technology* **9**: 562–566.

Bisaria, V. S. and Mishra, S. 1989. Regulatory aspects of cellulase biosynthesis and secretion. *CRC Crit. Rev. Biotechnol.* **9**: 61–103.

Chen, C. M., Gritzali, M., and Stafford, D. W. 1987. Nucleotide sequence and deduced primary structure of cellobiohydrolase II from *Trichoderma reesei*. *Bio/Technology* **5**: 274–278.

Chen, H., Hayn, M., and Esterbauer, H. 1992. Purification and characterization of two extracellular *β*-glucosidases from *Trichoderma reesei*. *Biochim. Biophys. Acta* **1121**: 54–60.

Cheng, C., Tsukagoshi, N., and Udaka, S. 1990. Nucleotide sequence of the cellobiohydrolase gene from *Trichoderma viride*. *Nucl. Acids Res.* **18**: 5559.

Cubero, B. and Scazzochio, C. 1994. Two different, adjacent and divergent zinc finger binding sites are necessary for CREA-mediated carbon catabolite repression in the proline gene cluster of *Aspergillus nidulans*. *EMBO J.* **13**: 407–415.

Dowzer, C. E. A. and Kelly, J. M. 1989. Cloning of the *creA* gene from *Aspergillus nidulans*: a gene involved in carbon catabolite repression. *Curr. Genet.* **15**: 457–459.

Dowzer, C. E. A. and Kelly, J. M. 1991. Analysis of the *creA* gene, a regulator of carbon catabolite repression in *Aspergillus nidulans*. *Mol. Cell. Biol.* **9**: 5701–5709.

Durand, H., Baron, M., Calmels, T., and Tiraby, G. 1988. Classical and molecular genetics applied to *Trichoderma reesei* for the selection of improved cellulolytic industrial strains. In J.-P. Aubert, P. Béguin, and J. Millet (eds), *Biochemistry and Genetics of Cellulose Degradation*. FEMS Symposium No. 43. Academic Press, London, pp. 135–152.

El-Gogary, S., Leite, A., Crivellaro, O., Eveleigh, D. E., and El-Dorry, H. 1989. Mechanism by which cellulose triggers cellobiohydrolase I gene expression in *Trichoderma reesei*. *Proc. Natl. Acad. Sci. USA* **86**: 6138–6141.

Ellouz, S., Durand, H., and Tiraby, G. 1987. Analytical separation of *Trichoderma reesei* cellulases by ion-exchange fast protein liquid chromatography. *J. Chromat.* **396**: 307–317.

FOWLER, T. and BROWN, R. D. JR. 1992. The *bgl1* gene encoding extracellular β-glucosidase from *Trichoderma reesei* is required for rapid induction of the cellulase complex. *Mol. Microbiol.* **6**: 3225–3235.

FOWLER, T., GRITZALI, M., and BROWN, R. D. JR. 1993. Regulation of the cellulase genes of *Trichoderma reesei*. In P. Suominen and T. Reinikainen (eds), *Proceedings of the second TRICEL symposium on* Trichoderma reesei *cellulases and other hydrolases*, Espoo, Finland, 1993, Vol. 8, pp. 199–210. Foundation for Biotechnical and Industrial Fermentation Research.

FRITSCHER, C., MESSNER, R., and KUBICEK, C. P. 1990. Cellobiose metabolism and cellobiohydrolase I biosynthesis in *Trichoderma reesei*. *Exp. Mycol.* **14**: 451–461.

GONG, C.-S., LADISCH, M. R., and TSAO, G. T. 1979. Biosynthesis, purification and mode of action of cellulases of Trichoderma reesei. In R. D. Brown, JR. and L. Jurasek (eds), Hydrolysis of Cellulose: Mechanism of Enzymatic and Acid Catalysis. *Adv. Chem. Ser.* **181**: 261–288.

GONZALEZ, R., PEREZ-GONZALEZ, J. A., GONZALEZ-CANDELAS, L., and RAMON, D. 1994. Transcriptional regulation of the *Trichoderma longibrachiatum egl1* gene. *FEMS Microbiol. Lett.* **122**: 303–308.

GRITZALI, M. and BROWN, R. D. JR. 1979. The cellulase system of *Trichoderma*: relationships between purified extracellular enzymes from induced or cellulose–grown cells. *Adv. Chem. Ser.* **181**: 237-260.

HAYN, M. and ESTERBAUER, H. 1985. Separation and partial characterization of *Trichoderma reesei* cellulase by fast liquid chromatography. *J. Chromat.* **329**: 379–387.

HENRIQUE-SILVA, F., EL-GOGARY, S., CARLE-URIOSTE, J. C., MATHEUCCI, E. JR., CRIVELLARO, O., and EL-DORRY, H. 1996. Two regulatory regions controlling basal and cellulose-induced expression of the gene encoding cellobiohydrolase I of *Trichoderma reesei* are adjacent to its TATA box. *Biochem. Biophys. Res. Comm.* **228**: 229–237.

HENRISSAT, B., DRIGUEZ, H., VIET, C., and SCHÜLEIN, M. 1985. Synergism of cellulases from *Trichoderma reesei* in the degradation of cellulose. *Bio/Technology* **3**: 722–726.

HERZOG, P., TÖRRÖNEN, A., HARKKI, A. M., and KUBICEK, C. P. 1992. Mechanism by which xylan and cellulose trigger the biosynthesis of endo-xylanase I by *Trichoderma reesei*. In J. Visser, G. Beldman, M. A. Kusters-van Someren, and A. G. J. Voragen (eds), *Xylans and Xylanases*. Elsevier Science, Amsterdam, pp. 289–293.

HRMOVÁ, M., BIELY, P., and VRSANSKA, M. 1986. Specificity of cellulase and β-xylanase induction in *Trichoderma reesei* QM 9414. *Arch. Microbiol.* **144**: 307–311.

HRMOVÁ, M., PETRAKOVA, E., and BIELY, P. 1991. Induction of cellulose- and xylan-degrading enzyme systems in *Aspergillus terreus* by homo- and heterodisaccharides composed of glucose and xylose. *J. Gen. Microbiol.* **137**: 541–547.

ILMÉN, M. 1997. *Molecular mechanisms of glucose repression in the filamentous fungus* Trichoderma reesei. Ph.D. thesis. To be published in the VTT Publication Series.

ILMÉN, M., THRANE, C., and PENTTILÄ, M. 1996a. The glucose repressor gene *cre1* of *Trichoderma*: isolation and expression of a full length and a truncated mutant gene. *Mol. Gen. Genet.* **251**: 451–460.

ILMÉN, M., SALOHEIMO, A., ONNELA, M.-L., and PENTTILÄ, M. 1997. Regulation of cellulase expression in the filamentous fungus *Trichoderma reesei*. *Appl. Envir. Microbiol.* **63**: 1298–1306.

ILMÉN, M., ONNELA, M.-L., KLEMSDAL, S., KERÄNEN, S., and PENTTILÄ, M. 1996b. Functional analysis of the cellobiohydrolase I promoter of the filamentous fungus *Trichoderma reesei*. *Mol. Gen. Genet.* **253**: 303–314.

IYAYI, C. B., BRUCHMANN, E.-E., and KUBICEK, C. P. 1989. Induction of cellulase formation in *Trichoderma reesei* by cellobiono-1,5-lactone. *Arch. Microbiol.* **151**: 326–330.

JOHNSON, P. F. and MCKNIGHT, S. L. 1989. Eucaryotic transcriptional regulatory proteins. *Annu. Rev. Biochem.* **58**: 799–839.

KUBICEK, C. P. 1987. Involvement of a conidial bound endoglucanase and a plasma membrane bound β-glucosidase in the induction of endoglucanase synthesis by cellulose in *Trichoderma reesei. J. Gen. Microbiol.* **133**: 1481–1487.

KUBICEK, C. P. 1993. From cellulose to cellulase inducers: facts and fiction. In P. Suominen and T. Reinikainen (eds), *Proceedings of the second TRICEL symposium on* Trichoderma reesei *cellulases and other hydrolases*, Espoo, Finland. Foundation for Biotechnical and Industrial Fermentation Research, Vol. 8, pp. 181–188.

KUBICEK, C. P., MESSNER, R., GRUBER, F., MACH, R. L. and KUBICEK PRANZ, E. M. 1993a. The *Trichoderma* cellulase regulatory puzzle: from the interior life of a secretory fungus. *Enzyme Microb. Technol.* **15**: 90–99.

KUBICEK, C. P., MESSNER, R., GRUBER, F., MANDELS, M., and KUBICEK-PRANZ, E. M. 1993b. Triggering of cellulase biosynthesis by cellulose in *Trichoderma reesei*: involvement of a constitutive, sophorose-inducible and glucose-inhibited β-diglucoside permease. *J. Biol. Chem.* **268**: 19364–19368.

KUBICEK, C. P., MÜHLBAUER, G., GROTZ, M., JOHN, E., and KUBICEK-PRANZ, E. M. 1988. Properties of the conidial-bound cellulase system of *Trichoderma reesei. J. Gen. Microbiol.* **134**: 1215–1222.

KULMBURG, P., MATHIEU, M., DOWZER, C., KELLY, J., and FELENBOEK, B. 1993. Specific binding sites in the *alcR* and *alcA* promoters of the ethanol regulon for the CreA repressor mediating carbon catabolite repression in *Aspergillus nidulans. Mol. Microbiol.* **7**: 847–857.

LUTFIYYA, L. L. and JOHNSTON, M. 1996. Two zinc-finger-containing repressors are responsible for glucose repression of SUC2 expression. *Mol. Cell. Biol.* **16**: 4790–4797.

MACH, R. L., STRAUSS, J., ZEILINGER, S., SCHINDLER, M., and KUBICEK, C. P. 1996. Carbon catabolite repression of xylanase I (*xyn1*) gene expression in *Trichoderma reesei. Mol. Microbiol.* **21**: 1273–1281.

MACH, R. L., SEIBOTH, B., MYASNIKOV, A., GONZALEZ, R., STRAUSS, J., HARKKI, A. M., and KUBICEK, C. P. 1995. The *bgl1* gene of *Trichoderma reesei* QM 9414 encodes an extracellular, cellulose-inducible β-glucosidase involved in cellulase induction by sophorose. *Mol. Microbiol.* **16**: 687-697.

MANDELS, M. and REESE, E. T. 1960. Induction of cellulase in fungi by cellobiose. *J. Bacteriol.* **79**: 816–826.

MANDELS, M., PARRISH, F. W., and REESE, E. T. 1962. Sophorose as an inducer of cellulases in *Trichoderma reesei. J. Bacteriol.* **83**: 400–408.

MARCHLER, G., SCHÜLLER, C., ADAM, G., and RUIS, H. 1993. *Saccharomyces cerevisiae* UAS element controlled by protein kinase A activates transcription in response to a variety of stress conditions. *EMBO J.* **12**: 1997–2003.

MARGOLLES-CLARK, E., ILMÉN, M., and PENTTILÄ, M. 1997. Expression patterns of ten hemicellulase genes of the filamentous fungus *Trichoderma reesei* on various carbon sources. *J. Biotechnol.*, **57**: 167–179.

MARGOLLES-CLARK, E., SALOHEIMO, M., SIIKA-AHO, M., and PENTTILÄ, M. 1996c. The α-glucuronidase-encoding gene of *Trichoderma reesei. Gene* **172**: 171–172.

MARGOLLES-CLARK, E., TENKANEN, M., NAKARI-SETÄLÄ, T., and PENTTILÄ, M. 1996a. Cloning of genes encoding α-L-arabinofuranosidase and β-D-xylosidase from *Trichoderma reesei* by expression in *Saccharomyces cerevisiae. Appl. Env. Microbiol.* **62**: 3840–3846.

MARGOLLES-CLARK, E., TENKANEN, M., LUONTERI, E., and PENTTILÄ, M. 1996d. Three α-galactosidase genes of *Trichoderma reesei* cloned by expression in yeast. *Eur. J. Biochem.* **240**: 104–111.

MARGOLLES-CLARK, E., TENKANEN, M., SÖDERLUND, H., and PENTTILÄ, M. 1996b. Acetyl xylan esterase from *Trichoderma reesei* contains an active site serine residue and a cellulose binding domain. *Eur. J. Biochem.* **237**: 553–560.

MATHIEU, M. and FELENBOK, B. 1994. The *Aspergillus nidulans* CREA protein mediates glucose repression of the ethanol regulon at various levels through competition with the ALCR-specific transactivator. *EMBO J.* **13**: 4022–4027.

MESSNER, R. and KUBICEK, C. P. 1991. Carbon source control of cellobiohydrolase I and II formation by *Trichoderma reesei. Appl. Environ. Microbiol.* **57**: 630–635.

MESSNER, R., KUBICEK-PRANZ, E. M., GSUR, A., and KUBICEK, C. P. 1991. Cellobiohydrolase II is the main conidial-bound cellulase in *Trichoderma reesei* and other *Trichoderma* strains. *Arch. Microbiol.* **155**: 601–606.

MISHRA, S., RAO, S., and DEB, J. K. 1989. Isolation and characterization of a mutant of *Trichoderma reesei* showing reduced levels of extracellular β-glucosidase. *J. Gen. Microbiol.* **135**: 3459–3465.

MONTENECOURT, B. S. and EVELEIGH, D. E. 1979. Selective screening methods for the isolation of high yielding cellulase mutants of *Trichoderma reesei. Adv. Chem. Ser.* **181**: 289–301.

MORAWETZ, R., GRUBER, F., MESSNER, R., and KUBICEK, C. P. 1992. Presence, transcription and translation of cellobiohydrolase genes in several *Trichoderma* species. *Curr. Genet.* **21**: 31–36.

MORIKAWA, Y., OHASHI, T., MANTANI, O., and OKADA, H. 1995. Cellulase induction by lactose in *Trichoderma reesei* PC-3-7. *Appl. Microbiol. Biotechnol.* **44**: 106–111.

NAGATA, O., TAKASHIMA, T., TANAKA, M., and TSUKAGOSHI, N. 1993. *Aspergillus nidulans* nuclear proteins bind to a CCAAT-element and the adjacent upstream sequence in the promoter region of the starch-inducible Taka-amylase A gene. *Mol. Gen. Genet.* **237**: 251–260.

NEHLIN, J. O. and RONNE, H. 1990. Yeast MIG1 repressor is related to the mammalian early growth response and Wilms' tumour finger proteins. *EMBO J.* **9**: 2891–2898.

NEVALAINEN, K. M. H. and PALVA, E. T. 1978. Production of extracellular enzymes in mutants isolated from *Trichoderma viride* unable to hydrolyze cellulose. *Appl. Environ. Microbiol.* **35**: 11–16.

PENTTILÄ, M., SALOHEIMO, A., ILMÉN, M., and ONNELA, M.-L. 1993. Regulation of the expression of *Trichoderma* cellulases at mRNA and promoter level. In P. Suominen and T. Reinikainen (eds), *Proceedings of the second TRICEL symposium on* Trichoderma Reesei *Cellulases and Other Hydrolases,* Espoo, Finland. Foundation for Biotechnical and Industrial Fermentation Research, Vol. 8, pp. 189–197.

PENTTILÄ, M., LEHTOVAARA, P., NEVALAINEN, H., BHIKHABHAI, R., and KNOWLES, J. 1986. Homology between cellulase genes of *Trichoderma reesei*: complete nucleotide sequence of the endoglucanase I gene. *Gene* **45**: 253–263.

ROYER, J. C. and NAKAS, J. P. 1990. Interrelationship of xylanase induction and cellulase induction of *Trichoderma longibrachiatum. Appl. Envir. Microbiol.* **56**: 2535–2539.

SAKAI, A., SHIMIZU, Y., KONDOU, S., CHIBAZAKURA, T., and HISHINUMA, F. 1990. Structure and molecular analysis of *RGR1*, a gene required for glucose repression of *Saccharomyces cerevisiae. Mol. Cell. Biol.* **10**: 4130–4138.

SALOHEIMO, M., NAKARI-SETÄLÄ, T., TENKANEN, M., and PENTTILÄ, M. 1997. cDNA cloning of a *Trichoderma reesei* cellulase and demonstration of endoglucanase activity by expression in yeast. *Eur. J. Biochem.* **249**: 584–591.

SALOHEIMO, A., HENRISSAT, B., HOFFRÉN, A.-M., TELEMAN, O., and PENTTILÄ, M. 1994. A novel, small endoglucanase gene, *egl5*, from *Trichoderma reesei* isolated by expression in yeast. *Mol. Microbiol.* **13**: 219–228.

SALOHEIMO, A., ILMÉN, M., MARGOLLES-CLARK, E., ARO, N., and PENTTILÄ, M. Regulatory mechanisms involved in expression of extracellular hydrolytic enzymes of *Trichoderma reesei. Proceedings of the TRICEL meeting: Carbohydrases from* Trichoderma reesei *and other microorganisms,* Ghent, Belgium, August 1997. The Royal Society of Chemistry Publishing, U.K. in press.

SALOHEIMO, M., LEHTOVAARA, P., PENTTILÄ, M., TEERI, T. T., STAHLBERG, J., JOHANSSON, G., PETTERSSON, G., CLAEYSSENS, M., TOMME, P., and KNOWLES,

J. K. C. 1988. EG III, a new endoglucanase from *Trichoderma reesei*: the characterization of both gene and enzyme. *Gene* **63**: 11–21.

Seiboth, B., Messner, R., Gruber, F., and Kubicek, C. P. 1992. Disruption of the *Trichoderma reesei cbh2* gene coding for cellobiohydrolase II leads to a delay in the triggering of cellulase formation by cellulose. *J. Gen. Microbiol.* **138**: 1259–1264.

Seiboth, B., Hakola, S., Mach, R. L., Suominen, P. L., and Kubicek, C. P. 1997. Role of four major cellulases in the triggering of cellulase gene expression by cellulose in *Trichoderma reesei. J. Bacterial.* **179**: 5318–5320.

Senior, D. J., Mayers, P. R., and Saddler, J. N. 1989. Production and purification of xylanases. *ACS Symp. Ser.* **309**: 641–655.

Sestak, S. and Farkaš, V. 1993. Metabolic regulation of endoglucanase synthesis in *Trichoderma reesei*: participation of cyclic AMP and glucose-6-phosphate. *Can. J. Microbiol.* **39**: 342–347.

Sheir-Neiss, G. and Montenecourt, B. S. 1984. Characterization of the secreted cellulases of *Trichoderma reesei* wild type and mutants during controlled fermentations. *Appl. Microbiol. Biotechnol.* **20**: 46–53.

Shoemaker, S., Schweickart, V., Ladner, M., Gelfand, D., Kwok, S., Myambo, K., and Innis, M. 1983. Molecular cloning of exo-cellobiohydrolase from *Trichoderma reesei* strain L27. *Bio/Technology* **1**: 691–696.

Shoemaker, S. P., Raymond, J. C., and Bruner, R. 1981. Cellulases: diversity amongst improved *Trichoderma* strains. In A. Hollaender, R. Rabson, Rogers, Pietro, Valentine and Wolfe (eds), *Trends in the Biology of Fermentations for Fuels and Chemicals.* Plenum, New York, pp. 89–109.

Stangl, H., Gruber, F., and Kubicek, C. P. 1993. Characterization of the *Trichoderma reesei cbh2* promoter. *Curr. Genet.* **23**: 115–122.

Sternberg, D. and Mandels, G. R. 1979. Induction of cellulolytic enzymes in *Trichoderma reesei* by sophorose. *J. Bacteriol.* **139**: 761–767.

Strauss, J. and Kubicek, C. P. 1990. *β*-Glucosidase and cellulase formation by a *Trichoderma reesei* mutant defective in constitutive cellulase formation. *J. Gen. Microbiol.* **136**: 1321–1326.

Strauss, J., Mach, R. L., Zeilinger, S., Stöffler, G., Wolschek, M., Hartler, G., and Kubicek, C. P. 1995. Cre1, the carbon catabolite repressor protein from *Trichoderma reesei. FEBS Lett.* **376**: 103–107.

Suominen, P. L., Mäntylä, A. L., Karhunen, T., Hakola, S., and Nevalainen, K. M. H. 1993. High frequency one-step gene replacement in *Trichoderma reesei.* II: Effects of deletions of individual cellulase genes. *Mol. Gen. Genet.* **241**: 523–530.

Szakmary, K., Wotawa, A., and Kubicek, C. P. 1991. Origin of oxidized cellulose degradation products and mechanism of their promotion of cellobiohydrolase I biosynthesis. *J. Gen. Microbiol.* **137**: 2873–2878.

Takashima, S., Iikura, H., Nakamura, A., Masaki, H., and Uozumi, T. 1996b. Analysis of Cre1 binding sites in the *Trichoderma reesei cbh1* upstream region. *FEMS Microbiol. Lett.* **145**: 361–366.

Takashima, S., Nakamura, A., Iikura, H., Masaki, H., and Uozumi, T. 1996a. Cloning of a gene encoding a putative carbon catabolite repressor from *Trichoderma reesei. Biosci. Biotechnol. Biochem.* **60**: 173–176.

Teeri, T., Salovuori, I., and Knowles, J. 1983. The molecular cloning of the major cellulase gene from *Trichoderma reesei. Bio/Technology* **1**: 696–699.

Teeri, T. T., Lehtovaara, P., Kauppinen, S., Salovuori, I., and Knowles, J. 1987. Homologous domains in *Trichoderma reesei* cellulolytic enzymes: gene sequence and expression of cellobiohydrolase II. *Gene* **51**: 43–52.

Torigoi, E., Henrique-Silva, F., Escobar-Vera, J., Carle-Urioste, J. C., Crivellaro, O., El-Dorry, H., and El-Gogary, S. 1996. Mutants of *Tricho-*

derma reesei are defective in cellulose induction but not basal expression of cellulase-encoding genes. *Gene* **173**: 199–203.

UMILE, C. and KUBICEK, C. P. 1986. A constitutive, plasma-membrane bound β-glucosidase in *Trichoderma reesei*. *FEMS Microbiol. Lett.* **34**: 291–295.

VAHERI, M. P. 1982a. Oxidation as a part of degradation of crystalline cellulose by *Trichoderma reesei*. *J. Appl. Biochem.* **4**: 356–363.

VAHERI, M. P. 1982b. Acidic degradation products of cellulose during enzymatic hydrolysis by *Trichoderma reesei*. *J. Appl. Biochem.* **4**: 153–160.

VAHERI, M. P., LEISOLA, M., and KAUPINNEN, V. 1979b. Transglycosylation products of the cellulase system of *Trichoderma reesei*. *Biotech. Lett.* **1**: 41–46.

VAHERI, M. P., VAHERI, M. E. O., and KAUPPINEN, V. S. 1979a. Formation and release of cellulolytic enzymes during growth of *Trichoderma reesei* on cellobiose and glycerol. *Eur. J. Appl. Microbiol. Biotechnol.* **8**: 73–80.

VAN DER VEEN, P., ARST, H. N., FLIPPHI, M. J. A., and VISSER, J. 1994. Extracellular arabinases in *Aspergillus nidulans*: the effect of different *cre* mutations on enzyme levels. *Arch. Microbiol.* **162**: 433–440.

VAN DER VEEN, P., FLIPPHI, M. J. A., VORAGEN, A. G. J., and VISSER, J. 1993. Induction of extracellular arabinases on monomeric substrates in *Aspergillus niger*. *Arch. Microbiol.* **159**: 66–71.

VAN HEESWIJCK, R. and HYNES, M. J. 1991. The *amdR* product and a CCAAT-binding factor bind to adjacent, possibly overlapping DNA sequences in the promoter region of the *Aspergillus nidulans amdS* gene. *Nucl. Acid Res.* **19**: 2655–2660.

VAN PEIJ, N. N. M. E., VISSER, J., and DE GRAFIFF, L. H. 1998. Isolation and analysis of *Xln*R, encoding a transcriptional activator co-ordinating xylanolytic expression in *Aspergillus niger*. *Molec. Microbiol.* **27**: 131–142.

WANG, P. and NUSS, D. L. 1995. Induction of a *Cryphonectria parasitica* cellobiohydrolase I gene is suppressed by hypovirus infection and regulated by a GTP-binding-protein-linked signalling pathway involved in fungal pathogenesis. *Proc. Natl. Acad. Sci. USA* **92**: 11529–11533.

WARD, M., WU, S., DAUBERMANN, J., WEISS, G., LARENAS, E., BOWER, B., REY, M., CLARKSON, K., and BOTT, R. 1993. Cloning, sequence and preliminary structural analysis of a small, high pI endoglucanase (EG III) from *Trichoderma reesei*. In P. Suominen and T. Reinikainen (eds), Trichoderma Reesei *Cellulases and Other Hydrolases. Enzyme structures, Biochemistry, Genetics and Applications*. Fagepaino Oy, Helsinki, Finland, pp. 153–158.

WEY, T.-T., HSEU, T.-H., and HUANG, L. 1994. Molecular cloning and sequence analysis of the cellobiohydrolase I gene from *Trichoderma koningii* G-39. *Curr. Microbiol.* **28**: 31–39.

WITTEVEEN, C. F. B., BUSINK, R., VAN DE VONDERVOORT, P., DIJKEMA, C., SWART, K., and VISSER, J. 1989. L-Arabinose and D-xylose catabolism in *Aspergillus niger*. *J. Gen. Microbiol.* **135**: 2163–2171.

ZEILINGER, S., SCHINDLER, M., HERZOG, P., MACH, R. L., and KUBICEK, C. P. 1996. Differential induction of two xylanases in *Trichoderma reesei*. *J. Biol. Chem.* **271**: 25624–25629.

4

Chitinolytic enzymes and their genes

M. LORITO

Istituto di Patologia Vegetale, Università degli Studi di Napoli "Federico II" and Centro di Studio CNR sulle Tecniche di Lotta Biologica (CETELOBI), Portici, Italy

4.1 Introduction

Chitin, the (1-4)-β-linked homopolymer of *N*-acetyl-D-glucosamine, is one of the most abundant polymers in the biosphere, and chitinolytic enzymes are found among all kingdoms, i.e., protista, bacteria, fungi, plants, invertebrates and vertebrates, including humans. Enzymatic degradation of chitin is involved in many biological processes, such as autolysis, morphogenesis and nutrition, and plays a role in relationships between organisms, including plant–fungus, insect–fungus and fungus–fungus interactions. (Note that, throughout this chapter, "chitinolytic enzymes" refers to enzymes secreted extracellularly, unless otherwise indicated.)

Fungal strains assigned to the genera *Trichoderma* and *Gliocladium* are well-known producers of chitinolytic enzymes and are used commercially as sources of these proteins. Additional interest in these enzymes is stimulated by the fact that chitinolytic strains of *Trichoderma* and *Gliocladium* are among the most effective agents for biological control of plant diseases and can be serious pathogens for mushroom farming (Chet, 1987; Harman, 1990; Harman *et al.*, 1993a; Komatsu, 1976; Lo *et al.*, 1996; Muthumeenakshi *et al.*, 1994; Samuels, 1996; Speranzini *et al.*, 1995; Tronsmo, 1991; this volume). Evidence is presented in this chapter that supports the correlation between extracellular production of chitinolytic enzymes and biocontrol or mycoparasitic ability of these fungi.

Research on chitinolytic enzymes from *Trichoderma* and *Gliocladium* has flourished in recent years, moving from the purification and characterization of the active proteins to the cloning of the encoding nucleotide (nt) sequences (Carsolio *et al.*, 1994; Chet *et al.*, 1993; Draborg *et al.*, 1995, 1996; Fekete *et al.*, 1996; Garcia *et al.*,

Abbreviations
nt = nucleotide(s);
GlcNAc = NAG = *N*-acetyl-β-D-glucosamine;
CWDE = cell wall degrading enzyme;
MAC = membrane-affecting compound;
PR-protein = pathogenesis-related protein;
DMI = sterol demethylation inhibitor.

1994; Hayes *et al.*, 1994; Limón *et al.*, 1995; Peterbauer *et al.*, 1996). To date, several laboratories around the world are applying these genes to a variety of biocontrol strategies and studying the mechanism of fungal antagonism and mycoparasitism. This chapter is a review of the present knowledge concerning the diversity, the characteristics and the functions of *Trichoderma* and *Gliocladium* chitinolytic enzymes and genes. It also addresses the intriguing yet unclear role of chitinolytic enzymes in the physiology of these fungi and discusses the potential of genetic approaches to bioremediation of chitin waste and biocontrol of fungal pathogens.

4.2 Chitinolytic enzymes from *Trichoderma* and *Gliocladium*

4.2.1 *Nomenclature and assay*

Nomenclature of chitinolytic enzymes is confusing (Sahai and Manocha, 1993) and inadequate to describe the chitinolytic system of *Trichoderma* and *Gliocladium*. The recommended enzyme nomenclature (Bielka *et al.*, 1984; Webb, 1992) ignores a distinct, well-recognized type of enzyme activity (chitobiosidase, see below) and is unclear about the partitioning of exo- and endo-acting enzymes (Cabib *et al.*, 1996; Robbins *et al.*, 1988; Sahai and Manocha, 1993) (Figure 4.1). The definition of chitinase activity as the *random* hydrolysis of *N*-acetyl-*β*-D-glucosaminide 1,4-*β*-linkages in chitin and chitodextrin (EC 3.2.1.14) describes only endochitinase activity (Figure 4.1A) and does not consider exo-acting enzymes that catalyze the progressive release *only* of diacetylchitobiose from the non-reducing ends of chitin chains

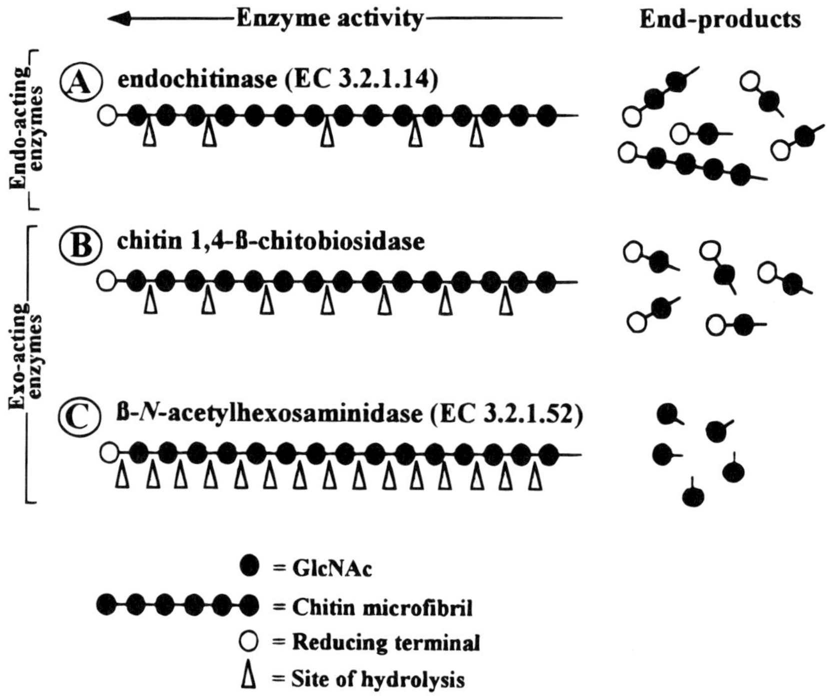

Figure 4.1 Mechanism of action of chitinolytic enzymes from *Trichoderma* and *Gliocladium*.

(Harman *et al.*, 1993b; Sahai and Manocha, 1993) (Figure 4.1B). In addition, enzymes that catalyze the progressive release *only* of *N*-acetylglucosamine (NAG) residues from the non-reducing end of the chain (Bielka *et al.*, 1984) (Figure 4.1C) should reasonably be considered as another type of exo-acting chitinolytic enzymes but have been catalogued first as chitobiases, then as β-D-acetylglucosaminidases and now as β-*N*-acetylhexosaminidases (EC 3.2.1.52). Thus, only two terms for chitinolytic enzymes are recommended today – endochitinase (EC 3.2.1.14) and β-*N*-acetylhexosaminidase (EC 3.2.1.52) – but there is enough evidence for at least three different types of chitinolytic activities (Sahai and Manocha, 1993) (Figure 4.1A–C). This is also the case for *Trichoderma* and *Gliocladium* chitinolytic enzymes (Harman *et al.*, 1993b), which either cut randomly inside the chitin chain or release only *N*-acetylglucosamine or only diacetylchitobiose from the non-reducing end (Figure 4.1).

The following terminology accounts for the observed diversities in the *Trichoderma* and *Gliocladium* chitinolytic system and includes the recommendations of Enzyme Nomenclature (Webb, 1992) (Figure 4.1). All the enzymes capable of catalyzing the degradation of chitin or chitooligomers by hydrolysis of the *N*-acetyl-β-D-glucosaminide 1,4-β-linkage are defined as *chitinolytic enzymes* and are differentiated according to the reaction end-products:

- *endochitinase* (EC 3.2.1.14) corresponds to the definition of "chitinase" in Enzyme Nomenclature (Webb, 1992). It *randomly* cleaves chitin and chitooligomers and releases a mixture of soluble low molecular mass end-products of different sizes, with diacetylchitobiose $(GlcNAc)_2$ being the main end-product because it is unable to cut further (Figure 4.1A). Thus, the recommended term "chitinase" should be used only to indicate endochitinase activity.
- *chitin 1,4-β-chitobiosidase* (chitobiosidase) is a new name (Harman *et al.*, 1993b; Tronsmo and Harman, 1993) for an already well-described exo-chitinase activity (Figure 4.1B) (Sahai and Manocha, 1993). It cleaves chitin and chitooligomers $[(GlcNAc)_{\geq 3}]$ progressively from the non-reducing end and releases *only* $(GlcNAc)_2$ (Figure 4.1B). The term chitobiosidase, not officially recognized with an EC number, is derived by analogy from the nomenclature of cellulolytic enzymes – cellulose 1,4-β-cellobiosidase (EC 3.2.1.91) (Webb, 1992) – and should be preferred to "chitobiohydrolase" which evokes the systematic classification of cellulases (Webb, 1992). The *Trichoderma* and *Gliocladium* chitobiosidases that have been characterized to date are different in many ways from other chitinolytic enzymes with regard to substrate specificity, amino acid sequence, antifungal activity, serological properties and other characteristics (Harman *et al.*, 1993b; Lorito *et al.*, 1993a; Tronsmo *et al.*, 1996).
- *β-N-acetylhexosaminidase* (EC 3.2.1.52) (chitobiase 3.2.1.29 and *N*-acetyl-β-D-glucosaminidases 3.2.1.30 are now deleted entries) includes enzyme activity that cleaves chitooligomers and also chitin progressively from the non-reducing end and releases *only* *N*-acetylglucosamine monomers (GlcNAc) (Figure 4.1C); this term also refers to enzymes with similar activity on *N*-acetylgalactosides (Webb, 1992). Thus, it is a type of exo-chitinase distinct from chitobiosidase (Figure 4.1B,C) and the only enzyme able to cut $(GlcNAc)_2$. The synonymous term "chitobiase" should be avoided because it may be confused with chitobiosidase and, like the term *N*-acetyl-β-D-glucosaminidases, is no longer recommended by Enzyme Nomenclature (Webb, 1992).

The substrates used to assay, purify and characterize chitinolytic enzymes from *Trichoderma* and *Gliocladium* include *p*-nitrophenyl chitooligomers, differently purified chitins or fungal cell walls, 4-methylumbelliferyl derivatives, blue substrates and a few others (Bertagnolli *et al.*, 1996; de la Cruz *et al.*, 1992; Di Pietro *et al.*, 1993; Harman *et al.*, 1993b; Koga *et al.*, 1991; Lorito *et al.*, 1994a; Ridout *et al.*, 1988; Tronsmo and Harman, 1993; Ulhoa and Peberdy, 1991a, 1992; Wirth and Wolf, 1992). However, the use of purified chitins and synthetic substrates may affect the specificity of the enzyme activity because of their "unnatural" interaction with catalytic sites, thus giving misleading indications about enzyme modes of action (Diekman and Breves, 1993). Perhaps the mechanism of action of *Trichoderma* and *Gliocladium* chitinolytic enzymes requires further analysis, since most of the studies completed to date employed these methods. For instance, chitins retaining their native structures may be prepared from cell walls of antagonist or host fungi (Bade *et al.*, 1988) and used as substrates to examine the end-products of the reaction (de la Cruz *et al.*, 1992; Tronsmo *et al.*, 1996). This would determine whether the indicated endochitinases, chitobiosidases or *β*-*N*-acetylhexosaminidases exert in nature the same modes of action showed *in vitro* and should help us to understand their role in the biology of these fungi.

4.2.2 *Multiplicity*

The multiplicity of enzymes secreted by *Trichoderma* and *Gliocladium* is somewhat astonishing. Wild type strains secrete endochitinase, chitobiosidase, *β*-*N*-acetylhexosaminidase, *N*-acetyl-*β*-galactosaminidase, *β*-1,3-glucanase, *β*-1,6-glucanase, protease, DNase, α-amylase, cellulase, lipase, mannanase, xylanase, urease, RNase, pectinase, pectin lyase, laccase, peroxidase and mutanase (Bertagnolli *et al.*, 1996; Freitag *et al.*, 1991; Koga *et al.*, 1991; Quivey and Kriger, 1993). The ability of *Trichoderma* to produce fungal cell wall degrading enzymes (CWDEs), such as chitinolytic, cellulolytic and glucanolytic enzymes, was reported several years ago (Chet *et al.*, 1979; Elad *et al.*, 1982; Hadar *et al.*, 1979; Reese and Mandels, 1959). But only recently has the chitinolytic system of *Trichoderma* been dissected and some of the enzymes characterized after being purified to homogeneity (de la Cruz *et al.*, 1993; Harman *et al.*, 1993b; Lorito *et al.*, 1993a, 1994a; Ulhoa and Peberdy, 1991a, 1992). The chitinolytic system of *Gliocladium* has received much less attention, probably because the biocontrol ability of these fungi is thought to rely mainly on production of antibiotics instead of CWDEs. Nevertheless, *Gliocladium* endochitinases, chitobiosidases and *β*-*N*-acetylhexosaminidases have been detected and purified from *G. roseum* and *G. virens* (now *T. virens*) (Di Pietro *et al.*, 1993; Pachenari and Dix, 1980; Roberts and Lumsden, 1990; Hayes, C. K., Lorito, M., and Harman, G. E., unpublished data).

Chitinolytic enzymes have been found in *T. harzianum*, *T. aureoviride*, *T. atroviride*, *T. koningii*, *T. longibrachiatum*, *T. viride*, *T. pseudokoningii*, *T. longipilis*, *T. minutisporum*, *T. hamatum* and *T. reesei*, but the chitinolytic system of *T. harzianum* is by far the most studied (Bruce *et al.*, 1995; de la Cruz *et al.*, 1993; Garcia *et al.*, 1994; Harman *et al.*, 1993b; Lorito *et al.*, 1994a; Turóczi *et al.*, 1996; Ulhoa and Peberdy, 1991a, 1992; Usui *et al.*, 1990). When cell-free extracts were carefully fractionated, a complex system composed of a variety of chitinolytic enzymes (up to six for a single strain) (Haran *et al.*, 1995) was found. Table 4.1 summarizes the chitinolytic system of *T. harzianum*, identifying each enzyme by its apparent molecular

Table 4.1 The chitinolytic system of *Trichoderma harzianum*

Enzyme[a] (reference)	Strain	Enzyme[b] activity	Substrate[c]	Non-substrate[c]	Reaction product	Purification	T_{opt} (°C)	pH opt.	Antifungal $(ED_{50})^d$	Synergist. with[e]	Gene cloned[f]	Other characteristics
Group I												
CHIT102 (Haran *et al.*, 1995)	TM	hexosam	MU, MU2, MU3	—	GlcNAc	—	—	—	—	—	—	heat sensitive
CHIT72 (CHIT73) (Lorito *et al.*, 1994a) (Haran *et al.*, 1995) (Ridout *et al.*, 1993)	 P1 TM IMI298372	hexosam	Np, Np2, MU, MU2, MU3 fungal cell-walls	—	GlcNAc	Lorito *et al.*, 1994a	60	5.0–5.5	(51 μg/ml)	CHIT42	*nag1* (Peterbauer *et al.*, 1996) *exc1* (Draborg *et al.*, 1995)	no affinity with CHIT32 and CHIT40; heat resistant; glycosylated; pI 4.6
CHIT64 (Ulhoa & Peberdy, 1991a)	39.1	hexosam	Np (GlcNAc)2	—	GlcNAc	Ulhoa & Peberdy, 1991a	50	5.5	—	—	—	insensitive to metal ions, EDTA, SDS, high ionic strength
Group II												
CHIT52 (Haran *et al.*, 1995)	TM	endochit	MU2, MU3	MU	mixture, mainly $(GlcNAc)_2$	—	—	—	—	—	—	highly heat sensitive
CHIT42 (Ulhoa & Peberdy, 1992) (de la Cruz *et al.*, 1993) (Harman *et al.*, 1993b) (Haran *et al.*, 1995) (Lorito *et al.*, 1993b)	 39.1 CECT2413 P1 TM	endochit	chitin, Np3, Np2 MU2, MU3 fungal cell-walls	MU, Np, $(GlcNAc)_2$	mixture, mainly $(GlcNAc)_2$	Ulhoa & Peberdy, 1992; de la Cruz *et al.*, 1992; Harman *et al.*, 1993b	40–45	4.0	(47 μg/ml)	CHIT72 CHIT40 CHIT33 CHIT37	*ThEn-42* (Hayes *et al.*, 1994); *chit42* (Garcia *et al.*, 1994; *ech-42* (Carsolio *et al.*, 1994; *chi1* (Draborg *et al.*, 1996)	heat resistant; no affinity with CHIT40 or CHIT72; pI 4.0 or 6.2; insensitive to high ionic strength
CHIT40 (Harman *et al.*, 1993b)	P1	chitobio	MU2, Np2	MU, Np	$(GlcNAc)_2$	Harman *et al.*, 1993b	—	4.0–7.0	(117 μg/ml)	CHIT42	—	no affinity with CHIT42 or CHIT72;
CHIT37 (de la Cruz *et al.*, 1993)	CECT2413	endochit	chitin, Np2, fungal cell-walls, MU2	Np	mixture, mainly $(GlcNAc)_2$	de la Cruz *et al.*, 1993	45–50	—	—	CHIT42	—	glycosylated pI 4.6–5.0
CHIT33 (de la Cruz *et al.*, 1993) (Haran *et al.*, 1995)	 CECT2413 TM	endochit	chitin, MU2, fungal cell-walls, MU3	Np, MU	mixture, mainly $(GlcNAc)_4$	de la Cruz *et al.*, 1993	45–50	—	—	CHIT42	*chit33* (Limon *et al.*, 1995)	heat resistant; pI 7.4–7.8
CHIT31 (Haran *et al.*, 1995)	TM	endochit	MU2, MU3	MU	mixture	—	—	—	—	—	—	heat resistant
CHIT28 (Deane *et al.*, 1995)	T198	hexosam	—	—	—	Deane *et al.*, 1995	—	—	—	—	—	glycosylated

[a] Enzyme names are based on apparent molecular weight on SDS-PAGE; Koga *et al.* (1991) purified a 150 kDa *β-N*-acetylhexosaminidase from *T. harzianum*, but its MW was only determined by gel filtration and, therefore, it is not clear if it is a novel *β-N*-acetylhexosaminidase. It has been suggested that CHIT102 is CHIT64 and that CHIT37 is CHIT31 (Haran *et al.*, 1995).

[b] Enzyme activity based on the end-products; hexosam = *β-N*-acetylhexosaminidase; endochit = endochitinase; chitobio = chitobiosidase; the end-product of CHIT40 to be confirmed by HPLC or TLC analysis.

[c] Substrates reported to be cleaved or non-cleaved by the given enzyme: Np = p-nitrophenyl-*β*-D-*N*-acetylglucosaminide; Np2 = p-nitrophenyl-*β*-D-*N*,*N*′-diacetylchitobiose; Np3 = p-nitrophenyl-*β*-D-*N*,*N*′,*N*″-triacetylchitotriose; MU = 4-methylumbelliferyl-*N*-acetyl-*β*-D-glucosaminide; MU2 = 4-methylumbelliferyl-*β*-D-*N*,*N*′-diacetylchitobiose; MU3 = 4-methylumbelliferyl-*β*-D-*N*,*N*′,*N*″-triacetylchitotriose; GlcNAc = *N*-acetylglucosamine.

[d] Inhibition of spore germination *in vitro*.

[e] Synergistic in the inhibition of spore germination or degradation of fungal cell walls.

[f] In addition, Draborg *et al.* (1995) cloned *exc2* from *T. harzianum* strain T 25-t, which perhaps codes for a novel *β-N*-acetylhexosaminidase.

weight (MW) on SDS-PAGE. The following two main groups may be designated: *group I*, including enzymes of MW around 60–70 kDa or higher (CHIT102, CHIT64, CHIT72 and CHIT73), all of which are *β-N*-acetylhexosaminidases, and *group II* including enzymes with lower MW (CHIT52, CHIT41, CHIT42, CHIT40, CHIT37, CHIT33, CHIT31 and CHIT28), mostly endochitinases. Perhaps, *group II* may be further separated into *group IIa* containing endochitinases and chitobiosidases with MW higher than 40 kDa (CHIT52-40) and *group IIb* with MW lower than 40 kDa (CHIT37-28), consisting mostly of endochitinases.

Some apparently different enzymes may actually be the same proteins, and differences in MW values may be due to the use of diverse separation protocols. For instance, CHIT72 is probably CHIT73 and CHIT41 is probably CHIT42 since they have identical properties when compared with one another. Furthermore, Haran *et al.* (1995) have suggested that CHIT102 is CHIT64 and CHIT37 is CHIT31. However, other enzymes of very similar MW (e.g., CHIT41 and CHIT40) are clearly different as shown by N-terminus sequencing and serological tests (de la Cruz *et al.*, 1992; Harman *et al.*, 1993b; Lorito *et al.*, 1993b). New enzymes will probably be added to this list, since more chitobiosidases (Harman G. E., unpublished data), *β-N*-acetylhexosaminidases (Draborg *et al.*, 1995; Koga *et al.*, 1991; Peterbauer *et al.*, 1996) and intracellular or wall-bound enzymes (Peterbauer *et al.*, 1996) have been detected.

The distribution of the various enzymes among different *Trichoderma* strains seems complex and difficult to interpret from the available data (Bruce *et al.*, 1995; Turóczi *et al.*, 1996), since some basic properties as well as the role of each enzyme are still unknown. For instance, *T. harzianum* strain TM secretes at least six different chitinolytic enzymes (Haran *et al.*, 1995) but does not produce chitobiosidases or other enzymes found in other strains (de la Cruz *et al.*, 1992; Harman *et al.*, 1993b). The most commonly detected chitinolytic enzymes are *β-N*-acetylhexosaminidases of group I and some endochitinases (CHIT42 and CHIT33) of group II (de la Cruz *et al.* 1993; Garcia *et al.*, 1994; Haran *et al.*, 1995; Harman *et al.*, 1993b; Lorito *et al.*, 1994a; Turóczi *et al.*, 1996; Ulhoa and Peberdy, 1991a, 1992). Other enzymes such as the CHIT40 (chitobiosidase) and the CHIT52 have been found more rarely thus far. However, a comprehensive study on the presence, characteristics and distribution of different chitinolytic enzymes among strains and species of *Trichoderma* and *Gliocladium* is generally lacking.

4.2.3 *Mechanism of induction*

The *Trichoderma* and *Gliocladium* mechanism of chitinolytic enzyme induction is still a matter of speculation in many respects. High-level induction of extracellular chitinolytic enzymes is usually obtained by growing *Trichoderma* and *Gliocladium* on purified chitin, fungal cell walls or mycelia as sole carbon sources. No or much less induction was attained when related compounds such as chitosan, cellulose, unpurified chitin or laminarin were used, indicating a rather specific induction *in vitro* (de la Cruz *et al.*, 1993; Harman *et al.*, 1993b; Lorito *et al.*, 1994a; Ulhoa and Peberdy, 1991b). However, different enzymes may not have the same mechanism of induction. For example, *N*-acetyl-glucosamine specifically induced *β-N*-acetylhexosaminidase (*N*-acetyl-glucosaminidase) but not endochitinase or chitobiosidase production in *T. harzianum* (de la Cruz *et al.*, 1993; Peterbauer *et al.*, 1996; Ulhoa and Peberdy, 1991b). During the mycoparasitic activity of *T. harzianum*, induction and production levels of various enzymes, mainly CHIT102, CHIT72 and some group II endochitinases, were affected differently if the fungus was confronted with different hosts (Haran *et al.*, 1996; Inbar and Chet, 1995), which suggests that a strain responds to distinct environmental conditions by the production of a specific set of chitinolytic enzymes. Further, great differences were found when purified cell walls from different basidiomycetes were used to induce endochitinase activity in different *Trichoderma* strains (Bruce *et al.*, 1995). Thus, a specific chitinolytic response may support distinct behaviors and various phases of saprophytism, mycoparasitism and antagonism, with substantial differences occurring among strains.

In vitro formation of most chitinolytic enzymes is proportionally repressed by increasing amounts of glucose, sucrose and end-products (Carsolio *et al.*, 1994; de la Cruz *et al.*, 1993; Garcia *et al.*, 1994; Margolles-Clark *et al.*, 1996b; Peterbauer *et al.*, 1996; Ulhoa and Peberdy, 1991b), suggesting that enzyme synthesis is specifically regulated by catabolic repression. This has been studied in more detail for the *T. harzianum* endochitinase CHIT42 and the encoding gene *ThEn-42*, which is triggered during the mycoparasitic interaction (Carsolio *et al.*, 1994; Hayes *et al.*, 1994; Inbar and Chet, 1995). Many authors have reported glucose repression of CHIT42 formation also at the mRNA level (Carsolio *et al.*, 1994; de la Cruz *et al.*, 1993; Garcia *et al.*, 1994; Margolles-Clark *et al.*, 1996b), and it was suggested that the carbon catabolite repressor Cre1 was involved (Carsolio *et al.*, 1994). Recently, Lorito *et al.* (1996a) have cloned and analyzed the 5′ non-coding region of *ThEn-42* from the mycoparasitic strain P1 and looked for sequences binding to regulatory elements. They found strong evidence for Cre1-based carbon catabolite repression of *ThEn-42*, which was modulated by the interaction with the host *Botrytis cinerea* possibly through a "mycoparasitic DNA-binding regulator" (Lorito *et al.*, 1996a) (see section 4.3.3). Further, the formation of CHIT72, CHIT42 and CHIT33 is regulated at the level of transcription (Garcia *et al.*, 1994; Limón *et al.*, 1995; Peterbauer *et al.*, 1996) and the induction or de-repression of several *T. harzianum* enzymes requires *de novo* transcription and translation (de la Cruz *et al.*, 1993; Inbar and Chet, 1995; Ulhoa and Peberdy, 1991b).

Although there are some insights into the mechanism of repression, very little is known about factors that turn on the chitinolytic system or positively regulate its induction. As occurs in other systems, the production of chitinolytic enzymes could be induced in a concentration-dependent manner by soluble chitooligomers (Blaiseau *et al.*, 1992; Monreal and Reese, 1969; Reyes *et al.*, 1989; Roby and Esquerre-Tugaye, 1987; Smith and Grula, 1983; St Leger *et al.*, 1986), as seems to be the case with CHIT72 and CHIT42 (de la Cruz *et al.*, 1993; Peterbauer *et al.*, 1996; Ulhoa and Peberdy, 1991b). Moreover, there is evidence that at least trace quantities of constitutive CHIT102, CHIT42 and CHIT33 are produced, even on glucose (Bruce *et al.*, 1996; Garcia *et al.*, 1994; Haran *et al.*, 1995; Inbar and Chet, 1995; Limón *et al.*, 1995; Margolles-Clark *et al.*, 1996b; Ulhoa and Peberdy, 1991b; Lorito M., Hayes C. K., and Harman G. E., unpublished data). These enzymes, which may be secreted continuously or only under starvation or stress conditions, would release oligomers from chitins present in the soil to locate substrates or fungal hosts in the environment and induce the hydrolytic system.

Interestingly, Inbar and Chet (1992, 1995) demonstrated that formation of chitinolytic enzymes in a *T. harzianum* mycoparasitic strain is elicited by a lectin-mediated interaction with the host, which may be an event that precedes the induction by chitooligomers. Their lectin-based recognition system nicely mimicked the presence of the host and may be valuable for studying early phases of the mycoparasitic process. However, it may be limiting in the study of chitinolytic enzyme induction because mycelia-released molecules may be required to fully activate the chitinolytic system. In fact, autoclaved mycelia were more powerful *in vitro* inducers of CHIT33 and CHIT42 than purified chitin (Garcia *et al.*, 1994; Limón *et al.*, 1995), and the presence of a living host but not chitin was sufficient to overcome glucose repression of CHIT42 (Lorito *et al.*, 1996a).

Other possibilities so far unsubstantiated are that induction is stimulated by light, sporulation (Carsolio *et al.*, 1994) and/or physical contact between the cell and the insoluble substrate, as suggested for cellulase biosynthesis (Berg and Pettersson, 1977). It has also been suggested that starvation itself is a strong inducer of CHIT42 (Margolles-Clark *et al.*, 1996b), since nt-motifs for binding of stress response elements may be found on the regulatory region of *ThEn-42* (Kubicek C. K., Mach R. L., and Lorito M., unpublished data). In addition, chitinolytic enzyme formation may be affected by the activity of other co-induced compounds such as other cell wall degrading enzymes (proteinases and glucanases) (de la Cruz *et al.*, 1993, 1995; Geremia *et al.*, 1991; Goldman, 1993; Lora *et al.*, 1995; Lorito *et al.*, 1994a), cell wall bound proteins (Lora *et al.*, 1994), permeases (Vasseur *et al.*, 1995) and antibiotics (Schirmböck *et al.*, 1994).

In conclusion, a unique mechanism of induction does not appear to occur, but formation of chitinolytic enzymes may be concurrently modulated by specific (e.g., lectin recognition, contact with the substrate, chitooligomer formation or specific interactions with other microorganisms) as well as non-specific stimuli (e.g., light, sporulation, stress due to nutrient depletion or antagonism, etc.). In addition, nothing is known about mechanisms of induction in the natural environment of *Trichoderma* and *Gliocladium*. This aspect may be difficult to explore, due to the complexity of the chitinolytic system, although the recent cloning of some regulatory sequences (Lorito *et al.*, 1996a) allows preparation of appropriate gene reporter systems to study chitinolytic enzyme induction *in situ* (Inbar and Chet, 1991).

4.2.4 *Production, purification and characterization*

Chitinolytic enzymes from *Trichoderma* and *Gliocladium* strains are readily produced on a variety of substrates, mainly defined mineral salts media with different carbon and nitrogen sources (e.g., purified fungal cell walls, autoclaved mycelia and mushroom extracts). In the past few years, enzymes have been purified to homogeneity and characterized from the culture filtrates of a few strains, mostly *Trichoderma* (Table 4.1). Combinations of standard techniques such as gel chromatography, chromatofocusing, affinity chromatography, isoelectric focusing, preparative electrophoresis, HPLC, ammonium sulfate precipitation, etc., have been successfully used (de la Cruz *et al.*, 1993; Harman *et al.*, 1993b; Lorito *et al.*, 1994a, unpublished data; Ulhoa and Peberdy, 1991a, 1992). To date, purification has been accomplished for most of the chitinolytic enzymes detected in *T. harzianum* with the exception of CHIT52, CHIT31 and CHIT102 (de la Cruz *et al.*, 1993; Harman *et al.*, 1993b; Lorito *et al.*, 1994a; Ulhoa and Peberdy, 1991a, 1992).

The physio-chemical properties of purified chitinolytic enzymes from *T. harzianum* and *T.* (*Gliocladium*) *virens* are typical for these kinds of enzymes (Table 4.1), although in many cases complete characterization is lacking. Reported features include the following: (a) MW from about 20 to about 100 kDa, which is comparable to the size of isozymes from other sources (Collinge *et al.*, 1993; Gooday, 1990); (b) occurrence of N-glycosylation, which is possibly associated with the high stability of enzymes; (c) acidic pH optimum from 4.0 to 5.5 and temperature optimum from 40 to 60°C; (d) sensitivity to some proteinases; (e) enzyme activity that does not require co-factors and is not greatly affected by metal ions, except Zn^{2+}, EDTA or high ionic strength; (f) heat resistance, with the exception of CHIT52 and CHIT102; and (g) general stability, as noted by the retention of enzyme activity when enzymes are dried, frozen, left in water solution at room temperature or expressed in different hosts, including fungi, yeast, bacteria and plants (de la Cruz *et al.*, 1993; Di Pietro *et al.*, 1993; Draborg *et al.*, 1995, 1996; Inbar and Chet, 1995; Haran *et al.*, 1995; Harman *et al.*, 1993b; Lorito *et al.*, 1994a, 1995, 1996b, unpublished data; Margolles-Clark *et al.*, 1996a,b; Ridout *et al.*, 1988; Ulhoa and Peberdy, 1991a, 1992).

Substrate specificity of chitinolytic enzymes from *Trichoderma* and *Gliocladium* is complex. Endochitinases, chitobiosidases and *β-N*-acetylhexosaminidases may act on colloidal, glycol or swollen chitin, purified cell walls, chitooligomers and less so on raw chitin extracted from crab shells. Chitins and chitinous cell walls are better substrates for endochitinases than for chitobiosidases or *β-N*-acetylhexosaminidases. Chitooligomers $(GluNAc)_3$ and higher analogues are cleaved by all three enzymes, but *β-N*-acetylhexosaminidases may act more slowly in reducing the turbidity of a colloidal chitin suspension; $(GluNAc)_2$ is a good substrate for *β-N*-acetylhexosaminidases but is not cleaved by endochitinases or chitobiosidases and, therefore, can be used to distinguish between endochitinase or chitobiosidase and *β-N*-acetylhexosaminidase activity. Moreover, the release of only GluNAc, only $(GluNAc)_2$ or a mixture of different sized end-products can distinguish between *β-N*-acetylhexosaminidase, chitobiosidase and endochitinase activity, respectively (see section 4.2.1) (Figure 4.1).

Purification yields of extracellular chitinolytic enzymes from *Trichoderma* or *Gliocladium* are usually a few milligrams per liter of culture, although the addition of V8 juice, yeast extract, proteose peptone or polyvinylpyrrolidone to chitin/salt media may increase enzyme levels (Tronsmo and Harman, 1992). Such minute quantities limit the direct testing of enzymes in applications for treatment of fungal diseases or processing of chitin-waste. More recently, molecular strategies have been designed to overcome the limitation in enzyme production. Margolles-Clark and co-workers (1996a,b) over-expressed *ThEn-42* in a strain of *T. reesei* selected for large-scale production of extracellular enzymes and reported at least a 20-fold improvement in the extracellular level of CHIT42. Draborg *et al.* (1995, 1996) have expressed genes encoding *β-N*-acetylhexosaminidases and CHIT42 in a yeast strain, which can be readily mutagenized for high production of extracellular hydrolases. Such studies, which are ongoing in several laboratories, should (a) provide additional data on amino acid sequence, substrate specificity, reaction products, molecular structure and mechanism of hydrolysis, which are required to fully understand the multiplicity and complexity of this chitinolytic system, and (b) help to find useful applications of these enzymes (see below).

4.2.5 *Lytic activity, antifungal properties and synergism*

Chitinolytic enzymes have cell wall lytic activity and antifungal effect that can be synergistically improved by making combinations of diverse enzymes or combining enzymes with other compounds (Collinge *et al.*, 1993; Roberts *et al.*, 1988; Sahai and Manocha, 1993). Even from the same organism, great variation has been found in the level of antifungal activity of different enzymes, such that some chitinases and glucanases from plants have lytic but not antifungal activity (Sela-Buurlage *et al.*, 1993). In fact, both antifungal and lytic activity have been reported only for some *Trichoderma* and *Gliocladium* chitinolytic enzymes (Table 4.1) (de la Cruz *et al.*, 1993; Di Pietro *et al.*, 1993; Lorito *et al.*, 1993b,c, 1994a,b, 1996c; Schirmböck *et al.*, 1994), which may in part be dependent on the procedure used for the assay. Regardless, *in vitro* results on enzyme lytic and antifungal effects require confirmation by *in vivo* experiments, which should be carried out when larger quantities of enzymes are available.

Lytic and antifungal effects of T. harzianum *chitinolytic enzymes*

CHIT42, CHIT40 and CHIT72 from *T. harzianum* strain P1 and *T. virens* strain 41 have a substantial inhibitory effect on the germination and hyphal elongation of several fungal pathogens, including *B. cinerea, Fusarium* spp., *Alternaria* spp., *Ustilago avenae, Uncinula necator* and other chitin-containing fungi, when they were incubated in an enzyme solution (Di Pietro *et al.*, 1993; Lorito *et al.*, 1993b,c, 1994a,b, 1996c; Schirmböck *et al.*, 1994; Harman *et al.*, unpublished data; Broekaert, W., personal communication). As reported for other chitinolytic systems, the endochitinases are among the most effective for both antifungal and lytic activities in comparison with other types of chitinolytic enzymes (de la Cruz *et al.*, 1993; Lorito *et al.*, 1996c,d). Moreover, the levels of inhibition for *Trichoderma* and *Gliocladium* enzymes were usually higher than those described for plant, bacterial or other fungal chitinolytic enzymes assayed under the same conditions (Lorito *et al.*, 1996d). This suggests that *Trichoderma* and *Gliocladium* enzymes have a distinct biological activity when compared with enzymes with similar functions from other sources and play a multiple role in both nutrition and defensive processes. In fact, *Trichoderma* and *Gliocladium* enzymes were able to lyse not only the "soft" structure of the hyphal tip but also the "hard" chitinous wall of mature hyphae, conidia, chlamydospores and sclerotia (Benhamou and Chet, 1993, 1996; Chérif and Benhamou, 1990; Lorito *et al.*, 1993b; Rousseau *et al.*, 1996). Additionally, many other morphological changes such as swelling, branching, necrosis and vacuolization were noted in hyphae treated with enzyme combinations (Lorito *et al.*, 1993b). These observed antifungal and morphological effects represent some of the best evidence to date supporting the involvement of the chitinolytic enzyme system, or at least part of it, in mycoparasitism by *Trichoderma* and *Gliocladium.* Moreover, certain chitinolytic enzymes may virtually affect any chitin-containing organism whether pathogenic or not, and this may have an impact on future research and technology development in specific sectors of agriculture and medicine.

Synergistic interactions

Synergistic relationships often occur between biologically active molecules that act together to perform specific tasks (e.g., antibiosis). Chitinolytic enzymes are no exception, and their synergism has been observed in the study of biological processes (e.g., plant defense against fungal attack) or used in biotechnological applications (e.g., improvement of plant disease resistance or biocontrol with microbes) (Broglie *et al.*, 1991; Collinge *et al.*, 1993; Haran *et al.*, 1993; Jach *et al.*, 1995; Lorito, 1995; Lorito *et al.*, 1994c, 1996e; Schirmböck *et al.*, 1994).

Trichoderma and *Gliocladium* chitinolytic enzymes with antifungal and lytic activity also interact synergistically with other chitinolytic or glucanolytic enzymes, which results in a great improvement of the lytic and inhibitory action even in cases where the enzymes have little to no activity when applied alone (de la Cruz *et al.*, 1993; Lorito *et al.*, 1993b,c, 1994a,b, 1996c,d). Moreover, if appropriate combinations of enzymes with different modes of action are made (e.g., endo- and exo-acting enzymes), the level of antifungal activity approaches that of chemical fungicides or naturally occurring antibiotics (Lorito *et al.*, 1994b, 1996c). For instance, an ED_{50} value (50% inhibition) of a few μg per ml, which is comparable to values reported for commonly used fungicides, is obtained when conidia of *B. cinerea* are left to germinate in a solution containing equal parts of CHIT42, CHIT40, CHIT72 and a glucan 1,3-β-glucosidase from *T. harzianum* strain P1 (Lorito *et al.*, 1994b). Interestingly, no or much less synergism was observed when two exo-acting enzymes were combined (Lorito *et al.*, 1996d). However, what makes *Trichoderma* and *Gliocladium* chitinolytic enzymes attractive for biocontrol applications is their ability to enhance the antifungal effect of non-enzymatic compounds or other microorganisms (Lorito *et al.*, 1993c, 1994b). For instance, chitinolytic enzymes were synergistic with cell membrane-affecting compounds (MACs) in the inhibition of fungi, probably due to a concerted effect of enzyme and MAC activity on the integrity of the cell wall (Lorito *et al.*, 1996c). In fact, peptaibol antibiotics were found to support *Trichoderma* mycoparasitism by inhibiting membrane-bound chitin and glucan synthase of *B. cinerea*, thus enhancing cell wall degradation and hyphal growth inhibition caused by chitinolytic enzymes (Lorito *et al.*, 1996f; Lorito, M., Farkaš, V., and Kubicek, C. P., unpublished data). However, this may not be the only explanation for the observed synergism between CWDEs and MACs.

Synergistic interactions have been found for different *T. harzianum* chitinolytic enzymes with the following:

- Plant PR-proteins such as osmotin, which acts as a MAC (Abad *et al.*, 1996), and a class I tobacco chitinase (Lorito *et al.*, 1996c, unpublished data). The target fungi were *B. cinerea* and *F. oxysporum*, and the level of synergism was the highest when CHIT42 from *T. virens* or *T. harzianum* were used (Lorito *et al.*, 1996c).
- *Trichoderma* and *Gliocladium* antibiotics such as trichorzianine A, trichorzianine B and gliotoxin, which all act as MACs and may be concurrently induced with the enzymes (Lorito *et al.*, 1996c; Schirmböck *et al.*, 1994).
- Metabolites and antibiotics from other microorganisms such as gramicidin, valinomycin and phospholipase, which all alter membrane permeability and structure (Lorito *et al.*, 1996c).
- Chemical fungicides, especially the DMI azole fungicides flusilazole and miconazole, which are widely used against fungal diseases of plants, animals and humans. These compounds also act on membrane structure and function and showed the highest levels of synergism with the enzymes (e.g., the addition of

only 10 μg/ml of CHIT42 improved the fungicidal activity of flusilazole almost one hundred-fold) (Lorito *et al.*, 1994b).

- Other microorganisms such as the biocontrol bacterium *Enterobacter cloacae.* CHIT42 and CHIT40 synergistically enhanced its binding to the hyphae of the pathogen and its ability to disrupt the target fungus (Lorito *et al.*, 1993c).

Two main conclusions can be drawn from the above points. First, the chitinolytic system of *Trichoderma* and *Gliocladium* is made up of different enzymes acting in concert to efficiently degrade fungal cell walls or any other chitinous substrate. This system is associated with the production of various metabolites, including other cell wall degrading enzymes, antibiotics and self-defense compounds (e.g., Qid3, see below). When all of these molecules act synergistically, they represent a powerful tool for antagonism and saprophytism and may provide an explanation for the paradox between low compound concentrations and high antifungal activity that occurs *in vivo.*

Second, a great number of applications may be envisaged for some of the enzymes and the encoding genes to improve plant disease resistance and biocontrol ability of microorganisms or to formulate new fungicides containing low levels of chemicals plus enzymes used as additives. It should also be noted that no other chitinolytic enzymes from plants, bacteria or fungi have shown comparable synergism with such a broad range of chemical and biological control agents as the enzymes of *Trichoderma* and *Gliocladium.*

The cell wall degrading system, of which chitinolytic enzymes are key components, is both efficacious and complex, and this raises the question of how *Trichoderma* and *Gliocladium* are able to protect their own cell walls from breaking down. Lora *et al.* (1994) found a putative cell wall protein (Qid3) co-induced with chitinolytic enzymes, which most likely acts as an inhibitor of chitinase activity on the walls of *Trichoderma.* Since different compounds (e.g., CWDEs, MACs, etc.) may act synergistically *in vivo* at very low concentrations, the specific inhibition of a key activity (e.g., chitinolytic activity interruption by Qid3) on the hyphae of the producing organism should disrupt this interaction and raise the effective dose of the components to a level not occurring *in vivo.* Thus, the synergistic mixture will be highly effective on another fungus or an appropriate substrate but not concentrated enough to substantially damage *Trichoderma* or *Gliocladium* tissues. In other words, synergism not only supports the lytic/antifungal effect but also allows efficient self-protection by using a few cell wall- or membrane-bound specific inhibitors. This hypothesis may explain the fact that *T. harzianum* P1 is resistant to its own chitinolytic enzymes, even in combinations (Lorito *et al.*, 1993b), and that hyphae of *Trichoderma* spp. penetrate inside the host by the action of CWDEs without suffering obvious damage to their own cell walls (Elad *et al.*, 1983).

4.3 Genes encoding chitinolytic enzymes

The first gene encoding a chitinolytic enzyme from *Trichoderma* and *Gliocladium* was cloned and sequenced only a few years ago (Garcia *et al.*, 1994; Hayes *et al.*, 1994). Since then, this topic has received much attention for the following reasons: (a) cloning of these genes is a relatively straightforward task once the enzymes have been purified and characterized; (b) coding sequences are relatively small and therefore easy to manipulate; (c) transcription may be induced at high levels by using appropriate substrates and/or fungal host tissues; (d) genes seem to be entirely located at one locus as a single copy and show some homology with similar genes from other organisms; and (e) cloned sequences can be expressed in other fungi, yeast, bacteria and plants for enzyme production or genetic manipulation.

4.3.1 Cloned genes

A small number of genes have been cloned thus far, mostly from *T. harzianum* (Table 4.1). The first one was the gene coding for CHIT42, probably because it is one of the most abundant and commonly found enzymes under chitin induction (Garcia *et al.*, 1994; Hayes *et al.*, 1994). This gene has been cloned from strains P1, CECT 2413, IMI206040, T 25-t and named *ThEn-42*, *chit42*, *ech-42* and *chi1*, respectively. A gene encoding another endochitinase, CHIT33, has also been cloned from strain CECT 2413 and named *chit33* (Limón *et al.*, 1995). Among exo-acting enzymes, the encoding sequence of CHIT72 has been recently cloned from strain P1 and named *nag1* (Peterbauer *et al.*, 1996) and from strain T 25-t and named *exc1* (Draborg *et al.*, 1995). Another *β-N*-acetylhexosaminidase-encoding sequence, named *exc2*, similar to *exc1*, has been cloned from T 25-t (Draborg *et al.*, 1995) and it perhaps codes for a novel hexosaminidase. Other cloned genes include *tham-ch* which codes for a CHIT42 from *T. hamatum* (Fekete *et al.*, 1996) and two endochitinase encoding genes from *T.* (*Gliocladium*) *virens* strain 41 (Harman, G. E. *et al.*, unpublished data). To date, no cloned sequence has been reported for a *Trichoderma* or *Gliocladium* chitobiosidase, such as CHIT40, although an N-terminus sequence is available (Harman *et al.*, 1993b).

4.3.2 Characteristics of cloned genes

Sequence features

Cloned *Trichoderma* and *Gliocladium* chitinolytic enzyme genes encoded proteins of lengths ranging from approximately 320 to 600 amino acids (Draborg *et al.*, 1995, 1996; Fekete *et al.*, 1996; Garcia *et al.*, 1994; Hayes *et al.*, 1994; Limón *et al.*, 1995; Peterbauer *et al.*, 1996). All sequences apparently contained a highly hydrophobic NH_2-terminus amino acid sequence (about 20 aa) that may correspond to a signal peptide cleaved upon secretion of the protein (Schrempf *et al.*, 1993; Yanai *et al.*, 1992). Interestingly, *ThEn-42*, *ech-42*, *chit42* and *tham-ch* each contained an additional 12 aa of a highly hydrophilic NH_2-terminal peptide, which is probably cleaved by proteolytic activation, perhaps correlated with the regulation of chitin synthase activity (Adams *et al.*, 1993). Two or three relatively short introns ranging from about 50 to 80 nt were found in endochitinase- and *β-N*-acetylhexosaminidase-encoding genes (Carsolio *et al.*, 1994; Peterbauer *et al.*, 1996). Typical motifs for filamentous fungal genes were present in the transcribed sequences, including signals for termination of transcription, polyadenylation, start of translation, intron splicing, etc. (Draborg *et al.*, 1995, 1996; Garcia *et al.*, 1994; Hayes *et al.*, 1994; Limón *et al.*, 1995; Peterbauer *et al.*, 1996). The length and the A/T content of leader sequences, 3′ untranslated sequences and the polyA tails were also similar to those of other fungal genes. The deduced aa sequence contained, in most of the cases, putative *N*-glycosylation and sometimes also phosphorylation sites, although glycosylation of the mature protein has only been demonstrated for CHIT72 encoded by *nag1* (Lorito *et al.*, 1994a; Peterbauer *et al.*, 1996). Several well-preserved domains were found in *ThEn-42*, *ech-42*, *chi1*, *chit42*, *chit33* and *nag1*. For instance, the endochitinase-encoding genes contain two conserved regions that especially resemble bacterial sequences and are associated with the active site, which also includes the amino acids Asp and Glu supposedly essential for catalytic action (Hayes *et al.*, 1994; Limón *et al.*, 1995; Watanabe *et al.*, 1993). However, these genes may lack typical domains such as those of chitin binding, Ser/Thr-rich and C-terminal processing (Hayes *et al.*, 1994; Limón *et al.*, 1995).

Sequence homologies

All *Trichoderma* and *Gliocladium* chitinolytic enzyme encoding genes cloned so far occurred as a single copy and did not cross-hybridize with other genome fragments even under low stringency conditions (except *exc1* and *exc2*) (Draborg *et al.*, 1995; Garcia *et al.*, 1994; Limón *et al.*, 1995). Nt-sequences coding for different enzymes showed no substantial levels of similarity and were also different with regard to homology with genes from other organisms (except *exc1* and *exc2* which share a 72% homology). *ThEn-42*, *ech-42*, *chit42* and *chi1* all coding for CHIT42 had no sequence identity with yeast and plant chitinases, but they showed the highest degree of homology with bacterial and filamentous fungal genes (Draborg *et al.*, 1996; Garcia *et al.*, 1994; Hayes *et al.*, 1994; Tronsmo *et al.*, 1993b). This may suggest that some fungal and bacterial chitinolytic enzymes evolved from an ancestral set of prokaryotic genes. However, the *chit33* endochitinase-encoding gene showed no similarity with prokaryotic or other *Trichoderma* genes, except for the two conserved regions associated with the active site; instead a high homology (up to 43% identity) was found with corresponding yeast, fungal and plant sequences (Limón *et al.*, 1995). The *β-N*-acetylhexosaminidase encoding genes *nag1*, *exc1* and *exc2* had considerable regional homologies with fungal and higher eukaryote genes, less similarity with a bacterial chitinase and no similarity with other *Trichoderma* genes.

Interestingly, genes coding for the same or similar enzymes showed unexpected differences when the sequences were compared. The gene encoding CHIT42 has always been detected by Southern analysis on one of the two largest chromosomes in mycolytic and non-mycolytic isolates of *T. harzianum*, *T. reesei*, *T. koningii*, *T. longibrachiatum*, *T. virens*, *T. viride*, *T. atroviride* and *T. hamatum* (Fekete *et al.*, 1996; Garcia *et al.*, 1994; Lorito, M., unpublished data). Most of the translated sequence was conserved among cloned CHIT42-encoding genes, but none was conserved in the region of the NH_2-terminal propeptide, which is putatively cleaved to release the mature protein (Figure 4.2). With regard to the overall sequence identity, *ThEn-42* from *T. harzianum* P1 was almost identical to *ech-42* from *T. harzianum* IMI206040 (99% identity) and to *tham-ch* from a *T. hamatum* strain (94.5% identity), but *ThEn-42* was significantly different than *chit42* from *T. harzianum* CECT 2413 and *chi1* from *T. harzianum* T 25-t (*chit42* and *chi1* are identical to each other at 97%). Furthermore, *nag1* from *T. harzianum* P1 was about 80% identical to *exc1* and much less similar to *exc2*. Theoretically, if genes commonly occurring among diverse species are considered, sequence differences may be found

```
                  10        20         30          40        50        60        70
                   +         +          +           +         +         +         +
         *******-------*******

ThEn-42  mlgflgksvallaalqatlisa spvtANDVsvekr asgyanavyftnwgiygr-nfqpqnlvasdithvi
ech-42   mlgflgksvallaalqatlisa spvtANDVsvekr asgyanavyftnwgiygr-nfqpqnlvasdithvi
chit42   mlsflgksvallaalqatlssp kpghRRA-svekr angyansvyftnwgiydrNnfqpadlvasdvthvi
chi1     mlsflgksvallaalqatlssa splaTEERsvekr angyansvyftnwgiydr-nfqpadlvasdvthvi
tham-ch  mlgflgksvallaalqatltsa splsTNDVtvekr asgyanavyftnwgiygr-nfqpqdlvasdtthvi
         1111111111111111111223 3334    11111 111111111111111111 1111111111111111
              pre-peptide       pro-peptide            mature protein
```

Figure 4.2 Comparison of N-terminus amino acid sequence deduced from endochitinase CHIT42-encoding genes of various *Trichoderma* strains. *ThEn-42, chit42, ech-42* and *chi1* were cloned from *T. harzianum* strain P1, CECT 2413, IMI206040 and T 25-t, respectively; *tham-ch* was cloned from *T. hamatum*. Amino acids in capital letters indicate a lack of sequence similarity. Where similarity occurs (small letters), a score related to the statistical significance of the alignment is given below each position. The lower the score the higher is the reliability of the alignment. Optimal multiple alignment with indices of reliability was obtained by using the MATCH-BOX_server 1.1 from the World Wide Web. Polar (asterisks) and hydrophobic (dashed line) sequences of the pre-peptide for protein secretion are indicated.

and used as molecular tools to assist in the taxonomy of *Trichoderma* and *Gliocladium* (see Volume 1, Chapter 2).

4.3.3 *Regulation of gene expression and promoters*

Little is know about the mechanisms that regulate gene expression of *Trichoderma* and *Gliocladium* chitinolytic enzymes. *De novo* transcription and regulation at the transcriptional level have been implicated in the production of these enzymes. Direct correspondence between enzyme and mRNA formation was found for *nag1*, *chit33*, *chit42* and *ech-42*, although additional post-transcriptional regulation was suggested for *chit33* (Carsolio *et al.*, 1994; Garcia *et al.*, 1994; Limón *et al.*, 1995). Transcription of all these genes was strongly or moderately co-induced by autoclaved mycelia, chitin, *N*-acetylglucosamine or glucose de-repression. However, some genes seem to be independently regulated (Haran *et al.*, 1996; Limón *et al.*, 1995). For instance, the highest transcription was obtained for *chit42* under induction with chitin or fungal cell walls and for *chit33* under starvation conditions (Limón *et al.*, 1995).

Expression of the CHIT42-encoding genes has been studied in more detail. For example, *ech-42* was strongly induced by chitin, light, and mycoparasitic interactions and effectively repressed by glucose, as noted for the other genes (Carsolio *et al.*, 1994). Putative binding sites for the protein BrlA, which controls light-induced sporulation in *Aspergillus nidulans*, were found on the 5′-regulatory sequence (Carsolio *et al.*, 1994), but there was no *in vitro* or *in vivo* evidence that these sites were functional or that binding actually occurred. More recently, Lorito *et al.* (1996a) have cloned and characterized a substantial part of the promoter region of *ThEn-42* from *T. harzianum* P1. Analysis based on DNA–protein binding indicated the presence of several putative sites for functional attachment of both known and unknown regulatory factors. Strong evidence was found to support catabolite repression of *ThEn-42* mediated by Cre1 for the following reasons: (a) Cre1 binds to the promoter region of *ThEn-42*; (b) its binding and release correspond, respectively, to the repression and de-repression of the gene; and (c) mycoparasitic interaction relieves Cre1 binding even in the presence of glucose, apparently allowing binding of a mycoparasitic-induced protein complex that possibly functions as a positive regulator of *ThEn-42* expression and CHIT42 synthesis. This study represents the first inside look at the regulation of the expression of a chitinase-encoding gene under mycoparasitic conditions and should stimulate new interest on this topic. While the *T. harzianum* Cre1-encoding gene has been cloned (Margolles-Clark *et al.*, 1996b), research is still in progress to identify and characterize the putative mycoparasitic inducer of *ThEn-42* and clone the gene. Moreover, *in vivo* and *in vitro* functional analyses of both the *ThEn-42* and *nag1* promoter region (Lorito *et al.*, 1996a; Peterbauer *et al.*, 1996) will identify active binding sites and regulatory elements controlling the expression of these two important genes under different conditions. Other aspects of the mechanism that oversees the expression of these sequences, such as signal surface recognition and transduction to the nucleus, are completely undiscovered and should also be investigated.

4.4 Roles of chitinolytic enzymes and their genes

The role of the chitinolytic enzyme system in the biology of *Trichoderma* and *Gliocladium* is still debated. It has been suggested that there is an involvement in morphogenesis, antagonism, mycoparasitism and saprophytism, but conclusive proof of the role of any of the enzymes, for instance from gene disruption experiments, is still lacking.

4.4.1 Role in morphogenesis

Fungal chitinases participate in hyphal elongation, cell separation, plasmogamy, branching, spore swelling, germination, sporangium formation, response to mechanical injuries and autolysis (for a review, see Gooday, 1990; Sahai and Manocha, 1993). Exo- and endo-acting enzymes localized intracellularly, in the periplasmic space or on the plasma membrane, may act synergistically to balance chitin synthase activity (Sahai and Manocha, 1993). The morphogenetic role of chitinolytic enzymes in *Trichoderma* and *Gliocladium* is still an unexplored field. A detailed study on intracellular, membrane-associated, or autolytic chitinolytic enzymes is lacking and there is no proof either in favor of or against the involvement in morphogenesis of any of the enzymes detected thus far. A certain amount of cell-bound endochitinase activity was detected by Ulhoa and Peberdy (1991b), while Tronsmo and Harman (1992) noted chitinolytic activity in liquid culture possibly associated with autolysis. Haran *et al.* (1995) reported that CHIT102 was detected only intracellularly under non-inducive conditions. Garcia *et al.* (1994) proposed a morphogenetic role for CHIT42 due to its generally low levels among different strains, while Limón *et al.* (1995) indicated that CHIT33 has no such role because its sequence lacks domains typical of cell-wall associated enzymes (Kuranda and Robbins, 1991; Yanai *et al.*, 1992). Finally, the presence of cell bound hexosaminidase activity has been suggested by Peterbauer *et al.* (1996). This research topic deserves more attention since *Trichoderma* and *Gliocladium* contain high proportions of chitin in their cell walls and, therefore, the production of chitinolytic enzymes with a morphogenetic role is expected.

4.4.2 Role in saprophytism, antagonism and biocontrol

Trichoderma and *Gliocladium* chitinolytic enzymes are secreted extracellularly to degrade chitinous substrates (e.g., fungal mycelia), repressed by simple nutrients, induced by starvation and co-produced or synergistic with other compounds involved in nutrition or antagonism. Therefore, some of these enzymes have a nutritional role and may be involved in biocontrol, as well as in causing mushroom diseases (Muthumeenakshi *et al.*, 1994; Speranzini *et al.*, 1995). However, it is not clear to what extent chitinolytic enzymes participate in *Trichoderma* and *Gliocladium* antagonism and mycoparasitism or if they are used mainly for saprophytic growth. Belanger *et al.* (1995) observed that an effective biocontrol strain of *T. harzianum* first killed the host with antibiotics and much later produced chitinolytic enzymes to saprophytically colonize the dead host tissue. According to this, biocontrol efficacy of *Trichoderma* and *Gliocladium* strains was associated with the production of non-volatile antibiotics or to other mechanisms rather than mycoparasitism (Caron, 1993; Howell, 1987; Lumsden and Lewis, 1989; Prokkola, 1992). On the other hand, the following evidence points to a major involvement of chitinolytic enzymes in mycoparasitism, antagonism and biocontrol.

- *In vitro* mycoparasitic interaction strongly induced expression of chitinolytic enzyme-encoding genes in the first few hours of contact, and enzymes seem to be involved both in penetration and killing of the host (Carsolio *et al.*, 1994; Chérif and Benhamou, 1990; Elad *et al.*, 1983; Inbar and Chet, 1995; Haran *et al.*, 1996; Lorito *et al.*, 1996a,f).
- Autoclaved mycelia induced a high level of *chit42* transcription in mycoparasitic but not in non-mycoparasitic *Trichoderma* and *Gliocladium* strains (Garcia *et al.*, 1994).
- Increased chitinolytic activity in *T. harzianum* augmented its antagonistic ability *in vitro* (Haran *et al.*, 1993; Limón *et al.*, 1996).
- A *T. harzianum* gene encoding a chitinolytic enzyme (probably CHIT42) expressed in *E. coli* conferred biocontrol ability to the bacterium (Chet *et al.*, 1993).
- Purified enzymes have inhibitory activity against a wide range of fungi (Di Pietro *et al.*, 1993; Lorito *et al.*, 1993b, 1994b).
- Chitinolytic enzymes and antibiotic peptaibols or gliotoxin are able to interact at low concentrations to produce powerful antifungal mixtures (Lorito *et al.*, 1996c,d). Further, some commercially utilized biocontrol strains of *Trichoderma* are not known to produce non-volatile antibiotics (Harman *et al.*, 1993b) and a direct correlation was found between *in vivo* chitinolytic activity and biocontrol capability in three *T. harzianum* strains (Elad *et al.*, 1982).

In conclusion, the role of the *Trichoderma* and *Gliocladium* chitinolytic enzyme system is complex and far from being completely understood. Substantial differences probably occur among strains that evolved in various habitats and developed different nutritional and antagonistic behaviors. In addition, this system could be involved in diverse phases of many physiological processes, which makes it less redundant than it may appear at first glance (Table 4.1). With the availability of genes, promoters and efficient transformation protocols (Gruber *et al.*, 1990; Herrera-Estrella *et al.*, 1990; Lorito *et al.*, 1993d), the role of key enzymes could be determined by analyzing the physiology of strains in which specific enzyme activities have been disrupted or the expression of key genes de-regulated. Such mutants may be constructed by using heterologous promoters (Margolles-Clark *et al.*, 1996b), augmenting gene copy numbers (Limón *et al.*, 1996), altering native promoters (Lorito, M., Kubicek, C. P., and Mach, R. L., unpublished data) or selectively disrupting "chitinolytic" genes (Harman, G. E., Hayes, C. K., and Lorito, M., unpublished data). In addition, promoter regions containing regulatory motifs may be fused with reporter genes coding for compounds easily detectable *in situ* (Spelligt *et al.*, 1996) and used, together with identified regulatory factors, for expression studies. These and other genetically-based approaches are now being applied in several laboratories in an attempt to define the role of chitinolytic enzymes in biocontrol and to set limits and directions for a full exploitation of *Trichoderma* and *Gliocladium* genetic potential.

4.5 Potential applications and commercial usefulness of chitinolytic enzymes and their genes

Fungi and insects, in contrast to plants and higher vertebrates, have large proportions of structural chitin. In addition, over a million tons of chitin wastes are discarded annually (Aloise *et al.*, 1996; Cosio *et al.*, 1982). Therefore, chitinolytic enzymes and their genes may have a number of potential applications for fungal and insect disease therapy, biomass degradation and preparation of useful biopolymers. In particular, *Trichoderma* and *Gliocladium* chitinolytic enzymes are stable, easy to handle, active on a wide range of substrates and pathogenic fungi, synergistic with other enzymes and non-enzymatic compounds, non-toxic and directly encoded by single, relatively small genes that can be used to produce active enzymes virtually in any organism (Coenen *et al.*, 1995; Harman and Hayes, 1996; Lorito *et al.*, 1994c). Although chitinolytic enzymes may exert an important influence in the biocontrol activity of *Trichoderma* strains commercialized as Trichodex™, Binab-T™, Root-Shield™, Bio-Trek22G™, Trichoject™, Tricho Minidowels™ and SoilGard™, no commercial use is reported thus far for these enzymes or their genes. On the other hand, some applications may soon be realized since genes are being exploited and enzymes are being produced transgenically in large quantities (Draborg *et al.*, 1996; Margolles-Clark, 1996a). Potential use and recent accomplishments are summarized below (see also other chapters in this book).

4.5.1 Industry and medicine

Trichoderma and *Gliocladium* enzymes may be applicable for degradation of chitin wastes from the shellfish industry (Cosio *et al.*, 1982) to obtain NAG, which is currently utilized as a food supplement, a growth substrate and a drug for treatment of gastrointestinal disorders in humans and animals (Aloise *et al.*, 1996). They can also be used for preparation of fabrics, detergents and protoplasts and, because of their antifungal activity, can be used as sterilizing solutions. In addition, *Trichoderma* endochitinase efficiently catalyzes the synthesis of $(GluNAc)_6$ and $(GluNAc)_7$ chitooligomers, which are being studied and commercialized as strong antitumor agents or inducers of plant resistance against pathogens (Usui *et al.*, 1990; Vander and Moerschbacher, 1993).

The recent discovery that chitinolytic enzyme activity in human serum and macrophages is elicited by fungal infection or other diseases (e.g., Gaucher disease) presents new possibilities for applying chitinolytic enzymes and their genes in enzyme or gene therapy (Aerts *et al.*, 1996; Muzzarelli, 1993; Overdijk and Van Steijn, 1994). In this respect, enzymes and genes from *Trichoderma* and *Gliocladium* may be particularly useful because they have strong, broad antifungal activity and show an impressive level of synergism with commonly used antifungal drugs (e.g., miconazole), which has never been reported for other chitinolytic enzymes (Lorito *et al.*, 1994b, 1996c). In addition, appropriate combinations of enzymes may be used as co-adjuvants of drugs applied topically against human and animal fungal diseases (Davies and Pope, 1978). This should lower the required dose of chemical fungicide in the formula, thereby reducing the risk of poisoning and the occurrence of resistant strains during prolonged treatments.

4.5.2 Plant disease control

Some chitinolytic enzymes from *Trichoderma* and *Gliocladium* have a long shelf life and may be used directly in combination with chemical fungicides to protect fruit from storage rots. Direct field or greenhouse application of enzyme-based preparations could also be considered, since enzymes may improve the efficacy of systemic fungicides and antagonism of biocontrol bacteria (Lorito *et al.*, 1993c, 1994b). Also promising for plant defense is the transfer of the encoding genes in many different species among plants, fungi and bacteria (Lorito *et al.*, 1996d). For instance, the endochitinase encoding gene *ThEn-42* has been constitutively expressed in tobacco, tomato, potato, petunia, apple and other plant species. In tobacco, the transgenic CHIT42 maintained its antifungal activity, was accumulated extracellularly in leaves, roots, stems and flowers and did not affect the interaction of the plant with mycorrhizal fungi (Lorito, 1995; Lorito, M. and Bonfante, P., unpublished data; Rousseau *et al.*, 1996). It may act synergistically with plant PR-proteins (e.g., osmotin and glucanases) (Lorito *et al.*, 1996c) to inhibit penetrating fungi and release oligomers that elicit plant defense response (Barcelo *et al.*, 1996). Consequently, transgenic lines of tobacco, potato and apple have been selected for further studies because they have shown a substantial improvement in resistance to aerial- and soil-borne phytopathogenic fungi, such as *Alternaria alternata*, *Rhizoctonia solani* and *Venturia inequalis* (Lorito *et al.*, 1996b, unpublished data; Harman, G. E., personal communication). In addition, high level expression of transgenic *Trichoderma* endochitinase did not significantly affect mycorrhiza formation and completion of the life cycle by *Gigaspora margarita* on tobacco roots (Lorito, M. and Bonfante, P., unpublished data). *ThEn-42* has also been expressed in the biocontrol bacterium *E. cloacae*; these transgenic bacteria seem to be more damaging to plant pathogenic fungi than the wild type (Harman, G. E., personal communication). Further, *ThEn-42* was constitutively overexpressed under the control of a *Trichoderma* cellulase promoter in a *T. reesei* strain capable of producing extracellular enzymes at an industrial level (Margolles-Clark *et al.*, 1996a). The same construct improved chitinolytic activity of the original *T. harzianum* strain P1 (Margolles-Clark *et al.*, 1996b) and the resulting mutants are being tested for biocontrol ability. Similarly, preliminary results of a constitutive overexpression of *chit33* in two *T. harzianum* strains indicated an improvement of antagonistic ability against *R. solani* (Limón *et al.*, 1996).

All of these results clearly support the use of *Trichoderma* and *Gliocladium* chitinolytic genes for plant disease control. Moreover, the spectrum and the level of activity could be substantially improved by co-expressing in transgenic plants different genes in synergistic combinations (e.g., genes encoding *Trichoderma* endochitinase, β-1,3-glucanase and tobacco osmotin) (Jach *et al.*, 1995; Zhu *et al.*, 1994; Lorito, M. and Woo, S. L., unpublished data). However, new crop varieties and strains genetically modified with these genes still need to be evaluated in field and greenhouse trials against pathogenic and nuisance fungi. Only extensive testing will assess whether or not these genes and their products have any commercial value and if they are of practical utility for reducing the application of chemical fungicides for a sustainable agriculture.

Acknowledgments

I am very grateful to Sheri Woo and also to Felice Scala for the critical revision of the manuscript and to Arne Tronsmo, Gary Harman and Clemens Peterbauer for useful discussions on enzyme terminology.

References

Abad, L. R., Durzo, M. P., Liu, D., Narasimhan, M. L., Reuveni, M., Zhu, J. K., Niu, X. M., Singh, N. K., Hasegawa, P. M., and Bressan, R. A. 1996. Antifungal activity of tobacco osmotin has specificity and involves plasma membrane permeabilization. *Plant Sci.* **118**: 11–23.

Adams, D. J., Causier, B. E., Mellor, K. J., Keer, V., Milling, R., and Dada, J. 1993. Regulation of chitin synthase and chitinase in fungi. In R. A. A. Muzzarelli (ed.), *Chitin Enzymology*. European Chitin Society, Lyon and Ancona, pp. 15–26.

Aerts, J. M. F. G., Boot, R. G., Renkeme, G. H., van Weely, S., Hollak, C. E. M., Donker-Koopman, W. E., Strijland, A., and Verhoek, M. 1996. Chitotriosidase: a human macrophage chitinase that is a marker for Gaucher disease manifestation. In R. A. A. Muzzarelli (ed.), *Chitin Enzymology II*. Atec, Grottammare (AP), Italy, pp. 3–10.

Aloise, P. A., Lumme, M., and Haynes, C. A. 1996. *N*-acetyl-D-glucosamine production from chitin-waste using chitinases from *Serratia marcescens*. In R. A. A. Muzzarelli (ed.), *Chitin Enzymology II*. Atec, Grottammare (AP), Italy, pp. 581–594.

Bade, M. C., Stinson, A., and Moneam, N. A. 1988. *Connective Tiss. Res.* **17**: 137–151.

Barcelo, A. R., Zapata, J. M., and Calderon, A. A. 1996. A basic peroxidase isoenzyme, marker of resistance against *Plasmopara viticola* in grapevines, is induced by an elicitor from *Trichoderma viride* in susceptible grapevines. *J. Phytopathol.* **144**: 309–313.

Belanger, R. R., Dufour, N., Caron, J., and Benhamou, N. 1995. Chronological events associated with the antagonistic properties of *Trichoderma harzianum* against *Botrytis cinerea*: indirect evidence for sequential role of antibiosis and parasitism. *Biocontrol Sci. Technol.* **5**: 41–53.

Benhamou, N. and Chet, I. 1993. Hyphal interactions between *Trichoderma harzianum* and *Rhizoctonia solani*: ultrastructure and gold cytochemistry of the mycoparasitic process. *Phytopathology* **83**: 1062–1071.

BENHAMOU, N. and CHET, I. 1996. Parasitism of sclerotia of *Sclerotium rolfsii* by *Trichoderma harzianum*: ultrastructural and cytochemical aspects of the interaction. *Phytopathology* **86**: 405–416.

BERG, B. and PETTERSSON, G. 1977. Location and formation of cellulases in *Trichoderma viride*. *J. Appl. Bacteriol.* **42**: 65–75.

BERTAGNOLLI, B. L., DALSOGLIO, F. K., and SINCLAIR, J. B. 1996. Extracellular enzyme profiles of the fungal pathogen *Rhizoctonia solani* isolate 2B-12 and of two antagonists, *Bacillus megaterium* strain B153-2-2 and *Trichoderma harzianum* isolate Th008. I: Possible correlations with inhibition of growth and biocontrol. *Physiol. Molec. Plant Pathol.* **48**: 145–160.

BIELKA, H., DIXON, H. B. F., KARLSON, P., LIEBECQ, C., SHARON, N., VAN LENTEN, S. F., VELICK, S. G., VLIEGENTHART, J. F. G., and WEBB, E. C. 1984. *Enzyme Nomenclature*. Academic Press, New York, 646 pp.

BLAISEAU, P. L., KUNZ, C., GRISON, R., BERTHEAU, Y., and BRYGOO, Y. 1992. Cloning and expression of a chitinase gene from the hyperparasitic fungus *Aphanocladium album*. *Curr. Genet.* **21**: 61–66.

BROGLIE, K., CHET, I., HOLLIDAY, M., CRESSMAN, R., BIDDLE, P., KNOWLTON, S., MAUVAIS, C. J., and BROGLIE, R. 1991. Transgenic plants with enhanced resistance to the fungal pathogen *Rhizoctonia solani*. *Science* **254**: 1194–1197.

BRUCE, A., SRINIVASAN, U., STAINES, H. J., and HIGHLEY, T. L. 1995. Chitinase and laminarinase production in liquid culture by *Trichoderma* spp. and their role in biocontrol of wood decay fungi. *Int. Biodeterior. Biodegrad.* **35**: 337–353.

CABIB, E., SHAW, J. A., MOL, P. C., BOWERS, B., and CHOI, W. J. 1996. Chitin biosynthesis and morphogenetic processes. In R. Brambl and G. A. Marzluf (eds), *Mycota*. Vol. III. Springer-Verlag, Berlin, pp. 243–267.

CARON, J. 1993. *Isolement et caractérisation de divers isolats de Trichoderma comme agent de lutte biologique contre la moisissure grise* (Botrytis cinerea) *dans la production de la fraise*. M.S. Thesis no. 12350, Université Laval, Québec.

CARSOLIO, C., GUTIÉRREZ, A., JIMÉNEZ, B., VAN MONTAGU, M., and HERRERA-ESTRELLA, A. 1994. Characterization of *ech-42*, a *Trichoderma harzianum* endochitinase gene expressed during mycoparasitism. *Proc. Natl. Acad. Sci. USA* **91**: 10903–10907.

CHÉRIF, M. and BENHAMOU, N. 1990. Cytochemical aspects of chitin breakdown during the parasitic action of a *Trichoderma* sp. on *Fusarium oxysporum* f. sp. *radicis-lycopersici*. *Phytopathology* **80**: 1406–1414.

CHET, I. 1987. *Trichoderma* – application, mode of action, and potential as a biocontrol agent of soilborne plant pathogenic fungi. In I. Chet (ed.), *Innovative Approaches to Plant Disease Control*. Wiley, New York, pp. 137–160.

CHET, I., BARAK, Z., and OPPENHEIM, A. 1993. Genetic engineering of microorganisms for improved biocontrol activity. In I. Chet (ed.), *Biotechnology in Plant Disease Control*. Wiley-Liss, New York, pp. 211–255.

CHET, I., HADAR, Y., ELAD, Y., KATAN, J., and HENIS, Y. 1979. Biological control of soilborne plant pathogens by *Trichoderma harzianum*. In B. Schippers and W. Gams (eds), *Soil-borne Plant Pathogens*. Academic Press, London, pp. 585–591.

COENEN, T. M. M., SCHOENMAKERS, A. C. M., and VERHAGEN, H. 1995. Safety evaluation of β-glucanase derived from *Trichoderma reesei*: summary of toxicological data. *Food Chem. Toxic.* **33**: 859–866.

COLLINGE, D. B., KRAGH, K. M., MIKKELSEN, J. D., NIELSEN, K., RASMUSSEN, U., and VAD, K. 1993. Plant chitinases. *Plant J.* **3**: 31–40.

COSIO, I. G., FISHER, R. A., and CARROAD, P. A. 1982. Bioconversion of shellfish chitin waste: waste treatment, enzyme production, process design, and economic analysis. *J. Food Sci.* **47**: 901–905.

DAVIES, D. A. L. and POPE, A. M. S. 1978. Mycolase, a new kind of systemic antimycotic. *Nature* **273**: 235–236.

DEANE, E. E., PEBERDY, J. F., WHIPPS, J. M., and LYNCH, J. M. 1995. Isolation of a partial chitinase clone from *Trichoderma harzianum* T198. Fifth International *Trichoderma* and *Gliocladium* Workshop, April 1995, Beltsville, MD, abstract.

DE LA CRUZ, J., PINTOR-TORO, J. A., BENITEZ, T., LLOBELL, A., and ROMERO, L. C. 1995. A novel endo-β-1,3-glucanase, *BGN13.1*, involved in the mycoparasitism of *Trichoderma harzianum*. *J. Bacteriol.* **177**: 6937–6945.

DE LA CRUZ, J., HIDALGO-GALLEGO, A., LORA, J. M., BENITEZ, T., PINTOR-TORO, J. A., and LLOBELL, A. 1992. Isolation and characterization of three chitinases from *Trichoderma harzianum*. *Eur. J. Biochem.* **206**: 859–867.

DE LA CRUZ, J., REY, M., LORA, J. M., HIDALGO-GALLEGO, A., DOMINGUEZ, F., PINTOR-TORO, J. A., LLOBELL, A., and BENITEZ, T. 1993. Carbon source control on β-glucanase, chitobiase and chitinase from *Trichoderma harzianum*. *Arch. Microbiol.* **159**: 316–322.

DIEKMAN, H. and BREVES, R. 1993. Multiplicity of chitinases and substrate specificity. In R. A. A. Muzzarelli (ed.), *Chitin Enzymology*. European Chitin Society, Lyon and Ancona, pp. 295–302.

DI PIETRO, A., LORITO, M., HAYES, C. K., BROADWAY, R. M., and HARMAN, G. E. 1993. Endochitinase from *Gliocladium virens*: isolation, characterization and synergistic antifungal activity in combination with gliotoxin. *Phytopathology* **83**: 308–313.

DRABORG, H., KAUPPINEN, S., DALBØGE, H., and CHRISTGAU, S. 1995. Molecular cloning and expression in *Saccharomyces cerevisiae* of two exochitinases from *Trichoderma harzianum*. *Biochem. Molec. Biol. Int.* **36**: 781–791.

DRABORG, H., CHRISTGAU, S., HALKIER, T., RASMUSSEN, G., DALBØGE, H., and KAUPPINEN, S. 1996. Secretion of an enzymatically active *Trichoderma harzianum* endochitinase by *Saccharomyces cerevisiae*. *Curr. Genet.* **29**: 404–409.

ELAD, Y., CHET, I., and HENIS, Y. 1982. Degradation of plant pathogenic fungi by *Trichoderma harzianum*. *Can. J. Microbiol.* **28**: 719–725.

ELAD, Y., CHET, I., BOYLE, P., and HENIS, Y. 1983. Parasitism of *Trichoderma* spp. on *Rhizoctonia solani* and *Sclerotium rolfsii*: scanning electron microscopy and fluorescence microscopy. *Phytopathology* **73**: 85–88.

FEKETE, C., WESZELY, T., and HORNOK, L. 1996. Assignment of a PCR-amplified chitinase sequence cloned from *Trichoderma hamatum* to resolved chromosomes of potential biocontrol species of *Trichoderma*. *FEMS Microbiol. Lett.* **145**: 385–391.

FREITAG, M., MORREL, J. J., and BRUCE, A. 1991. Biological protection of wood: status and prospects. *Biodeterioration Abstracts* **5**: 1–12.

GARCIA, I., LORA, J. M., DE LA CRUZ, J., BENITEZ, T., LLOBELL, A., and PINTOR-TORO, J. A. 1994. Cloning and characterization of a chitinase (CHIT42) cDNA from the mycoparasitic fungus *Trichoderma harzianum*. *Curr. Genet.* **27**: 83–89.

GEREMIA, R., JACOBS, D., GOLDMAN, G. H., VAN MONTAGU, M., and HERRERA-ESTRELLA, A. 1991. Induction and secretion of hydrolytic enzymes by the biocontrol agent *Trichoderma harzianum*. In A. B. R. Beemster, G. J. Bollen, M. Gerlagh, M. A. Ruissen, B. Shippers and A. Tempel (eds), *Biotic Interactions and Soil-borne Diseases*. Elsevier, Amsterdam, pp. 181–186.

GOLDMAN, G. H. 1993. *Molecular genetic studies of mycoparasitism by* Trichoderma *spp.* Ph.D. Thesis, Universiten Gent, Belgium, 107 pp.

GOODAY, G. W. 1990. Physiology of microbial degradation of chitin and chitosan. *Biodegradation* **1**: 177–190.

GRUBER, F., VISSER, J., KUBICEK, C. P., and DE GRAAFF, L. H. 1990. The development of a heterologous transformation system for the cellulolytic fungus *Trichoderma reesei* based on a pyrG-negative mutant strain. *Curr. Genet.* **18**: 71–76.

HADAR, Y., HENIS, Y., and CHET, I. 1979. Biological control of *Rhizoctonia solani* damping off with wheat bran culture of *Trichoderma harzianum*. *Phytopathology* **69**: 64–68.

HARAN, S., SCHICKLER, H., OPPENHEIM, A., and CHET, I. 1995. New components of the chitinolytic system of *Trichoderma harzianum*. *Mycol. Res.* **99**: 441–446.

HARAN, S., SCHICKLER, H., OPPENHEIM, A., and CHET, I. 1996. Differential expression of *Trichoderma harzianum* chitinases during mycoparasitism. *Phytopathology* **86**: 980–985.

HARAN, S., SCHICKLER, H., PE'ER, S., LOGEMANN, S., OPPENHEIM, A., and CHET, I. 1993. Increased constitutive chitinase activity in transformed *Trichoderma harzianum*. *Biol. Control* **3**: 101–108.

HARMAN, G. E. 1990. Deployment tactics for biocontrol agents in plant pathology. In R. R. Baker and P. E. Dunn (eds), *New Directions in Biological Control: Alternatives for Suppressing Agricultural Pests and Diseases*. Alan R. Liss, New York, pp. 779–792.

HARMAN, G. E. and HAYES, C. K. 1996. Biologically-based technologies for pest control: pathogens and pests of agriculture. *Report to the Office Technology Assessment*, US Congress, April 1996.

HARMAN, G. E., HAYES, C. K., and LORITO, M. 1993a. Genome of biocontrol fungi: modification and genetic components for plant disease management strategies. In G. C. Marten (ed.), *Pest Management: Biologically Based Technologies*. US Department of Agriculture, Beltsville, MD, pp. 205–228.

HARMAN, G. E., HAYES, C. K., LORITO, M., BROADWAY, R. M., DI PIETRO, A., PETERBAUER, C., and TRONSMO, A. 1993b. Chitinolytic enzymes of *Trichoderma harzianum*: purification of chitobiosidase and endochitinase. *Phytopathology* **83**: 313–318.

HAYES, C. K., KLEMSDAL, S., LORITO, M., DI PIETRO, A., PETERBAUER, C., NAKAS, J. P., TRONSMO, A., and HARMAN, G. E. 1994. Isolation and sequence of an endochitinase-encoding gene from a cDNA library of *Trichoderma harzianum*. *Gene* **138**: 143–148.

HERRERA-ESTRELLA, A., GOLDMAN, G. H., and VAN MONTAGU, M. 1990. High-efficiency transformation system for the biocontrol agents *Trichoderma* spp. *Molec. Microbiol.* **4**: 839–843.

HOWELL, C. R. 1987. Relevance of mycoparasitism in the biological control of *Rhizoctonia solani* by *Gliocladium virens*. *Phytopathology* **77**: 992–994.

INBAR, J. and CHET, I. 1991. Detection of chitinolytic activity in the rhizosphere using image analysis. *Soil Biol. Biochem.* **23**: 239–242.

INBAR, J. and CHET, I. 1992. Biomimics of fungal cell–cell recognition by use of lectin-coated nylon fibers. *J. Bacteriol.* **174**: 1055–1059.

INBAR, J. and CHET, I. 1995. The role of recognition in the induction of specific chitinases during mycoparasitism of *Trichoderma harzianum*. *Microbiology* **141**: 2823–2829.

JACH, G., GÖRNHARDT, B., MUNDY, J., LOGEMANN, J., PINSDORF, E., LEACH, R., SCHELL, J., and MAAS, C. 1995. Enhanced quantitative resistance against fungal diseases by combinatorial expression of different barley antifungal proteins in transgenic tobacco. *Plant J.* **8**: 97–109.

KOGA, K., IWAMOTO, Y., SAKAMOTO, H., HATANO, K., SANO, M., and KATO, I. 1991. Purification and characterization of a *β-N*-acetylhexosaminidase from *Trichoderma harzianum*. *Agric. Biol. Chem.* **55**: 2817–2823.

KOMATSU, M. 1976. Studies on *Hypocrea*, *Trichoderma* and allied fungi antagonistic to shiitake, *Lentinus edodes*. *Report of the Tottori Mycological Institute* **13**: 55–61.

KURANDA, M. J. and ROBBINS, P. W. 1991. Chitinase is required for cell separation during growth of *Saccharomyces cerevisiae*. *J. Biol. Chem.* **266**: 19758–19767.

LIMÓN, M. C., LLOBELL, A., PINTOR-TORO, J. A., and BENITEZ, T. 1996. Over-expression of chitinase by *Trichoderma harzianum* strains used as biocontrol fungi. In R. A. A. Muzzarelli (ed.), *Chitin Enzymology II*. Atec, Grottammare (AP), Italy, pp. 245–252.

LIMÓN, M. C., LORA, J. M., GARCIA, I., DE LA CRUZ, J., LLOBELL, A., BENITEZ, T., and PINTOR-TORO, J. A. 1995. Primary structure and expression pattern of the

33-kDa chitinase gene from the mycoparasitic fungus *Trichoderma harzianum. Curr. Genet.* **28**: 478–483.

LO, C.-T., NELSON, E. B., and HARMAN, G. E. 1996. Biological control of turfgrass diseases with a rhizosphere competent strain of *Trichoderma harzianum. Plant Dis.* **80**: 736–741.

LORA, J. M., DE LA CRUZ, J., BENITEZ, T., LLOBELL, A., and PINTOR-TORO, J. A. 1994. A putative catabolic-repressed cell wall protein from the mycoparasitic fungus *Trichoderma harzianum. Molec. Gen. Genet.* **242**: 461–466.

LORA, J. M., DE LA CRUZ, J., BENITEZ, T., LLOBELL, A., and PINTOR-TORO, J. A. 1995. Molecular characterization and heterologous expression of an endo-β-1,6-glucanase gene from the mycoparasitic fungus *Trichoderma harzianum. Molec. Gen. Genet.* **247**: 639–645.

LORITO, M. 1995. Expression of genes from *Trichoderma harzianum* in transgenic plants. Fifth International *Trichoderma* and *Gliocladium* Workshop, April 1995, Beltsville, MD, abstract.

LORITO, M., HAYES, C. K., DI PIETRO, A., and HARMAN, G. E. 1993d. Biolistic transformation of *Trichoderma harzianum* and *Gliocladium virens* using plasmid and genomic DNA. *Curr. Genetics* **24**: 349–356.

LORITO, M., PETERBAUER, C., HAYES, C. K., and HARMAN, G. E. 1994b. Synergistic interaction between fungal cell wall-degrading enzymes and different antifungal compounds enhances inhibition of spore germination. *Microbiology* **140**: 623–629.

LORITO, M., WOO, S. L., DONZELLI, B., and SCALA, F. 1996d. Synergistic antifungal interactions of chitinolytic enzymes from fungi, bacteria and plants. In R. A. A. Muzzarelli (ed.), *Chitin Enzymology II.* Atec, Grottammare (AP), Italy, pp. 157–164.

LORITO, M., DI PIETRO, A., HAYES, C. K., WOO, S. L., and HARMAN, G. E. 1993c. Antifungal, synergistic interaction between chitinolytic enzymes from *Trichoderma harzianum* and *Enterobacter cloacae. Phytopathology* **83**: 721–728.

LORITO, M., FARKAŠ, V., REBUFFAT, S., BODO, B., and KUBICEK, C. P. 1996f. Cell-wall synthesis is a major target of mycoparasitic antagonism by *Trichoderma harzianum. J. Bacteriol.* **178**: 6382–6385.

LORITO, M., HAYES, C. K., DI PIETRO, A., WOO, S. L., and HARMAN, G. E. 1994a. Purification, characterization and synergistic activity of a glucan 1,3-β-glucosidase and an *N*-acetyl-β-glucosaminidase from *Trichoderma harzianum. Phytopathology* **84**: 398–405.

LORITO, M., WOO, S. L., FILIPPONE, E., COLUCCI, G., and SCALA, F. 1996b. Expression in plants of genes from mycoparasitic fungi – a new strategy for biological control of fungal diseases. International Union of Microbiological Societies (IUMS) Congresses, August 1996, Jerusalem, Israel, abstract.

LORITO, M., HAYES, C. K., PETERBAUER, C., TRONSMO, A., KLEMSDAL, S., and HARMAN, G. E. 1993a. Antifungal chitinolytic enzymes from *T. harzianum* and *G. virens*: purification, characterization, biological activity and molecular cloning. In R. A. A. Muzzarelli (ed.), *Chitin Enzymology.* European Chitin Society, Lyon and Ancona, pp. 383–392.

LORITO, M., PETERBAUER, C., SPOSATO, P., MACH, R. L., STRAUSS J., and KUBICEK, C. P. 1996a. Mycoparasitic interaction relieves binding of the Cre1 carbon catabolite repressor protein to promoter sequences of the *ech-42* (endochitinase-encoding) gene in *Trichoderma harzianum. Proc. Natl. Acad. Sci. USA* **93**: 14868–14872.

LORITO, M., WOO, S. L., HARMAN, G. E., SPOSATO, P., MUCCIFORA, S., and SCALA, F. 1996e. Genes encoding for chitinolytic enzymes from biocontrol fungi – applications for plant disease control. In R. A. A. Muzzarelli (ed.), *Chitin Enzymology II.* Atec, Grottammare (AP), Italy, pp. 95–102.

LORITO, M., HARMAN, G. E., HAYES, C. K., BROADWAY, R. M., TRONSMO, A., WOO, S. L., and DI PIETRO, A. 1993b. Chitinolytic enzymes produced by

Trichoderma harzianum: antifungal activity of purified endochitinase and chitobiosidase. *Phytopathology* **83**: 302–307.

LORITO, M., HAYES, C. K., ZOINA, A., SCALA, F., DEL SORBO, G., WOO, S. L., and HARMAN, G. E. 1994c. Potential of genes and gene products from *Trichoderma* sp. and *Gliocladium* sp. for the development of biological pesticides. *Molec. Biotechnol.* **2**: 209–217.

LORITO, M., WOO, S. L., D'AMBROSIO, M., HARMAN, G. E., HAYES, C. K., KUBICEK, C. P., and SCALA, F. 1996c. Synergistic interaction between cell wall degrading enzymes and membrane affecting compounds. *Molec. Plant–Microbe Interact.* **9**: 206–213.

LUMSDEN, R. D. and LEWIS, J. A. 1989. Selection, production, formulation and commercial use of plant disease biocontrol fungi: problems and progress. In J. M. Whipps and R. D. Lumsden (eds), *Biotechnology of Fungi for Improving Plant Growth.* Cambridge University Press, Cambridge, pp. 171–190.

MARGOLLES-CLARK, E., HARMAN, G. E., and PENTTILÄ, M. 1996b. Enhanced expression of endochitinase in *Trichoderma harzianum* with the *cbh1* promoter of *Trichoderma reesei. Appl. Environ. Microbiol.* **62**: 2152–2155.

MARGOLLES-CLARK, E., HAYES, C. K., HARMAN, G. E., and PENTTILÄ, M. 1996a. Improved production of *Trichoderma harzianum* endochitinase by expression in *Trichoderma reesei. Appl. Environ. Microbiol.* **62**: 2145–2151.

MONREAL, J. and REESE, E. T. 1969. The chitinase of *Serratia marcescens. Can. J. Microbiol.* **15**: 689–696.

MUTHUMEENAKSHI, S., MILLS, P. R., BROWN, A. E., and SEABY, D. A. 1994. Intraspecific molecular variation among *Trichoderma harzianum* isolates colonizing mushroom compost in the British Isles. *Microbiology* **140**: 769–777.

MUZZARELLI, R. A. A. 1993. Advances in *N*-acetyl-β-D-glucosaminidase. In R. A. A. Muzzarelli (ed.), *Chitin Enzymology.* European Chitin Society, Lyon and Ancona, pp. 357–374.

OVERDIJK, B. and VAN STEIJN, G. J. 1994. Chitinase in humans. *Glycobiology* **4**: 797–803.

PACHENARI, A. and DIX, N. J. 1980. Production of toxins and wall degrading enzymes by *Gliocladium roseum. Trans. Br. Mycol. Soc.* **74**: 561–566.

PETERBAUER, C., LORITO, M., HAYES, C. K., HARMAN, G. E., and KUBICEK, C. P. 1996. Molecular cloning and expression of the *nag1* (*N*-acetyl-β-D-glucosaminidase-encoding) gene from *Trichoderma harzianum* P1. *Curr. Genet.* **30**: 325–331.

PROKKOLA, S. 1992. Antagonistic properties of *Trichoderma* species against *Mycocentrospora acerina.* In D. F. Jensen, J. Hockenhull, and N. Fokkema (eds), *New Approches in Biological Control of Soil-borne Diseases.* IOBC: WPRS Bulletin, pp. 76–79.

QUIVEY, R. G. and KRIGER, P. S. 1993. Raffinose-induced mutanase production from *Trichoderma harzianum. FEMS Microbiol. Lett.* **11**: 307–312.

REESE, E. T. and MANDELS, M. 1959. β-1,3-glucanases in fungi. *Can. J. Microbiol.* **5**: 173–185.

REYES, F., CALATAYUD, J., and MARTINEZ, M. J. 1989. Endochitinase from *Aspergillus nidulans* implicated in the autolysis of its cell wall. *FEMS Microbiol. Lett.* **60**: 119–124.

RIDOUT, C. J., COLEY-SMITH, J. R., and LYNCH, J. M. 1988. Fractionation of extracellular enzymes from a mycoparasitic strain of *Trichoderma harzianum. Enzyme Microb. Technol.* **10**: 180–187.

ROBBINS, P. W., ALBRIGHT, C., and BENFIELD, B. 1988. Cloning and expression of a *Streptomyces plicatus* chitinase (chitinase-63) in *Escherichia coli. J. Biol. Chem.* **263**: 443–447.

ROBERTS, D. P. and LUMSDEN, R. D. 1990. Effect of extracellular metabolites from *Gliocladium virens* on germination of sporangia and mycelial growth of *Pythium ultimum. Phytopathology* **80**: 461–465.

ROBERTS, W. K., LAUE, B. E., and SELITRENNIKOFF, C. P. 1988. Antifungal proteins from plants. *Annals N.Y. Acad. Sci.* **544**: 144–149.

ROBY, D. and ESQUERRE-TUGAYE, M.-T. 1987. Induction of chitinases and of translatable mRNA for these enzymes in melon plants infected with *Colletotrichum lagenarium. Plant Sci.* **52**: 175–185.

ROUSSEAU, A., BENHAMOU, N., CHET, I., and PICHE, Y. 1996. Mycoparasitism of the extramatrical phase of *Glomus intraradices* by *Trichoderma harzianum. Phytopathology* **86**: 434–443.

SAHAI, A. S. and MANOCHA, M. S. 1993. Chitinases of fungi and plants: their involvement in morphogenesis and host–parasite interaction. *FEMS Microbiol. Rev.* **11**: 317–338.

SAMUELS, G. J. 1996. *Trichoderma*: a review of biology and systematics of the genus. *Mycol. Res.* **100**: 923–935.

SCHIRMBÖCK, M., LORITO, M., WANG, Y.-L., HAYES, C. K., ARISAN-ATAC, I., SCALA, F., HARMAN, G. E., and KUBICEK, C. P. 1994. Parallel formation and synergism of hydrolytic enzymes and peptaibol antibiotics, molecular mechanisms involved in the antagonistic action of *Trichoderma harzianum* against phytopathogenic fungi. *Appl. Environ. Microbiol.* **60**: 4364–4370.

SCHREMPF, H., BLAAK, H., SCHNELLMANN, J., and STOCH, S. 1993. Chitinases from *Streptomyces olivaceoviridens* cloning and sequencing of two genes and analysis of their overproduced gene products. In R. A. A. Muzzarelli (ed.), *Chitin Enzymology*. European Chitin Society, Lyon and Ancona, pp. 409–416.

SELA-BUURLAGE, M. B., PONSTEIN, A. S., BRES-VLOEMANS, S. A., MELCHERS, L. S., VAN DEN ELZEN, P. J. M., and CORNELISSEN, B. J. C. 1993. Only specific tobacco chitinases and *β*-1,3-glucanases exhibit antifungal activity. *Plant Physiol.* **101**: 857–863.

SMITH, R. J. and GRULA, E. A. 1983. Chitinase is an inducible enzyme in *Beauveria bassiana. J. Invert. Pathol.* **42**: 319–326.

SPELLIGT, T., BOTTIN, A., and KAHMANN, R. 1996. Green fluorescent protein (GFP) as a new vital marker in the phytopathogenic fungus *Ustilago maydis. Molec. Gen. Genet.* **252**: 503–509.

SPERANZINI, D., CASLE, A., RINKER, D., and ALM, G. 1995. Genetic variation among *Trichoderma* isolates from North American mushroom farms. Fifth International *Trichoderma* and *Gliocladium* Workshop, April 1995, Beltsville, MD, abstract.

ST LEGER, R. J., COOPER, R. M., and CHARNLEY, A. K. 1986. Cuticle-degrading enzymes of entomopathogenic fungi: regulation of production of chitinolytic enzymes. *J. Gen. Microbiol.* **132**: 1509–1517.

TRONSMO, A. 1991. Biological and integrated controls of *Botrytis cinerea* on apple with *Trichoderma harzianum. Biol. Control* **1**: 59–62.

TRONSMO, A. and HARMAN, G. E. 1992. Coproduction of chitinases and biomass for biological control by *Trichoderma harzianum* on media containing chitin. *Biol. Control* **2**: 272–277.

TRONSMO, A. and HARMAN, G. E. 1993. Detection and quantification of *N*-acetyl-*β*-D-glucosaminidase, chitobiosidase and endochitinase in solution and on gels. *Anal. Biochem.* **208**: 74–79.

TRONSMO, A., KLEMSDAL, S. S., HAYES, C. K., LORITO, M., and HARMAN, G. E. 1993. The role of hydrolytic enzymes produced by *Trichoderma harzianum* in biological control of plant diseases. In P. Suominen and T. Reinikainen (eds), Trichoderma reesei *Cellulases and Other Hydrolases: Enzyme Structure, Biochemistry, Genetics and Applications*, Vol. 8, Foundation for Biotechnical and Industrial Fermentation Research, pp. 159–168.

TRONSMO, A., HJELJORD, L., KLEMSDAL, S. S., VAARUM, K. M., NORDTVEIT-HJERDE, R., and HARMAN, G. E. 1996. Chitinolytic enzymes from the biocontrol agent *Trichoderma harzianum*. In R. A. A. Muzzarelli (ed.), *Chitin Enzymology II*. Atec, Grottammare (AP), Italy, pp. 235–244.

TURÓCZI, G., FEKETE, C., KERÉNYI, Z., NAGY, R., POMAZI, A., and HORNOK, L. 1996. Biological and molecular characterisation of potential biocontrol strains of *Trichoderma*. *J. Basic. Microbiol.* **36**: 63–72.

ULHOA, C. J. and PEBERDY, J. F. 1991a. Purification and characterization of an extracellular chitobiase from *Trichoderma harzianum*. *Curr. Microbiol.* **23**: 285–289.

ULHOA, C. J. and PEBERDY, J. F. 1991b. Regulation of chitinase synthesis in *Trichoderma harzianum*. *J. Gen. Microbiol.* **137**: 2163–2169.

ULHOA, C. J. and PEBERDY, J. F. 1992. Purification and some properties of the extracellular chitinase produced by *Trichoderma harzianum*. *Enzyme Microb. Technol.* **14**: 236–240.

USUI, T., MATSUI, H., and ISOBE, K. 1990. Enzymic synthesis of useful chito-oligosaccharides utilizing transglycosylation by chitinolytic enzymes in a buffer containing ammonium sulfate. *Carbohydr. Res.* **203**: 65–77.

VANDER, P. and MOERSCHBACHER, M. 1993. Chitin oligomers produced by HF-solvolysis induce resistance reactions in higher plants. In R. A. A. Muzzarelli (ed.), *Chitin Enzymology*. European Chitin Society, Lyon and Ancona, pp. 437–440.

VASSEUR, V., VAN MONTAGU, M., and GOLDMAN, G. H. 1995. *Trichoderma harzianum* genes induced during growth on *Rhizoctonia solani* cell wall. *Microbiology* **141**: 767–774.

WATANABE, T., KOBORI, K., MIYASHITA, K., FUJII, T., SAKAI, H., UCHIDA, M., and TANAKA, H. 1993. Identification of glutamic acid 204 and aspartic acid 200 in chitinase A1 of *Bacillus circulans* WL-12 as essential residues for chitinase activity. *J. Biol. Chem.* **268**: 18567–18572.

WEBB, E. C. 1992. *Enzyme Nomenclature*. Academic Press, San Diego, 863.

WIRTH, S. J. and WOLF, G. A. 1992. Micro-plate colourimetric assay for endo-acting cellulase, xylanase, chitinase, 1,3-β-glucanase and amylase extracted from forest soil horizons. *Soil Biol. Biochem.* **24**: 511–519.

YANAI, K., TAKAYA, N., KOJIMA, N., HORIUCHI, H., OHTA, A., and TAKAGI, M. 1992. Purification of two chitinases from *Rhizopus oligosporus* and isolation and sequencing of the encoding gene. *J. Bacteriol.* **174**: 7398–7406.

ZHU, Q., MAHER, E. A., MASOUD, S., DIXON, R. A., and LAMB, C. J. 1994. Enhanced protection against fungal attack by constitutive co-expression of chitinase and glucanase genes in transgenic tobacco. *Bio/Technology* **12**: 807–812.

5

Glucanolytic and other enzymes and their genes

T. BENÍTEZ, C. LIMÓN, J. DELGADO-JARANA and M. REY

Departamento de Genética, Universidad de Sevilla, Sevilla, Spain

5.1 Introduction

Most of the *Trichoderma* and *Gliocladium* strains described in the scientific literature have been isolated either as potential biocontrol agents against fungal plant pathogens or because of their cellulolytic and hemicellulolytic activities. Chitinases and glucanases have been frequently studied because of their ability to degrade fungal cell walls during biocontrol (see Chapter 3), although other cell wall degrading enzymes such as proteases, lipases, and phosphatases may also be involved. This chapter deals with the properties, enzymology and genetics of the glucanases and other lytic enzymes produced by *Trichoderma* and *Gliocladium* spp.

Glucans are homopolymers of D-glucose linked in an α and/or β configuration. Some are simple molecules consisting of linear chains of glucosyl residues joined by a single linkage type, while others are more complex and consist of linkages in either linear or branched chains (Pitson *et al.*, 1993). However, many other glucans and glucan-degrading enzymes are produced by different organisms.

Both α- and β-glucans are widespread in nature. The enzymes degrading the polymers that accumulate in the largest amounts, cellulose and hemicellulose, are dealt with in Chapters 1 and 2 and are therefore not covered here. Among the others, glycogen – an α-1,4-glucan that serves as the principal cellular storage polysaccharide – appears to be universally distributed in fungi, except for oomycetes, which have β-1,3-glucans as their main storage carbohydrate (Table 5.1; Pitson *et al.*, 1993). Glucans also occur as constituents of cell walls and as extracellular polysaccharides (Bartnicki-García, 1968; Sentandreu *et al.*, 1994).

Glucanolytic enzymes are widely distributed among higher plants, bacteria, and fungi and have various physiological functions depending on their source. In plants, they are considered to act as part of the defense system against fungal pathogens (Mauch *et al.*, 1988), but additional roles, such as in cell differentiation, also have been suggested (Fincher, 1989). In bacteria, a nutritional function has been documented (Watanabe *et al.*, 1992). A number of different functions have been discussed for fungal β-glucanases: mobilization of cell wall glucans and storage carbohydrates under starvation conditions (Kuhn *et al.*, 1990); degradation of callose in plants by

Table 5.1 Role of fungal glucanases

Type	Origin	Role	Example
β-1,3-glucanase	Oomycetes	nutritional	Mobilization of β-1,3-glucan storage carbohydrate (Griffin, 1994)
		morphogenesis	Degradation of β-1,3-glucan cell wall components (Bartnicki-García, 1968)
	Ascomycetes	morphogenesis	Synthesis of *Neurospora crassa* cell walls (Chiba *et al.*, 1988)
			Saccharomyces cerevisiae sporulation (San Segundo *et al.*, 1993)
		nutritional	Mobilization of β-1,3-glucan in *Penicillium* (Santos *et al.*, 1978a)
	Basidiomycetes	nutritional	Mobilization of *Schizophyllum commune* storage carbohydrate (Griffin, 1994)
		morphogenesis	Differentiation of fruiting bodies *Agaricus bisporus* (Griffin, 1994)
			Synthesis *S. commune* cell walls (Kuhn *et al.*, 1990)
	Deuteromycetes	antifungal	Lysis of host cell walls in mycoparasitism: *Trichoderma* (De la Cruz *et al.*, 1993), *Stachybotrys* (Tweddell *et al.*, 1994)
		morphogenesis	Degradation of β-1,3-glucan of cell walls in *Trichoderma* (Benítez *et al.*, 1976), *Candida* (Chambers *et al.*, 1993)
α-1,3-glucanase	Ascomycetes	nutritional	Mobilization of *Aspergillus nidulans* α-1,3-glucan storage carbohydrate (Zonneveld, 1972)
		morphogenesis	Degradation of *A. nidulans* α-1,3-glucan from cleistothecium cell walls (Zonneveld, 1972)
	Basidiomycetes	nutritional	
		morphogenesis	Degradation of *S. commune* cell walls (Kuhn *et al.*, 1990)
	Deuteromycetes	antifungal	Lysis of host cell walls in mycoparasites (De Vries and Wessels, 1973)

Table 5.1 *(Cont)*

Type	Origin	Role	Example
α-1,4-glucanase (α-amylase)	Oomycetes	nutritional	Mobilization of *Mucor* starch storage carbohydrate (Radford *et al.*, 1996)
	Ascomycetes	nutritional	Mobilization of starch storage carbohydrate (Yamamoto, 1988)
		morphogenesis	*Saccharomicopsys fibulligera* sporulation specific (Radford *et al.*, 1996)
	Basidiomycetes	nutritional	Mobilization of starch storage carbohydrate (Yamamoto, 1988)
glucoamylase (exo α-1,4; α-1,3; α-1,6)	Zygomycetes	nutritional	Mobilization of *Rhizopus* storage carbohydrate (Radford *et al.*, 1996)
	Ascomycetes	nutritional	Mobilization of *Aspergillus* spp. and *S. cerevisiae* storage carbohydrate (Radford *et al.*, 1996)
		morphogenesis	Synthesis of *Aspergillus* cell walls (Bobbit *et al.*, 1977)
	Deuteromycetes	nutritional	Mobilization of storage carbohydrate in *T. reesei* (Fagerström and Kalkkinen, 1995)
β-1,2-glucanase	Ascomycetes	nutritional	Degradation of β-1,2-glucan in *Aspergillus* spp. (Reese *et al.*, 1961)
	Deuteromycetes	nutritional	Degradation of β-1,2-glucan in *Fusarium* spp. (Reese *et al.*, 1961)
β-1,4-glucanase	Oomycetes	morphogenesis	Synthesis of *Achbya* cell walls (Griffin, 1994)
	Ascomycetes	nutritional	Degradation of cellulose in *Aspergillus* growing in wood (Radford *et al.*, 1996)
	Basidiomycetes	nutritional	Degradation of cellulose in *A. bisporus* growing in wood (Radford *et al.*, 1996)
	Deuteromycetes	nutritional	Degradation of cellulose in *T. reesei* (Penttilä *et al.*, 1986)
β-1,6-glucanase	Ascomycetes	morphogenesis	Synthesis of β-1,6-glucan of cell walls in *S. cerevisiae* (Cid *et al.*, 1995)
	Basidiomycetes	morphogenesis	Synthesis of *S. commune* cell walls (Kuhn *et al.*, 1990)
	Deuteromycetes	antifungal	Lysis of host cell walls in mycoparasites (De la Cruz *et al.*, 1995a)

phytopathogens; nutrition of saprophytes; and involvement in the mechanism of attack and nutrition of mycoparasites (Chet, 1987; see Table 5.1). Glucanases involved in morphogenesis may be intracellular or wall-bound, are usually characterized by a high K_m and are either formed constitutively or are developmentally regulated. Metabolic hydrolases are preferentially extracellular, display a low K_m, are regulated by nutrient parameters and usually do not accumulate within the mycelia (Friebe and Holldorf, 1975; Griffin, 1994).

5.2 Glucanolytic enzymes and their genes

5.2.1 *β-Glucanases*

β-Glucanases are usually distinguished according to the type of hydrolysis (e.g. exo vs. endo) and by the type of linkages hydrolysed (α- or β-linkage; and 1,2-, 1,3-, 1,4- or 1,6-linkage). Exo-activity is defined by the ability to hydrolyze β-glucans from the non-reducing end yielding mono- or dimers as sole end-products, whereas endo-activity is characterized by a higher specific activity than the former, and by the formation of oligosaccharides. β-Glucanases specific only for the β-1,3-linkage frequently are reported in the literature; β-1,6-glucanases have rarely been detected in microorganisms. Most exo-glucanases display β-1,3- as well as β-1,6- activities (Vázquez de Aldana *et al.*, 1991; Yamamoto and Nagasaki, 1975).

Purification and properties of β-glucanase enzymes

The β-glucanolytic system of *Trichoderma* and *Gliocladium* consists of endo- and exoglucanases which synergistically catalyze the degradation of β-glucans to oligomers. Final hydrolysis is aided by β-glucosidase. However, knowledge about the synergism and regulation of these enzymes is hampered by the fact that there are several components of each enzyme that differ in several properties, including their M_r (Table 5.2). It is not known whether these different enzymes are different gene products or results of post-translational modification, since different components have been derived from different strains (i.e., *T. harzianum* P1 and CECT 2413, which are not the same species) (C. P. Kubicek, personal communication), conditions for enzyme production were not the same, and different methods have been applied to estimate the M_r. However, sera raised against one endo-β-1,3- and one endo-β-1,6-glucanase from *T. harzianum* strain CECT 2413 reacted with β-glucanases from various other *Trichoderma* species (M. Rey, unpublished), showing that at least these two enzymes may commonly occur in *Trichoderma.* Isolation of the genes encoding the different β-glucanases is therefore a prerequisite to establish unequivocally how many different β-glucanases are produced by *Trichoderma.*

Most workers employed standard chromatographic techniques for β-glucanase purification. Dubordieu *et al.* (1985) separated two exo-β-1,3-glucanases, one endo-β-1,6-glucanase and a single β-glucosidase from a commercial enzyme preparation of *T. harzianum.* One of the two exo-β-1,3-glucanases with an M_r of 40 kDa constituted the major portion of activity and was able to digest the β-1,3/β-1,6

Table 5.2 Properties of purified glucanases from different *Trichoderma* species

Enzyme	Origin	M_r	pI	K_m [a]	V_{max} [b]	Type of enzyme	Hydrolysis
β-1,3-glucanase	*T. harzianum*	78	8	3.3	75	Endo β-1,3	G4 + G2 + G
	T. harzianum	78	6.2	ND	ND	Exo β-1,3	G1
	T. harzianum	36	ND	1.18	1.26	Endo β-1,3	ND
	T. harzianum	ND	ND	0.285	ND	β-1,3	ND
	T. harzianum	31.5	ND	2.08	620	Exo β-1,3	G1
	T. harzianum	40	7.8	0.015	0.3	Exo β-1,3	G1
	T. longibrachiatum	70	7.2	0.065 mM	3.17	Endo β-1,3	G5 + G4 + G3 + G1 G1
	T. reesei	70	4.2	0.28	ND	Exo β-1,3 β-1,6	G1
β-1,6-glucanase	*T. harzianum*	43	5.8	2.4	224	Endo β-1,6	G2
	T. harzianum	51	ND	0.8	3.2	Endo β-1,6	G2 + G
α-1,3-glucanase	*T. viride*	47	ND	0.046 M	0.16	Endo α-1,3	G2 + G1
	T. viride	47	ND	7.1 mM	ND	Exo α-1,3	G1
	T. harzianum	15	ND	ND	ND	Endo α-1,3	G1
glucoamylase	*T. reesei*	66	4.0	0.11	ND	Exo α-1,4, β-1,6	G1

[a] mg substrate per ml where molarity is not indicated.
[b] μmol glucose per min per mg of protein.
ND, non described; Gn indicates glucose oligomers of n chain length.

glucans (cinerean) of *Botrytis cinerea.* Fractionation of a crude preparation of *T. viride* by chromatography resulted in a preparation of β-glucanase-free amyloglucosidase and amyloglucosidase-free β-glucanases, which may be useful in the analysis of the β-glucans in barley and in malted barley. The β-glucanases consisted of two endo-β-1,3-glucanases and one exo-β-1,3-glucanase with isoelectric points (pI) of 5.25, 4.95 and 4.20, respectively (Thomas *et al.*, 1983). From another commercial enzyme preparation of *T. viride*, called "Onozuka", Bielecki and Galas (1977) purified two β-1,3-glucanases: an acidic non-lytic β-1,3-glucanase that hydrolyzed laminarin (β-1,3-glucan) and another that showed lytic activity on cell walls of *Saccharomyces cerevisiae.* Also, De Vries and Wessels (1973) partly purified an exo-β-1,3-glucanase produced by another strain of *T. viride* during growth on *Schizophillum commune* cell walls. However, only the capacity of the enzyme to liberate protoplasts was described. Tangarone *et al.* (1989) compared the β-glucanases produced by *T. longibrachiatum* grown on *Agaricus bisporus* with those secreted in the presence of glucose and found two different enzymes. The enzyme produced on glucose had a K_m for laminarin 80-fold lower than that produced on the basidiomycete cell walls, and the authors suggested therefore that the former may be involved in morphogenesis.

The β-glucanase enzyme system of the mycoparasite *T. harzianum* CECT 2413 has recently been investigated in more detail; it was shown to consist of one basic and at least three acidic isozymes, which could be separated by isoelectric focusing (De la Cruz *et al.*, 1995b). The basic β-1,3-glucanase (β-1,3-glucanase I) BGN13.1 was purified to homogeneity by adsorption to pustulan (β-1,6-/β-1,3-glucan). It had an M_r of 78 kDa (De la Cruz *et al.*, 1995b) and was not glycosylated. The enzyme showed maximal activity on *S. cerevisiae* cell walls and laminarin (K_m 3.3 mg/ml for the latter). The enzyme was specific for β-1,3-linkages and exhibited an exo-type of action. Antisera raised against β-glucanases from plants and fungi did not cross-react with any of the β-1,3-glucanase enzymes from *T. harzianum*, and antiserum raised against BGN13.1 did not react with any other β-glucanase, thus suggesting this enzyme is unique to *Trichoderma. T. harzianum* CECT 2413 also produced at least two extracellular endo-β-1,6-glucanases of 51 and 43 kDa that were also purified to homogeneity. The 43 kDa β-1,6-glucanase (BGN16.2) was specific for β-1,6-linkages, exhibited an endo-type of action, and was devoid of bound carbohydrates (De la Cruz *et al.*, 1995b). Western analysis with culture filtrates from several other *Trichoderma* species (e.g., *T. viride, T. longibrachiatum, T. reesei, T. koningii* and *T. virens*) indicated that BGN16.2 was formed by all of these and under the same conditions (M. Rey, unpublished).

Several other glucanases with different catalytic activities, molecular weights and substrate specificities have been found in supernatants from *T. harzianum* cultures, but it is not known whether they are differently processed gene products from the same gene or from separate genes. Noronha and Ulhoa (1996) also detected β-1,3-glucanases produced by another strain of *T. harzianum* when it was grown in the presence of chitin or isolated fungal cell walls. An acidic endo-β-1,3-glucanase of 36 kDa was purified and was active towards laminarin. Other authors described two exo-β-1,3-glucanases of 32 and 78 kDa from *T. harzianum* (Kitamoto *et al.*, 1987; Lorito *et al.*, 1994) and one extracellular exo-β-1,3-glucanase of 70 kDa from a strain of *T. reesei* (Bamforth, 1980). These β-1,3-glucanases from *T. harzianum* and *T. reesei* behaved as exo-enzymes that hydrolyzed β-1,3-linkages and, less efficiently, β-1,6-linkages (Table 5.2).

β-glucanase genes

Two genes encoding glucanolytic enzymes have been cloned from *T. harzianum* CECT 2413: one gene that encodes the 78 kDa endo-β-1,3-glucanase (β-1,3-glucanase I) and the other that encodes the 43 kDa endo-β-1,6-glucanase (β-1,6-glucanase II) (see above and Table 5.2). Cloning of the β-1,3-glucanase-encoding gene (*bgn13.1*), which represents the major β-1,3-glucanase activity of this strain, was carried out using amino acid sequence information of the purified protein. There appeared to be only one copy of *bgn13.1* in the genome of *T. harzianum* as shown by Southern hybridization. The mature protein consisted of 728 amino acids, which was preceded by an N-terminal 34 amino acid long preprosequence, terminating in a KR and thus most probably cleaved by a Kex2-like processing peptidase (De la Cruz *et al.*, 1995b). This preprosequence also was correctly processed and secreted when expressed in yeast (Lora *et al.*, 1995). A comparison of amino acid homologies with other β-glucanases (Chen *et al.*, 1993) also identified a glutamic residue at the putative active site. No evidence for the presence of a substrate binding domain, as present in cellulases and some glucanases (see Chapter 1), was obtained. In tobacco class I chitinases, the N-terminal cysteine-rich domain, which is essential for chitin binding (Iseli *et al.*, 1993), is flanked by imperfect direct repeats of 9–10 bp length. It has been suggested that these domains arose from a common ancestral gene and were introduced by transposition events (Shinshi *et al.*, 1990). In *Myxococcus xanthus*, evidence for independent acquisition of the binding and catalytic domains of the *celA* gene (encoding a β-1,4-endoglucanase) by horizontal gene transfer from actinomycetes, was presented (Quillet, 1995). This transpositional origin of the binding domain in some glucanases would explain its absence in others such as the β-glucanases from *Trichoderma*.

Comparison of the sequence of *bgn13.1* with those present in the databases showed few significant homologies with α- or β-glucanases, so the gene represented a new class of β-1,3-glucanases. Alignment of *bgn13.1* with sequences from bacteria, yeasts, filamentous fungi and plants indicated that this gene shared none of the conserved domains present in β-glucanases from other sources (De la Cruz *et al.*, 1995b). However, limited homology was detected with a β-1,3-glucanase gene from *Cochliobolus carbonum*. This was substantiated by a phylogenetic tree of more than 60 glucanase gene sequences, which documents that *bgn13.1* formed an outgroup distant from other β-glucosidase (Barnett *et al.*, 1991) or β-1,6-glucanase-encoding genes (De la Cruz *et al.*, 1995b), but was rather related to *Trichoderma* cellulases (Figure 5.1).

The gene encoding the major β-1,6-glucanase (*bgn16.2*) from *T. harzianum* CECT 2413 was isolated by the aid of degenerate oligonucleotides derived from the amino acid sequences of tryptic fragments of the purified protein. In contrast to the *bgn13.1* gene, Southern analysis revealed the presence of three gene copies with nearly identical sequences in the genome of *T. harzianum* (Lora *et al.*, 1995). The deduced protein was also N-terminally preceded by a 17 amino acid prepropeptide terminating in a KR motif (De la Cruz *et al.*, 1995a; Lora *et al.*, 1995). The protein sequence showed partial homology with EXG1 and SPR1 (a sporulation specific exo-β-1,3-glucanase that contributed to ascospore thermoresistance) from *S. cerevisiae* (Muthukumar *et al.*, 1993; San Segundo *et al.*, 1993) and with an exo-β-1,3-glucanase of *Candida albicans* (Chambers *et al.*, 1993). Several strongly conserved areas were detected, including one forming a putative catalytic site (IEVLNEP) that was present in endo- as well as exo-β-glucanases of different origin. The deduced sequence contained several putative glycosylation sites, but a comparison of the migration of the purified protein in SDS-PAGE with the deduced M_r and treatment of the enzyme with EndoH revealed only little covalently attached carbohydrate and sequencing of internal peptides of the second endo-β-1,6-glucanase purified from this strain (De la Cruz, unpublished).

Regulation and function of β-glucanases

In *Trichoderma* and *Gliocladium* spp. (De la Cruz *et al.*, 1993), as in many other filamentous fungi (Griffin, 1994), β-glucanases are mainly controlled by induction and carbon catabolite repression. The question of how insoluble molecules like β-1,3-glucan can induce enzymes, catalyzing their own breakdown, remains unanswered. By analogy to the cellulase system (Kubicek, 1987; Kubicek *et al.*, 1993; see also Chapter 3), biochemical and molecular biological approaches would be consistent with a model in which conidial-bound exoglucanases first attack the β-1,3-glucan molecule. The disaccharides liberated are taken up by the mycelia and promote further glucanase biosynthesis. Culture supernatants of *T. virens*, grown on different carbon sources including host fungal cell walls, contained β-glucanase, chitinase, lipase and proteinase activities. Chitin also induced β-1,3-glucanase from *T. longibrachiatum* (Tangarone *et al.*, 1989) and *Cochliobolus carbonum* (Schaeffer *et al.*, 1994). Substrates for β-1,3-glucanases and chitinases frequently occur together in nature (fungal cell walls), and the simultaneous induction of glucanases and chitinases by the substrate of only one of them is not unusual (Lorito *et al.*, 1994; Table 5.3). The relative activities can vary with the substrate, but the highest specific activ-

Table 5.3 Hydrolase activity (mU/mg protein) measured when *T. harzianum* CECT 2413 was cultivated in minimal medium with different carbon sources

Carbon source (%) (w/v)	β(1-3)	β(1-6)	Chitinase
Glucose 0.1%	10	15	0
Glucose 2%	75	10	0
Fructose 2%	160	0	0
Glycerol 2%	150	15	0
Chitin 1.5%	640	280	25
Laminarin 1%	5	0	0
Pustulan 1%	340	100	0
Nigeran 1%	540	315	8
B.c. C.W. 1%	1040	360	5
Yeast C.W. 1%	1000	185	5
P.c. C.W.	700	683	24
R.s. C.W.	748	600	17
T.h. C.W.	666	617	22

Modified from De la Cruz *et al.*, 1993.
C.W., cell walls; *B.c.*, *Botrytis cinerea*; *P.c.*, *Phytophthora citrophtora*; *R.s.*, *Rhizoctonia solani*; *T.h.*, *Trichoderma harzianum*.

ities of β-glucanases were observed during growth on host cell walls (van Tilburg and Thomas, 1993). SDS-PAGE of samples, harvested over an 8-day period, showed that the appearance of extracellular proteins was apparently influenced by induction/repression mechanisms, protein inactivation and protein degradation and that the detected enzyme activity was the result of different activities induced at different times throughout the process (van Tilburg and Thomas, 1993). In other fungi, the mechanism of protein inactivation was shown to depend on *de novo* protein synthesis and to be due to limited proteolysis and partial removal of the carbohydrate component (Friebe and Holldorf, 1975; Santos *et al.*, 1978b). Differential enzyme induction by cell walls from different fungi has been correlated with the capacity of *Trichoderma* to mycoparasitize these strains (Sivan and Chet, 1989). However, Ridout *et al.* (1986) showed that the protein profile secreted by *Trichoderma* during growth on isolated cell walls differed from the profile during direct antagonism with phytopathogenic fungi.

Some of the major β-glucanases have been investigated in detail. De la Cruz *et al.* (1993, 1995b) found that the production of extracellular β-1,3- and β-1,6-glucanases in different carbon sources by *T. harzianum* CECT 2413 was induced by chitin, nigeran (α-1,3-/α-1,4-glucan), pustulan (β-1,6-glucan) and fungal cell walls (Table 5.3) and repressed by glucose. Compounds interfering either with RNA or protein synthesis inhibited enzyme induction, thus suggesting pretranslational regulation of enzyme synthesis. A similar type of regulation was also observed in several other strains (*T. longibrachiatum*, *T. viride*, *T. koningii*, *T. reesei*, and *T. virens*). In most cases, however, basal β-glucanase activities were also observed in the presence of 2% glucose (M. Rey, unpublished; see also below) (Table 5.4).

A different picture of regulation was obtained, however, when the different β-glucanase isozymes were investigated separately (De la Cruz *et al.*, 1995b); all four isozymes were detected upon growth on chitin, laminarin and pustulan, whereas on fungal cell walls only the basic extracellular endo-β-1,3-glucanase was formed. Transcription of *bgn13.1*, which encodes this basic endo-β-1,3-glucanase, was not repressed by glucose, and its activity was detected during growth on 2% glucose (Table 5.5). Although this endo-β-1,3-glucanase alone was unable to form clearing halos when incubated with fungal cell walls, it inhibited the growth of phytopathogenic

Table 5.4 β-1,6-glucanase activity (mU/100 μg dry weight) when the indicated *Trichoderma* strains were cultivated in minimal medium with different carbon sources

Strains	Glucose 2%	Glucose 0.1%	Chitin 1%	*Bot.* C.W.	*P.* spp C.W.	*P.c.* C.W.	*S.c.* C.W.
T. harzianum CECT	3	3	35	42	30	30	32
T. viride	6	7	14	38	30	30	20
T. virens	1	3	7	35	19	33	15
T. harzianum IMI	0	4	7	20	23	23	16
T. longibrachiatum	1	4	1	25	13	10	11
T. koningii	3	8	2	26	15	20	12
T. reesei	0	4	2	21	10	50	12

M. Rey, unpublished.
Bot., *Botrytis cinerea*; *P.* spp, *Penicillium* spp; *P.c.*, *Phythophthora citrophtora*; *S.c.*, *Saccharomyces cerevisiae*; C.W., cell walls.

Table 5.5 Different regulation of the two chitinases (CHIT42 and CHIT33) and the two β-glucanases (BGN16.2 and BGN13.1) of *T. harzianum* CECT 2413 on various substrates

Enzyme		Glucose 2%	Glucose 0.1%	Chitin 1%	Pustulan	Laminarin	*R.s.* C.W.[a]	*S.c.* C.W.[b]
CHIT42	N–	+	++	ND	ND	+++	+++
	W–	++	++	ND	ND	+++	+++
CHIT33	N–	+++	+	–	ND	++	ND
	W–	+	+++	–	ND	++	ND
BGN16.2	N–	+++	++	–	ND	++	ND
	W–	–	+	++	ND	+	ND
BGN13.1	N–	–	++	–	+	++	+++
	A+	–	++	+++	++	+++	+++

N, northern experiments; W, western experiments; A, activity in gel.
[a] *R.s.*, *R. solani*; [b] *S.c.*, *S. cerevisiae*.
ND, not done.

fungi in combination with other hydrolases. Because of its specific induction and lytic activity, this enzyme has been thought to contribute as an antifungal enzyme during antagonism of *Trichoderma* against other fungi, although its implication in other saprophytic roles has not been discounted (De la Cruz *et al.*, 1995b).

The gene (*bgn16.2*) encoding the endo-β-1,6-glucanase of *T. harzianum* CECT 2413 was also repressed by glucose and induced by cell wall polymers. Only a single cDNA clone out of 80 000 was obtained for this endo-β-1,6-glucanase, and only very low levels of its mRNA were found, indicating that the corresponding gene is weakly expressed (Table 5.5). Endo-β-1,6-glucanase II produced clearing halos when incubated with yeast cell walls and inhibited the growth of other fungi in combination with other hydrolytic enzymes when used at very high concentrations (25 mg/ml for 70–80% inhibition) (De la Cruz *et al.*, 1995a).

Despite the regulatory patterns outlined above, some constitutively formed β-glucanases have also been decribed. Del Rey *et al.* (1979) reported on the presence of β-1,3- and β-1,6-glucanases both in cell-free extracts as well as in the culture medium of *T. viride* during growth on glucose. Resting cells, which were deprived of glucose, either failed to produce β-glucanases or produced them in low quantities. The authors interpreted these data as meaning that these β-glucanases are involved in morphogenesis (i.e., cell wall turnover). In support of this, Kubicek (1982) described the constitutive production of β-1,3-glucanases by strains of *T. pseudokoningii* and *T. aureoviride*; activity of β-1,3-glucanase correlated with the release of a cell wall bound β-glucosidase into the medium, suggesting that β-1,3-glucanase was involved in the mechanism of release of β-glucosidase from the cell walls during turnover of the β-glucan. However, in *T. pseudokoningii*, cell wall-associated β-glucanase activity was shown to be due to three isoenzymes, and it was unclear whether they all had the same or different functions (Kubicek, 1982). In *Penicillium italicum*, it was suggested that the wall-bound β-1,3-glucanases II and III may be involved in the growth and extension of the cell wall, whereas β-1,3-glucanase I could be directed towards a different function such as mobilization of reserve wall glucan or conidiation (Santos *et al.*, 1978a,b). It should be noted in this context that cell wall binding itself does not necessarily imply a morphogenetic function; β-glucosidase has been located in the outermost β-glucan layer of *T. reesei* cell walls and the plasma membranes, but its function is in the final step of cellulose hydrolysis (Cummings and Fowler, 1996).

Several different functions have been attributed to fungal β-glucanases, including nutritional (Table 5.6) and/or antifungal roles: the 78 kDa glucan-β-1,3-glucosidase (β-1,3-glucanase) of *T. harzianum* P1 (Lorito *et al.*, 1994) had antifungal activity that was synergistic with endochitinase and a chitin-β-1,4-chitobiase to inhibit spore germination and germ tube elongation in *B. cinerea.* Concentrations of 10–20 μg/ml of total protein was needed for complete inhibition of germination or germ tube elongation. Clarkson (1992) also described the production of antifungal β-1,3-glucanase by *T. harzianum.* In addition, isolates of *T. virens* produced β-1,3-glucanases and β-1,4-glucanases which affected cell wall integrity of the host and induced cytoplasmic leakage (Jeffries and Young, 1994). The role of β-glucanases and other lytic enzymes in mycoparasitism is dealt with in detail in Chapter 7.

β-Glucans form a substantial part of the cell walls of *Trichoderma* and *Gliocladium* spp. as well and are therefore subject to the action of the glucanolytic enzymes (De la Cruz *et al.*, 1993, 1995b). The question of how they overcome the lytic action of their own enzymes is not completely clear. Possible explanations include transport as inactive zymogenic forms to the outside (Adams *et al.*, 1993; Manocha and Balasubramaniam, 1988). Alternatively, the presence of structural components in the cell walls that inhibit these lytic enzymes has been discussed. For example, α-glucan displays the folding characteristics of a lamellar chain which renders it very resistant to enzymatic lysis (Kuhn *et al.*, 1990). Other components such as structural proteins could also play a protective role; Goldman *et al.* (1994) and Vasseur *et al.* (1995) isolated three gene clones encoding proteins of 69, 37 and 15.6 kDa by differential screening of a mycoparasitism-induced cDNA library of *T. harzianum.* The predicted amino acid sequence of the 15.6 kDa protein displayed motifs found in serine- and alanine-rich structural proteins, which can inhibit hydrolases. Also melanin provides protection from enzymatic lysis (Bull, 1970).

Several β-glucanases also seem to be relevant for the autolytic potential of fungi; their significance for processes such as turnover of cell wall components, survival of cells under conditions of nutrient deprivation, interaction with antimicrobial compounds that interfere with cell wall synthesis and related phenomena, has not been established yet. Morphogenetic changes in some filamentous fungi take place under carbon-limiting conditions and are accompanied by a sharp increase in several degradative enzymes of which β-glucanases have also been thought to supply energy from reserve β-glucans. Both β-1,3- and β-1,6-glucanases participate in the autolysis of the mycelium that takes place in aged hyphae and collaborate in the extension of the cell wall by acting on the structural glucan during normal growth (Santos *et al.*, 1977, 1978a,b).

5.2.2 α-Glucanases: purification, properties and function

α-1,3-glucanases

The hyphal walls of most fungi, including all basidio- and ascomycetes, contain an alkali-soluble α-1,3-glucan. Lytic enzyme preparations of *Trichoderma* strains grown

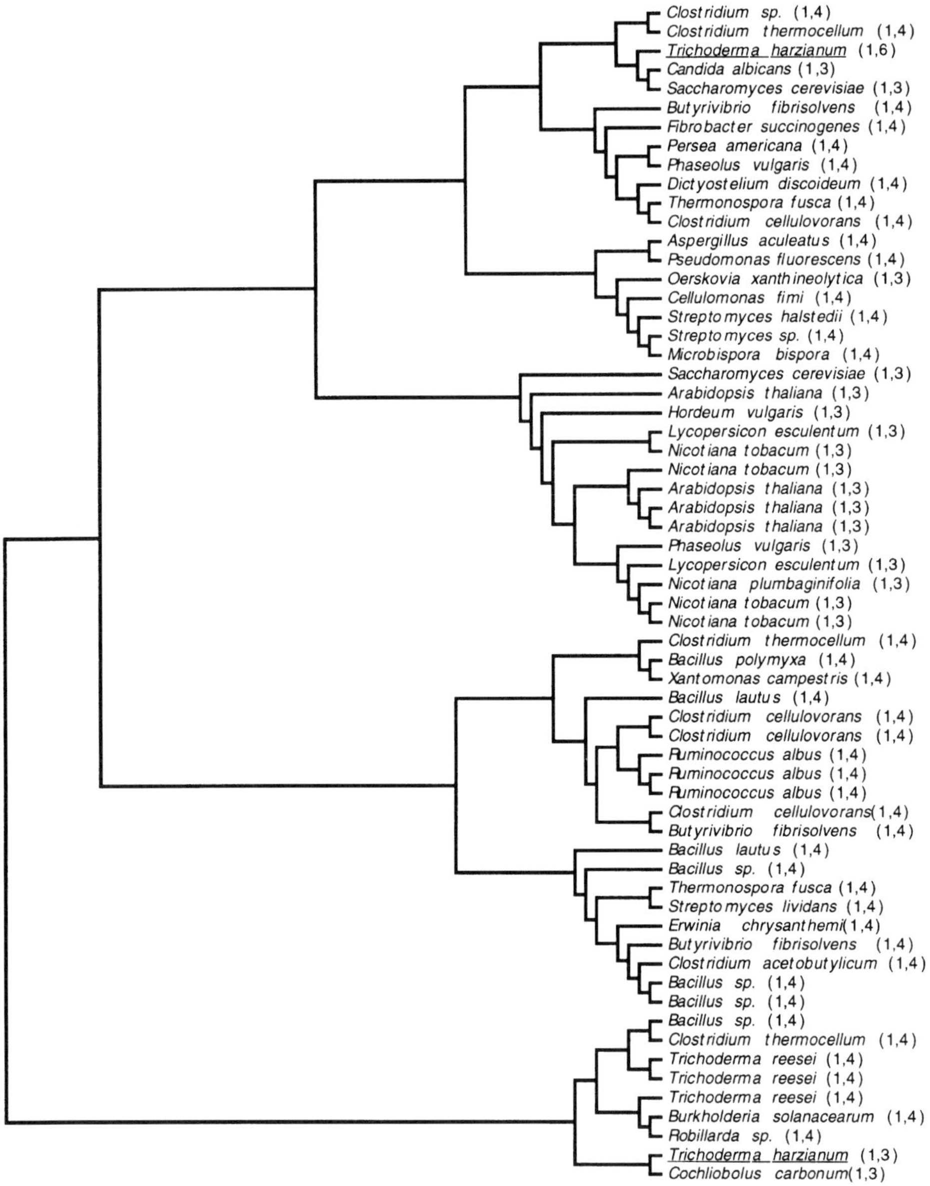

Figure 5.1 Phylogenetic tree formed by comparing *β*-glucanase sequences. Numbers in parentheses (1.3; 1.4; 1.6) indicate the type of linkage split by the enzyme. The *β*(1,3)- and *β*(1,6)-glucanases of *T. harzianum* are underlined.

on hyphal walls of some of these fungi are highly enriched in α-1,3-glucanases which, in combination with other hydrolases, release fungal protoplasts (De Vries and Wessels, 1973). These authors described purification of an endo-α-1,3-glucanase from *T. viride* BS754.33 grown on *S. commune* cell walls by CM-cellulose column chromatography at low pH and ionic strength. Tsunoda *et al.* (1977) described the purification of an exo-α-1,3-glucanase from a commercial cellulase preparation from

T. viride. The enzyme of 47 kDa had an optimal pH of 4.5, a K_m of 7.1 mM and liberated only glucose from pseudonigeran, an α-1,3-glucan from *Aspergillus niger*. Unfortunately, the activity of this enzyme on other substrates such as nigeran (α-1,3- plus α-1,4-glucan) or starch was not tested. A second peak, putatively representing an isozyme of the exo-α-1,3-glucanase, which appeared during DEAE-Sephadex anion exchange chromatography, was not investigated.

T. reesei also produces at least one extracellular endo-α-1,3-glucanase of 47 kDa when grown in a medium containing pseudonigeran. The enzyme had rather strict structural requirements hydrolyzing exclusively α-1,3- and α-1,4-linked polymers, but not starch or dextran, which establishes the enzyme as a highly specific tool for the analysis of polysaccharides. Its K_m (10^{-2} M) was unusually high for an enzyme involved in carbon nutrition (Hasegawa and Nordin, 1969). Formation of the enzyme was repressed by glucose.

Quivey and Kriger (1993) also described a 15 kDa endo-α-1,3-glucanase in *T. harzianum* strain OMZ779, induced by raffinose, mutan (a glucose polymer containing α-1,3- and α-1,6-linkages) and light, although its function is so far unknown.

Amylases

α- and β-amylases and glucoamylases catalyze the hydrolysis of α-1,4-linkages of starch and glycogen. The respective glucose monomers are α-1,4-linked, producing a linear chain which represents the soluble form of starch, i.e., amylose (Radford *et al*., 1996). A second component contains also α-1,6-branches, which results in a poorly soluble molecule, amylopectin. Fungal starch degradation makes use of a synergism of two types of enzymes, i.e., α-amylases, which are endo-acting enzymes, and glucoamylase, which is an exo-type enzyme that liberates glucose from the non-reducing end of starch. Glucoamylases also have some activity for α-1,6-bonds of amylopectin and α-1,3-bonds. Fungi also produce α-glucosidases hydrolyzing α-linked oligoglucosides, but most of them are rather involved in the processing reaction of glycoprotein biosynthesis. Most of the reported α-glucosidases hydrolyze both α-1,3- as well as α-1,4-glucosidic linkages (Yamasaki *et al*., 1995).

Table 5.6 *In vitro* antagonism (Anta) versus β-1,3 and β-1,6-glucanase activities (U/mg dry weight) of different *Trichoderma* strains cultivated in dual cultures with the phytopathogenic strains indicated

	Penicillium spp			*P. citropthora*			*R. solani*		
Strains	Anta.	β-1,3	β-1,6	Anta.	β-1,3	β-1,6	Anta.	β-1,3	β-1,6
T. harzianum CECT	−	0.9	0.3	+	0.9	0.3	±	0.8	0.3
T. viride	−	0.8	0.3	+	0.8	0.3	±	0.7	0.3
T. virens	−	1.0	0.2	+	0.7	0.3	+	0.8	0.2
T. harzianum IMI	−	1.0	0.2	+	0.8	0.2	+	0.8	0.2
T. longibrachiatum	−	1	0.1	+	0.6	0.04	+	0.7	0.2
T. koningii	−	0.5	0.1	+	0.1	0.05	+	0.5	0.2
T. reesei	−	0.6	0.2	+	0.3	0.03	−	0.6	0.2

+, Overgrowth and death of the pathogen was observed between 1–4 days.
±, Id. between 4–10 days.
−, Neither overgrowth nor death was observed after 10 days (M. Rey, unpublished).

Amylase production was investigated in a strain of *Trichoderma* spp. in chemostat culture using glucose-limited conditions (Aiba *et al.*, 1977). Formation of amylase activity depended on dilution rate and was subject to catabolite repression.

Several amylolytic enzymes were isolated and purified from *Trichoderma* strains. A glucoamylase from *T. reesei* was purified as a contaminant of a recombinantly produced glucoamylase from *Homocornis resinae.* The enzyme had an isoelectric point of 4.0 and a ratio of pullulan vs. starch degradation of 15%. Its aminoterminal sequence showed 60% identity with the amino terminal sequence of glucoamylases P and S from *H. resinae* (Fagerström, 1994; Fagerström and Kalkkinen, 1995) and 27% with that of an *A. niger* glucoamylase. Production of recombinant glucoamylase in *T. reesei* did not affect the specific activity of the native enzyme, but resulted in heterogeneous glycosylation that affected the enzyme's stability. Another glucoamylase was purified from *T. reesei*, which had an M_r of 66 kDa, contained <1% covalently-linked carbohydrate, and displayed optimal activity at a pH of 5.5 and at 70°C. The enzyme exhibited significant activity with α-1,6-glucosidically linked substrates, whereas β-cyclodextrin slightly inhibited the enzyme (Fagerström and Kalkkinen, 1995). The amino acid sequences of several tryptic peptides generated from the purified glucoamylase revealed about 60% identity with the *Aspergillus* glucoamylase. An α-glucosidase from *T. viride*, capable of hydrolyzing starch, maltooligosaccharides, and also α-1,3-linked oligoglucosides, has been purified by Yamasaki *et al.* (1995). The protein had an M_r of 65 kDa and exhibited optimal activity at pH 4.0 and at 65°C. The ratio of activities against α-1,3- to α-1,4-linked substrates was 1:16, and it was thus suggested that the enzyme primarily participates in the breakdown of starch.

5.3 Other hydrolases and their genes

5.3.1 Proteases

Protein and peptide utilization as nitrogen and sulfur sources by *Trichoderma* and *Gliocladium* requires exocellular digestion by proteases to hydrolyze the peptides to free amino acids. In general, proteases are strongly affected by catabolic repression by carbon, nitrogen and/or sulfur sources in the medium. In addition to extracellular proteases, intracellular proteases serve functions other than the procurement of nitrogen. Proteases can be classified into several categories according to their catalytic properties (amino peptidases, carboxypeptidases and endohydrolases). They are, in turn, divided into serine, thiol, carboxylic or metalloproteases.

Characterization and properties of proteases

Several examples of protease production from *Trichoderma* or *Gliocladium* grown in the presence of filamentous fungi or fungal cell walls have been reported. Two proteases – an aspartate protease from *T. reesei* and a serine protease from *T. harzianum* – have been purified to homogeneity. The gene encoding the serine protease has also been isolated and characterized. Haab *et al.* (1990) detected protease activity of *T. reesei* strain QM 9414 grown on cellulose, which was induced by the presence of organic nitrogen. An aspartate protease of 42.5 kDa, stable at pH 3–5 and pI of 4.3 was partly purified from this strain; it was insensitive to pepstatin A, an inhibitor of most fungal acidic proteases. The aspartic proteinase from *T. reesei*, which contained novel structural features, was further purified and crystallized (Pitts *et al.*, 1995). In addition, one mutant of *T. reesei* isolated by classical mutagenesis and defective in extracellular proteases lacked the 43 kDa chymosin-like pepstatin A inhibited aspartic protease (Mäntylä *et al.*, 1994).

Clarkson (1992) described the presence of a serine protease in *T. harzianum* which is inducible by β-glucans and fungal cell walls (Elad *et al.*, 1983). Another serine protease from *T. harzianum* was also purified by Geremia *et al.* (1993), which revealed the presence of at least three different proteases of 6.5, 7.0 and 9.2 pI in this strain. The alkaline (serine) protease from *T. harzianum* (PRB1) was biochemically characterized. Its molecular mass was 31 kDa, and it hydrolyzed substrates specific for chymotrypsin and aminopeptidase (Geremia *et al.*, 1993). The amino acid sequence of some peptides of the serine protease (PRB1) showed homology with that of protease K, particularly with the region containing the D presumably involved in catalysis.

Protease genes

A genomic *prb1* clone was isolated using a serine protease encoding cDNA clone as a probe (Geremia *et al.*, 1993). The gene (*prb1*) contained two introns and was present as a single copy gene in the *T. harzianum* genome. A highly homologous gene is also present in *T. viride*. *Pbr1* was transcriptionally regulated, subjected to catabolite repression and specifically expressed during the mycoparasitic process. For this reason it was suggested that the secretion of the protease by *T. harzianum* was triggered by a signal present in the host cell wall (Geremia *et al.*, 1993). The amino acid sequence of the putatively encoded protein showed extensive homology with several subtilisin-like proteases, comprised 409 amino acids, was synthesized as a preproenzyme and remained inactive until it was secreted (Geremia *et al.*, 1993; Goldman *et al.*, 1994).

Regulation and function of proteases

Proteases play various and frequently pivotal roles in cellular metabolism, nutrition and morphogenesis. They can also be responsible for post-secretional processing of multiple forms of extracellular hydrolytic enzymes from *Trichoderma*. Chen *et al.* (1993) detected three forms of CBHI cellobiohydrolase from *T. reesei* of 65, 58 and 54 kDa, the last two being two truncated forms of the intact CBHI, which possessed two specific proteolytic cleavage points. Haab *et al.* (1990) found that high levels of protease correlated with the appearance of proteolytic cellulase degradation products. Margolles-Clark *et al.* (1996) also found that the endochitinase encoded by *ThEn-42* of *T. harzianum* probably was sensitive to an acidic protease at the late stages of cultures of *T. reesei* transformants.

Of the three different proteases detected in *T. harzianum*, neutral proteases were present in all the supernatants analyzed, whereas growth of *T. harzianum* on fungal cell walls stimulated specifically the synthesis and/or secretion of the basic proteinase *Prb1*. Polymers such as laminarin, pustulan and cinerean failed to induce *prb1*, whereas colloidal chitin did induce protease secretion (Geremia *et al.*, 1993). High levels of expression of *prb1* were detected during the *T. harzianum–R. solani* interaction or when *T. harzianum* was cultivated on fungal cell walls (Flores *et al.*, 1997).

The antifungal activity of proteases is unclear. Mischke (1996) demonstrated *in vitro* that specific activity of a protease produced by a strain of *Trichoderma* did not correlate with its biocontrol ability, whereas Schirmböck *et al.* (1994) suggested that the parallel formation and synergism of several hydrolases (including protease activity) and antibiotics had an important role in the antagonistic action of *T. harzianum* against fungal pathogens. Transformants of *T. harzianum*, which carried from two to ten copies of *prb1*, were significantly improved in the biocontrol ability (Flores *et al.*, 1997); incorporation of these transformants into pathogen-infested soil significantly reduced the disease caused by *R. solani* in cotton plants. These results suggested roles for this serine protease that included activation of toxins or other enzymes during infection and the release of nutrients from the host in the form of amino acids.

5.3.2 *Nucleic acid hydrolases*

Nucleases are produced by numerous microorganisms, among which fungi are the most potent producers. Cleavage of RNA by RNases occurs by a two step mechanism which yields a 3′ nucleotide via 2′,3′-cyclic nucleotide intermediate (cyclizing RNases). These RNases are classified into three groups: guanine-specific, purine-specific and non-specific (Inada *et al.*, 1991). RNases are used in the food industry, for investigations on the structure and function of RNA, and for synthesis of oligonucleotides. In addition, some RNases possess antitumor, neurotoxic, or immunosuppressive properties.

Purification and properties of nucleic acid hydrolyzing enzymes

An RNase was purified from *T. viride* (RNase Trv) (Inada *et al.*, 1991) that belongs to a group of base non-specific and adenylic acid-preferential RNases (RNase T2 family) whose structure–function relationship is not well understood. Its enzymatic properties were very similar to typical T2-family RNases. The enzyme had a molecular mass of 26 kDa and the location of ten cysteine residues was consistent with that of other RNases of this family. In order to identify amino acid residues potentially important for catalysis, they compared the amino acid sequences of RNases of the same family from different species of fungi, including those of the genus *Trichoderma*. The sequences around three histidine residues were highly conserved in RNase Trv, and the enzyme presented homologies with RNase T (from *Aspergillus oryzae*), RNase M (from *Aspergillus saitoi*) and RNase Rh (from *Rhizopus niveus*). Chemical modification of other RNases showed the indispensability of two histidine residues and that a carboxylic group was essential for the catalysis.

Trichoderma strains are good producers of nucleases when they grow in media with low phosphate levels (Lin *et al.*, 1991). Vasileva-Tonkova and Bezborodova (1986) tested eleven strains of *Trichoderma* (including seven different species) for their ability to secrete extracellular RNases when grown in different synthetic media. Two *T. harzianum* strains showed highest levels of activity. An alkaline extracellular, guanine-specific RNase (RNase Th1) of one of these strains was purified to homogeneity; it displayed an M_r of 11 kDa and had an isoelectric point of 9.5, which sharply contrasted with the strongly acidic pI of other fungal RNases. Otherwise, a comparison of RNase Th1 with other guanyl-specific fungal RNases showed a significant degree of homology, indicating a common origin (Bezborodova *et al.*, 1988).

Finally, Lin *et al.* (1991) described isolation of an acidic extracellular protein of 14 kDa from *T. viride* (tricholin), which acted as a potent inhibitor of cell-free protein synthesis. The inhibition was due to a nucleolytic attack on the ribosome, which generated a specific cleavage product, an α-sarcin-like RNA fragment. Although a comparison of the amino acid composition of tricholin and α-sarcin did not reveal significant homology, antibodies against α-sarcin strongly cross-reacted with tricholin. Tricholin also appeared to act as a ribosome-inactivating toxin by attacking 28S rRNA. Whether this protein acts as an RNA N-glycosidase that modifies the 28S rRNA is not known, but this protein may be useful in the attack of other fungi during biocontrol. Transgenic tobacco plants that expressed a barley ribosome inactivating protein exhibited increased protection against *R. solani* infection (Logemann *et al.*, 1992).

No genes encoding any of the nucleases or ribosome-inactivating toxins of *Trichoderma* or *Gliocladium* have as yet been isolated.

5.3.3 Trehalases

Trehalose is a non-reducing disaccharide of D-glucose (α-D-glucopyranosyl-1,1-α-D-glucopyranoside), which is found in invertebrates and fungi and which functions as a reserve carbohydrate and a protectant against physical stress. Trehalases have been thoroughly studied in *Trichoderma*, particularly with respect to their physico-chemical properties and their kinetic properties such as the role of donor substrate configuration, stereochemistry of the reaction, and others (Weiser *et al.*, 1988). A partially purified trehalase from *T. reesei* (Vijayakumar *et al.*, 1978) had optimal activity at pH 4.4, exhibited an isoelectric point of 5.7 and yielded a K_m of 3.1 mM. The hydrolysis of different substrates by the enzymes suggested a high catalytic versatility, indicating that the catalytic groups of trehalase had the flexibility to catalyze different stereochemical reactions (Kasumi *et al.*, 1986). However, when α-glucosidases and trehalases from *T. reesei* were compared (Alabran *et al.*, 1983), trehalase was highly specific for α,α′-trehalose, whereas α-glucosidases acted on a wide variety of α-glucopyranosides. Furthermore, both enzymes catalyzed the hydration of gluco-octenitol, but only that catalyzed by α-glucosidase was strongly inhibited by the substrate (Weiser *et al.*, 1988). These data indicate several differences between trehalases and α-glucosidases of *T. reesei.*

Kelley and Rodríguez-Kabana (1976) noted that *T. harzianum* formed high activities of trehalase during competition with *Phytophthora cinnamomi*, which they interpreted as an indication that the mycoparasite utilized *P. cinnamomi* trehalose as a carbon source.

5.3.4 Pectinases

Pectinolytic enzymes degrade the pectin of plant cell walls, liberating polygalacturonates and galacturonates that can be used as nutritional sources. Whereas bacteria contain pectin-lyases (transeliminases), endo-polygalacturonide (pectic) lyases (endo-PL or PGL) and pectin methylesterases (pectin esterases; PE or PME), endo-polygalacturonases (endo-PG) and pectin-lyases (transeliminases; endo-polymethyl galacturonide (pectin) lyases; endo-PL or PML) have been reported from fungal sources. *Trichoderma* strains are known to produce high amounts of extracellular pectinases, but they have been only poorly characterized so far. Growth on substrates such as straw and corn cobs usually leads to the formation of xylanolytic, cellulolytic, proteolytic and pectinolytic enzymes (Kisielewska and Bujak, 1977). A polygalacturonase and a pectin-methylesterase were partly purified from *T. lignorum*; both enzymes had acidic isoelectric points with optimal activity between pH 4 and 4.45 (Mabrouk *et al.*, 1979).

5.3.5 *Mannanases, lipases and other lytic enzymes*

Many lytic enzymes are synthesized by *Trichoderma* and *Gliocladium*, including mannanases, lipases, phosphatases and others (Elad *et al.*, 1983; Labudova and Gogorova, 1988). In many cases, however, the number of the different proteins that integrate each activity are unknown. Lipases and phosphatases have neither been purified nor characterized, and only mannanases have received attention. Mannanases are more extensively described in other chapters of this book (see Chapter 4).

Due to their importance in the food industry, lactonases were also investigated in *Trichoderma* strains. Lactonases were components of a complete synergistic cellulase system. A lactonase was purified from a commercial cellulase preparation from *T. reesei.* The enzyme was active on glucono-1,5-lactone and cellobionolactone (Bruchmann *et al.*, 1987).

5.4 Biotechnological applications of hydrolytic enzymes

5.4.1 *Medical applications*

Insoluble glucan polymers play a major role in the development of dental plaque and the onset and progression of dental disease. The 15 kDa α-1,3-glucanase (also called mutanase) from *T. harzianum* was capable of degrading this water-insoluble glucan (Quivey and Kriger, 1993). This α-1,3-glucanase was used as a caries preventive agent and, when administered in chewing gum, reduced the disease potential of dental plaque and gingivitis (Kelstrup *et al.*, 1978).

5.4.2 *Agricultural applications*

Trichoderma strains are known to produce lyases, proteases, lipases, etc. which may be used for the degradation of cell walls of target fungal pathogens. The 78 kDa β-1,3-glucanase and the 72 kDa N-acetyl-glucosaminidase isolated from *T. harzianum* strain P1 had synergistic inhibitory effects on spore germination and germ tube elongation of *B. cinerea* (Lorito *et al.*, 1994). β-1,3-Glucanase from *T. harzianum* CECT 2413 induced morphological changes such as hyphal tip swelling, bursting, leakage of cytoplasm, formation of numerous septae and fungal growth inhibition on *R. solani* and *Fusarium* spp. hyphae (Llobell *et al.*, unpublished). The 43 kDa β-1,6-glucanase from *T. harzianum* CECT 2413 also produced hydrolytic halos on yeast cell walls and, when combined with β-1,3-glucanase and/or chitinase, hydrolyzed filamentous fungal cell walls and inhibited the growth of the fungi tested (Bruce *et al.*, 1995; De la Cruz *et al.*, 1995b). It can therefore be expected that inhibitory effects on fungal pathogens would be synergistically increased by mixing these fungal enzymes and bacterial biocontrol strains as described for chitinolytic enzymes and *Enterobacter* strains (Lorito *et al.*, 1993a,b).

Some researchers have concentrated on improving single antagonistic traits during strain development to obtain more effective biocontrol strains (Harman and Hayes, 1993). Haran *et al.* (1993) have examined ways of using genetic manipulation to increase constitutive hydrolase activity of *Trichoderma* and *Gliocladium* isolates that already possessed good biocontrol activity. Combinations of biocontrol agents and fungicides or even purified enzymes and fungicides have also been proposed as alternative ways of exerting biological control (Lorito *et al.*, 1996; Rice, 1996). When the competitive saprophytic ability of strains of *Trichoderma* was determined in the rhizosphere, the cellulolytic activity of these strains directly correlated with the ability to colonize roots (Ahmad and Baker, 1987). Furthermore, addition of fungal cell walls to seed coats increased the ability of *T. hamatum* to protect seeds against *Pythium* (Harman *et al.*, 1981).

Transformants of *T. harzianum* CECT 2413 that possessed multiple copies of the gene encoding β-1,6-glucanase II have been produced. Increased levels of β-1,6-glucanase activity and lytic and antifungal activity against *R. solani* was also detected, probably due to cooperative and synergistic effects with other hydrolases (J. Delgado-Jarana, unpublished). In addition, a mutant from *T. harzianum* CECT 2413 with higher levels of β-1,3- and β-1,6-glucanases has been isolated. Preliminary results in experiments carried out *in vitro* in dual cultures with *R. solani* indicated that the mutant PS1 possessed higher antifungal activity than the wild type. Finally, Flores *et al.* (1997) reported an improvement in biocontrol efficacy by the introduction of multiple copies of *prb1* into *T. harzianum.*

5.4.3 ***Food and beverage applications***

Wine

Genetic modification of actively dried wine yeasts by introducing glucanases and pectinases opened the way for the construction of strains that express metabolic activities, resulting in superior organoleptic characteristics of the wine. The addition of exogenous enzymes to increase aroma is a frequent practice in wineries, although complex and undefined mixtures of enzymes are used. Therefore, the structural genes encoding enzymes of enological interest – e.g., the pectate lyase from *T. longibrachiatum* – have been expressed in wine yeasts, and it was observed in many cases that their expression correlated with an increase in the aroma of the wine. Other glucanases and glycosidases investigated were able to hydrolyze the glycosidic linkage of terpene-glycosides, and the released terpenes increased the flavor (Querol and Ramón, 1996). In addition, exo-β-1,3-glucanases isolated from *T. harzianum* (Dubordieu *et al.*, 1985), which are able to digest β-1,3-, β-1,6-glucans (cinereans) produced by *B. cinerea*, were proposed for use in wine-making to hydrolyze this polymer and to help wine clarification.

Beer

β-Glucanase preparations are used in beer production to degrade the barley β-glucans present in the wort, which can result in filtering problems and turbid appearance and even occurrence of precipitates in beer. The external addition of commercial β-glucanase preparations is possible, but they often contain undesirable side activities. To overcome this problem, a gene encoding an endo-β-1,4-glucanase from *T. reesei* has been expressed in brewing yeasts; the recombinant yeast strains were stable even under non-selective conditions, and the endoglucanase was detected in the extracellular medium. The recombinant yeast improved the filterability of the wort, and the quality of the beer was also good (Xie *et al.*, 1995).

Bread

Wheat flour contains an adequate level of β-amylase but is deficient in α-amylase, which is therefore usually added to bread dough for optimal starch degradation. The sources of α-amylase for the baking industry include malt, bacterial and fungal enzyme preparations. The latter most improve the rheological properties of bread dough. Transgenic baker's yeast expressing the fungal α-amylase under the control of the *ACT1* promoter gave an optimal level of enzyme; the fermented dough gave a bread with a higher loaf volume and a softer crumb than the control. In addition, the hardening of the bread was retarded, and its shelf life and freshness were increased (Rández-Gil *et al.*, 1995).

5.4.4 *Other applications*

Purified lytic enzymes have been used for elucidating structures of fungal glucans and fungal cell walls (Pitson *et al.*, 1993). The 47 kDa endo-α-1,3-glucanase from *T. viride* (Hasegawa and Nordin, 1969) has been used for the determination of the structure of polysaccharides and allowed the isolation of intact carbohydrate chain fragments from *Aspergillus* fungal cell walls. Also, the β-1,6-glucanase II from *T. harzianum* CECT 2413 aided the identification of a phosphodiester linkage between proteins and a β-1,3-/β-1,6-heteropolymer in *S. cerevisiae* cell walls, which is responsible for binding of cell wall proteins (Kapteyn *et al.*, 1996). Furthermore, these enzymes have been used for the characterization of β-glucan in cereals, an important parameter defining its nutritional quality (Brenes *et al.*, 1993). Barlican, a β-glucanase preparation from *T. reesei*, was proposed for use as a feed additive. Feeding of barlican at dietary levels did not give rise to any evidence of toxicity in rats or broiler chicks (Coenen, 1995). Another enzyme mixture, which included endo-β-1,3-, -β-1,4-endoglucanase from *T. viride*, had also been tested in a barley diet (Vranjes and Wenk, 1996). The enzyme positively influenced the organic matter utilization and fiber degradability.

Other possibilities for the application of purified fungal enzymes include the conversion of algal biomass to fermentable sugars, partial polysaccharide hydrolysis for clinical use, polysaccharide synthesis, and others (Pitson *et al.*, 1993).

In summary, studies of fungal hydrolases such as those of *Trichoderma* and *Gliocladium* have given some insight into their role in fungal growth, development, autolysis, attachment and invasion of hyphae into the host cell. These studies have also promoted the application of hydrolases in medicine, crop management, brewing and feed improvement. Finally, these enzymes are tools for structural elucidation of fungal cell walls.

Acknowledgments

We thank L. Romero and B. Cubero for helpful discussion, R. Espejo for correction of the manuscript and A. Llobell and colleagues for sharing their unpublished data with us. This work was supported by grants BIO94-0289 and PTR94-0068 (CICYT) and the Junta de Andalucía (PAI 3288).

References

Adams, D. J., Cansier, B. E., Mellor, K. J., Keer, V., Milling, R., and Dada, J. 1993. Relation of chitinase and chitin synthase in fungi. In R. Muzzarelli (ed.), *Chitin Enzymology*. Atec Edizioni, Senigallia, Italy, pp. 1–2.

Adams, P. B. and Ayers, W. A. 1983. Histological and physiological aspects of infection of sclerotia of two *Sclerotinia* species by two mycoparasites. *Phytopathology* **73**: 1072–1076.

Ahmad, J. S. and Baker, R. 1987. Rhizosphere competence of *Trichoderma harzianum*. *Phytopathology* **77**: 182–189.

Aiba, S., Nagai, S., and Onodera, M. 1977. Induction and repression of exo-enzymes of cellulase and amylase from *Trichoderma sp.* Proc. Bioconversion Symp., IIT Delhi, 179.

Alabran, D. M., Ball, D. H., and Reese, E. T. 1983. Comparison of the trehalase of *Trichoderma reesei* with those from other sources. *Carbohydr. Res.* **123**: 179–181.

Bamforth, Ch. W. 1980. The adaptability, purification and properties of exo-β-1,3-glucanase from the fungus *Trichoderma reesei*. *Biochem. J.* **191**: 863–866.

Barnett, C. C., Berka, R. M., and Fowler, T. 1991. Cloning and amplification of the gene encoding an extracellular β-glucosidase from *Trichoderma reesei*: evidence for improved rates of saccharification of cellulosic substrates. *Bio/Technology* **9**: 562–567.

Bartnicki-Garcia, S. 1968. Cell wall chemistry, morphogenesis, and taxonomy of fungi. *Annu. Rev. Microbiol.* **22**: 87–108.

Benítez, T., Villa, T. G., and García-Acha, I. 1976. Effects on polioxin D on germination, morphological development and biosynthesis of the cell wall of *Trichoderma viride*. *Arch. Microbiol.* **108**: 183–188.

Bezborodova, S. I., Vasileva-Tonkova, E. S., Poliakov, K. M., and Schliapnikov, S. V. 1988. Isolation, analysis of amino acid sequence and crystallization of the extracellular ribonuclease Th1 from *Trichoderma harzianum*-01. *Bioorg. Khim.* **14**: 453–466.

Bielecki, S. and Galas, D. 1977. *β-1,3-Glucanase from cellulase "Onozuca" SS and its lytic capability*. Proc. Bioconversion Sym., IIT Delhi, 203–225.

Bobbit, T. F., Nordin, J. H., Roux, M., Revol, J. F., and Marchessault, R. H. 1977. Distribution and conformation of crystalline nigeran in hyphal walls of *Aspergillus niger* and *Aspergillus awamori*. *J. Bacteriol.* **132**: 691–703.

Brenes, A., Guenter, W., Marquardt, R. R., and Rotter, B. A. 1993. Effect of β-glucanase/pentosanase enzyme supplementation on the performance of chickens and laying hens fed wheat, barley, naked oats and rye diets. *Can. J. Anim. Sci.* **73**: 941–951.

Bruce, A., Srinivasan, U., Staines, H. J., and Highley, T. L. 1995. Chitinase and laminarinase production in liquid culture by *Trichoderma* spp. and their role in biocontrol of wood decay fungi. *Int. Biodeter. Biodegr.* **10**: 337–353.

BRUCHMANN, E. E, SCHACH, H., and GRAF, H. 1987. Role and properties of lactonase in a cellulase system. *Biotechnol. Appl. Biochem.* **9**: 146–159.

BULL, A. T. 1970. Inhibition of polysaccharases by melanin: enzyme inhibition in relation to mycolysis. *Arch. Biochem. Biophys.* **137**: 345–356.

CHAMBERS, R. S., WALDEN, A. R., BROOKE, G. S., CUTFIELD, J. G., and SULLIVAN, P. A. 1993. Identification of a putative active site residue in the exo-β-(1,3)-glucanase of *Candida albicans. FEBS Lett.* **327**: 366–369.

CHEN, L., FINCHER, G., and HOJ, P. B. 1993. Evolution of polysaccharide hydrolase substrate specificity. *J. Biol. Chem.* **268**: 13318–13326.

CHET, I. 1987. *Trichoderma* – applications, mode of action and potential as a biocontrol agent of soilborne plant pathogenic fungi. In I. Chet (ed.), *Innovative Approaches to Plant Diseases.* Wiley, New York, pp. 137–160.

CHIBA, Y., NAKAJIMA, T., and MATSUDA, K. 1988. A morphological mutant of *Neurospora crassa* with defects in the cell wall β-glucan structure. *Agric. Biol. Chem.* **52**: 3105–3111.

CID, V. J., DURAN, A., DEL REY, F., SNYDER, M. P., NOMBELA, C., and SÁNCHEZ, M. 1995. Molecular basis of cell integrity and morphogenesis in *Saccharomyces cerevisiae. Microbiol. Rev.* **59**: 345–386.

CLARKSON, J. M. 1992. Molecular biology of filamentous fungi used for biological control. In J. R. Kinghorn, and G. Turner (eds), *Applied Molecular Genetics of Filamentous Fungi.* Blackie Academic & Professional, Cambridge, pp. 175–190.

COENEN, T. M., SCHOENMAKERS, A. C., and VERHAGEN, H. 1995. Safety evaluation of β-glucanase derived from *Trichoderma reesei*: summary of toxicological data. *Food Chem. Toxicol.* **33**: 859–866.

CUMMINGS, C. and FOWLER, T. 1996. Secretion of *Trichoderma reesei* β-glucosidase by *Saccharomyces cerevisiae. Curr. Genet.* **29**: 227–233.

DE LA CRUZ, J., PINTOR-TORO, J. A., BENÍTEZ, T., and LLOBELL, A. 1995a. Purification and characterization of an endo-β-1,6-glucanase from *Trichoderma harzianum* related to its mycoparasitism. *J. Bacteriol.* **177**: 1864–1871.

DE LA CRUZ, J., PINTOR-TORO, J. A., BENÍTEZ, T., LLOBELL, A., and ROMERO, L. C. 1995b. A novel endo-β-1,3-glucanase, BGN13.1, involved in the mycoparasitism of *Trichoderma harzianum. J. Bacteriol.* **177**: 6937–6945.

DE LA CRUZ, J., REY, M., LORA, J. M., HIDALGO-GALLEGO, A., DOMÍNGUEZ, F., PINTOR-TORO, J. A., LLOBELL, A., and BENÍTEZ, T. 1993. Carbon source control on β-glucanase, chitibiosidase and chitinase from *Trichoderma harzianum. Arch. Microbiol.* **159**: 316–322.

DEL REY, F., GARCÍA-ACHA, I. and NOMBELA, C. 1977. The regulation of β-glucanase synthesis in fungi and yeast. *J. Gen. Microbiol.* **110**: 88–89.

DE VRIES, O. M. H. and WESSELLS, J. G. H. 1973. Release of protoplasts from *Schizophyllum commune* by combined action of purified α-1,3-glucanase and chitinase derived from *Trichoderma viride. J. Gen. Microbiol.* **76**: 319–330.

DRABORG, H., CHRISTGAU, S., HALKIER, T., RASMUSSEN, G., DALBOG, H., and KAUPPINEN, S. 1996. Secretion of an enzymatically active *Trichoderma harzianum* endochitinase by *Saccharomyces cerevisiae. Curr. Genet.* **29**: 404–409.

DUBORDIEU, D., DESPLANQUES, C., VILLETTAZ, J. C., and RIBEREAU-GAYON, P. 1985. Investigations of an industrial β-D-glucanase from *Trichoderma harzianum. Carbohydr. Res.* **144**: 277–287.

ELAD, Y., CHET, I., BOYLE, L., and HENIS, Y., 1983. Parasitism of *Trichoderma* spp. on *Rhizoctonia solani* and *Sclerotium rolfsii*: scanning electron microscopy and fluorescence microscopy. *Phytopathology* **73**: 85–88.

FAGERSTRÖM, R. 1994. Evidence for a polysaccharide-binding domain in *Hormoconis resinae* glycoamylase P: effects of its proteolytic removal on a substrate specificity and inhibition by β-cyclodextrin. *Microbiology* **140**: 2399–2407.

FAGERSTRÖM, R. and KALKKINEN, N. 1995. Characterization, subsite mapping and partial amino acid sequence of glucoamylase from the filamentous fungus *Trichoderma reesei*. *Biotechnol. Appl. Biochem.* **21**: 223–231.

FINCHER, G. B. 1989. Molecular and cellular biology associated with endosperm mobilization in germinating cereal grains. *Annu. Rev. Plant Physiol. Plant Mol. Biol.* **40**: 305–346.

FLORES, A., CHET, I., and HERRERA-ESTRELLA, A. 1997. Improved biocontrol activity of *Trichoderma harzianum* by over-expression of the proteinase-encoding gene *prb1*. *Curr. Genet.* **31**: 30–37.

FRIEBE, B. and HOLLDORF, A. W. 1975. Control of extracelllular β-1,3-glucanase activity in a basidiomycete species. *J. Bacteriol.* **122**: 818–825.

GARCÍA, I., LORA, J. M., DE LA CRUZ, J., BENÍTEZ, T., LLOBELL, A., and PINTOR-TORO, J. A. 1994. Cloning and characterization of a chitinase (CHIT42) cDNA from the mycoparasitic fungus *Trichoderma harzianum*. *Curr. Genet.* **27**: 83–89.

GEREMIA, R. A., GOLDMAN, G. H., JACOBS, D., ARDILES, W., VILA, S. B., VAN MONTAGU, M., and HERRERA-ESTRELLA, A. 1993. Molecular characterization of the proteinase-encoding gene, *prb1*, related to mycoparasitism by *Trichoderma harzianum*. *Mol. Microbiol.* **8**: 603–613.

GLENN, M., GHOSH, A., and GHOSH, B. K. 1985. Subcellular fractionation of a hypercellulolytic mutant, *Trichoderma reesei* Rut-C30: localization of endoglucanase in microsomal fraction. *Appl. Environ. Microbiol.* **50**: 1137–1143.

GOLDMAN, G. H., HAYES, C., and HARMAN, G. E. 1994. Molecular and cellular biology of biocontrol by *Trichoderma* spp. *TIBTECH* **12**: 478–482.

GRIFFIN, D. H. 1994. *Fungal Physiology*, 2nd edn. Wiley-Liss, New York.

HAAB, D., HAGSPIEL, K., SZAKMARY, K., and KUBICEK, C. P. 1990. Formation of the extracellular proteases from *Trichoderma reesei* QM 9414 involved in cellulase degradation. *J. Biotechnol.* **16**: 187–198.

HARAN, S., SCHICKLER, H., and CHET, I. 1996. Molecular mechanisms of lytic enzymes involved in the biocontrol activity of *Trichoderma harzianum*. *Microbiology UK* **142**: 2321–2331.

HARAN, S., SCHICKLER, H., PE'ER, S., LOGEMANN, S., OPPENHEIM, A., and CHET, I. 1993. Increased constitutive chitinase activity in transformed *Trichoderma harzianum*. *Biol. Control* **3**: 101–108.

HARMAN, G. E. and HAYES, C. K. 1993. The genetic nature and biocontrol ability of progeny from protoplast fusion in *Trichoderma*. In I. Chet (ed.), *Biotechnology in Plant Disease Control*. Wiley, New York, pp. 237–255.

HARMAN, G. E., CHET, I., and BAKER, R. 1981. Factors affecting *Trichoderma hamatum* applied to seeds as a biocontrol agent. *Phytopathology* **71**: 569–572.

HASEGAWA, S. and NORDIN, J. H. 1969. Enzyme that hydrolyze fungal cell wall polysaccharides. I: Purification and properties of an endo-α-D-(1-3)-glucanase from *Trichoderma viride*. *J. Biol. Chem.* **244**: 5460–5470.

INADA, Y., WATANABE, H., OHGI, K., and IRIE, M. 1991. Isolation, characterization, and primary structure of a base non-specific and adenylic acid preferential ribonuclease with higher specific activity from *Trichoderma viride*. *J. Biochem.* **110**: 896–904.

INBAR, J. and CHET, I. 1995. The role of recognition in the induction of specific chitinases during mycoparasitism by *Trichoderma harzianum*. *Microbiology UK* **141**: 2823–2829.

ISELI, B., BOLLER, T., and NEUHAUS, J. M. 1993. The N-terminal cysteine-rich domain of tobacco class I chitinase is essential for chitin binding but not for catalytic or antifungal activity. *Plant Physiol.* **103**: 221–226.

JEFFRIES, P. and YOUNG, T. W. K. 1994. *Interfungal Parasitic Relationships*. CAB International, Cambridge.

KAPTEYN, J. C., MONTIJN, R. C., VINK, E., DE LA CRUZ, J., LLOBELL, A., DOUWES, J. E., SHIMOI, H., LIPKE, P. N., and KLIS, F. M. 1996. Retention of

Saccharomyces cerevisiae cell wall proteins through a phosphodiester-linked β-1,3-/β-1,6-glucan heteropolymer. *Glycobiology* **6**: 337–345.

KASUMI, T., BREWER, C. F., REESE, E. T., and HEHRE, E. J. 1986. Catalytic versatility of trehalase: synthesis of α-D-glucosyl fluoride and α-D-xylose. *Carbohydr. Res.* **146**: 39–49.

KELLEY, W. D. and RODRÍGUEZ-KABANA, R. 1976. Competition between *Phytophthora cinnamoni* and *Trichoderma* spp. in autoclaved soil. *Can. J. Microbiol.* **22**: 1120–1127.

KELSTRUP, J., HOLM-PEDERSEN, P., and POULSEN, S. 1978. Reduction of the formation of dental plaque and gingivitis in humans by crude mutanase. *Scand. J. Dent. Res.* **86**(2): 93–102.

KISIELEWSKA, E. and BUJAK, S. 1977. Characterization of a complex of hemicellulolytic enzymes produced by some strains of the lower fungi. *Acta Microbiol. Polon.* **26**: 169–375.

KITAMOTO, Y., KONO, R., SHIMOTORI, A., MORI, N., and ICHIKAWA, Y. 1987. Purification and some properties of an exo-β-1,3-glucanase from *Trichoderma harzianum. Agric. Biol. Chem.* **51**: 3385–3386.

KUBICEK, C. P. 1982. β-Glucosidase excretion by *Trichoderma pseudokoningii*: correlation with cell wall bound β-1,3-glucanase activities. *Arch. Microbiol.* **132**: 349–354.

KUBICEK, C. P. 1987. Involvement of a conidial endoglucanase and a plasma-membrane-bound β-glucosidase in the induction of endoglucanse synthesis by cellulose in *Trichoderma reesei. J. Gen. Microbiol.* **133**: 1481–1487.

KUBICEK, C. P., MESSNER, R., GRUBER, F., MACH, R. L., and KUBICEK-PRANZ, E. M. 1993. The *Trichoderma* cellulase regulatory puzzle: from the interior life of a secretory fungus. *Enzyme Microbiol. Technol.* **15**: 90–99.

KUHN, P. J., TRINCI, A. P. J., JUNG, M. J., GOOSEY, M. W., and COPPING, L. G. 1990. *Biochemistry of cell walls and membranes in fungi.* Springer-Verlag, Berlin/Heidelberg.

LABUDOVA, I. and GOGOROVA, L. 1988. Biological control of phytopathogenic fungi through lytic action of *Trichoderma* species. *FEMS Microbiol. Lett.* **52**: 193–198.

LIMÓN, M. C., LORA, J. M., GARCÍA, I., DE LA CRUZ, J., LLOBELL, A., BENÍTEZ, T. and PINTOR-TORO, J. A. 1995. Primary structure and expression pattern of the 33-kDa chitinase gene from the mycoparasitic fungus *Trichoderma harzianum. Curr. Genet.* **28**: 478–483.

LIN, A., CHEM, C.-K., and CHEN, Y.-J. 1991. Molecular action of tricholin, a ribosome-inactivating protein isolated from *Trichoderma viride. Mol. Microbiol.* **5**: 3007–3013.

LOGEMANN, J., JACH, G., TOMMERUP, H., MUNDY, J., and SCHELL, J. 1992. Expression of a barley ribosome-inactivating protein leads to increased fungal protection in transgenic tobacco plants. *Bio/Technology* **10**: 305–308.

LORA, J. M., DE LA CRUZ, J., BENÍTEZ, T., LLOBELL, A., and PINTOR-TORO, J. A. 1995. Molecular characterization and heterologous expression of an endo-β-1,6-glucanase gene from the mycoparasitic fungus *Trichoderma harzianum. Mol. Gen. Genet.* **247**: 639–645.

LORITO, M., HAYES, C. K., DI PIETRO, A., WOO, S. L., and HARMAN, G. E. 1994. Purification, characterization, and synergistic activity of a glucan-1,3-β-glucosidase and an N-acetyl-β-glucosaminidase from *Trichoderma harzianum. Phytopathology* **84**: 398–405.

LORITO, M., DI PIETRO, A., HAYES, C. K., WOO, S. L., and HARMAN, G. E. 1993a. Antifungal, synergistic interaction between chitinolytic enzymes from *Trichoderma harzianum* and *Enterobacter cloacae. Phytopathology* **83**: 721–728.

LORITO, M., HARMAN, G. E., HAYES, C. K., BROADWAY, R. M., TRONSMO, A., WOO, S. L. and DI PIETRO, A. 1993b. Chitinolytic enzymes produced by *Trichoderma harzianum*: antifungal activity of purified endochitinase and chitobiosidase. *Phytopathology* **83**: 302–307.

LORITO, M., WOO, S. L., D'AMBROSIO, M., HARMAN, G. E., HAYES, C. K., KUBICEK, C. P., and SCALA, F. 1996. Synergistic interaction between cell wall degrading enzymes and membrane affecting compounds. *Mol. Plant Microb. Interact.* **9**: 206–213.

MABROUK, S. S., ABDEL-FATTAH, A. F., and ISMAIL, A. M. 1979. Preparation and properties of pectic enzymes produced by *Trichoderma lignorum. Zbl. Bakteriol.* **134**: 282–286.

MANOCHA, M. S. and BALASUBRAMANIAM, R. 1988. *In vitro* regulation of chitinase and chitin synthase activity of two mucoraceous hosts of a mycoparasite. *Can. J. Microbiol.* **34**: 1116–1121.

MÄNTYLÄ, A., SAARELAINEN, R., FAGERSTRÖM, R., SUOMINEN, P., and NEVALAINEN, H. 1994. Cloning of the aspartic protease gene from *Trichoderma reesei.* 2nd *European conference on fungal genetics, Lunteren,* The Netherlands. Abstract B 52.

MARGOLLES-CLARK, E., HAYES, C., HARMAN, G. E., and PENTTILÄ, M. 1996. Improved production of *Trichoderma harzianum* endochitinase by expression in *Trichoderma reesei. Appl. Environ. Microbiol.* **62**: 2145–2151.

MAUCH, F., MAUCH-MANI, B., and BOLLER, T. 1988. Antifungal hydrolases in pea tissue. *Plant Physiol.* **88**: 936–942.

MELDGAARD, M. and SVENDSEN, I. 1994. Different effects of *N*-glycosylation on the thermostability of highly homologous bacterial (1,3-1,4)-β-glucanases secreted from yeast. *Microbiology UK* **140**: 159–166.

MESSNER, R. and KUBICEK, C. P. 1991. Carbon source control of cellobiohydrolase I and II formation by *Trichoderma reesei. Appl. Environ. Microbiol.* **57**: 630–635.

MISCHKE, S. 1996. Evaluation of chromogenic substrates for measurements of protease production by biocontrol strains of *Trichoderma. Microbios* **87**: 175–183.

MUTHUKUMAR, G., SUHNG, S., MAGEE, P. T., JEWELL, R. D., and PRIMERANO, D. A. 1993. The *Saccharomyces cerevisiae SPR1* gene encodes a sporulation-specific exo-1,3-β-glucanase which contributes to ascospore thermoresistence. *J. Bacteriol.* **175**: 386–394.

NORONHA, E. F. and ULHOA, C. J. 1996. Purification and characterization of an endo-β-glucanase from *Trichoderma harzianum. Can. J. Microbiol.* **42**: 1039–1044.

PENTTILÄ, M., LEHTOVAARA, P., NEVALAINEN, H., BHIKHABHAI, R., and KNOWLES, J. 1986. Homology between cellulase genes of *Trichoderma reesei*: complete nucleotide sequence of the endoglucanase I gene. *Gene* **45**: 253–263.

PITSON, S. M., SEVIOUR, R. J., and MCDOUGALL, B. M. 1993. Noncellulolytic fungal β-glucanases: their physiology and regulation. *Enzyme Microbiol. Technol.* **15**: 178–192.

PITTS, J. E., CRAWFORD, M. D. NUGENT, P. G., WESTER, R. T., COOPER, J. B., MÄNIYLÄ, A., FAGERSTROM, R., and NEVALAINEN, H. 1995. The three-dimensional X-ray crystal structure of the aspartic proteinase native to *Trichoderma reesei* complexed with a renin inhibitor CP-80794. *Adv. Exp. Med. Biol.* **362**: 543–547.

QUEROL, A. and RAMÓN, D. 1996. The application of molecular techniques in wine microbiology. *TIFST* **7**: 73–77.

QUILLET, L. 1995. The gene encoding the β-endoglucanase (*CelA*) from *Mixococcus xanthus*: evidence for independent acquisition by horizontal transfer of binding and catalytic domains from actinomycetes. *Gene* **158**: 23–29.

QUIVEY, R. G. and KRIGER, P. S. 1993. Raffinose-induced mutanase production from *Trichoderma harzianum. FEMS Microbiol. Lett.* **112**: 307–312.

RADFORD, A., STONE, P. J., and TALEB, F. 1996. Cellulase and amylase complexes. In R. Brambl, R. and G. A. Marzluf (eds), *The Mycota III: Biochemistry and Molecular Biology.* Springer-Verlag, Berlin/Heidelberg, pp. 269–291.

RÁNDEZ-GIL, F., PRIETO, J. A., MURCIA, A., and SANZ, P. 1995. Construction of baker's yeast strains that secrete *Aspergillus oryzae* α-amylase and their use in bread making. *J. Cereal Sci.* **21**: 185–193.

REESE, E. T., PARRISH, F. W., and MANDELS, M. 1961. β-D-1,2-glucanases in fungi. *Can. J. Microbiol.* **7**: 309–317.

REESE, E. T., PARRISH, F. W., and MANDELS, M. 1962. β-D-1,6-glucanases in fungi. *Can. J. Microbiol.* **8**: 327–334.

RICE, E. L. 1996. *Biological Control of Weeds and Plant Diseases.* University of Oklahoma Press, Oklahoma.

RIDOUT, C. J., COLEY-SMITH, J. R., and LYNCH, J. M. 1986. Enzyme activity and electrophoretic profile of extracellular protein induced in *Trichoderma* spp. by cell walls of *Rhizoctonia solani. J. Gen. Microbiol.* **132**: 2345–2352.

SAN SEGUNDO, P., CORREA, J., VÁZQUEZ DE ALDANA, C. R., and DEL REY, F. 1993. *SSG1*, a gene encoding a sporulation-specific 1,3-β-glucanase in *Saccharomyces cerevisiae. J. Bacteriol.* **175**: 3823–3837.

SANTOS, T., VILLANUEVA, J. R., and NOMBELA, C. 1977. Production and catabolite repression of *Penicillium italicum. J. Bacteriol.* **129**: 52–58.

SANTOS, T., VILLANUEVA, J. R., and NOMBELA, C. 1978a. Regulation of β-1,3-glucanase synthesis in *Penicillium italicum. J. Bacteriol.* **133**: 542–548.

SANTOS, T., SÁNCHEZ, M., VILLANUEVA, J. R., and NOMBELA, C. 1978b. Region of the β-1,3-glucanase system in *Penicillium italicum*: glucose repression of the various enzymes. *J. Bacteriol.* **133**: 465–471.

SCHAEFFER, H. J., LEYKAM, J., and WALTON, J. D. 1994. Cloning and targeted gene disruption of *EXG1*, encoding exo-β-1,3-glucanase, in the phytopathogenic fungus *Cochliobolus carbonum. Appl. Environ. Microbiol.* **60**: 594–598.

SCHIRMBÖCK, M., LORITO, M., WANG, Y.-L., HAYES, C. K., ARISAN-ATAC, I., SCALA, F., HARMAN, G. E., and KUBICEK, C. P. 1994. Parallel formation and synergism of hydrolytic enzymes and peptaibol antibiotics, molecular mechanisms involved in the antagonistic action of *Trichoderma harzianum* against phytopathogenic fungi. *Appl. Environ. Microbiol.* **60**: 4364–4370.

SENTANDREU, R., MORMENEO, S., and RUIZ-HERRERA, J. 1994. Biogenesis of the fungal cell wall. In J. G. H. Wessels, and F. Meinhardt (eds), *The Mycota I: Growth, Differentiation and Sexuality.* Springer-Verlag, Berlin/Heidelberg, pp. 111–124.

SHINSHI, H., NEUHAUS, J. M., RYALS, J., and MEINS, F., JR. 1990. Structure of a tobacco endochitinase gene: evidence that different chitinase genes can arise by transposition of sequences encoding a cysteine-rich domain. *Plant Mol. Biol.* **14**: 357–368.

SIVAN, A. and CHET, I. 1989. Degradation of fungal cell walls by lytic enzymes from *Trichoderma harzianum. J. Gen. Microbiol.* **135**: 675–682.

TANGARONE, B., ROYER, J. C., and NAKAS, J. P. 1989. Purification and characterization of an endo-(1-3)-β-D-glucanase from *Trichoderma longibrachiatum. Appl. Environ. Microbiol.* **55**: 177–184.

THOMAS, D. A., STARK, J. R., and PALMER, G. H. 1983. Purification of glucan hydrolases from a commercial preparation of *Trichoderma viride* by chromatofocusing. *Carbohydr. Res.* **114**: 343–345.

TSUNODA, A., NAGARI, T., SAKANO, Y., and KOBAYASHI, T. 1977. Purification and properties of an exo-α-1,3-glucanase from *Trichoderma viride. Agric. Biol. Chem.* **41**: 939–943.

TWEDDELL, R. J., JABAJI-HARE, S. H., and CHARES, P. M. 1994. Production of chitinases and β-1,3-glucanases by *Stachybotrys elegans*, a mycoparasite of *Rhizoctonia solani. Appl. Environ. Microbiol.* **60**: 489–495.

VAN TILBURG, A. U. B. and THOMAS, M. D. 1993. Production of extracellular proteins by the biocontrol fungus *Gliocladium virens. Appl. Environ. Microbiol.* **59**: 236–242.

VASILEVA-TONKOVA, E. S. and BEZBORODOVA, S. I. 1986. Purification, physiochemical properties, and specificity of a ribonuclease produced by *Trichoderma harzianum. Enzyme Microb. Technol.* **18**: 147–152.

VASSEUR, V., VAN MONTAGU, M., and GOLDMAN, G. H. 1995. *Trichoderma harzianum* genes induced during growth on *Rhizoctonia solani* cell walls. *Microbiology UK* **141**: 767–774.

VÁZQUEZ DE ALDANA, C. R., CORREA, J., SAN SEGUNDO, P., BUENO, A., NEBREDA, A. R., MÉNDEZ, E., and DEL REY, F. 1991. Nucleotide sequence of the exo-β-D-1,3-glucanase-encoding gene, *EXG1*, of the yeast *Saccharomyces cerevisiae*. *Gene* **97**: 173–182.

VIJAYAKUMAR, P., ROSS, W., and REESE, E. T. 1978. α,α′-Trehalase of *Trichoderma reesei*. *Can. J. Microbiol.* **24**: 1280–1283.

VRANJES, V. and WENK, C. 1996. Influence of *Trichoderma viride* enzyme complex on nutrient utilization and performance of laying hens in diets with and without antibiotic supplementation. *Poult. Sci.* **75**: 551–555.

WATANABE, T., OYANAGI, W., SUZUKI, K., OHNISHI, K., and TNAKA, H. 1992. Structure of the gene encoding chitinase D of *Bacillus circulans* WL-12 and possible homology of the enzyme to other prokaryotic chitinases and class III plant chitinases. *J. Bacteriol.* **174**: 408–414.

WEISER, W., LEHMANN, J., CHIBA, S., MATSUI, H., BREWER, C. F., and HEHRE, E. J. 1988. Steric course of the hydratation of D-gluco-octenitol catalyzed by α-glucosidases and by trehalase. *Biochem.* **27**: 2294–2300.

XIE, Q., JIMÉNEZ, A., RAMÓN, D., and PÉREZ-GONZÁLEZ, J. A. 1995. Construction of an endoglucanolytic brewing yeast strain. *J. Inst. Brew.* **101**: 459–461.

YAMAMOTO, T. 1988 *Handbook of Amylases and Related Enzymes*. Pergamon Press, Oxford.

YAMAMOTO, S. and NAGASAKI, S. 1975. Purification, crystallization, and properties of endo-β-1,3-glucanase from *Rhizopus chinensis* R-69. *Agric. Biol. Chem.* **39**: 2163–2169.

YAMASAKI, Y., ELBEIN, A. D., and KONNO, H. 1995. Purification and properties of α-glucosidase from *Trichoderma harzianum*. *Biosci. Biotech. Biochem.* **59**: 2181–2182.

ZONNEVELD, J. M. 1972. Morphogenesis in *Aspergillus nidulans*: the significance of α-1,3-glucan of the cell wall and α-1,3-glucanase for cheistothecium development. *Biochem. Biophys. Acta* **273**: 174–187.

PART TWO

Application of *Trichoderma* and *Gliocladium* in Agriculture

6

Trichoderma and *Gliocladium* in biological control: an overview

L. HJELJORD and A. TRONSMO

Department of Biotechnological Sciences, Agricultural University of Norway, Aas, Norway

6.1 The need for biologically based fungicides

Modern agriculture, based on growing one or a few crop cultivars over large areas, is an ecologically unbalanced system which invites disease epidemics. Prevention of such epidemics has traditionally been achieved through use of chemical fungicides. However, consumers are becoming increasingly concerned about chemical pollution of the environment and pesticide residues on food, and farmers are more often being faced with pathogens resistant to available chemical fungicides. Furthermore, there is a need for efficient measures to combat soilborne disease and inoculum buildup. Thus both consumers and industry are anxious to find alternative methods of disease control.

Replacement or reduction of chemical application has been achieved through use of biologically based fungicides, a concept included in the broad definition of biological control proposed by Cook and Baker (1983): "Biological control is the reduction of the amount of inoculum or disease-producing activity of a pathogen accomplished by or through one or more organisms other than man." This broad definition includes use of less virulent variants of the pathogen, more resistant cultivars of the host, and microbial antagonists "that interfere with the survival or disease-producing activities of the pathogen". This section will concern the use of *Trichoderma* and *Gliocladium* spp. as microbial antagonists, i.e., biological control agents, of plant pathogenic fungi.

As living organisms, biological control agents will react to changes in their habitat in a manner unattainable by chemical fungicides. As fast-growing saprophytes, established *Trichoderma* and *Gliocladium* isolates can compete ecologically over the long term as well as at the time of application and are able to colonize potential infection courts, e.g., growing roots, wounds, or senescent tissue, as they become available. As aggressive mycoparasites they can also attack already-established pathogens as well as those arriving later. On the other hand, like any other organism, a biological control agent will be affected by abiotic and biotic factors such as weather, disease pressure, and competition from the indigenous microflora. Incomplete and variable disease control will result unless the antagonists are used in formulations and application schedules that give them a competitive advantage over the pathogens they are intended to control.

Perhaps we should reconsider what we hope to achieve with biological control measures. Although the severity of disease on the current crop is the most obvious measure of effectiveness of a treatment, reduction of disease in future crops is at least as important. Biological control measures are unlikely to eradicate a pathogen or its damage. However, the antagonism of an established biological control agent can reduce a pathogen's ability to produce and maintain high levels of inoculum. In the absence of other sources of inoculum, future disease levels will eventually be reduced by such biological control measures (Fokkema, 1995). For example, when applications are timed appropriately, *Trichoderma* and *Gliocladium* spp. are able to suppress disease caused by pathogens such as *Botrytis* and *Sclerotinia* by reducing the pathogen's sporulation in the primary infection cycle as well as by preventing accumulation of sclerotia, through a combination of nutrient competition and direct mycoparasitism (Dubos, 1987).

As discussed in Volume 1 (Chapters 1, 2, and 11), isolates of *Trichoderma* and *Gliocladium* spp. show a great degree of genetic variation. Biological control strains have been found that are effective in a variety of habitats and against an assortment of pathogens. *Trichoderma* and *Gliocladium* isolates are being used against diseases in many different crops, e.g., cotton, grapes, sweet corn, lettuce, onion, peas, plum, apples, and carrots, caused by pathogens such as *Pythium*, *Phytophthora*, *Rhizoctonia*, *Sclerotinia*, *Botrytis*, and *Fusarium* (see Nelson, 1991, and references therein). Various isolates have been used with success in greenhouse and field applications in soil and in the phyllosphere, as well as in cold storage (reviewed in Chet, 1987; Papavizas, 1985; Tronsmo, 1986b).

The genetic and ecological aspects of the variation between and within *Trichoderma* and *Gliocladium* species are discussed in detail in Volume 1. This part will focus on the use of these genera in agriculture, primarily in disease control and plant growth stimulation. This chapter presents an overview of characteristics affecting the use of *Trichoderma* and *Gliocladium* as biological control agents, with emphasis on recent research. The remaining chapters in this part describe in more detail the current state of knowledge on most of the central topics related to the use of *Trichoderma* and *Gliocladium* in biological control.

6.2 Suitability of *Trichoderma* and *Gliocladium* spp. as biological control agents

The high degree of ecological adaptability shown by strains within the genus *Trichoderma* is reflected in the fact that these fungi are common in soils all over the world, under differing environmental conditions, and living on various substrates. This considerable variation, coupled with their amenability to cultivation on inexpensive substrates, makes *Trichoderma* isolates attractive candidates for a variety of biological control applications. However, genetic characterization of this genus is as yet incomplete, and morphological classification is insufficient to predict behaviour of an unknown isolate. Therefore, although potential biocontrol agents with suitable antagonistic characteristics may be readily found, they must be screened carefully for other traits relevant to their use in a given application.

Even after careful screening, bioprotection candidates often show variable performance in field trials. The reliability of biological control systems can be improved by the use of formulations that provide conducive environments for the bioprotectant and fermentation systems that economically produce propagules of high quality (Harman, 1991). To become a successful bioprotectant, a *Trichoderma* or *Gliocladium* strain must therefore not only show desirable antagonistic and ecological behaviour but must also be amenable to manipulation in relevant formulations and delivery systems. Nonetheless, the most appropriate criterion for selection of a biocontrol agent may ultimately be consideration of environmental requirements, both of the antagonist and of the pathogen which it is intended to control (Marois and Coleman, 1995).

6.2.1 Sensitivity to abiotic environmental factors

Trichoderma and *Gliocladium* strains have been isolated from habitats of varying moisture, temperature and nutrient status (Danielson and Davey, 1973a; Papavizas, 1985; Roiger *et al.*, 1991). The ecological adaptability shown by members of these genera makes them ideal candidates for biological control applications in a variety of habitats, but the sensitivity of different isolates to abiotic environmental factors must be considered.

Temperature

The natural distribution of *Trichoderma* and *Gliocladium* spp. is greatly influenced by habitat temperature, and the species groups vary in their temperature optima and tolerances (Domsch *et al.*, 1980). Isolates within a species-group may also vary in their antagonistic activity at different temperatures (Köhl and Schlösser, 1989; Tronsmo and Dennis, 1978).

The thermal activity spectrum of the pathogen is another relevant factor. Dubos (1987) noted that most strains of *Trichoderma* would be unable to cover the wide thermal activity spectrum of *Botrytis* spp. and emphasized the importance of selecting strains that are actively antagonistic at the temperatures most conducive to disease. The temperature tolerance of the biocontrol isolate relative to that of the pathogen could be critical to the success of a number of applications.

Moisture

Other than temperature, moisture is considered the environmental parameter most influencing the natural distribution of the various *Trichoderma* spp. in soil (Danielson and Davey, 1973a). It is also considered a factor limiting the ability of introduced antagonists to colonize the phyllosphere and thus to compete with and control foliar pathogens (Elad and Kirshner, 1992; Hannusch and Boland, 1996a,b; Köhl *et al.*, 1995a). Environmental moisture affects growing fungi physiologically, e.g., through the increased energy costs of osmoregulation under dry conditions or decreased oxygen availability in wet habitats. Furthermore, moisture affects nutrient availability, through solute transport and presence of water films. To a certain degree, *Trichoderma* spp. can adapt to unfavourable habitat moisture conditions by cytoplasmic translocation. Cell wall synthesis is prioritized over cytoplasm synthesis, and the cytoplasm is relocated to the growing tips as the hyphae extend; in this manner the mycelium can extend over areas of low osmotic potential or nutrient content towards more suitable areas (Paustian and Schnürer, 1987a,b). Although hyphal growth of *T. harzianum* in soil thus may be relatively insensitive to low soil matric potential (Knudsen and Bin, 1990), initiation of growth from dried mycelial biomass may require addition of an osmoregulant to the formulation (Knudsen *et al.*, 1991).

Conidial germination is moisture sensitive, and *Trichoderma* isolates colonize habitats in the phyllosphere more effectively under moist conditions (Elad and Kirshner, 1993; Gullino, 1992; Gullino *et al.*, 1989; Knudsen and Bin, 1990; Köhl *et al.*, 1995a,b). Changes of as little as 5% relative humidity may significantly affect the ability of *T. viride* or *G. roseum* to control *S. sclerotiorum* and *B. cinerea* on bean leaves; both agents perform best at 100% relative humidity (Hannusch and Boland, 1996a,b). In a greenhouse application, relative humidity was found to affect efficacy of the biological control agent more than temperature did (Elad *et al.*, 1993).

Despite the environmental limitations imposed by temperature and moisture, appropriately-timed field applications of *Trichoderma* and *Gliocladium* have been shown to control disease in phyllosphere applications. During periods in which humidity and temperature are unsuitable for the antagonists, limited use of chemical fungicides has proven to be an effective strategy for increasing the consistency of biological disease control in the phyllosphere (Dubos, 1987; Elad *et al.*, 1993; Gullino, 1992; Harman *et al.*, 1996; Tronsmo, 1991), as well as in soil (Chet, 1987).

Nutrients

Conidia of *Trichoderma* spp. require exogenous nutrients in order to germinate (Danielson and Davey, 1973b). Lockwood and coworkers have studied the ability of fungal conidia to germinate in soil and concluded that the availability of utilizable nutrients was the critical factor. They found that, as a rule, small-spored fungi with a relatively long germination time, such as *Trichoderma* spp., were more sensitive to a lack of exogenous nutrients than were larger-spored, nutritionally independent fungi (Ko and Lockwood, 1967; Steiner and Lockwood, 1969). *Trichoderma* conidia must take up water and swell considerably before the germ tubes emerge, and swelling will not occur without a given period of exposure to sufficient nutrients (carbon and nitrogen sources) (Danielson and Davey, 1973c; Hawker, 1966; Martin and Nicolas, 1970; Steiner and Lockwood, 1969).

Propagule type affects the dependency on exogenous nutrients (Lockwood, 1981). Both hyphae and chlamydospores of *Trichoderma* and *Gliocladium* spp. are less sensitive than conidia to soil fungistasis (Beagle-Ristaino and Papavizas, 1985b; Hsu and Lockwood, 1971; Lewis *et al.*, 1993). Propagule age also affects the nutrient requirement. Danielson and Davey (1973b) found that *Trichoderma* spores become more sensitive to a lack of nutrients as they become older. These authors caution that "in natural systems the ability of spores to germinate may be constantly changing and predictions of fungistatic sensitivity based on young spores from nutrient-rich cultures may be erroneous".

Many formulations have been designed to accommodate the nutrient dependency of *Trichoderma* and *Gliocladium* isolates. Although these genera are able to utilize a wide variety of food bases, the challenge lies in designing a formulation that will not enhance activity of undesirable organisms (Harman *et al.*, 1981). Strategies based on physically enclosing the food base, e.g., solid matrix priming of seeds (Harman *et al.*, 1989) or choosing food bases more or less exclusively available to the antagonists (Nelson *et al.*, 1988; Tronsmo *et al.*, 1993), are described in Chapter 11.

6.2.2 *Interactions with other microorganisms*

The specificity of *Trichoderma* and *Gliocladium* isolates has been debated. An ideal antagonist would control all plant pathogens but not negatively affect beneficial saprophytic or mycorrhizal fungi. Although it would be unrealistic to expect such specificity, intra- and interspecific variation is seen in the degree and nature of the interactions of *Trichoderma* biocontrol agents with other microorganisms.

Pathogen specificity

Trichoderma strains are usually considered to be effective only against specific pathogens. Evidence is accumulating that this may be related not only to the ecological characteristics described above, but also to the pathogen-specificity of antagonistic weapons such as antibiotics (Howell and Stipanovic, 1995) and cell wall degrading enzymes (Haran *et al.*, 1996). In fact, Bell and coworkers (1982) detected in paired cultures on agar that a biocontrol isolate could be highly effective against one isolate of a pathogen and yet have only minimal effect on other isolates of the same species. It was feared that a single biocontrol strain would be too specific, or too localized, and several authors have proposed combining several strains in a formulation to widen the range of control (Cook, 1993; Duffy *et al.*, 1996; Sivan *et al.*, 1984).

However, Harman and coworkers (1989) have shown that complex mixtures are not necessary to achieve satisfactory disease control. In numerous trials in several different systems (soil, seed treatment, and phyllosphere), the antagonist *T. harzianum* strain 1295-22, a product of protoplast fusion (Stasz *et al.*, 1988), provided good control against a range of pathogens, including *Pythium ultimum*, *Rhizoctonia solani*, *Fusarium* spp., *Sclerotium rolfsii* and *Botrytis cinerea*, if properly formulated (Harman, 1990; Chapter 11).

Mushrooms

Unfortunately, we cannot expect that all strains of *Trichoderma* and *Gliocladium* will attack only pathogens. Shiitake mushroom bedlogs have often been reported damaged by *Trichoderma* and *Gliocladium* spp. (Badham, 1991; Komatsu, 1976; Tokimoto, 1985). However, resistant strains of shiitake have now been selected (Tokimoto and Komatsu, 1995).

Serious epidemics of green mould on mushroom compost caused by *T. harzianum* isolates have been reported in Europe and North America, causing concern among growers about the use of *T. harzianum* in biological control. It has recently been proven that not all *T. harzianum* isolates from mushroom compost are harmful. The use of cultural (Seaby, 1996) and molecular (Muthumeenakshi *et al.*,

1994) methods has resulted in unambiguous identification of the virulent taxa, allowing assessment of risk to mushroom compost as well as analysis of biological variation within these groups (Muthumeenakshi *et al.*, 1994). The effect of *Trichoderma* spp. on mushroom culture is treated more fully in Chapter 12.

Mycorrhizae

Mycorrhizal fungi promote plant growth in a number of species, e.g., by increasing the efficiency of mineral and water uptake through their symbiotic colonization of the roots. By altering root exudation, mycorrhizal fungi also affect microorganisms in the rhizosphere (Linderman, 1988). However, the effects of saprophytic microorganisms in the rhizosphere on root colonization by mycorrhizal fungi are not well understood, and it is hypothesized that arbuscular mycorrhizal fungi may be subjected to antibiosis, cell wall degrading enzymes, and direct mycoparasitism by other microorganisms (Fitter and Garbaye, 1994). It is obviously necessary to clarify the effects on mycorrhizal fungi of biocontrol agents intended for use against soilborne pathogens, and a number of recent investigations have been devoted to this subject. There have been conflicting reports regarding the effects of *Trichoderma* spp. on mycorrhizal fungi.

Most studies of the effects of biocontrol agents on mycorrhizae have been concerned with arbuscular mycorrhizal (AM) fungi. Wyss and coworkers (1992) tested the effect of *T. harzianum* on the formation of arbuscular mycorrhizal mycelium in soybean and found that mycorrhizal formation with *Glomus mosseae* was significantly depressed in the presence of the biocontrol agent, possibly because *T. harzianum* induced accumulation of the phytoalexin glyceollin in the roots, which inhibited root mycorrhizal colonization.

McAllister *et al.* (1994a,b) found that when *T. koningii* was inoculated before or at the same time as *G. mosseae*, mycorrhizal formation was reduced. This inhibition was not seen when *T. koningii* was inoculated two weeks after *G. mosseae*, suggesting that the negative effect was due to a competitive interaction of the saprophytes with the extramatrical phase of the mycorrhizal fungus. Paulitz and Linderman (1991) suggested that since AM fungi do not grow saprophytically before colonizing root tissue, and after colonization derive their nutrition directly from the host, they would not be affected by nutrient competition in the rhizosphere. They saw no detrimental effects of *Gliocladium virens* on primary colonization by *Glomus mosseae*, but they pointed out that possible effects of competing rhizosphere microorganisms on secondary colonization and growth of external hyphae need clarification.

Rousseau and coworkers (1996), using an *in vitro* system, demonstrated that an aggressive *T. harzianum* isolate penetrated and grew within and through the cells of the mycorrhizal fungus *G. intraradices*. The authors noted that they used a *Trichoderma* strain selected for its superior biocontrol potential and that such antagonistic mechanisms might not be shared by all *Trichoderma* isolates. We would add that although superior biocontrol isolates are just those of interest, interactions between "trapped" organisms *in vitro* do not necessarily reflect those occurring in a natural habitat.

Mycorrhizal fungi are also regarded as biocontrol agents. Datnoff and coworkers (1995) tested the efficiency of commercial formulations of *G. intraradices* and *T. harzianum*, alone or in combination, in controlling Fusarium crown and root rot of tomato. Either biocontrol agent alone showed significant disease control, but better results were obtained when both agents were used together. Similar results were reported by Siddiqui and Mahmood (1996), who found that although *T. harzianum* adversely affected both *G. mosseae* and *Verticillium chlamydosporium*, these three biocontrol agents applied together controlled wilt disease in pigeonpea better than when applied individually.

T. aureoviride has been reported to stimulate *in vitro* spore germination of *G. mosseae* and development of AM mycelium (Calvet *et al.*, 1993). Furthermore, inoculation of *T. aureoviride* together with the AM fungus produced a synergistic enhancement of growth in the host plant (Calvet *et al.*, 1993; Camprubi *et al.*, 1995). On the other hand, *T. koningii* inhibited germination of *G. mosseae* spores *in vitro*, suggesting effects of both volatile and soluble compounds produced by the antagonist (McAllister *et al.*, 1994b).

The somewhat conflicting results mentioned above may be due to the variation shown by *Trichoderma* isolates in their interactions with mycorrhizal fungi (Summerbell, 1987), as well as due to differing effects of the interactions, even on species of the same genus (McAllister *et al.*, 1994b). A number of reports attribute mycorrhizal inhibition by *Trichoderma* spp. to volatile metabolites. Recent results indicate that the inhibitory effects may be temporary (Burla *et al.*, 1996), suggesting that some of the discrepancy in reported results may be due to the differing time frames of the investigations. Furthermore, some effects on mycorrhizal colonization have been attributed to edaphic conditions, e.g., pH changes due to the inoculum substrate (Paulitz and Linderman, 1991). Conflicting results may also reflect the specificity of mycorrhizal associations: mycorrhizal infection of different crop species has been shown to be differentially affected by biocontrol agents (Dhillion, 1994).

Clearly, more research is needed on the effects of *Trichoderma* and *Gliocladium* biocontrol isolates on the beneficial microflora in the rhizosphere. It would seem natural to test those isolates of *T. harzianum* that have been selected for their rhizosphere competence, as these are the most likely candidates for use in seed treatments and would logically be those most likely to affect mycorrhizal fungi.

6.2.3 Effects on plants

An important aspect of the use of *Trichoderma* and *Gliocladium* in biological control is how the agents themselves affect plants, apart from the effects of disease reduction. Some isolates of these fungi have been implicated as the causal agents of disease, while others have been acclaimed as plant growth stimulators. Between these extremes is the role of some isolates in inducing resistance reactions in plants. The effects of *Trichoderma* and *Gliocladium* on plant growth and resistance are treated more thoroughly in Chapters 9 and 11.

Over the years there have been scattered reports that isolates of *Trichoderma* can cause plant diseases under conducive conditions. Storage rots caused by *Trichoderma* isolates have been found in sweet potatoes (Cook and Taubenhaus, 1911), citrus fruit (Knösel and Schickedanz, 1976), and apples (Conway, 1983; English, 1944). Farr *et al.* (1989) have listed 32 genera of plants that can be attacked by *T. viride*. Menzies (1993) detected a *T. viride* isolated from a healthy tomato root that was pathogenic to seedlings of cucumber, pepper and tomato in laboratory and greenhouse experiments.

Investigating reports of damage in maize crops, McFadden and Sutton (1975) found several isolates of *T. koningii*, *T. harzianum*, and *T. hamatum* that were potentially pathogenic on maize. These researchers pointed out that maize residues incorporated into soil could support population levels of *T. koningii* high enough to cause both lesions and reduced growth in the following crop. In a later study, however, these isolates were found not to be pathogenic, the disparity perhaps being due to differing environmental conditions or maize genotypes (Windham *et al.*, 1989).

The above results point out two important aspects of the pathogenicity of *Trichoderma* spp.: (1) isolates of the same species-groups differ in pathogenicity, and (2) disease or damage is affected by environmental conditions. The step from saprophytic colonization of senescing or necrotic tissue to pathogenicity on living tissue may be taken by some isolates under conducive conditions, and once again we are reminded of the importance of incorporating relevant tests when screening for biological control agents.

6.2.4 *Risk assessment*

To be suitable for use in biological control, *Trichoderma* and *Gliocladium* isolates must not only be effective and controllable antagonists toward fungal phytopathogens but must also be harmless towards vertebrates, including humans (see Volume 1, Chapter 8). Risk assessment should include not only residuals on edible plant products but also effects of exposure to propagules of the biocontrol agent during field or greenhouse applications. For example, systems involving frequent spraying with large amounts of conidia need to be evaluated with respect to toxic or allergenic effects on workers. This important aspect of biological control in practice is the subject of a current investigation in Denmark on health aspects of biological control measures in greenhouses (B. Løschenkohl, personal communication).

The genetics and taxonomy of *Trichoderma* and *Gliocladium* are of more than purely academic interest. With respect to biological control applications, both patenting interests and risk assessment analyses are dependent on positive identification and classification of the biocontrol strains of interest. Clarification of genetic relationships will also be useful in sorting out the wide range of inter- and intraspecific variation in the biological character of these genera (Muthumeenakshi *et al.*, 1994). Unfortunately, according to Samuels (1996), "We still do not know what a species of *Trichoderma* is." This author has recently reviewed the current status of the systematics and taxonomy of *Trichoderma* (Samuels, 1996; Volume 1, Chapter 1).

Molecular identification of *Trichoderma* and *Gliocladium* isolates will facilitate monitoring of biological control strains in field trials (Hjeljord, 1996). Selectable markers have been used for several years to determine the survival of specific *T. harzianum* isolates in soil (Pe'er *et al.*, 1991) and on the phylloplane (Migheli *et al.*, 1994). Recent advances now allow monitoring of activity of released isolates as well. Transformation of a *T. harzianum* isolate with the β-glucuronidase reporter gene (Green and Jensen, 1995; Thrane *et al.*, 1995) has allowed studies of its activity in mixed populations in the soil (Green and Jensen, 1996). Monoclonal antibodies recognizing *T. harzianum* conidia (Thorton and Dewey, 1996) and mycelia (Thorton *et al.*, 1994) have allowed investigation of its interactions with *Rhizoctonia solani* in the soil (Thorton, 1996). Such techniques should prove invaluable to mechanistic studies of the interactions of these antagonists with both target and non-target species, as well as in investigating the fate of isolates introduced into various habitats.

6.3 Antagonistic mechanisms

Antagonistic interactions between *Trichoderma* and *Gliocladium* spp. and other fungi have traditionally been classified as antibiosis, mycoparasitism, and competition for nutrients. These mechanisms are not mutually exclusive, and a given antagonistic mechanism can fall into several of these categories. For example, the control of *Botrytis* on grapes by *Trichoderma* involves both nutrient competition and mycoparasitism of sclerotia, both of which result in suppression of the pathogen's ability to cause and perpetuate disease (Dubos, 1987). Both antibiosis and mycoparasitism may be involved in competition for nutrients; indeed, production of toxic metabolites is known to be affected by the nutrient status of the growth medium (Ghisalberti and Sivasithamparam, 1991; Howell and Stipanovic, 1995). Recent evidence has shown that antibiotics and hydrolytic enzymes are not only produced together but act synergistically in mycoparasitic antagonism (Di Pietro *et al.*, 1993; Schirmböck *et al.*, 1994).

Some antagonistic interactions do not fall readily into any of the classical categories. For example, it has recently been suggested that the biocontrol agent *T. harzianum* T39 reduces the pathogenicity of *B. cinerea* by reducing the amount of pectin-degrading enzymes produced by the pathogen (Zimand *et al.*, 1996).

The importance of a given antagonistic mechanism has been shown in many studies to be dependent on the antagonist strain, the target organism, and the environmental conditions. Thus selection of biological control agents should take into consideration the intended application as well as the target pathogen.

6.3.1 *Antibiosis*

Trichoderma and *Gliocladium* strains produce a variety of volatile and non-volatile secondary metabolites, some of which inhibit other microorganisms with which they are not physically in contact. Such inhibitory substances are considered antibiotics. The best known of the antifungal metabolites produced by these genera are gliotoxin, viridin, and gliovirin (Howell and Stipanovic, 1983), which are produced by some *Gliocladium* isolates, and the coconut-scented 6-*n*-pentyl-2H-pyran-2-one (PPT) (Claydon *et al.*, 1987), which is characteristic of some *Trichoderma* isolates. Although the extracellular cell wall degrading enzymes produced by many strains are also capable of "killing at a distance", these are traditionally included in the concept of mycoparasitism, due to their integral role in direct physical interactions.

Trichoderma and *Gliocladium* species groups differ in their abilities to produce various antibiotics, and the environment also affects production, both qualitatively and quantitatively; further, specific antibiotics affect various pathogens differently (Claydon *et al.*, 1987; Dennis and Webster, 1971a,b; Ghisalberti and Sivasithamparam, 1991; Howell and Stipanovic, 1995). There is also conflicting evidence regarding the role of antibiotic production in disease suppression. The role of antibiosis in biological control by *Trichoderma* and *Gliocladium* isolates is thoroughly discussed in Volume 1, Chapter 7 and in Chapter 8 of this volume.

6.3.2 Mycoparasitism

The four stages distinguished in the direct interactions between *Trichoderma* spp. and other fungi are collectively termed mycoparasitism (Chet, 1990): (a) chemotropic growth, in which a chemical stimulus from the target fungus attracts the antagonist; (b) specific recognition, probably mediated by lectins on the cell surfaces of both pathogen and antagonist; (c) attachment and coiling of the *Trichoderma* hyphae around its host; and (d) secretion of lytic enzymes that degrade the host wall. The vast amount of research in this field is described in Chapters 4, 5, and 7.

6.3.3 Competition

Competition, an important aspect of biological control, occurs when two or more microorganisms demand more of the same resource than is immediately available. Competition between a biocontrol agent and a pathogen may lead to disease control, if the antagonist's growth results in reduction of the pathogen's population or inoculum production. On the other hand, competition between an introduced antagonist and the indigenous microflora may make long-term establishment of an introduced antagonist difficult.

Lockwood (1981, 1992) and Wicklow (1992) have applied the ecological concepts of *exploitation competition* and *interference competition* to interactions between fungal populations. Interference competition involves behavioural or chemical mechanisms by which one organism limits another organism's access to the substrate and results from both interspecific and intraspecific mycelial interactions (Wicklow, 1992).

Exploitation competition regards depletion of a resource by a given organism or population without directly limiting another organism's access to that resource (Lockwood, 1992). As nutrients are the resource most likely to be lacking in fungal habitats, exploitation competition is often referred to as nutrient competition and may occur at any stage in the fungal life cycle, affecting quiescence of persistent structures, germination, or exploitation of a substrate (Lockwood, 1981). Nutrient competition is the most effective antagonistic mechanism by which the natural microflora prevents infection of the leaf surfaces (Blakeman, 1975; Fokkema, 1973). It is also likely to be the most effective use of introduced biocontrol agents (Blakeman, 1993).

As soil saprophytes, *Trichoderma* and *Gliocladium* spp. are biologically adapted to aggressive colonization of available nutrient bases and to quiescent persistence as chlamydospores and conidia when nutrients are lacking. The saprophytic characteristics of these genera are reflected in their use as biological control agents: when induced to commence growth by the presence of nutrients, these isolates colonize the substrate rapidly, sometimes employing antibiosis or direct mycoparasitism against their competitors. Their rapid growth rate, prolific conidiation, and range of variation in substrate utilization make *Trichoderma* and *Gliocladium* isolates very efficient saprophytes, and successful application of these isolates as biocontrol agents usually involves some form of nutrient competition. Severity of necrosis is directly correlated with the extent of the pathogen's mycelium on the plant surface. Competition is most likely to be a successful biological control strategy against pathogens dependent upon rapid colonization of the site or nutrient base if *Trichoderma* colonizes the site heavily before other microbes become established (Fokkema, 1993; Hannusch and Bolland, 1996b).

Competition for necrotic tissue

The best-known phyllosphere applications of *Trichoderma* and *Gliocladium* involve biocontrol of *Botrytis* and *Sclerotinia* spp., pathogens that opportunistically invade senescing or dead plant tissue as a nutritive base from which to invade healthy tissue. When sprayed onto grape flowers during blossoming, appropriately selected *Trichoderma* isolates are able to colonize tissue as it senesces, delaying colonization by *Botrytis* and reducing later disease levels in the fruit (Dubos, 1987; Gullino, 1992; Harman *et al.*, 1996). The importance of timing of application indicates that the inhibition is due to competitive colonization: repeated applications of the antagonist during flowering give the best results, but the most important spray is that during late flowering, when the petals begin to senesce. Application of the antagonist only at the time of berry ripening is not sufficient to prevent disease (Dubos, 1987). *Trichoderma* and *Gliocladium* isolates have also been used successfully to competitively control colonization by *Botrytis* and *Sclerotinia* on other fruits and vegetables, e.g., strawberry (Tronsmo and Dennis, 1977), apple (Tronsmo and Raa, 1977), and cucumber (Elad *et al.*, 1993) (see also Chapter 11).

A related strategy based on nutrient competition is disease control through suppression of inoculum production by *Sclerotinia* or *Botrytis* spp. on senescent plant tissue (Braun and Sutton, 1988; Fokkema, 1995). Sporulation suppression can be based on manual removal of plant residues after harvest (Köhl *et al.*, 1995b) but has been more practically achieved through pre-emptive colonization of the necrotic tissue by saprophytic antagonists such as *Trichoderma* and *Gliocladium* (Köhl *et al.*, 1995a; Sutton and Peng, 1993a; Whipps, 1987). As the degree of disease caused by facultative pathogens such as *Botrytis* and *Sclerotinia* spp. is directly related to the extent of their colonization (Hannusch and Boland, 1996b), environmental factors affecting the ability of a biocontrol agent to successfully compete with the pathogen for nutrients will determine the success of the biological control measures.

Competition for plant exudates

Damping-off diseases caused by *Pythium ultimum* in a number of cereal and vegetable crops is initiated by the rapid response of the pathogen to exudates from seeds. Sporangia germinate and infect the seed within hours of planting in Pythium-infested soil (Nelson, 1987). Seed treatment with *Trichoderma* isolates results in reduced sporangia germination (Ahmad and Baker, 1988b), a phenomenon attributed to competition for germination stimulants (Harman and Nelson, 1994; Chapter 11).

Not all isolates capable of protecting seeds from attack are able to control root rot or post-emergence damping off. To be effective against root diseases, a biocontrol agent must be able to establish itself in the immediate vicinity of the root, i.e., the rhizosphere, and grow fast enough to maintain a high population density in the root tip zone (Ahmad and Baker, 1988b), unless it is to be mixed with the entire soil volume. The biological characteristics responsible for the disease protection afforded by such rhizosphere competent *Trichoderma* isolates have not been fully elucidated; however, improved utilization of root exudates and cellulose substrates on or near the root plane appear to be advantageous (Ahmad and Baker, 1988a,c; Cotes *et al.*, 1996). A rhizosphere competent *T. harzianum* isolate, made by protoplast fusion between the two biocontrol agents *T. harzianum* T-95 and *T. harzianum* T-12 (Stasz *et al.*, 1988), has proven to be a better biocontrol agent than the parent strains, probably due to an improved ability to colonize the whole root surface.

Nutrient competition also seems to be the most potent mechanism employed by *T. harzianum* T-35 in the control of *Fusarium oxysporum* in the rhizosphere of cotton and melon (Sivan and Chet, 1989).

Competition on wound sites

Yet another form for nutrient competition is that seen in early colonization of fresh wound sites, which, because they are poorly colonized and vulnerable to pathogen attack, may be the most promising sites for biological control in the phyllosphere by introduced antagonists (Fokkema, 1995).

One of the first examples of successful biological control in pruning wounds was the use of *T. viride*, applied in sprays or via pruning shears, to control the silver leaf pathogen, *Chondrostereum purpureum* (Corke, 1974; Grosclaude *et al.*, 1973). Introduced isolates of *Trichoderma* spp. have also proven able to colonize freshly-cut stumps and prevent infection by the root pathogen, *Amillaria luteobubalina* (Nelson *et al.*, 1995).

Stem rot often follows *Botrytis* infection of pruning wounds in greenhouse tomato plants; this disease is difficult to control by cultural methods (O'Neill *et al.*, 1996). *Trichoderma* and *Gliocladium* isolates have proven able to control stem rot when inoculated together with or before *Botrytis*, but not when inoculated afterwards, suggesting that competitive colonization of the wound site is the determinative factor in disease reduction (Koning and Köhl, 1995; O'Neill *et al.*, 1996).

In a study of *Pythium* infection of cucumber roots, it was shown that although *T. harzianum* strain T3 did not colonize the whole root, it did actively colonize wound sites (Thrane *et al.*, 1997). The antagonist's competition for nutrients leaking from the wound was apparently the cause of reduced infection by *Pythium.*

6.4 Integrated control

The great interest in *Trichoderma* and *Gliocladium* as bioprotectants is due to the need for alternatives to chemical plant protection. Chemical fungicides have undesirable effects on the environment and, when used regularly, encourage development of fungicide resistance in pathogens.

Replacement of some of the chemical fungicide treatments with biological control agents not only reduces the input of chemicals into agricultural soils but can also result in improved disease control. By using chemical and biological control measures together (integrated control), the duration of active disease control will be extended. Chemical protectants are effective under climatic conditions or levels of disease pressure in which the biological antagonist is less effective, while an active biological control agent can prophylactically colonize wounds or senescing plant tissue.

Integrated control measures may also be synergistic. Even reduced amounts of the fungicide can stress and weaken the pathogen and render its propagules more susceptible to subsequent attack by the antagonist (Lorito *et al.*, 1996). Fungicide-resistant or -tolerant isolates for use in integrated control are usually readily obtained by selection on pesticide-containing media (Abd-El Moity *et al.*, 1982; Tronsmo, 1986a, 1991). *Trichoderma* and *Gliocladium* strains differ in their sensitivities to different pesticides (Koomen *et al.*, 1993). Tronsmo (1989) showed that, on the average, insecticides used at recommended concentrations were more inhibitory to *Trichoderma* spp. than were fungicides; however, the compatibility of a biological control antagonist with a given pesticide should be confirmed before its use in integrated control. Non-resistant isolates may also be used in integrated control programmes, e.g., following fumigation measures that reduce population levels of the indigenous microflora, thus facilitating colonization by the antagonist.

An integrated control programme consisting of two to three sprays with a fungicide-resistant antagonist and one spray with a benzimidazole or dicarboximide fungicide has been found to provide more consistent control than use of the antagonist alone (Gullino *et al.*, 1995). A similar alternation schedule has been recommended for greenhouse cucumbers (Elad *et al.*, 1993; Chapter 11).

6.5 Formulation and delivery

A significant advantage of *Trichoderma* and *Gliocladium* spp. is that they are culturable on a wide range of carbon and nitrogen sources and can be induced to produce hyphal biomass, conidia, or chlamydospores suitable for a variety of applications. The choice of propagule for the biological control preparation depends on both production system and intended application. Although readily produced and quicker to commence growth upon exposure to nutrients (Hsu and Lockwood, 1971), hyphal fragments are more sensitive to environmental stresses such as drying. Conidia are also easily produced in many fermentation systems and are resistant to drying, although their physiological state strongly influences this characteristic (Harman *et al.*, 1991). However, conidia are sensitive to soil fungistasis and thus are dependent on exogenous nutrients to commence growth (Danielson and Davey, 1973c). Chlamydospores of *Trichoderma* and *Gliocladium* are larger than conidia and are less dependent on added nutrients (Beagle-Ristaino and Papavizas, 1985b). The choice of propagule type has varied with application and researcher. Some workers have found hyphal biomass and chlamydospores superior for a number of applications (Beagle-Ristaino and Papavizas, 1985a; Lewis and Papavizas, 1983; Papavizas *et al.*, 1984), while others have found conidia practical to produce in large quantities and suitable for their applications (Harman *et al.*, 1991; Jin *et al.*, 1991, 1996; Tronsmo and Harman, 1992). Either liquid or semi-solid fermentation may be used to produce biomass for biological control applications; these are discussed in Chapter 10.

It is well known among biological control researchers that apparently very suitable bioprotection candidates can be incapable of controlling disease consistently in field tests. Varying environmental conditions of temperature, humidity, disease pressure, and nutrient availability affect the antagonistic abilities of *Trichoderma* and *Gliocladium* spp. An important step in developing a reliable biocontrol agent is discovering the proper formulation to overcome the environmental limitations and give the antagonist a competitive advantage over the pathogens and other microflora (Connick *et al.*, 1990; Harman, 1991).

Given a suitably antagonistic isolate and a formulation that enables it to efficiently colonize the target site, the remaining barrier to a successful biological control product is an efficient production and delivery system. The time is past when biological control tests were best described as "spray and pray". It is now recognized that timing of the introduction of an antagonist is vital to its success. Sutton and Peng (1993b) encourage a liberation from application strategies imitating those developed for fungicides ("nozzle-head syndrome") toward "biological systems" strategies taking into consideration population dynamics of biocontrol organisms as well as the infection cycle of the pathogen. These authors also question the efficiency of sprays, particularly where fruits or flowers are the target, and encourage consideration of direct application alternatives, such as bee vectors (Sutton and Peng, 1993b) or contact applicators (James and Sutton, 1996).

6.6 Conclusions

We have presented an overview of aspects of the biology and antagonism of *Trichoderma* and *Gliocladium* isolates relevant to their use as biological control agents. Many of these topics will be described in more detail in the remaining chapters in this part.

As research in this field progresses, increasing knowledge of the biological mechanisms and environmental factors governing the interactions between antagonists and phytopathogenic fungi is clarifying the variation in disease control traditionally associated with biological control measures. We are now able to screen isolates for traits relevant not only to their antagonism but also to their ability to colonize habitats to which they are not naturally adapted.

Colonization is the key to superior disease control by biological agents in niches unexploitable by chemical measures. For example, rhizosphere competent *Trichoderma* isolates can colonize the entire root zone and afford localized protection unattainable through chemical control. Furthermore, successful colonization by *Trichoderma* and *Gliocladium* isolates adapted to specific applications will provide persistent protection only achievable by repeated doses of chemicals. Biological control agents colonize competitively through a variety of antagonistic mechanisms and thus avoid the development of resistance in pathogens that so often renders chemical fungicides obsolete.

Applied in biologically based formulations and delivery systems, appropriately selected *Trichoderma* and *Gliocladium* isolates should provide consistent and effective disease control without the dangers of environmental pollution and fungicide resistance so often associated with the use of chemical fungicides.

References

Abd-El Moity, T. H., Papavizas, G. C., and Shatla, N. N. 1982. Induction of new isolates of *Trichoderma harzianum* tolerant to fungicides and their experimental use for control of white rot on onion. *Phytopathology* **72**: 396–400.

Ahmad, J. S. and Baker, R. 1988a. Growth of rhizosphere-competent mutants of *Trichoderma harzianum* on carbon substrates. *Can. J. Microbiol.* **34**: 807–814.

Ahmad, J. S. and Baker, R. 1988b. Implications of rhizosphere competence of *Trichoderma harzianum*. *Can. J. Microbiol.* **34**: 229–234.

Ahmad, J. S. and Baker, R. 1988c. Rhizosphere competence of benomyl-tolerant mutants of *Trichoderma* spp. *Can. J. Microbiol.* **34**: 694–696.

Badham, E. 1991. Growth and competition between *Lentinus edodes* and *Trichoderma harzianum* on sawdust substrates. *Mycologia* **83**: 455–463.

Beagle-Ristaino, J. E. and Papavizas, G. C. 1985a. Biological control of Rhizoctonia stem canker and black scurf of potato. *Phytopathology* **75**: 560–563.

Beagle-Ristaino, J. E. and Papavizas, G. C. 1985b. Survival and proliferation of propagules of *Trichoderma* spp. and *Gliocladium virens* in soil and in plant rhizospheres. *Phytopathology* **75**: 729–732.

Bell, D. K., Wells, H. D., and Markham, C. R. 1982. *In vitro* antagonism of *Trichoderma* species against six fungal plant pathogens. *Phytopathology* **72**: 379–382.

Blakeman, J. P. 1975. Germination of *Botrytis cinerea* conidia *in vitro* in relation to nutrient conditions on leaf surfaces. *Trans. Br. Mycol. Soc.* **65**: 239–247.

Blakeman, J. P. 1993. Pathogens in the foliar environment. *Plant Pathol.* **42**: 479–493.

Braun, P. G. and Sutton, J. C. 1988. Infection cycle and population dynamics of *Botrytis cinerea* in strawberry leaves. *Can. J. Plant Pathol.* **10**: 133–141.

Burla, M., Goverde, M., Schwinn, F. J., and Wiemken, A. 1996. Influence of biocontrol organisms on root pathogenic fungi and on the plant symbiotic microorganisms *Rhizobium phaseolii* and *Glomus mosseae*. *J. Plant Dis. Prot.* **103**: 156–163.

Calvet, C., Barea, J. M., and Pera, J. 1992. *In vitro* interactions between the vesicular-arbuscular mycorrizal fungus *Glomus mosseae* and some saprophytic fungi isolated from organic substrates. *Soil Biol. Biochem.* **24**: 775–780.

Calvet, C., Pera, J., and Barea, J. M. 1993. Growth response of marigold (*Tagetes erecta* L.) to inoculation with *Glomus mosseae*, *Trichoderma aureoviride* and *Pythium ultimum* in a peat-perlite mixture. *Plant and Soil* **148**: 1–6.

Camprubi, A., Calvet, C., and Estaun, V. 1995. Growth enhancement of *Citrus reshni* after inoculation with *Glomus intraradices* and *Trichoderma aureoviride* and associated effects on microbial populations and enzyme activity in potting mixes. *Plant and Soil* **173**: 233–238.

Chet, I. 1987. *Trichoderma* – application, mode of action, and potential as a biocontrol agent of soilborne plant pathogenic fungi. In I. Chet (ed.), *Innovative Approaches to Plant Disease Control*. Wiley, New York, pp. 137–160.

Chet, I. 1990. Mycoparasitism – recognition, physiology and ecology. In R. R. Baker and P. E. Dunn (eds), *New Directions in Biological Control: Alternatives for Suppressing Agricultural Pests and Diseases*. Alan Liss, New York, pp. 725–733.

Claydon, N., Allan, M., Hanson, J. R., and Avent, A. G. 1987. Antifungal alkyl pyrones of *Trichoderma harzianum*. *Trans. Br. Mycol. Soc.* **88**: 503–513.

Connick, W. J., Lewis, J. A., and Quimby, P. C. 1990. Formulation of biocontrol agents for use in plant pathology. In R. R. Baker and P. E. Dunn (eds), *New Directions in*

Biological Control: Alternatives for Suppressing Agricultural Pests and Diseases. Alan Liss, New York, pp. 345–372.

CONWAY, W. S. 1983. *Trichoderma harzianum*: a possible cause of apple decay in storage. *Plant Dis.* **67**: 916–917.

COOK, M. T. and TAUBENHAUS, J. J. 1911.*Trichoderma köningii* the cause of a disease of sweet potatoes. *Phytopathology* **1**: 184–189.

COOK, R. J. 1993. The role of biological control in pest management in the 21st century. In R. D. Lumsden and J. L. Vaughn (eds), *Pest Management: Biologically Based Technologies.* American Chemical Society, Washington, DC, pp. 10–20.

COOK, R. J. and BAKER, K. F. 1983. *The Nature and Practice of Biological Control of Plant Pathogens.* American Phytopathological Society, St Paul, MN, 539 pp.

CORKE, A. T. K. 1974. The prospect for biotherapy in trees infected by silver leaf. *J. Hort. Sci.* **49**: 391–394.

COTES, A. M., LEPOIVRE, P., and SEMAL, J. 1996. Correlation between hydrolytic enzyme activities measured in bean seedlings after *Trichoderma koningii* treatment combined with pregermination and the protective effect against *Pythium splendens. Eur. J. Plant Pathol.* **102**: 497–506.

DANIELSON, R. M. and DAVEY, C. B. 1973a. The abundance of *Trichoderma* propagules and the distribution of species in forest soils. *Soil Biol. Biochem.* **5**: 485–494.

DANIELSON, R. M. and DAVEY, C. B. 1973b. Non-nutritional factors affecting the growth of *Trichoderma* in culture. *Soil Biol. Biochem.* **5**: 495–504.

DANIELSON, R. M. and DAVEY, C. B. 1973c. Effects of nutrients and acidity on phialospore germination of *Trichoderma in vitro. Soil Biol. Biochem.* **5**: 517–524.

DATNOFF, L. E., NEMEC, S., and PERNEZNY, K. 1995. Biological control of Fusarium crown and root rot of tomato in Florida using *Trichoderma harzianum* and *Glomus intraradices. Biol. Control* **5**: 427–431.

DENNIS, C. and WEBSTER, J. 1971a. Antagonistic properties of species-groups of *Trichoderma.* I: Production of non-volatile antibiotics. *Trans. Br. Mycol. Soc.* **57**: 25–39.

DENNIS, C. and WEBSTER, J. 1971b. Antagonistic properties of species-groups of *Trichoderma.* II: Production of volatile antibiotics. *Trans. Br. Mycol. Soc.* **57**: 41–48.

DHILLION, S. S. 1994. Effect of *Trichoderma harzianum, Beijerinckia mobilis* and *Aspergillus niger* on arbuscular mycorrhizal infection and sporulation in maize, wheat, millet, sorghum, barley and oats. *J. Plant Dis. Prot.* **101**: 272–277.

DI PIETRO, A., LORITO, M., HAYES, C. K., BROADWAY, R. M., and HARMAN, G. E. 1993. Endochitinase from *Gliocladium virens*: isolation, characterization and synergistic antifungal activity in combination with gliotoxin. *Phytopathology* **83**: 308–313.

DOMSCH, K. H., GAMS, W., and ANDERSON, T.-H. 1980. *Compendium of Soil Fungi.* Academic Press, London, 859 pp.

DUBOS, B. 1987. Fungal antagonism in aerial agrobiocenoses. In I. Chet (ed.), *Innovative Approaches to Plant Disease Control.* Wiley, New York, pp. 107–135.

DUFFY, B. K., SIMON, A., and WELLER, D. M. 1996. Combination of *Trichoderma koningii* with fluorescent pseudomonads for control of take-all on wheat. *Phytopathology* **86**: 188–194.

ELAD, Y. and KIRSHNER, B. 1992. Establishment of an active *Trichoderma* population in the phylloplane and its effect on grey mould (*Botrytis cinerea*). *Phytoparasitica* **20**: 137–141.

ELAD, Y. and KIRSHNER, B. 1993. Survival in the phylloplane of an introduced biocontrol agent (*Trichoderma harzianum*) and populations of the plant pathogen *Botrytis cinerea* as modified by abiotic conditions. *Phytoparasitica* **21**: 303–313.

ELAD, Y., ZIMAND, G., ZAQS, Y., ZURIEL, S., and CHET, I. 1993. Use of *Trichoderma harzianum* in combination or alternation with fungicides to control cucumber grey mould (*Botrytis cinerea*) under commercial greenhouse conditions. *Plant Pathol.* **42**: 324–332.

ENGLISH, H. 1944. Notes on apple rots in Washington. *Plant Dis. Rep.* **28**: 610–622.

FARR, D. F., BILLS, G. F., CHAMURIS, G. P., and ROSSMAN, A. Y. 1989. *Fungi on Plants and Plant Products in the United States.* APS Press, St Paul, MN, 999 pp.

FITTER, A. H. and GARBAYE, J. 1994. Interactions between mycorrhizal fungi and other soil organisms. *Plant Soil* **159**: 123–132.

FOKKEMA, N. J. 1973. The role of saphrophytic fungi in antagonism against *Drechslera sorokiniana* (*Helminthosporium sativum*) on agar plates and on rye leaves with pollen. *Physiol. Plant Pathol.* **3**: 195–205.

FOKKEMA, N. J. 1995. Strategies for biocontrol of foliar fungal diseases. In M. Manka (ed.), *Environmental Biotic Factors in Integrated Plant Disease Control.* The Polish Phytopathological Society, Poznan, pp. 69–79.

GHISALBERTI, E. L. and SIVASITHAMPARAM, K. 1991. Antifungal antibiotics produced by *Trichoderma* spp. *Soil Biol. Biochem.* **23**: 1011–1020.

GREEN, H. and JENSEN, D. F. 1995. A tool for monitoring *Trichoderma harzianum.* II: The use of a GUS transformant for ecological studies in the rhizosphere. *Phytopathology* **85**: 1436–1440.

GREEN, H. and JENSEN, D. F. 1996. Monitoring of a GUS-transformed strain of *Trichoderma harzianum* in soil and rhizosphere. In D. F. Jensen, H. B. Jansson and A. Tronsmo (eds), *Monitoring Antagonistic Fungi Deliberately Released Into the Environment.* Kluwer, Dordrecht, pp. 77–83.

GROSCLAUDE, C., RICARD, J., and DUBOS, B. 1973. Inoculation of *Trichoderma viride* spores via pruning shears for biological control of *Stereum purpureum* on plum tree wounds. *Plant Dis. Rep.* **57**: 25–28.

GULLINO, M. L. 1992. Control of Botrytis rot of grapes and vegetables with *Trichoderma* spp. In E. C. Tjamos, G. C. Papavizas and R. J. Cook (eds), *Biological Control of Plant Diseases.* Plenum Press, New York, pp. 125–132.

GULLINO, M. L., ALOI, C., and GARIBALDI, A. 1989. Evaluation of the influence of different temperatures, relative humidities and nutritional supports on the antagonistic activity of *Trichoderma* spp. against grey mould. In R. Cavalioro (ed.), *Influence of Environmental Factors on the Control of Grape Pests, Diseases and Weeds.* A. A. Balkema, Rotterdam, pp. 231–236.

GULLINO, M. L., MONCHIERO, M., and GARIBALDI, A. 1995. Biological control of *Botrytis cinerea* of grapevine: critical analysis of the results. In J. M. Whipps and T. Gerlagh (eds), *Biological Control of Sclerotium-forming Pathogens*, Vol. 18(3). IOBC/WPRS Bulletin, Warwick, UK, pp. 84–86.

HANNUSCH, D. J. and BOLAND, G. J. 1996a. Influence of air temperature and relative humidity on biological control of white mold of bean (*Sclerotinia sclerotiorum*). *Phytopathology* **86**: 156–162.

HANNUSCH, D. J. and BOLAND, G. J. 1996b. Interactions of air temperature, relative humidity and biological control agents on grey mold of bean. *Eur. J. Plant Pathol.* **102**: 133–142.

HARAN, S., SCHICKLER, H., OPPENHEIM, A., and CHET, I. 1996. Differential expression of *Trichoderma harzianum* chitinases during mycoparasitism. *Phytopathology* **86**: 980–985.

HARMAN, G. E. 1990. Deployment tactics for biocontrol agents in plant pathology. In R. R. Baker and P. E. Dunn (eds), *New Directions in Biological Control: Alternatives for Suppressing Agricultural Pests and Diseases*, Alan Liss, New York, pp. 779–792.

HARMAN, G. E. 1991. Seed treatments for biological control of plant disease. *Crop Prot.* **10**: 166–171.

HARMAN, G. E. and NELSON, E. B. 1994. Mechanisms of protection of seed and seedlings by biological seed treatments: implications for practical disease control. In T. Martin (ed.), *Seed Treatment: Progress and Prospects.* British Crop Protection Council Monograph No. 57, Canterbury, UK, pp. 283–292.

HARMAN, G. E., CHET, I., and BAKER, R. 1981. Factors affecting *Trichoderma hamatum*

applied to seeds as a biocontrol agent. *Phytopathology* **71**: 569–572.
HARMAN, G. E., TAYLOR, A. G., and STASZ, T. E. 1989. Combining effective strains of *Trichoderma harzianum* and solid matrix priming to improve biological seed treatment. *Plant Dis.* **73**: 631–637.
HARMAN, G. E., JIN, X., STASZ, T. E., PERUZZOTTI, G., LEOPOLD, A. C., and TAYLOR, A. G. 1991. Production of conidial biomass of *Trichoderma harzianum* for biological control. *Biol. Control* **1**: 23–28.
HARMAN, G. E., LATORRE, B., AGOSIN, E., MARTIN, R. S., RIEGEL, D. G., NIELSEN, P. A., TRONSMO, A., and PEARSON, R. C. 1996. Biological and integrated control of Botrytis bunch rot of grape using *Trichoderma* spp. *Biol. Control* **7**: 259–266.
HAWKER, L. E. 1966. Germination: morphological and anatomical changes. In M. F. Madelin (ed.), *The Fungus Spore.* Butterworths, London, pp. 151–161.
HJELJORD, L. 1996. Techniques for monitoring *Trichoderma* in the phyllosphere. In D. F. Jensen, H. B. Jansson, and A. Tronsmo (eds), *Monitoring Antagonistic Fungi Deliberately Released into the Environment.* Kluwer, Dordrecht, pp. 159–166.
HOWELL, C. R. and STIPANOVIC, R. D. 1983. Gliovirin, a new antibiotic from *Gliocladium virens*, and its role in the biological control of *Pythium ultimum. Can. J. Microbiol.* **29**: 321–324.
HOWELL, C. R. and STIPANOVIC, R. D. 1995. Mechanisms in the biocontrol of *Rhizoctonia solani*-induced cotton seedling disease by *Gliocladium virens*: antibiosis. *Phytopathology* **85**: 469–472.
HSU, S. C. and LOCKWOOD, J. L. 1971. Responses of fungal hyphae to soil fungistasis. *Phytopathology* **61**: 1355–1362.
JAMES, T. D. W. and SUTTON, J. C. 1996. Biological control of Botrytis leaf blight of onion by *Gliocladium roseum* applied as sprays and with fabric applicators. *Eur. J. Plant Pathol.* **102**: 265–275.
JIN, X., HARMAN, G. E., and TAYLOR, A. G. 1991. Conidial biomass and desiccation tolerance of *Trichoderma harzianum* produced at different medium water potentials. *Biol. Control* **1**: 237–243.
JIN, X., TAYLOR, A. G., and HARMAN, G. E. 1996. Development of media and automated liquid fermentation methods to produce desiccation-tolerant propagules of *Trichoderma harzianum. Biol. Control* **7**: 267–274.
KNÖSEL, D. and SCHICKEDANZ, F. 1976. Temperaturanspruche und extracellulare enzymatiche aktivitat einiger aus Citrus-importen isolierte Pilze. *Phytopathol. Z.* **85**: 217–226.
KNUDSEN, G. R. and BIN, L. 1990. Effects of temperature, soil moisture, and wheat bran on growth of *Trichoderma harzianum* from alginate pellets. *Phytopathology* **80**: 724–727.
KNUDSEN, G. R., ESCHEN, D. J., DANDURAND, L. M., and WANG, Z. G. 1991. Method to enhance growth and sporulation of pelletized biocontrol fungi. *Appl. Environ. Microbiol.* **57**: 2864–2867.
KO, W. and LOCKWOOD, J. L. 1967. Soil fungistasis: relation to fungal spore nutrition. *Phytopathology* **57**: 894–901.
KÖHL, J. and SCHLÖSSER, E. 1989. Decay of sclerotia of *Botrytis cinerea* by *Trichoderma* spp. at low temperatures. *J. Phytopathol.* **125**: 320–326.
KÖHL, J., MOLHOEK, W. M. L., PLAS, C. H. v. d., and FOKKEMA, N. J. 1995a. Effects of antagonists on sporulation of *Botrytis cinerea* on dead lily leaves under different field conditions. In M. Manka (ed.), *Environmental Biotic Factors in Integrated Plant Disease Control.* The Polish Phytopathological Society, Poznan, pp. 309–311.
KÖHL, J., MOLHOEK, W. M. L., PLAS, C. H. v. d., and FOKKEMA, N. J. 1995b. Suppression of sporulation of *Botrytis* spp. as a valid biocontrol strategy. *Eur. J. Plant Pathol.* **101**: 251–259.
KOMATSU, M. 1976. Studies on *Hypocrea, Trichoderma* and allied fungi antagonistic to shiitake, *Lentinus edodes* (Berk.) *Sing. Rep. Tottori Mycol. Inst. (Japan)* **13**: 1–113.
KONING, G. P. and KÖHL, J. 1995. Wound protection by antagonists against Botrytis

stem rot in cucumber and tomato. In M. Manka (ed.), *Environmental Biotic Factors in Integrated Plant Disease Control*, The Polish Phytopathological Society, Poznan, pp. 313–316.

KOOMEN, I., CROSS, J. V., and BERRIE, A. M. 1993. Effects of pesticides on growth and spore germination of selected *Trichoderma* species. *Ann. Appl. Biol.* **122** (Suppl.): 36–37.

LEWIS, J. A. and PAPAVIZAS, G. C. 1983. Production of chlamydospores and conidia by *Trichoderma* spp. in liquid and solid growth media. *Soil Biol. Biochem.* **15**: 351–357.

LEWIS, J. A., PAPAVIZAS, G. C., and HOLLENBECK, M. D. 1993. Biological control of damping-off of snapbeans caused by *Sclerotium rolfsii* in greenhouse and field with formulation of *Gliocladium virens*. *Biol. Control* **3**: 109–115.

LINDERMAN, R. G. 1988. Mycorrhizal interactions with the rhizosphere microflora: the mycorrhizosphere effect. *Phytopathology* **78**: 366–371.

LOCKWOOD, J. L. 1981. Exploitation competition. In D. T. Wicklow and G. C. Carroll (eds), *The Fungal Community: Its Organization and Role in the Ecosystem.* Marcel Dekker, New York, pp. 319–349.

LOCKWOOD, J. L. 1992. Exploitation competition. In G. C. Carroll and D. T. Wicklow (eds.), *The Fungal Community: Its Organization and Role in the Ecosystem*, 2nd edn. Marcel Dekker, New York, pp. 243–263.

LORITO, M., WOO, S. L., D'AMBROSIO, M. D., HARMAN, G. E., HAYES, C. K., KUBICEK, C. P., and SCALA, F. 1996. Synergistic interaction between cell wall degrading enzymes and membrane affecting compounds. *Molec. Plant–Microbe Interact.* **9**: 206–213.

MAROIS, J. J. and COLEMAN, P. M. 1995. Ecological succession and biological control in the phyllosphere. *Can. J. Bot.* **73** (Suppl. 1): S76–S82.

MARTIN, J. F. and NICOLAS, G. 1970. Physiology of spore germination in *Penicillium notatum* and *Trichoderma lignorum*. *Trans. Br. Mycol. Soc.* **55**: 141–148.

MCALLISTER, C. B., GARCIA-ROMERA, I., GODEAS, A., and OCAMPO, J. A. 1994a. Interactions between *Trichoderma koningii*, *Fusarium solani* and *Glomus mosseae*: effects on plant growth, arbuscular mycorrhizas and the saprophyte inoculants. *Soil Biol. Biochem.* **26**: 1363–1367.

MCALLISTER, C. B., GARCIA-ROMERA, I., GODEAS, A., and OCAMPO, J. A. 1994b. *In vitro* interactions between *Trichoderma koningii*, *Fusarium solani* and *Glomus mosseae*. *Soil Biol. Biochem.* **26**: 1369–1374.

MCFADDEN, A. G. and SUTTON, J. C. 1975. Relationships of populations of *Trichoderma* spp. in soil to disease in maize. *Can. J. Plant Sci.* **55**: 579–586.

MENZIES, J. G. 1993. A strain of *Trichoderma viride* pathogenic to germinating seedlings of cucumber, pepper and tomato. *Plant Pathol.* **42**: 784–791.

MIGHELI, Q., HERRERA-ESTRELLA, A., AVATANEO, M., and GULLINO, M. L. 1994. Fate of transformed *Trichoderma harzianum* in the phylloplane of tomato plants. *Molec. Ecol.* **3**: 153–159.

MUTHUMEENAKSHI, S., MILLS, P. R., BROWN, A. E., and SEABY, D. A. 1994. Intraspecific molecular variation among *Trichoderma harzianum* isolates colonizing mushroom compost in the British Isles. *Microbiology* **140**: 769–777.

NELSON, E. B. 1987. Rapid germination of sporangia of *Pythium* species in response to volatiles from germinating seeds. *Phytopathology* **77**: 1108–1112.

NELSON, E. B. 1991. Current limits to biological control of fungal phytopathogens. In D. K. Arora, B. Rai, D. G. Mukerji and G. R. Knudsen (eds), *Handbook of Applied Mycology: Soil and Plants.* Marcel Dekker, New York, pp. 327–355.

NELSON, E. B., HARMAN, G. E., and NASH, G. T. 1988. Enhancement of *Trichoderma*-induced biological control of Pythium seed rot and pre-emergence damping-off of peas. *Soil Biol. Biochem.* **20**: 145–150.

NELSON, E. E., PEARCE, M. H., and MALAJCZUK, N. 1995. Effects of *Trichoderma* spp. and ammonium sulphamate on establishment of *Armillaria luteobubalina* on stumps of

Eucalyptus diversicolor. Mycol. Res. **99**: 957–962.

O'NEILL, T. M., NIV, A., ELAD, Y., and SHTIENBERG, D. 1996. Biological control of *Botrytis cinerea* on tomato stem wounds with *Trichoderma harzianum. Eur. J. Plant Pathol.* **102**: 635–643.

PAPAVIZAS, G. C. 1985. *Trichoderma* and *Gliocladium*: biology, ecology, and potential for biocontrol. *Annu. Rev. Phytopathol.* **23**: 23–54.

PAPAVIZAS, G. C., DUNN, M. T., LEWIS, J. A., and BEAGLE-RISTAINO, J. 1984. Liquid fermentation technology for experimental production of biocontrol fungi. *Phytopathology* **74**: 1171–1175.

PAULITZ, T. C. and LINDERMAN, R. G. 1991. Lack of antagonism between the biocontrol agent *Gliocladium virens* and vesicular arbuscular mycorrhizal fungi. *New Phytol.* **117**: 303–308.

PAUSTIAN, K. and SCHNÜRER, J. 1987a. Fungal growth response to carbon and nitrogen limitation: a theoretical model. *Soil Biol. Biochem.* **19**: 613–620.

PAUSTIAN, K. and SCHNÜRER, J. 1987b. Fungal growth response to carbon and nitrogen limitation: application of a model to laboratory and field data. *Soil Biol. Biochem.* **19**: 621–629.

PE'ER, S., BARAK, Z., YARDEN, O., and CHET, I. 1991. Stability of *Trichoderma harzianum amd*S transformants in soil and rhizosphere. *Soil Biol. Biochem.* **23**: 1043–1046.

ROIGER, D. J., JEFFERS, S. N., and CALDWELL, R. W. 1991. Occurrence of *Trichoderma* species in apple orchard and woodland soils. *Soil Biol. Biochem.* **23**: 353–359.

ROUSSEAU, A., BENHAMOU, N., CHET, I., and PICHÉ, Y. 1996. Mycoparasitism of the extramatrical phase of *Glomus intraradices* by *Trichoderma harzianum. Phytopathology* **86**: 434–442.

SAMUELS, G. J. 1996. *Trichoderma*: a review of biology and systematics of the genus. *Mycol. Res.* **100**: 923–935.

SCHIRMBÖCK, M., LORITO, M., WANG, Y.-L., HAYES, C. K., ARISAN-ATAC, I., SCALA, F., HARMAN, G. E., and KUBICEK, C. P. 1994. Parallel formation and synergism of hydrolytic enzymes and peptaibol antibiotics, molecular mechanisms involved in the antagonistic action of *Trichoderma harzianum* against phytopathogenic fungi. *Appl. Environ. Microbiol.* **60**: 4364–4370.

SEABY, D. A. 1996. Differentiation of *Trichoderma* taxa associated with mushroom production. *Plant Pathol.* **45**: 905–912.

SIDDIQUI, Z. A. and MAHMOOD, I. 1996. Biological control of *Heterodera cajani* and *Fusarium udum* on pigeonpea by *Glomus mosseae, Trichoderma harzianum,* and *Verticillium chlamydosporium. Israel J. Plant Sci.* **44**: 49–56.

SIVAN, A. and CHET, I. 1989. The possible role of competition between *Trichoderma harzianum* and *Fusarium oxysporum* on rhizosphere colonization. *Phytopathology* **79**: 198–203.

SIVAN, A., ELAD, Y., and CHET, I. 1984. Biological control effects of a new isolate of *Trichoderma harzianum* on *Pythium aphanidermatum. Phytopathology* **74**: 498–501.

STASZ, T. E., HARMAN, G. E., and WEEDEN, N. F. 1988. Protoplast preparation and fusion in two biocontrol strains of *Trichoderma harzianum. Mycologia* **80**: 141–150.

STEINER, G. W. and LOCKWOOD, J. L. 1969. Soil fungistasis: sensitivity of spores in relation to germination time and size. *Phytopathology* **59**: 1084–1092.

SUMMERBELL, R. C. 1987. The inhibitory effect of *Trichoderma* species and other soil microfungi on formation of mycorrhiza by *Laccaria bicolor in vitro. New Phytol.* **105**: 437–448.

SUTTON, J. C. and PENG, G. 1993a. Biocontrol of *Botrytis cinerea* in strawberry leaves. *Phytopathology* **83**: 615–621.

SUTTON, J. C. and PENG, G. 1993b. Manipulation and vectoring of biocontrol organisms to manage foliage and fruit diseases in cropping systems. *Annu. Rev. Phytopathol.* **31**: 473–493.

THORTON, C. R. 1996. Development of monoclonal antibody-based immunoassays for the quantification of *Rhizoctonia solani* and *Trichoderma harzianum* in soil. In D. F. Jensen, H. B. Jansson and A. Tronsmo (eds.), *Monitoring Antagonistic Fungi Deliberately Released into the Environment*. Kluwer, Dordrecht, pp. 147–153.

THORTON, C. R. and DEWEY, F. M. 1996. Detection of phialoconidia of *Trichoderma harzianum* in peat-bran by monoclonal antibody-based enzyme-linked immunosorbent assay. *Mycol. Res.* **100**: 217–222.

THORTON, C. R., DEWEY, F. M., and GILLIGAN, C. A. 1994. Development of a monoclonal antibody-based enzyme-linked immunosorbent assay for the detection of live propagules of *Trichoderma harzianum* in a peat-bran medium. *Soil Biol. Biochem.* **26**: 909–920.

THRANE, C., TRONSMO, A., and JENSEN, D. F. 1997. Endo 1,3-β-glucanase and cellulase from *Trichoderma harzianum*: Purification and partial characterization, induction of and biological activity against plant pathogenic *Pythium* spp. *Eur. J. Plant Pathol.* **103**: 331–344.

THRANE, C., LÜBECK, M., GREEN, H., DEGEHFU, Y., ALLERUP, S., THRANE, U., and JENSEN, D. F. 1995. A tool for monitoring *Trichoderma harzianum*. I: Transformation with the GUS gene by protoplast technology. *Phytopathology* **85**: 1428–1435.

TOKIMOTO, K. 1985. Physiological studies on antagonism between *Lentinus edodes* and *Trichoderma* spp. in bedlogs of the former. *Rep. Tottori Mycol. Inst.* (*Japan*) **23**: 1–54.

TOKIMOTO, K. and KOMATSU, M. 1995. Selection and breeding of shiitake strains resistant to *Trichoderma* spp. *Can. J. Bot.* **73**(Suppl.1): 962–966.

TRONSMO, A. 1986a. *Trichoderma* used as a biocontrol agent against *Botrytis cinerea* rots on strawberry and apple. *Sci. Rep. Agric. Univ. Norway* **65**: 1–22.

TRONSMO, A. 1986b. Use of *Trichoderma* spp. in biological control of necrotropic pathogens. In N. J. Fokkema, and J. v. d. Heuvel (eds), *Microbiology of the Phyllosphere*. Cambridge University Press, Cambridge, pp. 348–362.

TRONSMO, A. 1989. Effect of fungicides and insecticides on growth of *Botrytis cinerea*, *Trichoderma viride* and *T. harzianum*. *Norw. J. Agric. Sci.* **3**: 151–156.

TRONSMO, A. 1991. Biological and integrated controls of *Botrytis cinerea* on apple with *Trichoderma harzianum*. *Biol. Control* **1**: 59–62.

TRONSMO, A. and DENNIS, C. 1977. The use of *Trichoderma* species to control strawberry fruit rots. *Neth. J. Plant Pathol.* **83** (Suppl. 1): 449–455.

TRONSMO, A. and DENNIS, C. 1978. Effect of temperature on antagonistic properties of *Trichoderma* species. *Trans. Br. Mycol. Soc.* **71**: 469–474.

TRONSMO, A. and HARMAN, G. E. 1992. Coproduction of chitinolytic enzymes and biomass for biological control by *Trichoderma harzianum* on media containing chitin. *Biol. Control* **2**: 272–277.

TRONSMO, A. and RAA, J. 1977. Antagonistic action of *Trichoderma pseudokoningii* against the apple pathogen *Botrytis cinerea*. *Phytopathol. Z.* **89**: 216–220.

TRONSMO, A., SKAUGRUD, Ø., and HARMAN, G. E. 1993. Use of chitin and chitosan in biological control of plant diseases. In R. A. A. Muzzarelli (ed.), *Chitin Enzymology*. European Chitin Society, Lyon and Ancona, pp. 265–270.

WHIPPS, J. M. 1987. Behaviour of fungi antagonistic to *Sclerotinia sclerotiorum* on plant tissue segments. *J. Gen. Microbiol.* **133**: 1495–1501.

WICKLOW, D. T. 1992. Interference competition. In G. C. Carroll and D. T. Wicklow (eds), *The Fungal Community: Its Organization and Role in the Ecosystem*, 2nd edn. Marcel Dekker, New York, pp. 265–274.

WINDHAM, G. L., WINDHAM, M. T., and WILLIAMS, W. P. 1989. Effects of *Trichoderma* spp. on maize growth and *Meloidogyne arenaria* reproduction. *Plant Dis.* **73**: 493–495.

WYSS, P., BOLLER, T., and WIEMKEN, A. 1992. Testing the effect of biological control agents on the formation of vesicular arbuscular mycorrhiza. *Plant and Soil* **147**: 159–162.

ZIMAND, G., ELAD, Y., and CHET, I. 1996. Effect of *Trichoderma harzianum* on *Botrytis cinerea* pathogenicity. *Phytopathology* **86**: 1255–1260.

7

Mycoparasitism and lytic enzymes

I. CHET*, N. BENHAMOU† and S. HARAN*

* *Otto Warburg Center for Agricultural Biotechnology, The Hebrew University of Jerusalem, Rehovot, Israel, and*
† *Recherche en Sciences de la vie et de la santé, Université Laval, Sainte-Foy, Québec, Canada*

Mycoparasitism is the direct attack of one fungus on another and is only one of three main antagonistic interactions between microorganisms, the other two being antibiosis and competition (Chet, 1987). *Trichoderma* spp. are mycoparasitic biocontrol agents of several economically important plant-pathogenic fungi. Parasitism by *Trichoderma* spp. is destructive, causing the death of the host fungus (Barnett and Binder, 1973). There are many research reports dealing with mycoparasitism as it relates to biological control (Baker and Cook, 1974; Chet, 1987). Here we concentrate on the molecular and biochemical events involved in this phenomenon and on the cellular mechanisms of the mycoparasitic interaction.

7.1 Sequential events involved in mycoparasitism

7.1.1 Chemotropic growth

Positive chemotropism is directed growth towards a chemical stimulus (Chet, 1990). As early as 1981, Chet *et al.* showed that *Trichoderma* can detect its host from a distance. The mycoparasite begins to branch in an atypical way, with the branches growing towards the target fungus. Although *Trichoderma* appears to grow according to a chemical gradient, no specific stimuli other than amino acids and sugars have been reported to induce this growth. It is, therefore, still unclear whether this phenomenon is host specific. Although chemotropism has been suggested to hold some advantages for the antagonist, it is not considered an essential step for mycoparasitism (Chet, 1990).

7.1.2 Recognition

In all experiments dealing with *Trichoderma* spp., specificity of the antagonist for a range of host fungi has been reported. This has led to the idea that molecular recognition between both partners is an essential event preceding the antagonistic process. The recognition may be physical, as in thigmotropism, or chemical (chemotropism). The latter involves hydrophobic interactions, or interactions between complementary molecules present at the surface of both the host and the parasite. Such recognition processes include lectin–carbohydrate interactions. Lectins are sugar-binding proteins or glycoproteins which agglutinate cells and are involved in the interactions between the cell surface components and their extracellular environment (Barondes, 1981). Elad *et al.* (1983a) demonstrated the role of lectins in the host–mycoparasite relationship between *T. harzianum* (isolate 203) and the plant-pathogenic fungus *Rhizoctonia solani*. The authors found that *R. solani* hyphae contain a lectin that specifically agglutinates erythrocytes of type O but not those of types A or B. This specific interaction was inhibited by pre-incubation of *R. solani* hyphae with fucose or galactose but was not modified by the other sugars tested. These findings led the authors to assume that a lectin on *R. solani* was involved in the early interaction and that the *Trichoderma* cell walls contained suitable binding sites, such as fucose or galactose, for this lectin. Indeed, they found that *Trichoderma* cell walls contain galactose and suggested that the lectin present in *R. solani* hyphae displays the ability to recognize the galactose residues on the *Trichoderma* cell walls and that this binding plays a key role in prey recognition by the predator. Barak *et al.* (1986) further investigated the potential role of lectins in the *Trichoderma–Rhizoctonia* interaction, using bacterial cells instead of erythrocytes for the agglutination assays. The authors compared the effects of various sugars on the adherence of *Escherichia coli* to *R. solani* hyphae with their effects on the fungus–fungus interaction. They found that methyl-L-fucoside, an inhibitor of *Rhizoctonia* agglutinin, prevents coiling of the biocontrol agent around *Rhizoctonia* hyphae. They also demonstrated the presence of L-fucosyl residues on the *Trichoderma* cell wall surface and suggested that these could serve as receptors for the *Rhizoctonia* agglutinin. They hypothesized that this recognition was a very early event in the fungus–fungus interactions subsequently leading to mycoparasitism. The viability of the interacting fungi during the mycoparasitic process was determined using fluorescein diacetate (FDA). This substrate penetrates the fungal cells as a non-fluorescent substrate and is then hydrolyzed by a number of different enzymes to produce fluorescein (Barak and Chet, 1986). The authors compared the vital staining of *R. solani* and *T. hamatum* when grown alone with that of each of the fungi during the interaction. They found that the attacked *Rhizoctonia* hyphae lose their viability faster than the unattacked mycelium. On the other hand, the parasitizing *Trichoderma* hyphae retained their fluorescence longer than the non-parasitizing hyphae, probably because of the abundant food source supporting the growth of the parasitic fungus. This work demonstrated that the initial attack of *T. hamatum* on *R. solani* occurs while the host is still alive and that, as early as 24 h after contact is established, the host is damaged.

Barak *et al.* (1985) found that another important plant pathogen, *Sclerotium rolfsii*, produces a lectin that is associated with its extracellular polysaccharide. The agglutination activity of this lectin was specifically inhibited by D-glucose and D-mannose. The authors reported that the ability of different isolates of *Trichoderma* spp. to attack *S. rolfsii* is associated with the agglutination of *Trichoderma* conidia by the pathogen and suggested that the agglutinin produced by *S. rolfsii* is involved in the specific recognition between *Trichoderma* isolates and this pathogen. Barak and Chet (1990) purified this lectin and found that it is strongly associated with wall-bound β-1,3-glucans from the pathogen cell wall. The purified lectin appeared as two protein bands of 55 and 60 kDa on an SDS-polyacrylamide gel. Using specific antibodies raised against this lectin, the authors provided evidence that the lectin is located at restricted sites along the *S. rolfsii* hyphae. They found that it adsorbs only to conidia of *T. hamatum* T-244, the antagonist of *S. rolfsii*, and not to those of non-antagonistic strains of *Trichoderma*, and they therefore suggested that this lectin determines the specificity of the parasitic interaction.

Another lectin from *S. rolfsii* was purified by Inbar and Chet (1994). This lectin was of 45 kDa and its agglutination activity was not inhibited by any of the mono- or disaccharides tested but was inhibited by the glycoproteins mucin and asialomucin. Proteases, as well as β-1,3-glucanases, totally destroyed the agglutination activity, indicating the necessity of both protein and β-glucan for agglutination. The authors developed a biomimetic system based on nylon fibers, designed to mimic the pathogen hyphae (Inbar and Chet, 1992). The system, based on the covalent binding of lectins to the surface of these fibers, enabled a dissection of the role of lectins in mycoparasitism. The purified lectin, bound to the surface of these nylon fibers, specifically induced mycoparasitic behavior in *T. harzianum*, a phenomenon which was not observed with the non-agglutinating protein-treated fibers or with untreated fibers (Inbar and Chet, 1994). These findings provided direct evidence for the involvement of lectins in mycoparasitism. The authors suggested that recognition and attachment of the mycoparasitic fungus *Trichoderma* to the *S. rolfsii* cell wall surface is mediated by a complexed agglutinating polymer that surrounds the host hyphae. This recognition initiates differentiation processes in *Trichoderma* which lead to the formation of infection structures and then to the sequence of events that eventually lead to host breakdown.

Recently, Gomez *et al.* (1997) investigated *Trichoderma–Trichoderma* interactions. They grouped *Trichoderma* isolates according to RAPD analyses and electrophoretic karyotypes and raised an interesting question: Would *Trichoderma* isolates of the same group recognize each other as the same type and form vegetative compatibility groups capable of undergoing anastomosis? Using confrontation assays they found that isolates of the same group exhibit compatible interactions that result in hyphal fusions and noted that these interactions are never associated with cell death. In contrast, interactions between isolates from different groups resulted in plasma membrane collapse and cell damage, indicating antagonistic behavior. These results suggested that a recognition event is also involved in the *Trichoderma–Trichoderma* interactions.

7.1.3 Attachment and coiling

Following recognition, *Trichoderma* hyphae attach to the host via the formation of hook-like structures and appressorium-like bodies and coil around the pathogen hyphae (Elad *et al.*, 1983a; Harman *et al.*, 1981). An example of this phenomenon is illustrated in Figure 7.1, which clearly shows coiling of mycoparasite hyphae around *R. solani* hyphae in a dual-culture experiment (Elad *et al.*, 1983b). The question of whether these hyphal interactions actually occur in the soil has only recently been investigated (Inbar *et al.*, 1996). *T. harzianum* conidia were found to germinate in sterile soil and form a mycelium that grew through the soil towards the mycelium of the host. When contact between both partners was established, the antagonist coiled around its host and developed typical appressorium-like structures. The intensity of these parasitic interactions in the soil was lower than that observed in the dual-

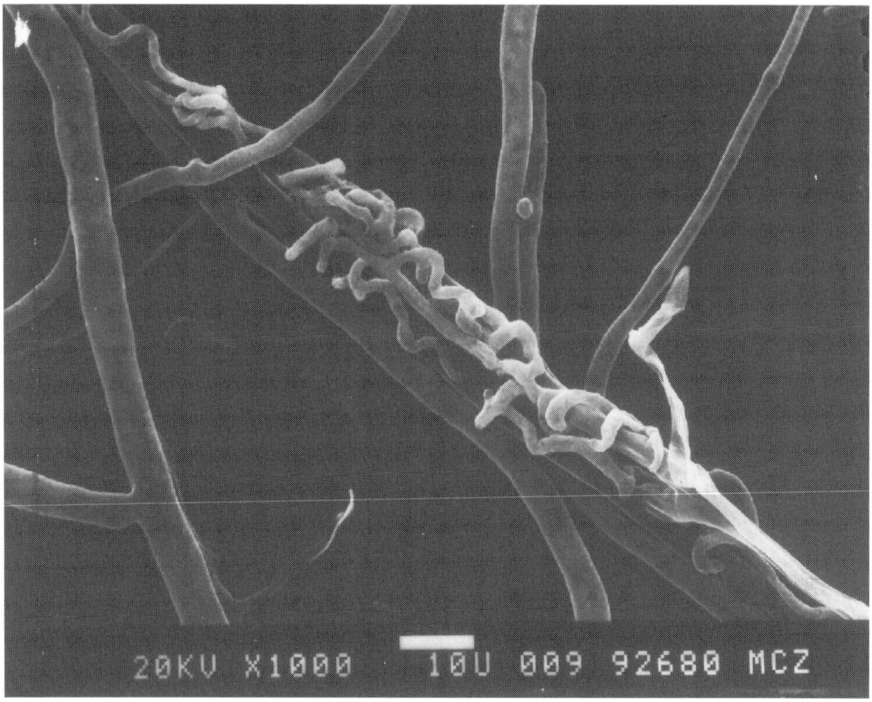

Figure 7.1 Mycoparasitic relationship. Scanning electron micrograph of *Trichoderma* hyphae coiled around *R. solani* hyphae (from Elad *et al.*, 1983b).

culture experiments, probably because of the lower concentration of nutrients in the soil supporting *Trichoderma* growth. This typical attachment and coiling appears to be the last step before lytic enzyme activity begins.

Attachment of another biocontrol agent, the antagonistic yeast *Pichia guilliermondii*, to the fungal plant pathogen *Botrytis cinerea* was observed by Wisniewski *et al.* (1991). This attachment could be blocked by agents that alter protein integrity (salts, proteases, etc.) as well as by certain sugars. The authors suggested that the mode of action of this postharvest biocontrol yeast combines tenacious attachment with secretion of cell wall degrading enzymes.

7.1.4 *Lytic activity*

Trichoderma spp. are known for their ability to degrade fungal cell walls (Baker, 1987; Chet, 1987). The molecular mechanisms via which lytic enzymes are involved in the biocontrol activity of *T. harzianum* have been recently reviewed by Haran *et al.* (1996a). Degradation of fungal cell walls is mainly due to chitinase, glucanase and protease activities (Elad *et al.*, 1982). After attachment and coiling of the antagonist around the phytopathogenic fungus, the parasitic hyphae were removed, revealing lysed sites and penetration holes in hyphae of the host fungus (Elad *et al.*, 1983b). The interaction sites exhibited intense fluorescence when stained with either fluorescein isothiocyanate-conjugated lectin, which binds to chitotriose, or with Calcofluor White, which binds to β-glucan and N-acetyl-D-glucosamine (GlcNAc) oligomers

(Elad *et al.*, 1983b). This phenomenon suggested that lytic enzymes excreted by *Trichoderma* degrade *R. solani* and *S. rolfsii* cell walls at these sites. In the presence of cycloheximide, antagonism was prevented and enzymatic activity was reduced. Chitinase and glucanase activities were also found in soil, when *Trichoderma* attacked *S. rolfsii* or *R. solani* (Elad *et al.*, 1982).

The possible connection between the enzymatic activity of *Trichoderma* and its ability to parasitize *S. rolfsii* sclerotia was examined by Elad *et al.* (1984). The *T. harzianum* strain capable of attacking the pathogen excreted more β-1,3-glucanase and chitinase than the strains that were incapable of attack. Baker and Cook (1974) suggested that a hyperparasite needs to be most effective against survival structures of plant pathogens, if it is to be a successful biological control agent. The ability of an antagonistic strain of *T. harzianum* to attack *S. rolfsii* survival structures was investigated by Elad *et al.* (1984). The authors found that *T. harzianum* hyphae invade *S. rolfsii* sclerotia and make holes in their surfaces. These sclerotia eventually lose their regular shape and their cell cytoplasm.

A considerable amount of recent research has been devoted to studying the lytic systems produced by *Trichoderma* spp. Most of the studies on the expression and regulation of these lytic enzymes have been performed in liquid cultures supplemented with different carbon sources (e.g., chitin, glucose, GlcNAc, fungal cell walls) and their antifungal effects determined *in vitro* (for further details, see Part 1). These growth conditions facilitated the identification of the lytic enzymes induced in *Trichoderma* to hydrolyze the polymers constituting the fungal cell walls; however, they did not accurately reflect the conditions existing during the antagonistic interactions between *Trichoderma* and its hosts. The works described here examined the expression and regulation of lytic enzymes during mycoparasitism on live mycelium of the attacked host.

Considering the complexity of the *in vivo* interaction between a mycoparasitic fungus and its host in the plant rhizosphere or in soil, dual-interaction assays provide the closest approximation and most feasible methodological approach for studying the biochemical events and molecular biology of mycoparasitism. Flores *et al.* (1996) used this method to investigate the expression of a *T. harzianum* basic proteinase gene (*pbr*1) during the antagonistic interaction between *T. harzianum* and *R. solani.* They found that the mRNA levels of *pbr*1 increase as the mycoparasitic interaction progresses. Carsolio *et al.* (1994) investigated the expression pattern of a 42 kDa endochitinase (CHIT 42) in *T. harzianum* IMI 206040 during direct confrontation assays with *R. solani.* The expression of CHIT 42 was strongly induced during the antagonistic interaction.

Inbar and Chet (1995) studied the induction of specific chitinases in *T. harzianum* during its parasitic interaction with *S. rolfsii.* In the early stages of the interaction, a *T. harzianum* 1,4-β-N-acetylglucosaminidase of 102 kDa (CHIT 102) was induced. As the interaction proceeded, its activity decreased, concomitant with the appearance of another 1,4-β-N-acetylglucosaminidase of 73 kDa (CHIT 73), which was expressed thereon. This phenomenon did not occur if the *S. rolfsii* mycelium was autoclaved prior to incubation with *T. harzianum,* suggesting its dependence on vital elements from the host. Cycloheximide inhibited this phenomenon, indicating that *de novo* synthesis of enzymes is taking place in *Trichoderma* during these stages of parasitism. To separate the effect of recognition from the mycoparasitic process, Inbar and Chet (1995) used the aforementioned biomimetic system, binding an *S. rolfsii*-purified lectin to the nylon fibers. When *Trichoderma* was grown on these lectin-coated fibers, the activity of CHIT 102 increased as compared with its activity when the fungus was grown on non-coated fibers. Note that no chitin was present in this biomimetic system. These results suggested that the induction of CHIT 102 in *Trichoderma* is a very early event that is elicited by the recognition signal (i.e., lectin–carbohydrate interactions).

Strains of *T. harzianum* have been found to differ in the levels of hydrolytic enzymes produced when mycelia of different soilborne pathogens are attacked in the soil. This phenomenon was correlated with the ability of each strain to control the respective soilborne pathogens (Elad *et al.*, 1982). However, the specific enzyme(s) actually involved in this activity was not identified. To find out which of the chitinolytic enzymes are involved in the parasitic interactions, Haran *et al.* (1996b) performed dual-culture experiments in which *T. harzianum* was separately confronted with *R. solani* or *S. rolfsii*, both containing chitin as a major cell wall component. The *T. harzianum* strain used in this study was capable of overgrowing *R. solani* hyphae in dual agar cultures but hardly overgrew *S. rolfsii* under the same conditions (Figure 7.2).

The expression of *T. harzianum* chitinases during these interactions was found to be regulated in a very specific and finely-tuned manner, which was affected by the attacked host. The efficient parasitic interaction against *R. solani* involved the expression of three endochitinases of 52 kDa (CHIT 52), 42 kDa (CHIT 42), and 33 kDa (CHIT 33), as well as the N-acetylglucosaminidase activity of CHIT 102 (Figure 7.3a,b). In contrast, during the non-efficient mycoparasitic interaction against *S. rolfsii*, only the exo-type activities of two N-acetylglucosaminidases (CHIT 102 and CHIT 73) were detected (Figure 7.4).

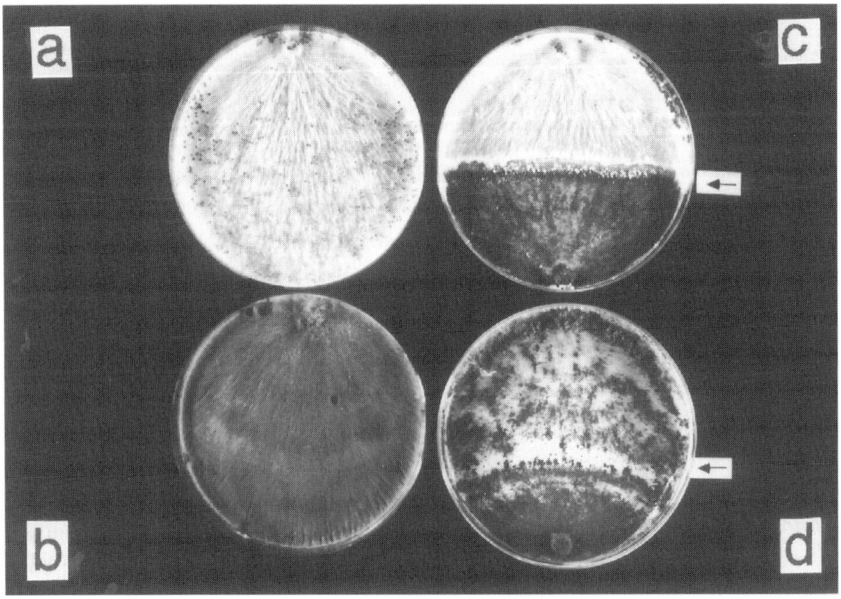

Figure 7.2 Dual cultures of *T. harzianum* with the plant pathogens *S. rolfsii* and *R. solani* 5 days after contact: **a**, *S. rolfsii* alone; **b**, *R. solani* alone; **c**, *S. rolfsii* (top) and *T. harzianum* (bottom); **d**, *R. solani* (top) and *T. harzianum* (bottom) (Haran, S., Schickler, H., Oppenheim, A., and Chet, I., unpublished).

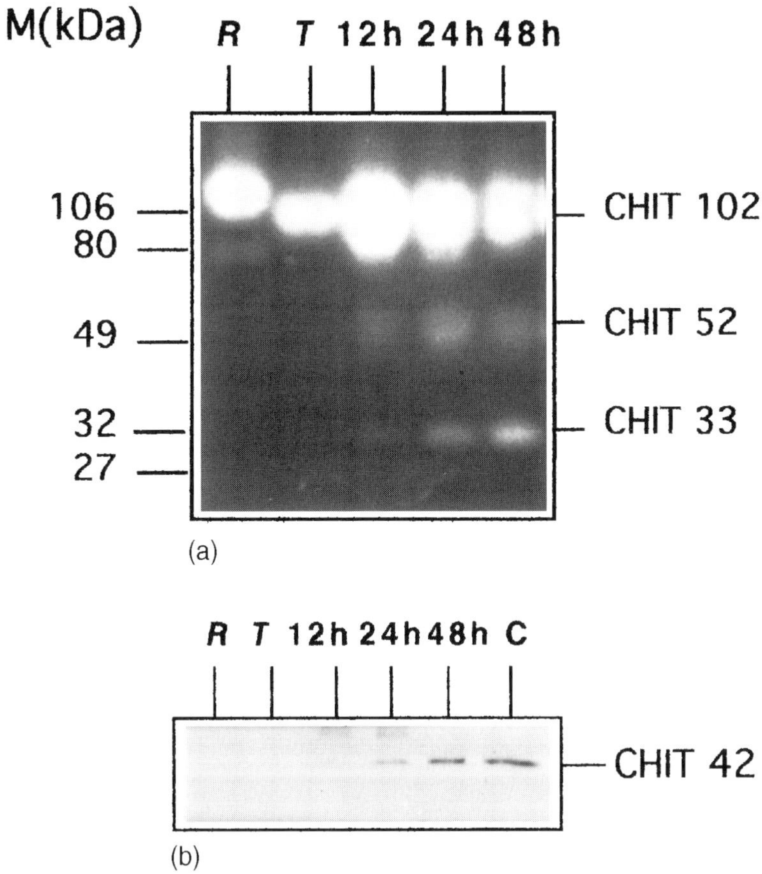

Figure 7.3 Expression of *T. harzianum* chitinases during the antagonistic interaction against *R. solani* (adapted from Haran *et al.*, 1996b). **a**, Activity detection. Lanes *R* and *T*: proteins produced by *R. solani* and by *T. harzianum* (respectively) before contact; 12, 24 and 48 h: proteins obtained from the interaction zone 12, 24 and 48 h after contact, respectively. **b**, Western blot analysis using antibodies raised against *T. harzianum* CHIT 42. Proteins were loaded as in Figure 7.3a with a positive control (C).

These results demonstrated that the chitinolytic system of *T. harzianum* is not regulated by a simple "on/off" mechanism that responds to the presence or absence of chitin. Synergistic effects of the antifungal chitinolytic enzymes *in vitro* were reported when combinations of enzymes with complementary modes of action were used (Lorito *et al.*, 1993, 1994). Indeed, Haran *et al.* (1996b) found that the efficient parasitic interaction with *R. solani* involves the expression of both the endochitinase activities and the exo-type N-acetylglucosaminidase activity, whereas during the non-efficient mycoparasitic interaction with *S. rolfsii* only the exo-type activities were detected. The authors therefore suggested that this differential expression of *T. harzianum* chitinases may influence the overall antagonistic ability of the fungus against a specific host, thereby determining its host range.

Most of the mycoparasitic interactions described thus far have been those involved in the biological control of soil plant pathogens by *Trichoderma*. However, positive results have also been reported for the use of *Trichoderma* as a biological

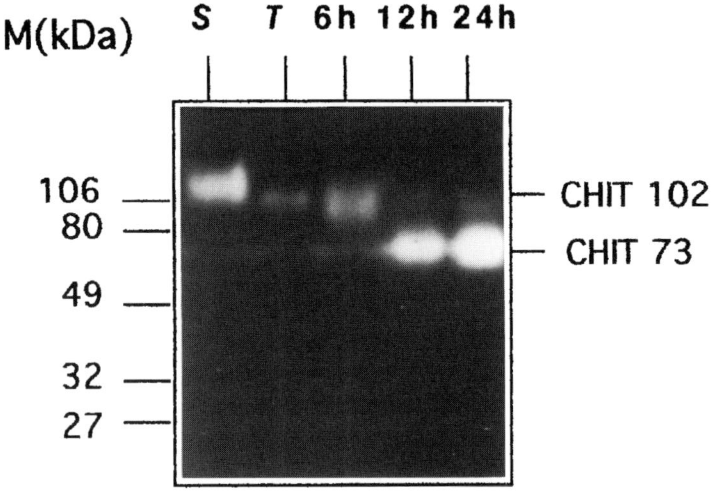

Figure 7.4 Expression of *T. harzianum* chitinases during the antagonistic interaction against *S. rolfsii* (adapted from Haran *et al.*, 1996b). Lanes *S* and *T*: proteins produced by *S. rolfsii* and by *T. harzianum* (respectively) before contact; 6 h, 12 h and 24 h: proteins obtained from the interaction zone 6, 12 and 24 h after contact, respectively.

control agent of aerial plant pathogens (Dubos, 1987). In the case of *Chondrostereum pupureum*, the agent of silver leaf in fruit trees, development of *Trichoderma* in the pruning wound and underlying wood led to a non-specific reaction by the tree in the form of vascular gummosis, which inhibited further development of the pathogen. In the case of the antagonistic interaction between *Trichoderma* and *B. cinerea*, a foliar and fruit pathogen, both competition and mycoparasitism were reported. Competition was found to be especially active during the colonization of floral debris, whereas mycoparasitism occurred from ripening on. Despite the fact that infected-bunch percentages did not differ between *Trichoderma*-treated and non-treated plots, the percentage of rot was significantly lower in the *Trichoderma*-treated vines. This indicated that when the inoculum is highly abundant (from the time of grape ripening onward), *Trichoderma* treatment does not reduce the number of infected foci but prevents their development. The antagonistic fungi fructified on the margin between necrotic and healthy areas on the rotting grapes, thus preventing the plant pathogen's spread to healthy grapes in the bunch. During this interaction, *Trichoderma* coiled around and penetrated the pathogen at numerous locations, reflecting typical mycoparasitic behavior (Dubos, 1987). Labudova and Gorgorova (1988) described *Trichoderma* isolates capable of producing lytic enzymes and implied that the nature of the antagonism of these isolates is based on mycoparasitism. The direct antifungal activity of *Trichoderma* lytic enzymes on *B. cinerea* was demonstrated *in vitro* by Lorito *et al.* (1993). However, Zimand *et al.* (1991) did not find any correlation between the levels of chitinase and β-1,3-glucanase activities in five *T. harzianum* isolates and their biological control abilities. Elad (1996) reviewed the mechanisms involved in the biological control of *B. cinerea*-incited diseases and concluded that the involvement of mycoparasitism and lytic enzymes in the biological control of *B. cinerea* has yet to be determined.

7.2 Heterologous lytic enzymes as a tool for enhancing biocontrol activity

Successful biocontrol is dependent upon several factors, including the development of superior biocontrol strains (Chet *et al.*, 1993). To this end, one of the most attractive approaches is the production of transgenic strains with genetic components conferring improved antifungal features.

Lytic enzymes, which degrade fungal cell walls, have been shown to be strong inhibitors of fungal growth and survival. They can therefore be used to improve biocontrol activity. Shapira *et al.* (1989) demonstrated the involvement of chitinase in the control of *S. rolfsii* using genetic-engineering techniques: the gene *Chi*A, coding for the major chitinase produced by *Serratia marcescens*, was cloned into *E. coli*. The enzyme produced by the cloned gene caused rapid and extensive bursting of *S. rolfsii* hyphal tips. This chitinase preparation was also effective in reducing the incidence of diseases caused by *S. rolfsii* in bean and by *R. solani* in cotton, under greenhouse conditions.

Sitrit *et al.* (1993) introduced the *Chi*A gene from *S. marcescens* into the plant symbiont *Rhizobium meliloti*, which colonizes alfalfa-root nodules. *Rhizobium* colonies harboring the chitinase gene exhibited antifungal activity during symbiosis on alfalfa roots. The importance of chitinolytic activity in biocontrol was demonstrated by Chernin *et al.* (1995). Chitinolytic strains of the bacterium *Enterobacter agglomerans* decreased the incidence of disease caused by *R. solani* in cotton by 64–86%. In contrast, Tn5 mutants of the bacterium, which were deficient in chitinolytic activity, were unable to protect the plants against this disease. The role of one of the *E. agglomerans* endochitinases in the antagonistic effect on plant pathogens was further investigated by Chernin *et al.* (1997). The *Chi*A gene encoding a 59 kDa endochitinase was cloned into *E. coli*. The transformed *E. coli*, which produced and secreted the *E. agglomerans* endochitinase, inhibited *R. solani* growth *in vitro* and in the greenhouse. The activity of this endochitinase also inhibited *Fusarium oxysporum* f. sp. *meloni* spore germination. This work directly demonstrated the antifungal effects of one of the *E. agglomerans* endochitinases.

Enhancement of biocontrol activity by *T. harzianum* (T-35) was attempted by Haran *et al.* (1993). An *S. marcescens* chitinase gene was introduced into the *T. harzianum* genome, under the control of a 35S constitutive promoter from cauliflower mosaic virus. The authors expected the constitutive elevation of extracellular lytic activity to improve *T. harzianum*'s natural ability to attack pathogens, thereby enabling its use as a superior biocontrol agent. When grown on synthetic medium containing glucose as the sole carbon source, the *Trichoderma* transformants expressed significantly higher chitinase activity than the recipient (wild type), suggesting that the signals for expression and regulation of the introduced gene had been recognized by *T. harzianum*. However, when grown in the presence of chitin as the sole carbon source, the transformants showed significantly lower chitinase activity than the wild type. These findings suggested that the transformation event results in constitutive expression of the foreign chitinase but interferes with the natural ability of *T. harzianum* to induce its own chitinolytic activity. This interference may be caused by the insertion of the foreign gene into regions in the *Trichoderma* genome involved in chitinase expression. When the transformants' antagonistic ability was evaluated against *S. rolfsii* in dual cultures containing glucose, the transformants produced significantly wider lytic zones along the contact front with the phytopathogenic fungus than those produced by wild-type *Trichoderma* under the same conditions. This phenomenon could be explained by the specific conditions used in these experiments: in the presence of glucose, the natural chitinolytic activity of *T. harzianum* was repressed and the transformants exhibited stronger hydrolysis of the pathogen's hyphae as a consequence of the constitutive expression of the foreign chitinase. These findings suggested that altering the regulation of *T. harzianum* chitinase expression has a significant influence on the efficiency of the mycoparasitic process.

The promoters of constitutive genes have proven themselves to be useful parts of expression vectors for genetic engineering in different organisms. Whereas promoters isolated from other organisms can sometimes prove useful, one can never be sure that they will work similarly in all physiological or developmental states of the heterologous host. Based on differential screening of an induced cDNA library, Goldman *et al.* (1994) isolated constitutively-expressed cDNA clones of *T. harzianum*. One of these cDNA clones corresponded to a gene (*cob*4) that encodes a novel serine + alanine-rich protein. Northern (RNA) blot analysis demonstrated that *cob*4 is expressed during growth when glucose or cell walls of a phytopathogenic fungus are provided as the carbon source. The authors suggested that constitutively expressed genes could provide reliable promoters useful for genetic manipulations.

Most of the work on *T. harzianum* as a biocontrol agent has been performed with phytopathogenic fungi that contain chitin and β-glucan as major cell wall components. However, the main cell wall component of plant-pathogenic oomycetes, such as *Pythium* spp., is cellulose (Bartnicki-Garcia, 1968). Migheli *et al.* (1994) attempted a different way of improving *Trichoderma* as a biocontrol agent by using the enzyme β-1,4-endoglucanase. Hypercellulolytic strains of *T. longibrachiatum* were obtained by co-transformation of plasmid pTLEG12, which contains the *egl*1 gene for the β-1,4-endoglucanase of *T. longibrachiatum*, and plasmid pAN7-1, which confers hygromycin-B resistance, for selection after transformation (Sanchez-Torres *et al.*, 1994). The transformants were tested for their ability to reduce Pythium damping-off on cucumber seedlings and were shown to be significantly more effective in controlling this disease than the wild type (disease incidence was reduced from 60% to 28%). These preliminary results suggest that cellulase activity may be involved in the biocontrol of *P. ultimum* by *T. longibrachiatum*.

Geremia *et al.* (1993) cloned a gene coding for the basic proteinase *prb*1 from *T. harzianum* strain IMI 206040 and suggested that it plays a key role in mycoparasitism. Flores *et al.* (1996) attempted to increase the *prb*1 gene dosage by introducing multiple copies of it into the *Trichoderma* genome. Integration occurred as multiple copies arranged in tandem, as previously described (Goldman *et al.*, 1990; Herrera-Estrella *et al.*, 1990). The number of copies was estimated at two to six in the various transformants obtained. Densitometric analysis of RNA expression by Northern analysis and protein expression by Western analysis indicated that the levels of *prb*1 mRNA and protease secreted by the transformants are up to 17 and 20 times higher than with the wild-type strain, respectively. However, analysis of *prb*1 mRNA and Prb1 protein production by each of the transformants indicated that high *prb*1 mRNA production does not always result in high protein production. The various transformants showed a gradient of proteinase protein production. The successful overexpression of the *prb*1 gene, driven by its own promoter, led the authors to test the effectiveness of the transformants as biocontrol agents (Flores *et al.*, 1996). Greenhouse experiments were performed in which the transformed *T. harzianum* strains were used to protect cotton seedlings from *R. solani*. An up to five-fold increase was observed in the biocontrol efficacy of the transformed strains, as compared with the wild type. However, the best protection was provided by a strain that produced only an intermediate level of proteinase protein. The authors suggested that extreme levels of proteinase protein can degrade other enzymes that are important in the mycoparasitic process. These results demonstrated that the introduction of multiple copies of *prb*1 improves the biocontrol activity of *T. harzianum* and showed the importance of the proteinase Prb1 in biological control.

The biocontrol activity of the same transformants against plant-parasitic nematodes was recently evaluated by Spiegel and Chet (unpublished data). Plants that were protected from nematodes with the transformed *T. harzianum*, overexpressing the proteinase-encoding gene *prb*1, showed increased top fresh weight, decreased galling indices, and decreased eggs per gram of root, as compared with plants protected by the *T. harzianum* wild-type strain. These results demonstrated that the transformation of *T. harzianum* to overexpress proteinase activity also improves its biocontrol activity against nematodes.

7.3 Ultrastructural changes and cellular mechanisms during the mycoparasitic process

In the last ten years, major advances have been made in understanding the sequence of events taking place in the regulation and expression of mycoparasitism (Shirmböck *et al.*, 1994). Progress in characterizing the mechanisms of cell-to-cell signalling and identifying the cascade of biochemical events leading to antagonist establishment in a particular fungal host has led to the consideration that mycoparasitism could provide a conceptual basis to confer enhanced plant protection from microbial attack (Chet, 1987). Recent studies have convincingly shown that, in many cases, the nature and spatiotemporal coordination of the events involved in the mycoparasitic process are crucial in defining the outcome of the interaction. However, in spite of the increasing attention focused on how molecular mechanisms are regulated and coordinated in fungus–fungus interactions, the spatiotemporal localization of key molecules had not been, until recently, fully investigated. If one considers that ultrastructural investigations of the host response to antagonist attack may provide key information on the mechanisms underlying mycoparasitism (Benhamou and Chet, 1993), it becomes clear that visualizing the cellular and molecular changes underlying the infection process is an essential complement to biochemical analyses (Chérif and Benhamou, 1990). The past few years have witnessed the implementation of new methods for the *in situ* localization of fungal molecules, and several papers have been published addressing the wide range of applications of cytochemistry in plant pathology (Benhamou, 1989; Benhamou and Bélanger, 1995). In the present review, some of the most recent findings associated with the ultrastructural aspects involved in the sequence of events leading to effective mycoparasitism are briefly summarized. To put this information in context, we concentrate on a host–pathogen system that has been the focus of much interest as a model in both basic and applied research. For more than two decades, *Trichoderma*–fungal pathogen interactions have received increasing attention, mainly because *Trichoderma* spp., which are easily isolated from soil and readily grown, are considered to be

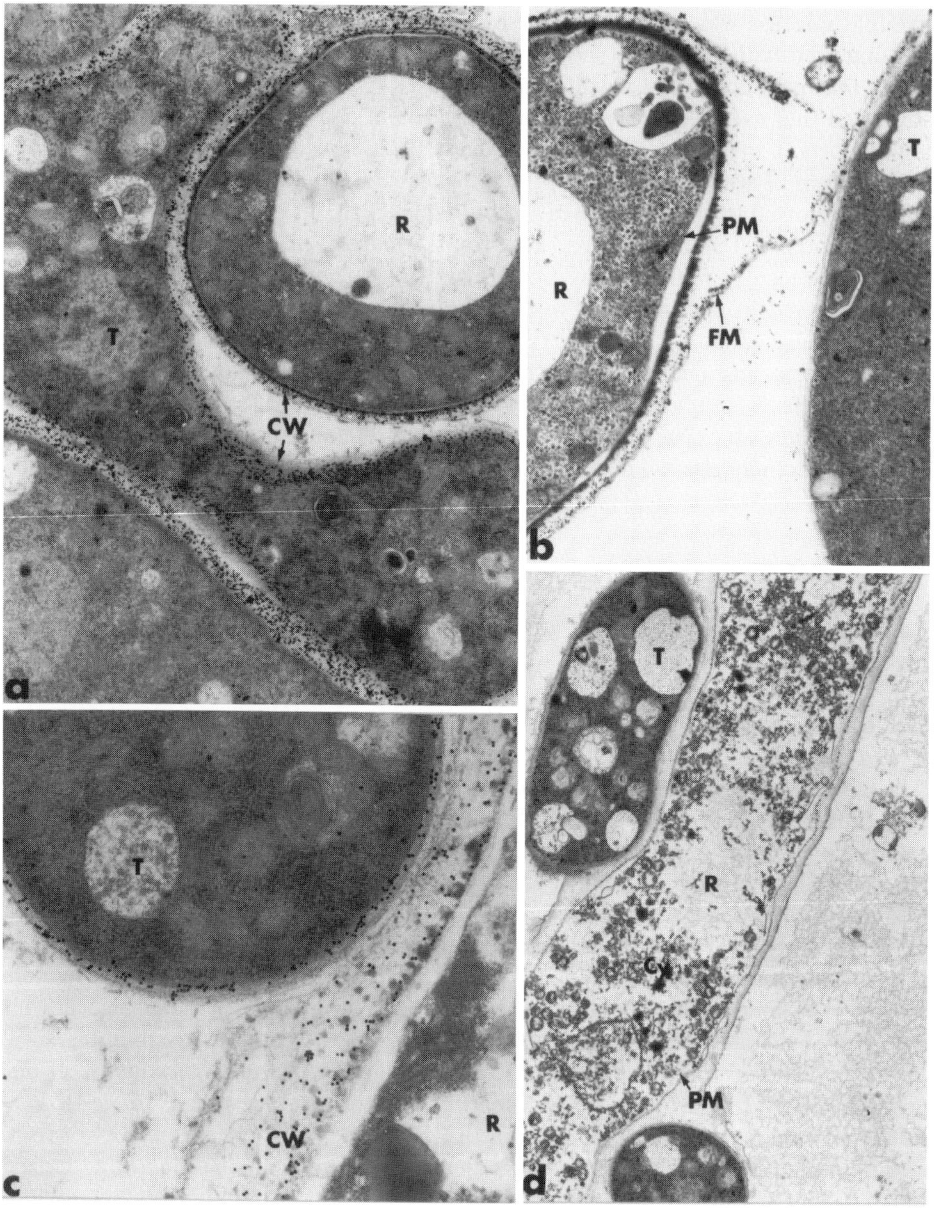

Figure 7.5 Transmission electron micrographs of dual cultures between *Trichoderma harzianum* (T) and *Rhizoctonia solani* (R). **a**, Two days after inoculation. A *T. harzianum* hypha encircling a *R. solani* cell. The cell walls (CW) of both fungi are evenly labelled by the WGA/ovomucoid–gold complex for the localization of chitin (× 15 000). **b**, Four days after inoculation. Early degradation of a *R. solani* hypha, mainly characterized by local retraction of the plasma membrane (PM). A fine external fibrillar matrix (FM), surrounding the *Rhizoctonia* cell wall, is apparently attracted by a cell of the antagonist and tends to stick on the antagonist cell surface (× 20 000). **c**, Five days after inoculation. An altered *R. solani* cell wall labelled with a few scattered gold particles following incubation with the WGA/ovomucoid–gold complex (× 35 000). **d**, At a more advanced stage of the interaction, *Rhizoctonia* hyphae show pronounced signs of alteration characterized by plasma membrane (PM) retraction and cytoplasm (Cy) disorganization (× 8000).

among the most promising biocontrol agents in terms of large-scale applications (Ahmad and Baker, 1988; Chet, 1987; Swan and Chet, 1986).

Ultrastructural investigations of the interactive regions between *Trichoderma* spp. and plant pathogens have proven useful in delineating the scheme of events underlying the mycoparasitic process. Results of cytological studies have firmly established the potential of some *Trichoderma* spp. such as *T. harzianum* to have adverse effects on various pathogens (Benhamou and Chet, 1993, 1996, 1997; Chérif and Benhamou, 1990). These studies have also provided clear ultrastructural and cytochemical evidence that antagonism is a multifaceted process requiring the synergistic contribution of several mechanisms, including cell-surface recognition between both partners and the production of specific hydrolytic enzymes by the antagonist. Benhamou and Chet (1993) reported that recognition events mediated by close contact between *T. harzianum* and *R. solani* (Figure 7.5a) preceded or coincided with major host-cell alterations including cell wall degradation, plasmalemma disorganization, and cytoplasm aggregation (Figure 7.5d). Scanning electron microscope (SEM) investigations of the interactive region between both fungi demonstrated that damage of *R. solani* hyphae appears soon after the coiling of *Trichoderma* around them (Benhamou and Chet, 1993). This observation confirmed that the outcome of the interaction is most likely to be determined by initial contact events that trigger firm binding of the antagonist to the host, ultimately leading to a series of events resulting in pathogen breakdown. More precise investigations with a transmission electron microscope (TEM) showed that cells of *Trichoderma* establish close contact with the host mycelium via a fine, fibrillar matrix originating from *R. solani* hyphae (Figure 7.5b). The polysaccharidic nature of this fine layer was demonstrated using a galactose-specific lectin, the *Ricinus communis* agglutinin (RcA) complexed to colloidal gold. The occurrence of significant amounts of galactose residues in the external matrix of *R. solani* suggested that receptors with galactose-binding affinity are present at the cell surface of *Trichoderma*. In recent years, the key role of lectins in early recognition events triggering subsequent responses, such as adhesion, coiling, and penetration of the antagonist into its host, has been well documented (Barak and Chet, 1990; Barak *et al.*, 1985; Nordbring-Hertz and Chet, 1986). The production of lytic enzymes by *T. harzianum* was easily shown by gold cytochemistry (Benhamou and Chet, 1993; Chérif and Benhamou, 1990). The observation that the amount of GlcNAc residues (chitin) significantly decreased in the outer wall layers of *R. solani* supported the concept that *Trichoderma* spp. display the ability to produce chitinases (Figure 7.5c) (Harman *et al.* 1993). Evidence of the gradual diffusion of hydrolytic enzymes lay in the observation that alteration of GlcNAc residues occurred earlier and at a higher level in the outer versus inner wall layers of the host. In that context, one may assume that chitin oligomers released from the outer wall layers of *R. solani* act as elicitors of further enzyme synthesis in a manner similar to fungal elicitors of plant chitinases (Roby *et al.*, 1988).

Indirect evidence of the production of hydrolytic enzymes by *T. harzianum* was recently provided by a study of the *T. harzianum–P. ultimum* interaction. Based on the cytological observations, Benhamou and Chet (1997) postulated that *Pythium* colonization by *T. harzianum* involves a chronological sequence of events: (1) coiling and local penetration of the antagonist into the pathogen hyphae (Figure 7.6a); (2) induction of a host structural response at sites of potential antagonist entry; (3) alteration of the host protoplasm; and (4) active multiplication of the antagonistic

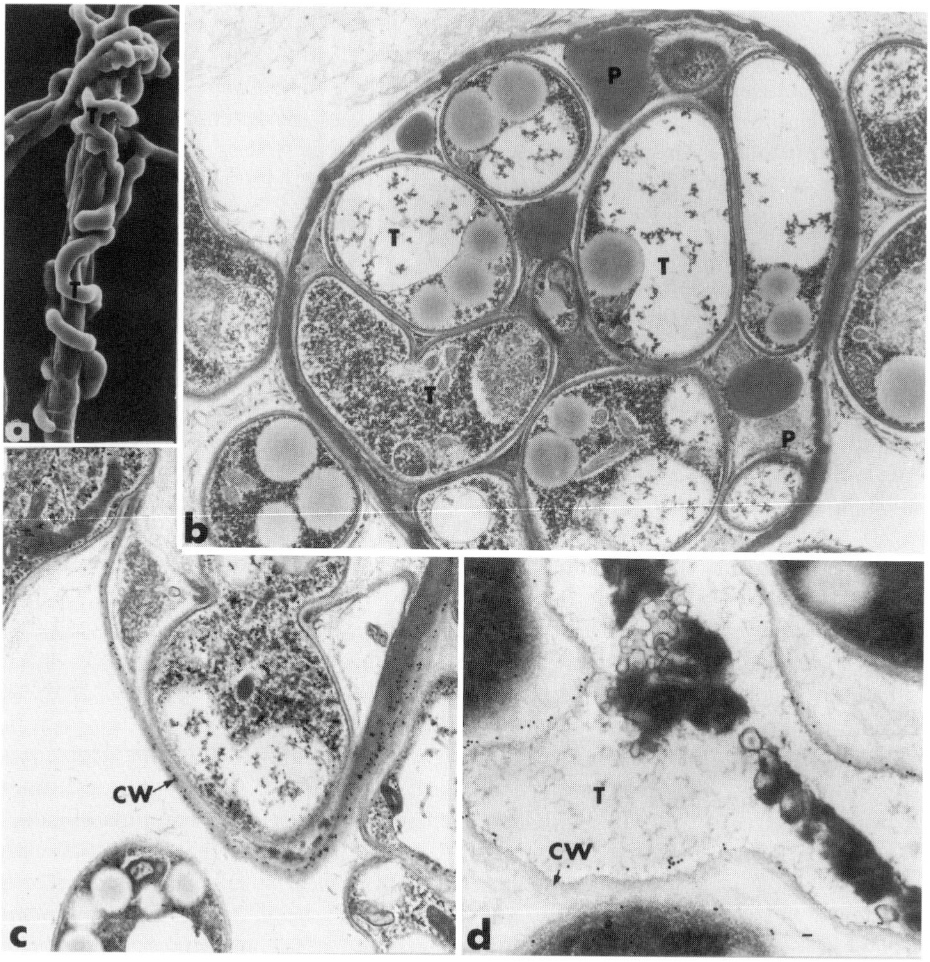

Figure 7.6 Scanning and transmission electron micrographs of dual cultures between *Trichoderma harzianum* (T) and *Pythium ultimum* (P). **a**, Three days after inoculation. *Trichoderma* hyphae form dense coils and tightly encircle hyphae of *P. ultimum*. A strong compression of the host cell is reflected by the wrinkled appearance of the cell surface (×1000). **b**, Four days after inoculation. Abundant multiplication of the antagonist in a *Pythium* cell (×15 000). **c**, Four days after inoculation. Labelling of cellulosic β-1,4-glucans with the exoglucanase–gold complex. The host cell wall (CW) is thicker than normal and irregularly labelled (×20 000). **d**, Five days after inoculation. Labelling of β-1,3-glucans with the tobacco β-1,3-glucanase. Labelling is reduced to a few gold particles over the outermost layers of the cell wall (CW) of severely damaged *Pythium* hyphae (×30 000).

cells in the pathogen hyphae leading to host-cell breakdown (Figure 7.6b). Ultrastructural localization of cellulose and β-1,3-glucan, the two main wall-bound compounds in *Pythium* cells, confirmed that the antagonist has the ability to penetrate the pathogen cell wall by altering its structure, as reflected by a decrease in the cellulosic wall-bound component (Figure 7.6c). In addition to altering the cellulosic component, *Trichoderma*'s ingress into *Pythium* hyphae was also associated with a marked alteration of the β-1,3-glucan component of the cell wall (Figure 7.6d).

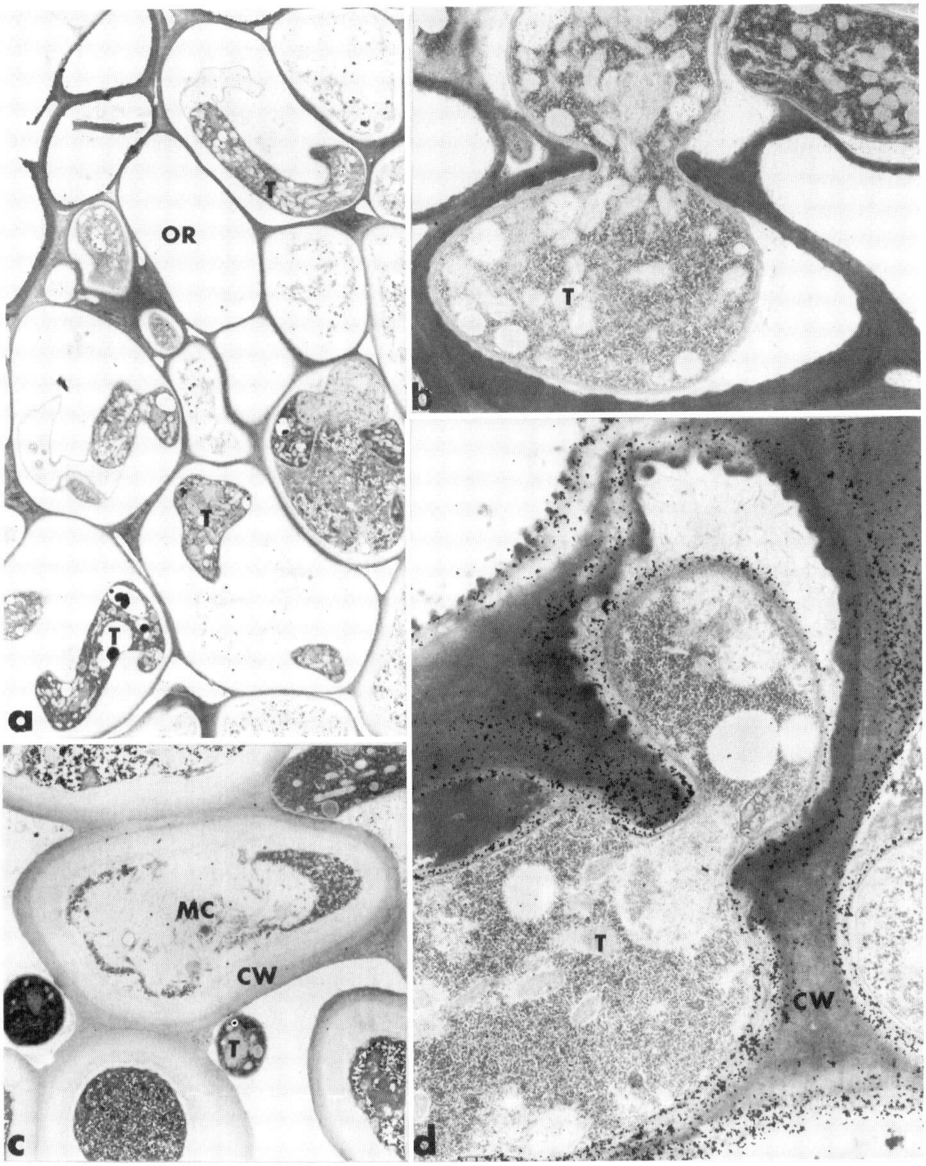

Figure 7.7 Transmission electron micrographs of sclerotia of *Sclerotium rolfsii* (S) parasitized by *Trichoderma harzianum* (T). **a** and **b**, Outer rind cells (OR). Invasion of the outermost rind cells occurs via direct penetration of the electron-dense cell walls (b). *Trichoderma* growth is mainly intracellular and coincides with extensive cell alterations such as retraction and aggregation of the cytoplasm and vacuolar breakdown (**a**, ×3000; **b**, ×13 000). **c**, Medullary cells. *Trichoderma* hyphae do not directly penetrate the medullary cells (MC) although these show pronounced alterations such as cytoplasmic disorganization and/or aggregation. The cell wall (CW) does not appear to be altered (×5000). **d**, Labeling of chitin with the WGA/ovomucoid–gold complex. Invasion of the rind cells by hyphae of *T. harzianum* proceeds via the formation of narrow channels of penetration. The electron-dense outermost rind cell wall (CW) is structurally preserved although gold particles appear irregularly distributed in areas of close contact with the antagonist (×20 000).

Alteration of β-1,3-glucans occurred not only in wall areas adjacent to *Trichoderma* cells but also at a distance from the sites of antagonist entry, so β-1,3-glucanases may have freely diffused extracellularly, probably facilitating *Trichoderma*'s ingress through loosened wall matrices. These cytochemical observations raised the following question: To what extent does weakening of the cell wall through the action of *Trichoderma* β-1,3-glucanases facilitate the diffusion of cellulolytic enzymes and/or toxic compounds by increasing the wall's permeability? Recently, the synergistic action of hydrolytic enzymes and antibiotics was found to be required in the biocontrol process mediated by *Gliocladium virens* (Di Pietro *et al.*, 1993). The authors suggested that both mechanisms are expressed in a controlled sequence and that slight alterations of the host cell walls are essential prerequisites for further antibiotic diffusion.

Hydrolytic enzymes produced by *T. harzianum* were also shown to be effective against sclerotial fungi (Elad *et al.*, 1983b, 1984). Benhamou and Chet (1996), studying the interaction between *T. harzianum* and sclerotia of the soilborne plant pathogen *S. rolfsii* by SEM and TEM, provided evidence that enzymes produced by the antagonist play a major role in breaching the host cell walls at sites of attempted penetration. Hyphae of *T. harzianum* grew abundantly on the sclerotial surface and formed a dense branched mycelium which developed in the sclerotial cells causing extensive host-cell alterations such as retraction and aggregation of the cytoplasm and vacuolar breakdown (Figure 7.7a,b). The synergistic action of chitinases and β-1,3-glucanases was found to be essential for successful sclerotial colonization. However, the finding that *Trichoderma* hyphae did not penetrate the medullary cells (Figure 7.7c), even though these showed pronounced alterations such as cytoplasmic disorganization and/or aggregation, was taken as an indication that lytic enzymes are not the only components involved in sclerotial mycoparasitism (Figure 7.7d). Whether antibiotics collaborate in the process remains to be elucidated.

The few examples in this review point to the interest in using ultrastructure and gold cytochemistry for a better understanding of the cellular and molecular events occurring in mycoparasitism. Exciting progress has been made in the understanding of the cellular events induced in response to parasitism, mainly because of the introduction of new biological probes (i.e., enzyme and lectin–gold complexes). In conjunction with biochemistry and molecular biology, cytochemistry of fungal cells has proven useful for elucidating some aspects of the relationships established between antagonists and pathogens. It is clear that both molecular and traditional approaches will continue to benefit from novel findings derived from the *in situ* localization of an increasing number of fungal molecules.

7.4 Concluding remarks

Mycoparasitism is an exciting phenomenon that plays a role in biological control processes. *Trichoderma* spp. appear to use this mode of action, along with competition and antibiosis, in the course of their biocontrol activity. However, it is very difficult to assess the relative importance of mycoparasitism in biological control when the fungus is used against plant pathogens under field conditions. As shown here, mycoparasitism itself is a complex process. It involves several successive steps, most of which are highly species specific. Recognition of several plant pathogens by *Trichoderma* spp. was shown to be mediated by specific lectins, since specific inhibition of the plant-pathogenic lectin prevented coiling of the mycoparasite around the pathogen's hyphae. Moreover, nylon fibers coated with one of these lectins initiated the formation of typical mycoparasitic infection structures in *T. harzianum.*

Evidence was provided that lytic enzymes, which have been shown to possess an antifungal effect, are involved in the mycoparasitic process. The expression of *T. harzianum* chitinases was affected by the attacked host. During an efficient antagonistic interaction against *R. solani,* both endochitinases and an N-acetylglucosaminidase were simultaneously expressed, whereas during a non-efficient interaction against *S. rolfsii,* only N-acetylglucosaminidases were detected. This differential expression is thought to influence the efficiency of *T. harzianum* biocontrol activity, thus determining its host range.

The possibilities of improving biocontrol activity by *T. harzianum* using genetic manipulation techniques are promising. Early studies with transgenic *T. harzianum,* transformed to produce increased amounts of specific proteins, are encouraging because, even with these preliminary systems, positive results have been obtained. Future research should focus on the development of strains expressing 'multigene' combinations while preserving the intrinsic vigor and ecological competence of the fungus. Once transgenic *T. harzianum* strains capable of producing highly efficient synergistic combinations of enzymes are obtained, a higher level of plant disease control should become possible. Transgenic *Trichoderma* therefore offers the potential of substantially reducing the amount of chemical fungicides required to produce disease-free plants.

References

AHMAD, J. S. and BAKER, R. 1988. Implications of rhizosphere competence of *Trichoderma harzianum. Can. J. Microbiol.* **34**: 229–234.

BAKER, K. F. 1987. Evolving concepts of biological control of plant pathogens. *Annu. Rev. Phytopathol.* **25**: 67–85.

BAKER, K. F. and COOK, R. J. 1974. Biological Control of Plant Pathogens. Am. Phytopath. Soc., St Paul, MN, 433 pp.

BARAK, R. and CHET, I. 1986. Determination, by fluorescein diacetate staining, of fungal viability during mycoparasitism. *Soil Biol. Biochem.* **18**: 315–319.

BARAK, R. and CHET, I. 1990. Lectin of *Sclerotium rolfsii*: its purification and possible function in fungal–fungal interactions. *Appl. Bacteriol.* **69**: 101–112.

BARAK, R., ELAD, Y., and CHET, I. 1986. The properties of L-fucose-binding agglutinin associated with the cell wall of *Rhizoctonia solani. Arch. Microbiol.* **144**: 346–349.

BARAK, R., ELAD, Y., MIRELMAN, D., and CHET, I. 1985. Lectins: a possible basis for specific recognition in the interaction of *Trichoderma* and *Sclerotium rolfsii. Phytopathology* **75**: 458–462.

BARNETT, H. L. and BINDER, F. L. 1973. The fungal host–parasite relationship. *Annu. Rev. Phytopathol.* **11**: 273–292.

BARONDES, S. H. 1981. Lectins: their multiple endogenous cellular functions. *Annu. Rev. Biochem.* **50**: 207–231.

BARTNICKI-GARCIA, S. 1968. Cell wall chemistry, morphogenesis, and taxonomy of fungi. *Annu. Rev. Microbiol.* **22**: 87–108.

BENHAMOU, N. 1989. Preparation and application of lectin–gold complexes. In M. A. Hayat (ed.), *Colloidal Gold: Techniques and Applications.* Academic Press, New York, pp. 95–143.

Benhamou, N. and Bélanger, R. R. 1995. Immuno-electron microscopy. In U. S. Singh and R. P. Singh (eds), *Advanced Methods in Plant Pathology*. CRC Press, London, pp. 15–29.

Benhamou, N. and Chet, I. 1993. Hyphal interactions between *Trichoderma harzianum* and *Rhizoctonia solani*: ultrastructure and gold cytochemistry of the mycoparasitic process. *Phytopathology* **83**: 1062–1071.

Benhamou, N. and Chet, I. 1996. Parasitism of sclerotia of *Sclerotium rolfsii* by *Trichoderma harzianum*: ultrastructural and cytochemical aspects of the interaction. *Phytopathology* **86**: 405–416.

Benhamou, N. and Chet, I. 1997. Cellular and molecular mechanisms involved in the interaction between *Trichoderma harzianum* and *Pythium ultimum*. *Appl. Environ. Microbiol.* **63**: 2095–2099.

Carsolio, C., Gutierrez, A., Jimenez, B., Van Montagu, M., and Herrera-Estrella, A. 1994. Characterization of *ech-42*, a *Trichoderma harzianum* endochitinase gene expressed during mycoparasitism. *Proc. Natl. Acad. Sci. USA* **91**: 10903–10907.

Chérif, M. and Benhamou, N. 1990. Cytochemical aspect of chitin breakdown during the parasitic action of a *Trichoderma* sp. on *Fusarium oxysporum* f. sp. *radicis-lycopersici*. *Phytopathology* **80**: 1406–1414.

Chernin, L., Ismailov, Z., Haran, S., and Chet, I. 1995. Chitinolytic *Enterobacter agglomerans* antagonistic to fungal plant pathogens. *J. Appl. Environ. Microbiol.* **61**: 1720–1726.

Chernin, L. S., de la Fuente, L., Sobolev, V., Haran, S., Vorgias, C. E., Oppenheim, A. B., and Chet, I. 1997. Molecular cloning, structural analysis and expression in *Escherichia coli* of a chitinase from *Enterobacter agglomerans*. *J. Appl. Environ. Microbiol.* **63**: 834–839.

Chet, I. 1987. *Trichoderma*: application, mode of action, and potential as a biocontrol agent of soilborne plant pathogenic fungi. In I. Chet (ed.), *Innovative Approaches to Plant Disease Control*. Wiley, New York, pp. 137–160.

Chet, I. 1990. Mycoparasitism – recognition, physiology and ecology. In R. R. Barker and P. E. Dunn (eds), *New Directions in Biological Control: Alternatives for Suppressing Agricultural Pests and Diseases*. Alan Liss, New York, pp. 725–733.

Chet, I., Barak, Z., and Oppenheim, A. 1993. Genetic engineering of microorganisms for improved biocontrol activity. In I. Chet (ed.), *Biotechnology of Plant Disease Control*. Wiley-Liss, New York, pp. 211–235.

Chet, I., Harman, G. E., and Baker, R. 1981. *Trichoderma hamatum*: its hyphal interactions with *Rhizoctonia solani* and *Pythium* spp. *Microb. Ecol.* **7**: 29–38.

Di Pietro, A., Lorito, M., Hayes, C. K., Broadway, R. M., and Harman, G. E. 1993. Endochitinase from *Gliocladium virens*: isolation, characterization, and synergistic antifungal activity in combination with gliotoxin. *Phytopathology* **83**: 308–313.

Dubos, B. 1987. Fungal antagonism in aerial agrobiocenoses. In I. Chet (ed.), *Innovative Approches to Plant Disease Control*. Wiley, New York, pp. 107–135.

Elad, Y. 1996. Mechanisms involved in the biological control of *Botrytis cinerea* incited diseases. *Eur. J. Plant Pathol.* **102**: 719–732.

Elad, Y., Barak, R., and Chet, I. 1983a. Possible role of lectins in mycoparasitism. *J. Bacteriol.* **154**: 1431–1435.

Elad, Y., Barak, R., and Chet, I. 1984. Parasitism of sclerotia of *Sclerotium rolfsii* by *Trichoderma harzianum*. *Soil Biol. Biochem.* **16**: 381–386.

Elad, Y., Chet, I., and Henis, Y. 1982. Degradation of plant pathogenic fungi by *Trichoderma harzianum*. *Can. J. Microbiol.* **28**: 719–725.

Elad, Y., Chet, I., Boyle, P., and Henis, Y. 1983b. Parasitism of *Trichoderma* spp. on *Rhizoctonia solani* and *Sclerotium rolfsii* – SEM studies and fluorescence microscopy. *Phytopathology* **73**: 85–88.

Flores, A., Chet, I., and Herrera-Estrella, A. 1996. Improved biocontrol activity of *Trichoderma harzianum* by over-expression of the proteinase-encoding gene *prb*1. *Curr. Genet.* **31**: 30–37.

GEREMIA, R., GOLDMAN, G., JACOBS, D., ARDILES, W., VILA, S., VAN MONTAGU, M., and HERRERA-ESTRELLA, A. 1993. Molecular characterization of the proteinase encoding gene, *prb*1, related to mycoparasitism by *Trichoderma harzianum. Molec. Microbiol.* **8**: 603–613.

GOLDMAN, G. H., VAN MONTAGU, M., and HERRERA-ESTRELLA, A. 1990. Transformation of *Trichoderma harzianum* by high-voltage electric pulse. *Curr. Genet.* **17**: 169–174.

GOLDMAN, G. H., VASSEUR, V., CONTRERAS, R., and VAN MONTAGU, M. 1994. Sequence analysis and expression studies of a gene encoding a novel serine + alanine-rich protein in *Trichoderma harzianum. Gene* **144**: 113–117.

GOMEZ, I., CHET, I., and HERRERA-ESTRELLA, A. 1997. Genetic diversity and vegetative compatibility in *Trichoderma harzianum. Molec. Gen. Genet.* **256**: 127–135.

HARAN, S., SCHICKLER, H., and CHET, I. 1996a. Molecular mechanisms of lytic enzymes involved in the biocontrol activity of *Trichoderma harzianum. Microbiology* **142**: 2321–2331.

HARAN, S., SCHICKLER, H., OPPENHEIM, A., and CHET, I. 1996b. Differential expression of *Trichoderma harzianum* chitinases during mycoparasitism. *Phytopathology* **86**: 980–985.

HARAN, S., SCHICKLER, H., PE'ER, S., LOGEMANN, S., OPPENHEIM, A., and CHET, I. 1993. Increased constitutive chitinase activity in transformed *Trichoderma harzianum. Biol. Control* **3**: 101–108.

HARMAN, G. E., CHET, I., and BAKER, R. 1981. Factors affecting *Trichoderma hamatum* applied to seeds as a biocontrol agent. *Phytopathology* **71**: 569–572.

HARMAN, G. E., HAYES, C. K., LORITO, M., BROADWAY, R. M., DI PIETRO, A., PETERBAUER, C., and TRONSMO, A. 1993. Chitinolytic enzymes of *Trichoderma harzianum*: purification of chitobiosidase and endochitinase. *Phytopathology* **83**: 313–318.

HERRERA-ESTRELLA, A., GOLDMAN, G. H., and VAN MONTAGU, M. 1990. High efficiency transformation system for the biocontrol agents, *Trichoderma* spp. *Molec. Microbiol.* **4**: 839–843.

INBAR, J. and CHET, I. 1992. Biomimics of fungal cell–cell recognition by use of lectin-coated nylon fibers. *J. Bacteriol.* **174**: 1055–1059.

INBAR, J. and CHET, I. 1994. A newly isolated lectin from the plant pathogenic fungus *Sclerotium rolfsii*: purification, characterization and its role in mycoparasitism. *Microbiology* **140**: 651–657.

INBAR, J. and CHET, I. 1995. The role of recognition in the induction of specific chitinases during mycoparasitism by *Trichoderma harzianum. Microbiology* **141**: 2823–2829.

INBAR, J., MENENDEZ, A., and CHET, I. 1996. Hyphal interaction between *Trichoderma harzianum* and *Sclerotinia sclerotiorum* and its role in biological control. *Soil Biol. Biochem.* **28**: 757–763.

LABUDOVA, I. and GOGOROVA, L. 1988. Biological control of phytopathogenic fungi through lytic action of *Trichoderma* species. *FEMS Microbiol. Lett.* **52**: 193–198.

LORITO, M., HAYES, C. K., DI PIETRO, A., WOO, S. L., and HARMAN, G. E. 1994. Purification, characterization, and synergistic activity of a glucan 1,3-β-glucosidase and an N-acetylglucosaminidase from *Trichoderma harzianum. Phytopathology* **84**: 398–405.

LORITO, M., HARMAN, G. E., HAYES, C. K., BROADWAY, R. M., WOO, S. L., and DI PIETRO, A. 1993. Chitinolytic enzymes produced by *Trichoderma harzianum.* II: Antifungal activity of purified endochitinase and chitobiase. *Phytopathology* **83**: 302–307.

MIGHELI, Q., FRIARD, O., RAMON-VIDAL, D., and GONZALEZ-CANDELAS, L. 1994. Hypercellulolytic transformants of *Trichoderma longibrachiatum* are active in reducing *Pythium* damping-off on cucumber. In M. J. Daniels (ed.), *Advances in Molecular Genetics of Plant–Microbe Interactions*, Vol. 3. Kluwer, Dordrecht, pp. 395–398.

NORDBRING-HERTZ, B. and CHET, I. 1986. Fungal lectins and agglutinins. In D. Mirelman (ed.), *Microbial Lectins and Agglutinins: Properties and Biological Activity.* Wiley, New York, pp. 393–407.

ROBY, D., TOPPAN, A., and ESQUERRÉ-TUGAYÉ, M. T. 1988. Systemic induction of chitinase activity and resistance in melon plants upon fungal infection or elicitor treatment. *Physiol. Molec. Plant Pathol.* **33**: 409–417.

SANCHEZ-TORRES, P., GONZALEZ, R., ESPEREZ-GONZALEZ, J. A., GONZALEZ-CANDELAS, L., and RAMON, D. 1994. Development of a transformation system for *Trichoderma longibrachiatum* and its use to construct multicopy transformants for *egl*1 gene. *Appl. Microbiol. Biotechnol.* **41**: 440–446.

SHAPIRA, R., ORDENTLICH, A., CHET, I., and OPPENHEIM, A. B. 1989. Control of plant diseases by chitinase expressed from cloned DNA in *Escherichia coli. Phytopathology* **79**: 1246–1249.

SHIRMBÖCK, M., LORITO, M., WANG, Y.-L., HAYES, C. K., ARISAN-ATAC, I., SCALA, F., HARMAN, G. E., and KUBICEK, C. P. 1994. Parallel formation and synergism of hydrolytic enzymes and peptaibol antibiotics, molecular mechanisms involved in the antagonistic action of *Trichoderma harzianum* against phytopathogenic fungi. *Appl. Environ. Microbiol.* **60**: 4364–4370.

SITRIT, Y., BARAK, Z., KAPULNIK, Y., OPPENHEIM, A., and CHET, I. 1993. Expression of *Serratia marcescens* chitinase gene in *Rhizobium meliloti* during symbiosis on alfalfa roots. *Molec. Plant Microbe Interact.* **6**: 293–298.

SWAN, A. and CHET, I. 1986. Possible mechanism for control of *Fusarium* spp. by *Trichoderma harzianum. Proc. Br. Crop Prot. Conf.* **2**: 865–872.

WISNIEWSKI, M., BILES, C., DROBY, S., MACLAUGHLIN, R. J., WILSON, C., and CHALUTZ, E. 1991. Mode of action of postharvest biocontrol yeast, *Pichia guilliermondii.* I. Characterization of attachment to *Botrytis cinerea. Physiol. Molec. Plant Pathol.* **39**: 245–258.

ZIMAND, G., ELAD, Y., and CHET, I. 1991. Biological control of *Botrytis cinerea* by *Trichoderma* spp. *Phytoparasitica* **19**: 252–253.

8

The role of antibiosis in biocontrol

C. R. HOWELL

USDA-ARS Cotton Pathology Research Unit, College Station, Texas, USA

8.1 Introduction

The genera *Trichoderma* and *Gliocladium* have long been associated with biological control phenomena, usually because of their presence in suppressive soils or on the root systems of surviving plants among diseased ones. Another conspicuous attribute is the ability of many species within these two genera to act as parasites of other fungi, some of which are plant pathogens (Baker, 1990; Chet, 1987; Cook and Baker, 1983; Howell, 1990; Papavizas, 1985). Although mycoparasitism is usually the first phenomenon that excites the attention of the observer, further investigation often reveals that secondary metabolites with antifungal and/or antibacterial activity are also produced by the biocontrol agent. The antimicrobial compounds produced by *Trichoderma* and *Gliocladium* constitute a rather diverse group with respect to structure and function (Taylor, 1986), and the group contains both volatile and nonvolatile compounds. A number of these antibiotic compounds have been related to biocontrol activity, but none has proven to be the sole arbiter of success or failure in a biocontrol system. More than likely, antibiosis is only one mechanism among the many that constitute a much more complex system, the end product of which is biological control. The entire range of secondary metabolites produced by these two genera and their structure and function are treated in detail in Volume 1, Chapter 8 and will not be addressed here. The purpose of this chapter is to acquaint the reader with the kinds of antibiotics produced by *Trichoderma* and *Gliocladium* spp. that have been associated with disease control, what the effects of environment are on their production and activity, how they act as mechanisms in the biocontrol process, and how genetic manipulation of these biocontrol agents can be used to enhance their antibiotic activity or reduce their phytotoxicity.

8.2 Antibiotics associated with disease control by *Trichoderma* and *Gliocladium* species

Before an association of antibiotic production with disease control within *Trichoderma* and *Gliocladium* can be discussed, a certain confusion in the early literature as to which species was producing what compound must be cleared up. The production of gliotoxin was first ascribed to *Trichoderma lignorum* (Weindling and Emerson, 1936) and that of gliotoxin and viridin to *T. viride* (Brian *et al.*, 1946; Brian and Hemming, 1945). Webster and Lomas (1964) subsequently demonstrated that these compounds were actually produced by *Gliocladium virens*. Moffatt *et al.* (1969) ascribed the production of viridiol to *T. viride*, but it was most likely a strain of *G. virens*. Recent results, however, indicate that *G. virens* is much more closely related to *Trichoderma* than was previously thought, and it has been transferred to that genus (von Arx, 1987; Rehner and Samuels, 1994). Thus we have essentially come full circle with respect to genus. The systematics used in this chapter are those of the authors whose papers are cited. A more modern approach to taxonomy can be found in Volume 1, Chapter 2.

Perhaps the first individuals to most comprehensively address the role of antibiotics produced by *Trichoderma* species in the biocontrol of plant pathogens were Dennis and Webster (1971a), who ascribed the antibiotic activity of nonvolatile compounds in extracts from *Trichoderma* spp. cultures to trichodermin and peptide antibiotics. They later found (Dennis and Webster, 1971b) that the antibiotic activity of some isolates was due to the production of volatile compounds, and they noted that active isolates emitted a strong coconut odor.

Pachenari and Dix (1980) isolated low molecular weight inhibitory substances from *Gliocladium roseum* by freeze-drying dialysates of malt extract cultures of the fungus. These compounds inhibited the germination and growth of conidia from *Botrytis allii*.

Maiti *et al.* (1991) reported the production of volatile inhibitors by *G. virens* (*T. virens*) that partially suppressed the growth of *Sclerotium rolfsii* in culture. Other researchers have isolated and identified many other antibiotics produced by members of the genus *Trichoderma*.

8.2.1 *Alkyl pyrones*

The main aroma constituent responsible for the coconut odor in *T. viride* has been identified as 6-pentyl-pyrone (Collins and Halim, 1972) and its antibiotic activity against a number of plant pathogens has been demonstrated (Claydon *et al.*, 1987; Ghisalberti *et al.*, 1990; Merlier *et al.*, 1984). This compound has been isolated from strains of *T. harzianum*, *T. viride*, *T. koningii*, and *T. hamatum* (Ghisalberti and Savasithamparam, 1991; Simon *et al.*, 1988), and it has shown phytotoxic activity to germinating lettuce seedlings (Claydon *et al.*, 1987).

8.2.2 *Isonitriles*

The antibiotics isonitrin A–D and isonitrinic acids E and F have been isolated from *T. hamatum*, *T. harzianum*, *T. koningii*, *T. polysporum*, and *T. viride* (Fujiwara *et al.*, 1982), and their production patterns correlated with species (Okuda *et al.*, 1982). Isonitrin A was broadly effective against both gram-positive and gram-negative bacteria and against fungi. Isonitrin D showed good activity against fungi but little or no activity against bacteria, and the others showed relatively weak antimicrobial activity.

8.2.3 *Polyketides*

Harzianolide, which was isolated from *T. harzianum*, inhibited growth of the take-all fungus, and the strain that produced it was the most effective in suppressing the disease in greenhouse tests (Almassi *et al.*, 1991). This same compound was isolated as a major constituent from another strain of *T. harzianum* that reduced the germination of *Fusarium oxysporum* f. spp. *melonis* and *vasinfectum* conidia and chlamydospores. This strain also reduced the disease incidence caused by these pathogens to melon and cotton plants (Ordentlich *et al.*, 1992). Harzianolide has also been isolated from *T. koningii* (Dunlop *et al.*, 1989).

8.2.4 *Peptaibols*

The peptide antibiotics trichopolyns A and B, isolated from *T. polysporum*, strongly inhibited the growth of a number of fungi and gram-positive bacteria (Fuji *et al.*, 1978). They were much less effective against gram-negative bacteria. Extraction of an isolate of *T. koningii*, selected for its antagonistic properties against other fungi, yielded the 19-residue peptaibol trikoningin KAV and two 11-residue lipopeptaibols trikoningins KBI and KBII (Auvin-Guette *et al.*, 1993). All three proved to be strongly active against gram-positive *Staphylococcus aureus* but inactive against gram-negative *Escherichia coli*. The trichorzianines, isolated from *T. harzianum*, have been characterized as peptides with 19 amino acid residues with a high proportion of a-aminoisobutyric acid (Bodo *et al.*, 1985; El Hajji *et al.*, 1987; Merlier *et al.*, 1984). When trichorzianines A and B were tested for activity against *Sclerotium rolfsii*, A reduced radial growth by 70% and B by 36% (Correa *et al.*, 1995). Strangely, trichorzianine A also reduced the radial growth of the *T. harzianum* strain from which it was isolated.

8.2.5 *Diketopiperazines*

With the transfer of *Gliocladium virens* to the genus *Trichoderma*, the list of antibiotics produced by members of this genus must be expanded to include the diketopiperazines. The antibiotic gliotoxin was first isolated by Weindling and Emerson (1936) and later designated as gliotoxin (Weindling, 1941). It was further characterized by Brian and Hemming (1945), and, as in all prior work, the producing fungus was misidentified as to either genus or species. The structure of gliotoxin was not elucidated until many years later (Bell *et al.*, 1958), and it was not until 1964 that the fungus was correctly identified as *G. virens* (Webster and Lomas, 1964). The first to associate gliotoxin production with biocontrol by *G. virens* (now *T. virens*) were Aluko and Hering (1970), who demonstrated that gliotoxin was the active factor in control of *Rhizoctonia solani* on seed potatoes. The association of gliotoxin production and control of Zinnia seedling diseases incited by *R. solani* and *Pythium ultimum* in soilless mix was demonstrated by Lumsden *et al.* (1992a). Gliotoxin has also been shown to inhibit the mycelial growth, sporangium formation, and zoospore motility of a number of *Phytophthora* species (Wilcox *et al.*, 1992). Howell *et al.* (1993) later reported that gliotoxin was produced only by certain strains (designated "Q" strains) within *T. virens*. They also showed that gliotoxin-producing strains gave better control of *R. solani*-incited disease than did nonproducers.

The diketopiperazine antibiotic gliovirin (Stipanovic and Howell, 1982) is also produced by *T. virens* (Howell and Stipanovic, 1983) but not by the strains that produce gliotoxin. Gliovirin, produced by "P" strains of the fungus, is very toxic to *P. ultimum*, and "P" strains are more effective biocontrol agents of this pathogen than are "Q" strains (Howell *et al.*, 1993).

8.2.6 *Sesquiterpenes*

Heptelidic acid is the one secondary metabolite that bridges the gap between *T. virens* and other *Trichoderma* species. It has been isolated from *T. viride*, *T. virens*, and *T. koningii* (Endo *et al.*, 1985; Itoh *et al.*, 1980; Stipanovic and Howell, 1983). It has been shown to have antibiotic activity against anaerobic bacteria (Itoh *et al.*, 1980), *P. ultimum*, and *R. solani* (Howell *et al.*, 1993) and to inhibit cholesterol biosynthesis (Endo and Kanbe, 1981). A carotane sesquiterpene, CAF-603, has also been isolated from *T. virens* and shown to have antifungal activity (Watanabe *et al.*, 1990).

8.2.7 *Steroids*

The antibiotic viridin was first isolated from *T. viride* (actually *G. virens*) by Brian and McGowan (1945) and shown to inhibit spore germination of a number of fungi (Brian and Hemming, 1945). Viridin, in combination with gliotoxin, has also been associated with the biocontrol of black scurf on potatoes by *T. virens* (Aluko and Hering, 1970). More recently, viridin production by *T. virens* has been associated with the suppression of growth by *R. solani* and *P. ultimum* and sclerotial germination by *Sclerotium rolfsii* (Lumsden *et al.*, 1992b). Viridiol (Moffatt *et al.*, 1969) is the end product of a pathway in which viridin is its precursor (Jones and Hancock, 1987). This compound was shown by Howell and Stipanovic (1984) to be a potent phytotoxin, causing severe necrosis of germinating seed radicles, and was suggested for use as a pre-emergence herbicide. The number of affected weeds was greatly expanded and less expensive means of producing the phytotoxin were reported by Jones *et al.* (1988).

8.3 Environmental parameters affecting antibiotic production and activity

Of the environmental factors that directly influence the production of antibiotics by *Trichoderma* and *Gliocladium* species, probably the most important is the substrate on which the fungus is grown. Early on, it was demonstrated that the production of gliotoxin by *T. virens* was favored by growing the fungus in Raulin–Thom medium compared with other media (Brian and Hemming, 1945), and that $Ca(NO_3)_2$ and glucose, galactose, starch or glycerol were the best sources of nitrogen and carbohydrate (Brian *et al.*, 1946). Park *et al.* (1991) later reported that the optimum medium for gliotoxin production contained glucose and phenylalanine at carbon:nitrogen ratios of 18:1, 31:1, or 42:1. Growth of *T. viride* on different carbohydrate sources was correlated with the ability of culture filtrates to protect wood blocks from *Fomes annosus* by Sierota (1977). Cultures utilizing d-xylose and cellulose gave the best results, and there was no correlation between mycelial growth of the antagonist and pathogen inhibition. Srinivasan *et al.* (1992) have also reported that soluble metabolites from *T. koningii* grown on malt medium were much more inhibitory to the wood decay fungus *Neolentinus lepideus* than those from a minimal medium. Another prime example is the effect of substrate on the production of viridiol by *T. virens*. Howell and Stipanovic (1984) showed that production on a rice substrate was much better than on other substrates with lower carbon:nitrogen ratios. This was confirmed by the work of Jones and Hancock (1987) who showed in defined media that high C:N ratios enhanced, and low C:N ratios suppressed, viridiol production.

The pH of the growth environment may also have a profound effect on antibiotic production and activity. Weindling (1941) demonstrated that production of gliotoxin was much higher at pH 3.5 than at 6.0. However, because gliotoxin was also shown to be unstable at high pH it is difficult to know whether the effect was on production or stability. Loss of activity at higher pH was also demonstrated for viridin (Brian and McGowan, 1945).

Surprisingly little attention has been paid to the effects of temperature on the production or activities of antibiotics produced by *Trichoderma* or *Gliocladium* species. Weindling (1941) did examine the effect of temperature on the stability and activity of gliotoxin. He found the compound to be stable at 123°C at pH 2.4, but that stability deteriorated quickly as the pH rose. He also found that gliotoxin activity was retarded by low temperature (3–18°C), but became progressively more active with rising temperature (21–32.5°C) against *R. solani*. Lumsden *et al.* (1992a) studied the effect of temperature on the production of gliotoxin in soilless medium by alginate prills of *T. virens*. Production of the antibiotic was low at 15°C, but it increased with rising temperature to a plateau between 25 and 30°C. Considering the low temperature of some soils at planting, this phenomenon could have an adverse effect on biocontrol activity.

A most interesting aspect of the effect of environment on antibiotic activity is the apparent synergism between the enzymes and antibiotics elaborated by *Trichoderma* species to effect antifungal activity. Di Pietro *et al.* (1993) studied the synergistic effects of an endochitinase and gliotoxin isolated from *T. virens* on conidia germination by *Botrytis cinerea*. They found that when 0.75 $\mu g\ ml^{-1}$ of gliotoxin or 75 $\mu g\ ml^{-1}$ of endochitinase were applied alone, they caused no inhibition and 20% inhibition, respectively. When the enzyme and antibiotic were applied in combination they caused 95% inhibition of conidia germination. The authors suggested that this may play a role in biocontrol by *T. virens*. Schirmböck *et al.* (1994) noted a similar phenomenon with metabolites of *T. harzianum*. They found that a concentration of 200 $\mu g\ ml^{-1}$ of the antibiotics trichorzianines A_1 or B_1 were required to completely inhibit germination of *B. cinerea* conidia. The addition of endochitinase (25 $\mu g\ ml^{-1}$), chitobiosidase (50 $\mu g\ ml^{-1}$), or β-1,3-glucanase (25 $\mu g\ ml^{-1}$) to the test medium containing the antibiotic reduced the required inhibitory concentration to 30 $\mu g\ ml^{-1}$. The same concentration of endochitinase alone resulted in only 35% inhibition. The authors also concluded that the synergism of hydrolytic enzymes and peptaibols contributed significantly to the antagonism of *T. harzianum* against phytopathogens. Subsequent work by Lorito *et al.* (1996b) has expanded this concept to include seven different antibiotics and two fungicides in all combinations with eight enzymes from fungi, bacteria, or plants. All enzyme plus antimicrobial compound combinations showed high levels of synergistic inhibition against *B. cinerea* and *Fusarium oxysporum*. Different levels of synergism were obtained among the combinations, depending on the antifungal activity of the enzyme. When the enzyme was added after the antibiotic, the level of synergism was lower than when the order was reversed, indicating that cell wall degradation was needed to establish the synergistic interaction.

8.4 Antibiotics as mechanisms in disease biocontrol, and their modes of action

Very little is known about the modes of action exerted by the antibiotic secondary metabolites of *Trichoderma* and *Gliocladium* species. Jones and Hancock (1988) studied the mechanism of gliotoxin action against *P. ultimum* and *R. solani*. They found that binding of radiolabeled thiol reagents to fungal thiol groups was inhibited by gliotoxin and that thiol reagents inhibited the uptake of radiolabeled gliotoxin. They also found that uptake of amino acids and glucose by these fungi was reduced 85% by gliotoxin. This suggests that the primary mode of gliotoxin action involves selective binding to cytoplasmic membrane thiol groups.

Trichorzianines and a number of closely related peptaibols form voltage-gated ion channels in black lipid membranes (Molle *et al.*, 1987), and they modify the membrane permeability of liposomes (El Hajji *et al.*, 1989; Le Doan *et al.*, 1986). They are active against fungi and gram-positive bacteria, and they act by perturbing the ionic balance of the cell (Goulard *et al.*, 1995; Schirmböck *et al.*, 1994).

The role of antibiotics produced by *Trichoderma* and *Gliocladium* in the biocontrol process has long been of interest (Dennis and Webster, 1971a,b), and there are numerous reports in the literature associating antibiosis with disease control (Fravel, 1988; Ghisalberti *et al.*, 1990; Worthington, 1988). In more recent years, evidence has begun to accumulate in support of a role for gliotoxin in the biocontrol of *P. ultimum*-induced seedling disease. Lumsden *et al.* (1992a) demonstrated that gliotoxin could be detected in soilless media suppressive to the pathogen and that its maximal accumulation corresponded to the time of greatest disease suppressive activity. This was followed by the production of gliotoxin-deficient mutants from *T. virens* that showed on average only 54% of the disease-suppressive activity of the wild-type isolate *in vivo* (Wilhite *et al.*, 1994). Antagonistic activity by the mutants against *P. ultimum in vitro* was almost totally lost. A similar situation was demonstrated by Howell and Stipanovic (1983) with mutants of *T. virens* deficient for the antibiotic gliovirin. Loss of gliovirin production was accompanied by a loss in biocontrol activity against *P. ultimum*-induced seedling disease. A mutant with increased gliovirin production gave disease control equivalent to that of the parent strain even though its growth rate was reduced. It has also been demonstrated that the biocontrol efficacy of gliovirin-producing strains of *T. virens* against *P. ultimum*-induced damping-off is far superior to that of the same strains against *R. solani*-induced seedling disease (Howell *et al.*, 1993). The pathogen in the latter case is not inhibited by gliovirin. Although gliotoxin has been implicated in the biocontrol of *R. solani* in several instances (Lumsden *et al.*, 1992a,b; Weindling, 1941), the role of this antifungal compound in the biocontrol phenomenon is much less clear. Howell and Stipanovic (1995) produced mutants of *T. virens* deficient for gliotoxin production by ultraviolet (UV) light mutagenesis. The mutants showed a total loss of inhibitory activity to *R. solani*, but assay of the mutants for biocontrol efficacy against *R. solani*-induced seedling disease showed them to be equal to the parental strains.

A similar phenomenon occurred with UV-induced mutants of *T. harzianum* that showed reduced or elevated levels of antibiotic production (Graeme-Cook and Faull, 1991). Assay of the mutants for interaction with *P. ultimum*, *R. solani*, and *Fusarium oxysporum* showed that there was no correlation between antibiotic production and ability to invade any of the three pathogens. In fact, increased antibiotic activity appeared to impede the interaction of *T. harzianum* with the test fungi.

8.5 Genetic manipulation of antibiotics and disease control

Although several excellent review articles have been written on the genetic manipulation of fungal biocontrol agents (Clarkson, 1992; Harman and Tronsmo, 1992), reports in the literature on the genetic manipulation of secondary metabolites in *Trichoderma* or *Gliocladium* are indeed sparse. The use of mutants induced with UV light to ascertain the role of antibiotics in the biocontrol activities of *T. virens* (Howell and Stipanovic, 1983, 1995; Wilhite *et al.*, 1994) and *T. harzianum* (Graeme-Cook and Faull, 1991) has already been outlined in this chapter. A UV-induced mutant of *T. harzianum* has also been shown to produce a heat-labile antifungal compound not found in the parent isolate (Papavizas *et al.*, 1982). The compound was inhibitory to *Sclerotium cepivorum*, and the strain that produced the highest amounts of both the heat-labile and another compound was the most effective in controlling white rot of onion.

Mutagenesis of *T. virens* strains with UV light and culture on media containing sterol inhibitors has resulted in the isolation of mutants deficient for production of the phytotoxin viridiol (Howell and Stipanovic, 1996). Suppression of viridiol production by these mutants virtually eliminated phytotoxicity of the biocontrol preparations to developing root systems of cotton seedlings, and the mutants retained the capacity to act as biocontrol agents of cotton seedling disease incited by *R. solani*.

There is virtually no information available on the transformation of genes coding for antibiotic synthesis within *Trichoderma* or *Gliocladium*.

Although not directly associated with the genetic manipulation of secondary metabolites, transformation of genes coding for hydrolytic enzymes that have been cloned (Carsolio *et al.*, 1994; Geremia *et al.*, 1991; Hayes *et al.*, 1993; Lorito *et al.*, 1994) could have profound effects on the activities of antibiotics produced by *Trichoderma* and *Gliocladium* species. As shown by Lorito *et al.* (1996b), there is strong synergism between cell wall degrading enzymes and antimicrobial compounds. This synergism may take the form of cell wall disruption by *T. harzianum*-produced β-1,3-glucanase and inhibition of host β-1,3-glucan synthase by the peptaibols trichorzianin TA and TB. Therefore, host cell wall disruption by hydrolytic enzymes is sustained by the peptaibols which prevent resynthesis of cell wall β-glucans (Lorito *et al.*, 1996a). It seems reasonable to expect that proper manipulation of genes coding for enzyme and antibiotic production could lead to better biocontrol strains.

8.6 Concluding remarks

The development of successful strategies for biological control of plant diseases requires intimate knowledge of the biocontrol agents, the mechanisms involved in the biocontrol process, and the effect of environment on both. Research to date on the role of antibiotics in the biocontrol process has shown them to be important in some systems but less so in others. Even where they appear to play a prominent role, it almost always is in concert with other mechanisms such as competition and/or mycoparasitism. Recent data indicates that antibiotics usually, if not always, act synergistically with cell wall degrading enzymes. The preponderance of recent research data also supports the concept that successful biocontrol is a matter of balance between a number of contributing factors. This must be taken into account when altering the genome of a biocontrol agent in order to optimize its disease control activity. Frequently, increased antibiotic production by a mutant or transformant results in reduced rather than increased biocontrol efficacy. In the final analysis it seems that successful biocontrol agents will be those that make the most efficient and balanced use of a number of mechanisms associated with antagonism and those that can function under a fairly wide range of environmental conditions.

References

ALMASSI, F., GHISALBERTI, E. L., NARBEY, M. J., and SIVASITHAMPARAM, K. 1991. New antibiotics from strains of *Trichoderma harzianum*. *J. Nat. Prod.* **54**: 396–402.

ALUKO, M. O. and HERING, T. F. 1970. The mechanisms associated with the antagonistic relationship between *Corticium solani* and *Gliocladium virens*. *Trans. Br. Mycol. Soc.* **55**: 173–179.

AUVIN-GUETTE, C., REBUFFAT, S., VUIDEPOT, I., MASSIAS, M., and BODO, B. 1993. Structural elucidation of trikoningins KA and KB, peptaibols from *Trichoderma koningii*. *J. Chem. Soc., Perkin Trans.* **1**: 249–255.

BAKER, R. 1990. An overview of current and future strategies and models for biological control. In D. Hornby (ed.), *Biological Control of Soil-borne Plant Pathogens*. Redwood Press, Melksham, Wiltshire, pp. 375–388.

BELL, M. R., JOHNSON, J. R., WILDI, B. S., and WOODWARD, R. B. 1958. The structure of gliotoxin. *J. Am. Chem. Soc.* **80**: 1001.

BODO, B., REBUFFAT, S., EL HAJJI, M., and DAVOUST, D. 1985. Structure of trichorzianines A IIIc, an antifungal peptide from *Trichoderma harzianum*. *J. Am. Chem. Soc.* **107**: 6011–6017.

BRIAN, P. W. and HEMMING, H. G. 1945. Gliotoxin, a fungistatic metabolic product of *Trichoderma viride*. *Ann. Appl. Biol.* **32**: 214–220.

BRIAN, P. W. and MCGOWAN, J. C. 1945. Viridin: a highly fungistatic substance produced by *Trichoderma viride*. *Nature (London)* **156**: 144.

BRIAN, P. W., CURTIS, P. J., HEMMING, H. G., and MCGOWAN, J. C. 1946. The production of viridin by pigment-forming strains of *Trichoderma viride*. *Ann. Appl. Biol.* **33**: 190–200.

CARSOLIO, C., GUTIERREZ, A., JIMENEZ, B., VAN MONTAGU, M., and HERRERA-ESTRELLA, A. 1994. Characterization of *ech-42*, a *Trichoderma harzianum* endochitinase gene expressed during mycoparasitism. *Proc. Natl. Acad. Sci. USA* **91**: 10903–10907.

CHET, I. 1987. *Trichoderma* – Application, mode of action and potential as a biocontrol agent of soilborne plant pathogenic fungi. In I. Chet (ed.), *Innovative Approaches to Plant Disease Control*. Wiley, New York, pp. 137–160.

CLARKSON, J. M. 1992. Molecular biology of filamentous fungi used for biological control. In J. R. King and G. Turner (eds), *Applied Molecular Genetics of Filamentous Fungi*. Blackie Academic & Professional, London, pp. 175–190.

CLAYDON, N., ALLAN, M., HANSON, J. R., and AVENT, A. G. 1987. Antifungal alkyl pyrones of *Trichoderma harzianum*. *Trans. Br. Mycol. Soc.* **88**: 503–513.

COLLINS, R. P. and HALIM, A. F. 1972. Characterization of the major aroma constituent of the fungus *Trichoderma viride* (Pers.). *J. Agr. & Food Chem.* **20**: 437–438.

COOK, R. J. and BAKER, K. F. 1983. The Nature and Practice of Biological Control of Plant Pathogens. The American Phytopathological Society, St. Paul, MN, 539 pp.

CORREA, A., REBUFFAT, S., BODO, B., ROQUEBERT, M. F., DUPONT, J., and BETTUCCI, L. 1995. *In vitro* inhibitory activity of trichorzianines of *Sclerotium rolfsii* Sacc. *Crypt. Mycol.* **16**: 185–190.

DENNIS, C. and WEBSTER, J. 1971a. Antagonistic properties of species-groups of *Trichoderma*. I: Production of non-volatile antibiotics. *Trans. Br. Mycol. Soc.* **57**: 25–39.

DENNIS, C. and WEBSTER, J. 1971b. Antagonistic properties of species-groups of *Trichoderma*. II: Production of volatile antibiotics. *Trans. Br. Mycol. Soc.* **57**: 41–48.

DI PIETRO, A., LORITO, M., HAYES, C. K., BROADWAY, R. M., and HARMAN, G. E. 1993. Endochitinase from *Gliocladium virens*: isolation, characterization, and synergistic antifungal activity in combination with gliotoxin. *Phytopathology* **83**: 308–313.

DUNLOP, R. W., SIMON, A., SIVASITHAMPARAM, K., and GHISALBERTI, E. L. 1989. An antibiotic from *Trichoderma koningii* active against soilborne plant pathogens. *J. Nat. Prod.* **52**: 67–74.

EL HAJJI, M., REBUFFAT, S., LECOMMANDEUR, D., and BODO, B. 1987. Isolation and sequence determination of trichorzianines A antifungal peptides from *Trichoderma harzianum*. *Int. J. Pept. & Prot. Res.* **29**: 207–215.

EL HAJJI, M., REBUFFAT, S., LE DOAN, T., KLEIN, G., SATRE, M., and BODO, B. 1989. Interaction of trichorzianines A and B with model membranes and with amoeba *Dictyostelium*. *Biochim. Biophys. Acta* **978**: 97–104.

ENDO, A. and KANBE, T. 1981. Isolation, chemical structure and mechanisms of action of koningic acid, a new inhibitor of cholesterol synthesis. *Proc. Japan. Conf. Biochem. Lipids* **23**: 172–174.

ENDO, A., HASUMI, K., SAKAI, K., and KANBE, T. 1985. Specific inhibition of glyceraldehyde-3-phosphate dehydrogenase by koningic acid (heptelidic acid). *J. Antib.* **38**: 920–925.

FRAVEL, D. R. 1988. Role of antibiosis in the biocontrol of plant diseases. *Annu. Rev. Phytopath.* **26**: 75–91.

FUJI, K., FUJITA, E., TAKAISHI, Y., FUJITA, T., ARITA, I., KOMATSU, M., and HIRATSUKA, N. 1978. New antibiotics, trichopolyns A and B: isolation and biological activity. *Experientia* **34**: 237–239.

FUJIWARA, A., OKUDA, T., MASUDA, S., SHIOMI, Y., MIYAMOTO, C., SEKINE, Y., TAZOE, M., and FUJIWARA, M. 1982. Fermentation, isolation and characterization of isonitrile antibiotics. *Agric. Biol. Chem.* **46**: 1803–1809.

GEREMIA, R. A., GOLDMAN, G. H., JACOBS, D., VAN MONTAGU, M., and HERRERA-ESTRELLA, A. 1991. Role and specificity of different proteinases in pathogen control by *Trichoderma harzianum*. *Petria* **1**: 131.

GHISALBERTI, E. L. and SIVASITHAMPARAM, K. 1991. Antifungal antibiotics produced by *Trichoderma* spp. *Soil Biol. Biochem.* **23**: 1011–1020.

GHISALBERTI, E. L., NARBEY, M. J., DEWAN, M. M., and SIVASITHAMPARAM, K. 1990. Variability among strains of *Trichoderma harzianum* in their ability to reduce take-all and to produce pyrones. *Plant and Soil* **121**: 287–291.

GOULARD, C., HLIMI, S., REBUFFAT, S., and BODO, B. 1995. Trichorzins HA and MA, antibiotic peptides from *Trichoderma harzianum*. I: Fermentation, isolation and biological properties. *J. Antib.* **48**: 1248–1253.

Graeme-Cook, K. A. and Faull, J. L. 1991. Effect of ultraviolet-induced mutants of *Trichoderma harzianum* with altered production on selected pathogens *in vivo*. *Can. J. Microbiol.* **37**: 659–664.

Harman, G. E. and Tronsmo, A. 1992. Methods of genetic manipulation for the production of improved bioprotectant fungi. In D. F. Jensen, J. Hockenhull, and N. J. Fokkema (eds), *New Approaches in Biological Control of Soil-Borne Diseases. IOBC/* WPRS Bulletin **XV/1**: 181–187.

Hayes, C. K., Klemsdal, S., Lorito, M., Di Pietro, A., Peterbauer, C., Nakas, J. P., Tronsmo, A., and Harman, G. E. 1993. Isolation and sequence of an endochitinase encoding gene from a cDNA library of *Trichoderma harzianum*. *Gene* **138**: 143–148.

Howell, C. R. 1990. Fungi as Biological Control Agents. In J. P. Nakas and C. Hagedorn (eds), *Biotechnology of Plant–Microbe Interactions.* McGraw-Hill, New York, pp. 257–286.

Howell, C. R. and Stipanovic, R. D. 1983. Gliovirin, a new antibiotic from *Gliocladium virens*, and its role in the biological control of *Pythium ultimum*. *Can. J. Microbiol.* **29**: 321–324.

Howell, C. R. and Stipanovic, R. D. 1984. Phytotoxicity to crop plants and herbicidal effects on weeds of viridiol produced by *Gliocladium virens*. *Phytopathology* **74**: 1346–1349.

Howell, C. R. and Stipanovic, R. D. 1995. Mechanisms in the biocontrol of *Rhizoctonia solani*-induced cotton seedling disease by *Gliocladium virens*: antibiosis. *Phytopathology* **85**: 469–472.

Howell, C. R. and Stipanovic, R. D. 1996. Mechanisms in cotton soreshin biocontrol by *Trichoderma virens*: viridiol production. Proc. Beltwide Cotton Conf., p. 271.

Howell, C. R., Stipanovic, R. D., and Lumsden, R. D. 1993. Antibiotic production by strains of *Gliocladium virens* and its relation to the biocontrol of cotton seedling diseases. *Biocontrol Sci. Tech.* **3**: 435–441

Itoh, Y., Kodama, K., Furuya, K., Takahashi, S., Haneishi, T., Takiguchi, Y., and Arai, M. 1980. A new sesquiterpene antibiotic, heptelidic acid. Producing organisms, fermentation, isolation and characterization. *J. Antib.* **23**: 468–473.

Jones, R. W. and Hancock, J. G. 1987. Conversion of viridin to viridiol by viridin-producing fungi. *Can. J. Microbiol.* **33**: 963–966.

Jones, R. W. and Hancock, J. G. 1988. Mechanism of gliotoxin action and factors mediating gliotoxin sensitivity. *J. Gen. Microbiol.* **134**: 2067–2075.

Jones, R. W., Lanini, W. T., and Hancock, J. G. 1988. Plant growth response to the phytotoxin viridiol produced by the fungus *Gliocladium virens*. *Weed Sci.* **36**: 683–687.

Le Doan, T., El Hajji, M., Rebuffat, S., Rajesvari, M. R., and Bodo, B. 1986. Fluorescence studies of the interaction of trichorzianine A IIIc with model membranes. *Biochim. Biophys. Acta* **858**: 1–5.

Lorito, M., Farkas, V., Rebuffat, S., Bodo, B., and Kubicek, C. P. 1996a. Cell wall synthesis is a major target of mycoparasitic antagonism by *Trichoderma harzianum*. *J. Bacteriol.* **178**: 6382–6385.

Lorito, M., Hayes, C. K., Zoina, A., Scala, F., Del Sorbo, G., Woo, S. L., and Harman, G. E. 1994. Potential of genes and gene products from *Trichoderma* sp. and *Gliocladium* sp. for the development of biological pesticides. *Molec. Biotech.* **2**: 209–217.

Lorito, M., Woo, S. L., D'Ambrosio, M. D., Harman, G. E., Hayes, C. K., Kubicek, C. P., and Scala, F. 1996b. Synergistic interaction between cell wall degrading enzymes and membrane affecting compounds. *Molec. Plant–Microbe Interact.* **9**: 206–213.

Lumsden, R. D., Locke, J. C., Adkins, S. T., Walter, J. F., and Ridout, C. J. 1992a. Isolation and localization of the antibiotic gliotoxin produced by *Gliocladium virens* from alginate prill in soil and soilless media. *Phytopathology* **82**: 230–235.

LUMSDEN, R. D., RIDOUT, C. J., VENDEMIA, M. E., HARRISON, D. J., WATERS, R. M., and WALTER, J. F. 1992b. Characterization of major secondary metabolites produced in soilless mix by a formulated strain of the biocontrol fungus *Gliocladium virens*. *Can. J. Microbiol.* **38**: 1274–1280.

MAITI, D., DASGUPTA, B., and SEN, C. 1991. Antagonism of *Trichoderma harzianum* and *Gliocladium virens* isolates to *Sclerotium rolfsii* and biological control of stem rot of ground nut and betelvine. *J. Biol. Control* **5**: 105–109.

MERLIER, A. M. O., BOIER, J. M., PONS, J. B., and RENAUD, M. C. 1984. European Patent Application EP 124388 (Chem. Abs. 183747r, 102, 1985).

MOFFATT, J. S., BU'LOCK, J. D., and YUEN, T. H. 1969. Viridiol, a steroid-like product from *Trichoderma viride*. *J. Chem. Soc. Chem. Comm.* **139**: 839.

MOLLE, G., DUCLOHIER, H., and SPACH, G. 1987. Voltage dependent and multistate ionic channels induced by trichorzianines, antifungal peptides related to alamethicin. *FEBS Letters* **224**: 208–212.

OKUDA, T., FUJIWARA, A., and FUJIWARA, M. 1982. Correlation between species of *Trichoderma* and production patterns of isonitrile antibiotics. *Agric. Biol. Chem.* **46**: 1811–1822.

ORDENTLICH, A., WIESMAN, Z., GOTTLIEB, H. E., COJOCARU, M., and CHET, I. 1992. Inhibitory furanone produced by the biocontrol agent *Trichoderma harzianum*. *Phytochemistry* **31**: 485–486.

PACHENARI, A. and DIX, N. J. 1980. Production of toxins and wall degrading enzymes by *Gliocladium roseum*. *Trans. Br. Mycol. Soc.* **74**: 561–566.

PAPAVIZAS, G. C. 1985. *Trichoderma* and *Gliocladium*: biology, ecology, and potential for biocontrol. *Annu. Rev. Phytopath.* **23**: 23–54.

PAPAVIZAS, G. C., LEWIS, J. A., and ABD-EL MOITY, T. H. 1982. Evaluation of new biotypes of *Trichoderma harzianum* for tolerance to benomyl and enhanced biocontrol capabilities. *Phytopathology* **72**: 126–132.

PARK, Y. H., STACK, J. P., and KENERLEY, C. M. 1991. Production of gliotoxin by *Gliocladium virens* as a function of source and concentration of carbon and nitrogen. *Mycol. Res.* **95**: 1242–1248.

REHNER, S. A. and SAMUELS, G. J. 1994. Taxonomy and phylogeny of *Gliocladium* analyzed by large subunit ribosomal DNA sequences. *Mycol. Res.* **98**: 625–634.

SCHIRMBÖCK, M., LORITO, M., WANG, Y. L., HAYES, C. K., ARISAN-ATAC, I., SCALA, F., HARMAN, G. E., and KUBICEK, C. P. 1994. Parallel formation and synergism of hydrolytic enzymes and peptaibol antibiotics, molecular mechanisms involved in the antagonistic action of *Trichoderma harzianum* against phytopathogenic fungi. *Appl. Environ. Microbiol.* **60**: 4364–4370.

SIEROTA, Z. H. 1977. Inhibitory effect of *Trichoderma viride* Pers. ex Fr. filtrates on *Fomes annosus* in relation to some carbon sources. *Eur. J. Forest Pathol.* **7**: 164–172.

SIMON, A., DUNLOP, R. W., GHISALBERTI, E. L., and SIVASITHAMPARAM, K. 1988. *Trichoderma koningii* produces a pyrone compound with antibiotic properties. *Soil Biol. Biochem.* **20**: 263–264.

SRINIVASAN, U., STAINES, H. J., and BRUCE, A. 1992. Influence of media type on antagonistic modes of *Trichoderma* spp. against wood decay basidiomycetes. *Material and Organismen* **27**: 301–321.

STIPANOVIC, R. D. and HOWELL, C. R. 1982. The structure of gliovirin, a new antibiotic from *Gliocladium virens*. *J. Antib.* **35**: 1326–1330.

STIPANOVIC, R. D. and HOWELL, C. R. 1983. The x-ray crystal structure determination and biosynthetic studies of the antibiotic, heptelidic acid. *Tetrahedron* **39**: 1103–1107.

TAYLOR, A. 1986. Some aspects of the chemistry and biology of the genus *Hypocrea* and its anamorphs, *Trichoderma and Gliocladium*. *Proc. Nova Scotia Inst. Sci.* **36**: 27–58.

VON ARX, J. A. 1987. Plant pathogenic fungi. *Beihefte zur Nova Hedwigia* **87**: 288.

WATANABE, N., YAMAGISHI, M., MIZUTANI, T., KONDOH, H., OMURA, S.,

HANADA, K., and KUSHIDA, K. 1990. CAF-603: a new antifungal carotane sesquiterpene: isolation and structure elucidation. *J. Nat. Prod.* **53**: 1176–1181.

WEBSTER, J. and LOMAS, N. 1964. Does *Trichoderma viride* produce gliotoxin and viridin? *Trans. Br. Mycol. Soc.* **47**: 535–540.

WEINDLING, R. 1941. Experimental consideration of the mold toxins of *Gliocladium* and *Trichoderma*. *Phytopathology* **31**: 991–1003.

WEINDLING, R. and EMERSON, O. H. 1936. The isolation of a toxic substance from the culture filtrate of *Trichoderma*. *Phytopathology* **26**: 1068–1070.

WILCOX, W. F., HARMAN, G. E., and DI PIETRO, A. 1992. Effect of gliotoxin on growth, sporulation, and zoospore motility of seven *Phytophthora* spp. *in vitro*. *Phytopathology* **82**: 1121.

WILHITE, S. E., LUMSDEN, R. D., and STRANEY, D. C. 1994. Mutational analysis of gliotoxin production by the biocontrol fungus *Gliocladium virens* in relation to suppression of Pythium damping-off. *Phytopathology* **84**: 816–821.

WORTHINGTON, P. A. 1988. Antibiotics with antifungal and antibacterial activity against plant diseases. *Nat. Prod. Rep.* **5**: 47–66.

9

Direct effects of *Trichoderma* and *Gliocladium* on plant growth and resistance to pathogens

B. A. BAILEY and R. D. LUMSDEN

Biocontrol of Plant Diseases Laboratory, USDA, ARS, Beltsville, Maryland, USA

9.1 Introduction

It is well established that microorganisms closely associated with the roots of plants can directly influence plant growth and development. The detrimental effects of plant pathogenic organisms have long been the subject of intensive study. The beneficial effects of microorganisms, such as rhizobia (Keister and Cregan, 1991), plant growth promoting rhizobacteria (Kloepper *et al*., 1991), and mycorrhizae (Linderman, 1988; Reid, 1990), on plant growth are well known and some interactions have been studied extensively. Species of *Trichoderma*, on the other hand, are primarily studied for their ability to control plant disease through mycoparasitism (Chapter 7) and/or the production of antimicrobial compounds (Chapter 8). Although the ability of species of *Trichoderma* to directly promote or inhibit plant growth has been noted for many years (Lindsey and Baker, 1967; Wright, 1956), efforts to define and exploit those influences have met with limited success. Determination of the beneficial or detrimental effects of *Trichoderma* species on plant growth is complicated by the many interactions that take place in the soil between *Trichoderma* spp. (both test isolates and background isolates), other microorganisms, changes in the soil environment, and the plant root. An example demonstrating the difficulty in interpreting results from such studies is the work by Thuy (1991). In this study, both beneficial and detrimental effects of *Trichoderma* spp. on germination of pepper seed were demonstrated and the influences of indigenous *Trichoderma* species and other microorganisms were brought into question. It would seem obvious that in the absence of disease, plant growth is enhanced. The tasks set forth in this chapter are to describe the direct detrimental and beneficial effects of *Trichoderma* species on plant growth and to discuss the proposed mechanisms involved.

9.2 Plant growth inhibition

9.2.1 *Phytotoxic compounds*

There has been a great deal of research concerning the antimicrobial compounds produced by *Trichoderma* species. These compounds have varying levels of both antifungal and herbicidal activities that influence our abilities to exploit their use in biocontrol. Gliotoxin and viridin are epidithiodiketopiperazine and sterol compounds, respectively. Gliotoxin and viridin inhibited the germination and root growth of mustard seed at 1 ppm (Wright, 1951) in petri dish culture. Gliotoxin and viridin concentrations of 1 ppm were not inhibitory to root growth of red clover or wheat seedlings, demonstrating the differences in tolerance to gliotoxin and viridin that exist among plant species. Swelling of the coleorhiza occurred at higher concentrations of the two compounds. The activity of gliotoxin and viridin tended to be greater at low pH (pH 4.0) than at a more neutral pH (pH 6.0), perhaps due to increased stability of the compounds. Wright (1956) observed that some strains produced predominantly gliotoxin or viridin while other strains produced no detectable quantities of either compound. The toxin-producing strains caused damage to white mustard seedlings at high inoculum concentrations. A common symptom was chlorosis of the cotyledons. Strains of *Gliocladium virens* (= *Trichoderma virens*) produced the compounds gliotoxin, viridin, and gliovirin in soilless growth medium (Lumsden *et al.*, 1992). The phytotoxic effects of gliotoxin and viridin are limited compared with their antifungal activities (Lumsden *et al.*, 1992; Lumsden and Walter, 1997; Wright, 1951).

The modes of action resulting in the toxicity of gliotoxin and viridin to plants are still under study. Haraguchi *et al.* (1996) observed that gliotoxin inhibited the enzyme acetolactate synthase (ALS) in plants. ALS is involved in the biosynthesis of branched chain amino acids. The addition of branched chain amino acids to the plant growth medium partially alleviated the toxicity of gliotoxin. Gliotoxin also inhibits the activity of other enzymes (Van Der Pyl *et al.*, 1992; De Clercq *et al.*, 1978).

Trichoderma species also produce viridiol, a dihydro-derivative of viridin (Moffatt *et al.*, 1969). Viridin is apparently enzymatically converted to viridiol by producer fungal strains (Jones and Hancock, 1987). *Gliocladium virens* isolates that did not produce viridin could not convert viridin to viridiol. Viridiol has very little antibiotic activity but has considerable herbicidal activity (Howell and Stipanovic, 1984). The phytotoxic effects of viridiol, active against a broad range of plant species, have been studied more extensively than other secondary metabolites produced by *Trichoderma* species. Viridiol, produced when positive strains are incorporated into soil, is most toxic to annual composite species (Table 9.1) and less toxic to monocot species (Jones *et al.*, 1988). Typical symptoms include reduced seedling emergence and reduced shoot and root weights. Jones and Hancock (1987) went as far as to suggest that viridin-producing strains of *Gliocladium virens* may be necrotrophic pathogens. Accumulation of viridiol in the soil peaked 5 to 6 days after inoculation and declined to undetectable levels by 2 weeks. Viridiol is too unstable for direct use as a herbicide and must be produced in the root zone to be effective (Howell and Stipanovic, 1984). A pulverized preparation of a viridiol-producing strain of *G. virens* inhibited pigweed germination when applied to pigweed-infested soil. Damage caused by viridiol included inhibition of seed germination and damage to seedling radicals (Howell and Stipanovic, 1984). The *G. virens* inoculum did not damage cotton seedlings planted below the area of inoculum incorporation, even though cotton was sensitive to viridiol. Damage to crop plants resulting from herbicidal applications of *G. virens* may be limited by directed placement of the inoculum. A suggested method and rate of application of a fungus–peat mixture was in a band requiring 815 kg/ha (Jones *et al.*, 1988). The high inoculum rate required for achiev-

Table 9.1 Effects of addition of the *Gliocladium virens* – peat mixture on emergence and dry weight values of crop and weed species at 14 days

Rate[a]	Plant	Seedling emergence (% of control)	Dry weight[b] (mg)			
			Root		Shoot	
			Control	Treated	Control	Treated
16.0	Cotton	75	44.1	3.4	150.1	41.1
	Sunflower	69	31.6	15.4	105.4	48.4
	Lettuce	32	1.4	0.2	5.6	0.4
	Soybean	41	59.0	34.6	176.9	86.7
	Cucumber	44	16.5	3.5	66.7	11.4
	Cantaloupe	25	13.9	0.6	61.7	1.6
	Corn	83	103.6	64.1	92.6	36.0
	Sudangrass	71	15.5	3.7	32.8	5.4
	Mustard	33	2.2	0.3	11.9	0.9
	Tomato	52	2.7	0.1	12.4	0.5
	Wild oat	89	30.7	17.8	13.7	6.8
8.7	Safflower	100	30.8	16.3	77.7	31.8
	Bean	100	189.6	143.1	423.3	257.1
	Alfalfa	75	6.3	2.1	17.6	3.2
	Sugarbeet	100	8.5	1.6	25.2	3.5
	Spinach	79	8.1	2.3	19.5	4.3
	Eastern black nightshade	23	0.9	0.9*	2.9	1.3
	Field bindweed	67	12.0	3.7	27.7	9.5
	Annual bluegrass	40	1.3	0.5	2.2	0.4
	Yellow foxtail	53	6.8	1.5	15.6	2.5
	Littleseed canarygrass	29	1.8	1.2	2.1	1.9*
	Curly dock	20	0.6	0.2	2.3	0.9
	Buckhorn plantain	46	0.9	0.5	1.5	0.8
4.5	Spotted catsear	58	1.6	0.1	2.7	0.7
	Redroot pigweed	71	1.5	0.1	5.7	0.1
	Common purslane	67	1.0	0.1	2.5	0.2
	Common fiddleneck	57	0.9	0.6	5.3	1.7
	Lambsquarters	69	0.2	0.1	0.7	0.4
2.3	Dandelion	60	1.9	0.9	2.6	1.2
	Bristly oxtongue	62	1.8	1.0	4.1	2.4
	Groundsel	96	0.5	0.3	1.8	1.1
	Annual sowthistle	58	1.0	0.6	3.1	1.0

[a] Fungus–peat mixture as percentage of total soil volume. Values represent treatment levels at which at least 10% of seedlings emerged relative to the control. Species that failed to emerge were seeded at the next lower concentration.

[b] Values represent mean of individual seedlings. Differences between treatment means for dry weight values were all significant ($P = 0.05$) using ANOVA-1, except where noted by asterisk (*).

Reproduced with permission from Jones *et al.* (1988).

ing significant herbicidal activity by *G. virens* and other *Trichoderma* spp. is one factor limiting their potential use as bioherbicides.

Cultures of *Trichoderma* spp. produce other metabolites with phytotoxic activities. Cutler *et al.* (1986) used a wheat coleoptile etiolation assay to demonstrate that the 6-pentyl-α-pyrone, produced by *T. viride* and *T. harzianum*, inhibits plant growth. The compound 6-pentyl-α-pyrone is a flavoring agent found in natural peach essence and has been synthesized for industrial purposes. This compound was non-toxic to greenhouse grown bean, corn, and tobacco. Cutler and Jacyno (1991) also used the wheat coleoptile etiolation assay to demonstrate the activity of (-)-harzianopyridone isolated from *T. harzianum*. The compound (-)-harzianopyridone was toxic to bean, corn, and tobacco plants at 10^{-2} M concentrations. A third group of compounds capable of inhibiting etiolation of wheat coleoptiles are the compounds Koninginin A (Cutler *et al.*, 1989) and Koninginin B (Cutler *et al.*, 1991) produced by *T. koningii*. Koninginin B appears to be slightly more phytotoxic than Koninginin A.

9.2.2 *Factors influencing growth inhibition*

The detrimental effects of *Trichoderma* metabolites on plant growth can limit their usefulness as biocontrol agents. The total amounts and type of inoculum used in protecting plants from disease must be limited to levels not producing phytotoxic effects (Howell, 1991; Howell and Stipanovic, 1994). Production of secondary metabolites is strain dependent (Howell *et al.*, 1993). Therefore, their detrimental effects can be limited in many cases by strain selection. Lumsden *et al.* (1992) identified strains of *G. virens* that produced high gliotoxin levels and low viridin levels or vice versa. The gliotoxin-producing strain GL-21 has been commercialized for use in the greenhouse production of bedding plants (Lumsden and Walter, 1997).

Growth conditions greatly influence the production of both epidithiodiketopiperazine and sterol compounds by *G. virens*. Howell and Stipanovic (1984) observed that whereas biomass of *G. virens* grown on rice produced large quantities of viridiol and was damaging to cotton seedlings, biomass grown on potato dextrose agar produced low levels of viridiol and was not phytotoxic. Production of gliotoxin, as opposed to viridin, is regulated by pH and C:N ratios in the growth medium. At C:N ratios of 150:1 gliotoxin production was accentuated at lower pH ($pH < 6.0$) and viridin production was accentuated at higher pH ($pH > 6.0$) (Lumsden and Ridout, 1991).

Other systems have been developed that limit the production of secondary metabolites by producing strains *in vivo*. The fungicides propiconazole and flusilazole inhibit cytochrome P-450 C_{14}-demethylation of sterols. Low concentrations of these triazole fungicides inhibited production of viridiol and viridin without inhibiting growth of the fungus (Howell and Stipanovic, 1994; Stipanovic and Howell, 1994). Production of the nonsteroid antimicrobial compounds gliotoxin and gliovirin was not influenced by the fungicide treatments (Table 9.2). The fungicides blocked production of viridin, allowing treatment of cotton seed with increased quantities of *G. virens*–millet inoculum (Howell and Stipanovic, 1994; Stipanovic and Howell, 1994). Thus, it may be possible to use a strain that normally produces

Table 9.2 Effect of sterol biosynthesis inhibitors on production ($\mu g\ ml^{-1}$) of the phytotoxin viridiol and other secondary metabolites by *Gliocladium virens* in culture

	Phytotoxin and antibiotics			
Strain/inhib.	Viridiol	Viridin	Gliotoxin	Gliovirin
P group				
G-4 + UT[a]	1510 + 100*[b]	110 + 30	NP	1560 + 100
G-4 + PC	0	47 + 12	NP	1430 + 60
G-8 + UT	1460 + 100*	37 + 6	NP	1680 + 50
G-8 + PC	0	56 + 6	NP	1460 + 100
G-9 + UT	1480 + 90*	37 + 6*	NP	1730 + 100
G-9 + PC	0	170 + 50	NP	1560 + 130
Q group				
G-6 + UT	2340 + 90*	30 + 17*	650 + 100	NP
G-6 + FL	0	0	890 + 150	NP
G-11 + UT	2460 + 20*	37 + 15*	700 + 80	NP
G-11 + FL	0	0	950 + 200	NP
G-20 + UT	2350 + 30*	30 + 10*	700 + 100	NP
G-20 + FL	0	0	800 + 70	NP

[a] UT = untreated control; PC = propiconazole; FL = flusilazole; NP = not produced. Sterol inhibitor concentrations = 1 $\mu g\ mol^{-1}$.
[b] Numbers followed by * within a strain are significantly different ($P < 0.05$) according to Student's t test.
Reproduced with permission from Howell and Stipanovic (1994).

high levels of phytotoxic compounds, such as viridin, but otherwise has good biocontrol characteristics.

9.2.3 *Diseases caused by* Trichoderma *species*

The term "disease" is defined by Agrios (1978) as "any disturbance of a plant that interferes with its normal structure, function, or economic value". In the same text, a "pathogen" is defined as "an entity that can incite disease". *Trichoderma* species have limited plant pathogenic capabilities. Perhaps the greatest pathogenic capability of *Trichoderma* species is not on plants at all but on mushrooms (Chapter 12). The primary exceptions are cases of post-harvest decay of fruits. An unidentified *Trichoderma* species was shown to cause rot of apples in storage (English, 1944). The infection was initiated at the calyx and progressed into the core of the apple. *T. lignorum* can cause post-harvest decay of citrus fruit (Knosel and Schickedanz, 1976). An isolate of *T. harzianum* was shown to cause brown firm rot on wounded apples in storage (Conway, 1983). The occurrence of brown firm rot of apple caused by *T. harzianum* is considered rare because of the limited competitive ability of *T. harzianum* compared with *Penicillium expansum*.

There have been incidences where *Trichoderma* spp. have been described as pathogens to field crops. An isolate of *T. viride* was reported to be pathogenic to alfalfa (Aube and Gagnon, 1970). Incorporation of *T. viride* inoculum into the soil reduced the number of alfalfa plants and dry weight of alfalfa shoots and roots.

T. viride stimulated the pathogenicity of *Pyrenochaeta terrestris* and *Rhizoctonia solani* when co-applied to alfalfa. Based on field isolations from stunted maize plants in Ontario, Canada, *T. koningii* was reported to be parasitic on maize (McIntosh and Sneh, 1972). Symptoms observed after inoculation of maize seedlings in the field with *T. koningii* included lesions on the first internodes, first nodal adventitious roots, and primary roots. Falloon (1985) provided evidence that *T. koningii* was pathogenic to ryegrass. It is quite possible that the reduction in seedling number observed by Falloon (1985) was the result of toxic metabolite production by the very high inoculum concentration used in the test. *Trichoderma* spp. sometimes build up to high populations (8×10^5 cfu/g soil) in native soils (Chet and Baker, 1981), but phytotoxic effects from these high populations have not been demonstrated in nature. In contrast to Falloon's (1985) results, Dewan and Sivasithamparam (1988) considered *T. koningii* to be nonpathogenic to ryegrass and wheat. They also found (Dewan and Sivasithamparam, 1988) that isolates of *T. hamatum* and *T. harzianum* taken from roots of wheat and ryegrass were nonpathogenic to wheat and ryegrass.

An isolate of *T. viride* was characterized by Menzies (1993) as a minor pathogen on cucumber, pepper, and tomato in hydroponics systems (Table 9.3). A 'minor pathogen' is defined by Salt (1979) as "a saprophyte or parasite damaging only meristematic and cortical cells and surviving in soil as a saprophyte, as resting spores or as sclerotia". Characteristics of minor pathogens include a wide distribution in cultivated soils, a strong environmental influence on their damaging effects, a lack of distinctive symptoms, and occurrence in mixed infections with other fungi.

Table 9.3 The effect of the inoculation of seed of cucumber cv. Corona, pepper cv. Bison, and tomatoes cv. Dombito with a water suspension of 10^6 conidia/ml of *Trichoderma viride* RF1 on seedling root, shoot, and total length and weight

Host seed treated	Root length[a] (mm)	Shoot length[a] (mm)	Total length[a] (mm)	Seedling fresh weight[a] (mg)
Cucumber				
Control	131.2	82.7	214.1	925
T. viride RF1	109.4	56.5	165.9	635
Cucumber[b]				
Control	121.3	74.2	195.5	909
T. viride RF1	83.8	31.5	115.3	480
Pepper				
Control	62.3	33.9	96.2	60
T. viride RF1	32.5	17.0	49.5	39
Tomato				
Control			155.9	
T. viride RF1			88.6	

[a] The values of the control and *T. viride* RF1 treated seeds are significantly different in all cases as determined using an F test ($P < 0.0001$ in all cases except tomato which had a $P < 0.0038$).

[b] The cucumber test was repeated, and both sets of data are presented.

Reproduced with permission from Menzies (1993).

Trichoderma species are excellent saprophytes and have been isolated from many different plant species. Some *Trichoderma* species can penetrate cortical tissues in what has been described as a mycorrhizal-like association (Kleifeld and Chet, 1992). Despite ample opportunity, damage to plants is rarely attributed to *Trichoderma* species and, when noted, occurs under very restricted or even artificial conditions.

9.3 Plant growth promotion

Plant growth promotion by strains of *Trichoderma* has been reported for many years (Lindsey and Baker, 1967). The results presented by Lindsey and Baker (1967) are significant in that isolates of *Chaetomium* sp., *Rhizopus nigricans*, and *Fusarium roseum* also promoted plant growth under gnotobiotic conditions, demonstrating that this phenomenon was not restricted to *Trichoderma* spp. (Table 9.4). Chang *et al.* (1986) observed plant growth promotion resulting in enhanced germination, more rapid flowering, increased flowering, and increased heights and fresh weights in pepper, periwinkle, chrysanthemums, and/or other plants after treatment of soil with peat/bran inoculum or conidial suspensions of *T. harzianum*. The study, which included data on rhizosphere competent isolates of *T. harzianum* from laboratories in the USA and Israel, did not assess the importance of pathogens or the mode(s) of action involved in the plant growth promotion. The isolates used were not effective on all plant species tested and some of the results presented were inconsistent with earlier results presented by these same laboratories using the same isolates (Baker *et al.*, 1984). Windham *et al.* (1986) used defined conditions including gnotobiotic conditions to demonstrate plant growth promotion by *T. harzianum* and *T. koningii* strains. Plant growth promotion was demonstrated in corn, tomato, tobacco, and radish and included increased germination rates, dry weights, and emergence. Harman *et al.* (1989) carried out field studies on growth and development of sweet corn treated with *T. harzianum* isolate T12 in soils both amended and non-amended with *Pythium ultimum*. In this study, T12 promoted higher seedling emergence, final stand, and dry weight in plots of sweet corn under conditions where *Pythium*

Table 9.4 Average height and weight of 42-day-old dwarf tomato plants grown in three different gnotobiotic environments

	Plant height and dry weight			
Environment	Plant[a] height (cm)	Shoot weight (g)	Root weight (g)	Total[a] weight (g)
Germ-free	14.1c	1.38	0.40	1.78c
Chaetomium sp.-infested	15.3c	1.37	0.41	1.78c
Trichoderma viride-infested	18.1d	1.50	0.41	1.91c

[a] Values followed by the same letter do not differ significantly at the 1% level, according to Duncan's multiple range test. Values are based on 20 plants/treatment.
Reproduced with permission from Lindsey and Baker (1967).

ultimum had no measurable effect. The possibility of the plant growth promotion's resulting from control of other undiagnosed diseases was not eliminated. Lynch *et al.* (1991) demonstrated plant growth promotion in lettuce after treatment with *T. harzianum* isolates grown on molasses medium which included increased emergence rates and increased dry weights after 25 days. Although not inoculated with a pathogen, the pathogen pressure in the growth medium was not determined. Kleifeld and Chet (1992) found that a peat/bran preparation of *T. harzianum* isolate T-203 was more effective at inducing plant growth promotion (germination, seedling length, dry weight, and leaf area) in pepper and other plant species than conidial suspensions or seed coatings, suggesting that the peat/bran served as a food reservoir for isolate T-203. The level of plant growth promotion was strongly dependent on the plant growth substrate. Plant growth promotion was demonstrated in plants grown in semi-sterile Hoagland's solution, suggesting the effect was not entirely associated with suppression of minor pathogens. Calvet *et al.* (1993) observed enhanced growth of marigold after treatment with *T. aureoviride* in combination with the mycorrhizal fungus *Glomus mosseae*. The *T. aureoviride* had no effect on disease caused by *Pythium ultimum* alone but did when combined with *G. mosseae*. This study emphasizes the potential importance of other microorganisms in the rhizosphere on plant growth promotion by *Trichoderma* species. Inbar *et al.* (1994) were able to demonstrate plant growth promotion by *T. harzianum* isolate T-203 in a commercial greenhouse production system. In this seedling production system, seedling height, leaf area, and dry weight were increased significantly at the time of marketing.

As mentioned previously, plant growth promotion by fungi is not restricted to *Trichoderma* species. Shivanna *et al.* (1996) have worked with various sterile fungi and isolates of *Penicillium* and *Trichoderma* species taken from the roots of zoysia grass. Many of these isolates, including a *Trichoderma* strain, were shown to promote growth of wheat and soybean under greenhouse conditions (Shivanna *et al.*, 1996). The plant growth promoting fungi were less effective when applied in field tests, but increased yield was observed in response to some isolates (Shivanna *et al.*, 1994).

9.3.1 ***Rhizosphere competence***

Rhizosphere competence (Harman, 1992) is a term that relates to the ability of an organism to establish itself in the rhizosphere of plants and is subject to influences of environment and competition with other organisms. Most wild-type *Trichoderma* strains have very limited ability to colonize the rhizosphere (Beagle-Ristaino and Papavizas, 1985; Chao *et al.*, 1986; Papavizas, 1982) and only increase in populations in the soil in close proximity to added food sources (Beagle-Ristaino and Papavizas, 1985; Kleifeld and Chet, 1992). Wild-type isolates with some level of rhizosphere competence have been described (Sivan and Chet, 1989).

The strategy to increase the rhizosphere competence of wild-type *Trichoderma* strains has resulted in the development of techniques for genetically altering *Trichoderma* species to improve their rhizosphere competence (Ahmad and Baker, 1987; Sivan and Harman, 1991; Stasz and Harman, 1990). The primary questions asked of rhizosphere competent *Trichoderma* strains are as follows: can the strains colonize the root surface along with the progression of root growth, to what depth can they colonize, what conditions are required for colonization, and what is their competitive ability with other microorganisms? Strains of *Trichoderma* with improved rhizosphere competence continue to be studied for their growth promotion and disease suppressing capabilities (Lynch *et al.*, 1991; Sivan and Harman, 1991; Björkman *et al.*, 1998; Chapter 4). One of the most detailed studies of rhizosphere competence among *Trichoderma* species was carried out with *T. harzianum* and other *Trichoderma* species by Ahmad and Baker (1987, 1988a,b,c). Strains of *T. harzianum, T. koningii, T. polysporum,* and *T. viride* were mutated using N-methyl-N-nitro-N-nitrosoguanidine (100 $\mu g\ ml^{-1}$) that produced benomyl-tolerant strains that also possessed improved rhizosphere competence (Ahmad and Baker, 1987, 1988b). The goal of the initial experiments was to develop benomyl-tolerant strains for co-application of *Trichoderma* spp. with benomyl to seedlings, but, unexpectedly, most of the benomyl-tolerant isolates possessed improved rhizosphere competence in the absence of benomyl (Ahmad and Baker, 1987). Rhizosphere competence was demonstrated by applying the *T. harzianum* inoculum as a seed coating and limiting watering to prevent movement of spores through the soil (Ahmad and Baker, 1987). While parental strains failed to colonize beyond 3 cm of the root, the mutant strains colonized the entire root system to the root tip (8 cm). The mutant strains of *T. harzianum* promoted seedling emergence and plant growth in several plant species and provided control of *Pythium ultimum* by significantly reducing populations of *P. ultimum* near the root tips (Ahmad and Baker, 1988a). The genetic control of tolerance to benomyl appears to be closely, but not completely, linked to rhizosphere competence (Ahmad and Baker, 1988b). Peterbauer *et al.* (1992) found that transformation of *T. reesei* with the *Neurospora crassa ben* gene, coding for benomyl-tolerant beta-tubulin, resulted in a phenotype similar to *Trichoderma* strains exposed to sublethal doses of benomyl. The rhizosphere competent strains were better able to utilize complex carbohydrates, including cotton linters, microcrystalline cellulose, wood cellulose, and xylan as carbon sources (Ahmad and Baker, 1988b,c). Thus Ahmad and Baker (1988c) proposed that the mutants have a better ability to utilize complex carbohydrates such as those associated with plant roots, resulting in improved hyphal extension rates for the resulting thalli, allowing the fungus to grow along with the root. Environmental factors such as pH, temperature, other microorganisms, and the host plant influenced the populations of *T. harzianum* in the rhizosphere (Ahmad and Baker, 1987). The populations of *T. harzianum* were highest around the root tip and in the upper root zone, producing a "C-shaped" population curve when graphed relative to root depth (Ahmad and Baker, 1987). It must be noted that seed treatment of peas with conidial suspensions of reported rhizosphere competent strain T95 (Ahmad and Baker, 1987) did not result in colonization of the root system in studies carried out by Chao *et al.* (1986), suggesting that the treatment methods and conditions are critical to demonstration of rhizosphere competence.

Improved rhizosphere competent strains that enhanced growth of sweet corn and cotton roots were produced by protoplast fusions of auxotrophic mutants (Sivan and Harman, 1991) derived from the *T. harzianum* strains T12 and T95. Strain T95 was produced by mutation and selection for tolerance to benomyl. The strain 1295-22 had an altered root colonization pattern, being able to colonize the rhizosphere of maize and cotton more uniformly than either parental strain (Table 9.5). Maize roots were colonized to a depth of 22 cm and cotton roots were colonized to

Table 9.5 Mean levels of the various strains of *Trichoderma* on the upper, middle, and lower rhizosphere and rhizoplane of maize and cotton roots

Root	Log_{10} (C.f.u.)[a]			
section	T95	1295-22	T12	MSD
Maize rhizosphere				
Upper portion	2.5	3.9	3.3	NS
Middle portion	1.2	2.9	1.6	1.08
Lower portion	2.8	3.2	1.8	NS
MSD[b]	NS	NS	1.52	
Maize rhizoplane				
Upper portion	2.8	2.4	2.6	NS
Middle portion	1.2	2.9	1.4	NS
Lower portion	1.5	3.1	2.4	NS
MSD[b]	NS	NS	NS	
Cotton rhizosphere				
Upper portion	1.3	2.7	2.5	1.13
Middle portion	0.9	2.0	0.6	0.80
Lower portion	0.6	0.7	0	NS
MSD[b]	0.45	1.66	1.14	
Cotton rhizoplane				
Upper portion	1.4	2.6	2.4	0.74
Middle portion	0.8	2.0	1.0	1.07
Lower portion	0.4	1.3	0	NS
MSD[b]	NS	NS	0.12	

[a] Units are per g dry weight of soil or root for rhizosphere or rhizoplane, respectively. In all cases, the upper portion of the roots was considered to be the first 6 cm below the hypocotyl, the lower portions were considered to be the lowest 4 cm, and the remainder was considered to be the middle portion. Values presented are means across two separate experiments.

[b] Minimum significant difference as determined by Waller and Duncan's test (GLM procedure, SAS Institute, Cary, NC, USA).

Reproduced with permission from Sivan and Harman (1991).

a depth of 16 cm by isolate 1295-22. A concurrent enhanced colonization of rhizoplane soil was observed for both plant species. Rhizosphere competence in *Trichoderma* species is a relative trait that has potential for continual enhancement. Although plant growth promotion was not described in the work by Lo *et al.* (1996), rhizosphere competent *T. harzianum* strain 1295-22 was able to persist at elevated levels in creeping bentgrass for over 8 months. This isolate increased root and shoot growth of the sweet corn variety Supersweet Jubilee in greenhouse pot studies an average of 66% over untreated controls (Björkman *et al.*, 1998). Studies designed to estimate the effect of minor pathogens indicated that plant growth promotion was a consequence of direct effects on seed and seedling vigor and not a result of the control of specific plant pathogens (Björkman *et al.*, 1998).

9.3.2 *The mechanism of plant growth promotion*

It is evident that the addition of specific *Trichoderma* isolates to the rhizosphere can result in plant growth promotion. The plant growth-promoting effects in some systems are prolonged even to the point of increasing yield (Ahmad and Baker, 1988a; Harman *et al.*, 1989; Lynch *et al.*, 1991). It is also apparent that various isolates of *Trichoderma* species differ in their abilities to maintain populations in the rhizosphere in the absence of outside food sources (Harman, 1992). The plant growth promoting *T. harzianum* isolate T-203 was able to penetrate and live in plant roots in a manner similar to mycorrhizal fungi (Kleifeld and Chet, 1992). By maintaining high populations of *Trichoderma* species in the rhizosphere, their influence on plant growth (i.e., promotion or inhibition) should be maximized. What is unclear at this point is how *Trichoderma* species promote plant growth and how the effect can be predicted and exploited. The mechanisms involved in plant growth promotion by *Trichoderma* species are only now being investigated.

Indirect effect by control of minor pathogens

It has been proposed that plant growth promotion in response to treatment with *Trichoderma* spp. is the result of control of minor pathogens. This hypothesis suggests that all plants grown in the open environment are to some extent diseased and unable to reach their maximum growth potentials. The addition of *Trichoderma* species to these "control" plants, with undefined disease potentials, results in disease suppression and enhanced plant growth. The primary minor pathogens are thought to be *Pythium* species (Ahmad and Baker, 1988a; Harman *et al.*, 1989) against which *Trichoderma* spp. have known activity (Harman *et al.*, 1989; Lumsden *et al.*, 1992). In several plant growth promotion studies, *T. harzianum* inhibited growth and development of *Pythium* species on the root (Ahmad and Baker, 1988a; Harman *et al.*, 1989).

Direct effects of Trichoderma *on plant growth promotion*

Of greater pertinence to this chapter are studies where the influence of minor pathogens are limited or presumably removed. The most obvious cases involve the use of gnotobiotic conditions (Lindsey and Baker, 1967; Windham *et al.*, 1986), where plant growth promotion was demonstrated under conditions where the only known organisms involved were the plant and the *Trichoderma* isolate. Gnotobiotic conditions are far removed from situations in native soil and so must be considered carefully (Ahmad and Baker, 1987). Attempts have been made to define factors responsible for growth promotion under gnotobiotic conditions. Evidence supporting the presence of a growth-promoting diffusable factor was obtained by separating germinating maize, tomato, and tobacco seeds from homogenized mycelial preparations of *T. koningii* or *T. harzianum* (Windham *et al.*, 1986) with a cellophane membrane. This diffusable factor was not identified. A more recent hypothesis has been put forth by Björkman *et al.* (1998). In these studies, the roots of maize plants treated with rhizosphere competent and plant growth promoting *T. harzianum* strain 1295-22 were more vigorous than roots of untreated plants. The evidence suggested that the *T. harzianum* isolate was limiting and even reversing the effects of oxidative damage to the roots. Addition of *T. harzianum* strain 1295-22 to hypochlorite-damaged seedlings restored seedling vigor (Figure 9.1). Hypochlorite is reported to cause oxidative damage to plant tissues (Schraufstaetter *et al.*, 1990).

Direct effects by induced resistance

Induced resistance involves activation of plant defense systems resulting in protection of plants from pathogens or pests (Ouchi, 1983; Boller, 1989). Host plant recognition of signals from potential pathogens is a key factor in the induction of resistance in plant–pathogen interactions. Are signals produced as a result of *Trichoderma*–pathogen–plant–environment interactions capable of inducing resistance in plants? The literature indicates that inducers or elicitors of the defense response are produced directly by the pathogen or result from the interaction between the pathogen and plant (Boller, 1989). It is important to note, especially with soilborne plant pathogens, that the pathogen and plant rarely, if ever, interact in the absence of other organisms. The environment in which pathogens interact with plants is teeming with other organisms, including *Trichoderma* species. The defense responses induced by plant–microbe interactions may remain localized or become systemic. Induction of systemic resistance can result in whole plants, both above- and below-ground parts, being resistant to a variety of plant pests (for a review of induced resistance see Ouchi, 1983; Ryals *et al.*, 1995; and Chen *et al.*, 1995).

Direct evidence implicating *Trichoderma* spp. in the induction of resistance in plants is limited. Wyss *et al.* (1992) observed that an isolate of *T. harzianum* induced accumulation of large amounts of the phytoalexin glyceollin in soybean roots but

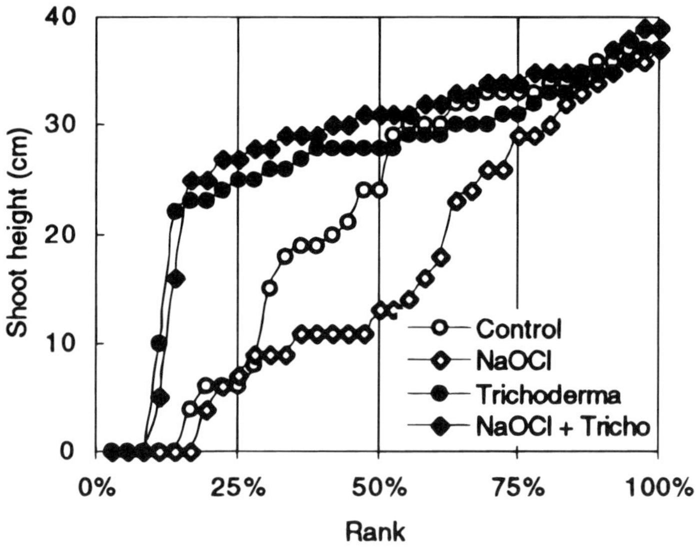

Figure 9.1 Distribution of sweet corn plant size as affected by hypochlorite treatment and *Trichoderma* colonization. The shoot height was measured after 21 days of growth in pots. Both hypochlorite and *Trichoderma* treatments affected primarily intermediate-vigor seedlings. High-vigor seedlings were not further enhanced by *Trichoderma* colonization. Reproduced with permission from Björkman *et al.* (1998).

did not provide protection against *Rhizoctonia solani*. Only now are more extensive studies concerning induced resistance being carried out with *Trichoderma* species (Zimand *et al.*, 1996; Hyakumachi, personal communication). Zimand *et al.* (1996) proposed an interesting mechanism whereby *T. harzianum* isolate T39 reduced disease caused by *Botrytis cinerea* on bean leaves by inhibiting pectolytic enzyme activity of the pathogen. These authors suggested that the reduced polygalacturonase activity might result in accumulation of large oligogalacturonides which can act as elicitors of plant defense. Plant growth promoting fungi are capable of inducing systemic resistance in plants (Meera *et al.*, 1994). Hyakumachi (personal communication) has found that many plant growth promoting fungi, including isolates of *T. harzianum* and *T. koningii*, applied as mycelial inocula on roots of cucumber seedlings, can provide protection to plants from foliar symptoms of anthracnose.

Trichoderma species produce elicitors that induce plant defense responses in plant tissue (Anderson *et al.*, 1993; Calderon *et al.*, 1993; Ishii, 1987). The responses of plant tissue to the 22 kDa xylanase of *Trichoderma viride* have been examined and a model system presented (Figure 9.2). This xylanase is produced by *T. viride* in culture during growth in liquid media containing D-xylose, xylan, or crude plant cell wall preparations as the primary carbon source (Dean *et al.* 1989). Other *Trichoderma* spp., including *T. harzianum*, *T. hamatum*, and *T.* (*Gliocladium*) *virens*, produce a related 22 kDa xylanase in culture (Baker *et al.*, 1977; Dean *et al.*, 1989). Further information on the chemistry and molecular biology of the 22 kDa xylanase is provided in Chapter 2. The 22 kDa xylanase induces plant defense responses including K^+, H^+, and Ca^{2+} channeling (Bailey *et al.*, 1992), pr-protein biosynthesis (Lotan and Fluhr, 1990), ethylene biosynthesis (Fuchs *et al.*, 1989), and glycosylation and fatty acylation of phytosterols (Moreau *et al.*, 1994). It is interesting to note that rhizosphere competent strains of *T. harzianum* have enhanced cellulytic enzyme

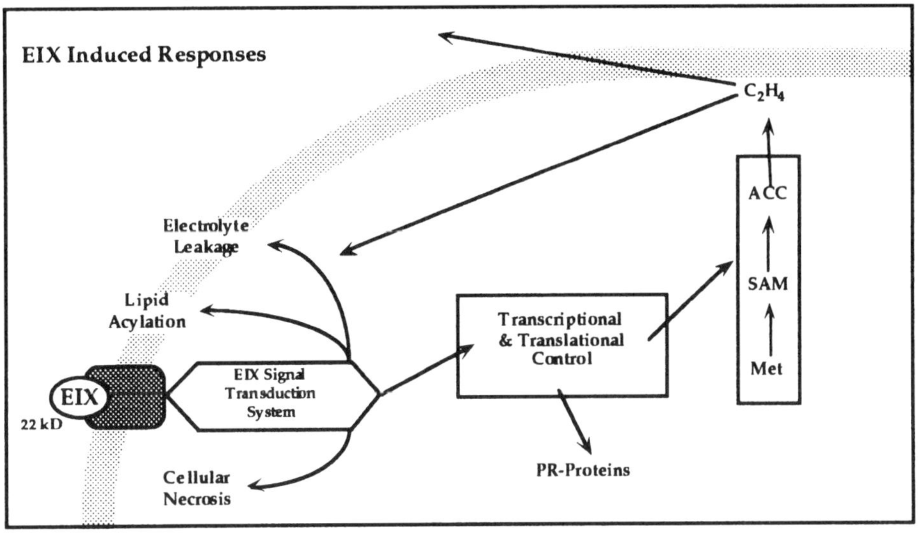

Figure 9.2 A model demonstrating the induced plant defense responses in Xanthi tobacco leaves after treatment with the *Trichoderma viride* 22 kDa xylanase (EIX). The model postulates that the 22 kDa xylanase binds to a membrane bound receptor and through a signal transduction pathway induces a hypersensitive-like response. The model was provided by Dr James D. Anderson, Weed Science Laboratory, ARS/USDA, Beltsville, Maryland 20705, USA.

activities in liquid culture and some are better able than wild-type strains to digest xylan and other complex carbohydrates even in the presence of simple sugars (Ahmad and Baker, 1988c). Simple sugars are known to repress cellulytic enzyme activity in wild-type strains of *Trichoderma* (Ahmad and Baker, 1988c; Ljundhal and Erriksson, 1985). A more detailed description of the influence of sugars on cellulytic enzyme activity is presented in Chapter 4. Sensitivity to the 22 kDa xylanase is not only dependent upon plant species (some solanaceous species are sensitive) but also upon the genotype of the cultivar being studied. Single genes have been identified in both tobacco (Bailey *et al.*, 1993) and tomato (Avni *et al.*, 1994) cultivars that control sensitivity to the 22 kDa xylanase. The 22 kDa xylanase must be injected into the intercellular spaces or xylem tissue (Bailey *et al.*, 1991). Once in the xylem, the protein migrates through the vascular system and out into the intercellular spaces by means of transpiration. Present work suggests the responses are not the result of the xylanase enzyme activity (Sharon *et al.*, 1993) but are mediated by membrane receptors interacting directly with the xylanase protein (A. Avni, personal communication). It is not known if this system functions *in vivo* to induce resistance in plant tissues. There are many other products, including cellulytic enzymes produced by *Trichoderma* species, capable of inducing plant responses, but how and if they function in the biocontrol of disease is unknown.

Trichoderma species produce many of the same or similar enzymes as plants that are capable of digesting walls and membranes of plant pathogens as well as plants (see Part 1 of this volume). For example, *T. harzianum* is capable of producing β-1,3-glucanases, chitinases, proteases, xylanases, and lipases, which are excreted into the growth medium in liquid culture (Elad *et al.*, 1982). These enzymes are also produced when *T. harzianum* attacks mycelia of other fungi. Similar enzymes have been credited with releasing compounds capable of eliciting defense responses (Boller, 1989; Ishii, 1987). It is possible that *Trichoderma* species release elicitor-active compounds from pathogenic fungi, plant cell walls, or organic matter in the soil or in the root, thereby inducing resistance in associated plant tissues. If this is true, and it must be tested, questions become critical concerning the fungal and plant growth medium, the soil environment, pathogen and biocontrol agent populations, plant species and age, and timing of the interactions. *Trichoderma* species can directly or indirectly produce compounds which stimulate defense responses in plants. The combination of this ability and the ability to colonize the rhizosphere and cortical tissue of plant roots makes some *Trichoderma* species excellent candidates for providing long-term induced resistance in plants.

9.4 Unraveling host plant interactions

Considerable work has been carried out demonstrating the interaction of *Trichoderma* spp. with plants, but unraveling the mechanisms involved in the interactions has been difficult. Until the mechanisms are understood, our ability to exploit these effects in commercial agricultural systems is limited. Of paramount importance to understanding the mechanisms involved is an understanding of the environment in which these interactions occur. Soil type, pH, temperature, nutrient availability, other microorganisms, and other environmental factors are critical in these studies (Ousley *et al.*, 1993).

At present, there is a great deal of inconsistency, especially in the expression of plant growth promoting effects of strains of *Trichoderma*, which is undoubtedly influenced by variations in the rhizosphere environment. Another requirement for understanding the plant growth promoting capabilities of *Trichoderma* spp. is the examination of more isolates. The vast majority of studies related to the plant growth promoting capabilities of *Trichoderma* spp. use *T. harzianum* strains related to isolates initially reported in the early 1980s (Chet and Baker, 1981). A larger number of plant growth promoting strains of *Trichoderma* spp. would not only lead to an improved understanding of the mechanisms involved but would also allow development of improved strains by genetic manipulation and recombination beyond what is available now.

The relationship between rhizosphere competence and plant growth promotion justifies the strategy of developing *Trichoderma* strains with improved ability to colonize and influence the rhizosphere environment. New strains of *Trichoderma* with improved rhizosphere competence have been developed by genetic alterations of these strains first by mutation and then by protoplast fusion (Sivan and Harman, 1991). The potential for rhizosphere competent strains to contribute to plant growth promotion in economically useful ways seems very promising. Still, there is some confusion as to exactly what rhizosphere competence as characterized in the literature means. Higher populations of colony forming units as measured by available techniques have no necessary relationship with microbial activity. In fact, it could be argued that colony forming units measured as production of conidia may reflect unfavorable conditions for the fungus, in that vegetative growth is shut down and survival spores are formed. The techniques used in rhizosphere competence studies are generally acknowledged to be the best that are available, but techniques that more directly monitor microbial activity are needed. By directly measuring microbial activity, the influences of environmental fluctuations could be examined in greater detail.

The newest area of research regarding the direct effects of *Trichoderma* strains on plants is the area of induced resistance. Based on the little we know at present, we can already postulate that induced resistance in response to strains of *Trichoderma* is likely to occur by several mechanisms. Very few strains of *Trichoderma* have been evaluated for their abilities to induce resistance in plants. It may require new isolation and screening techniques to identify strains with strong resistance-inducing capabilities. Here again, having isolates with the ability to thoroughly colonize plant roots might lead to a prolonged desirable effect.

It appears unlikely that *Trichoderma* spp. will be used to develop mycoherbicides. The herbicidal effect of *Trichoderma* strains is short-lived and requires very high rates of application to be effective. It may be possible to enhance phytotoxin production by a greater understanding of the biosynthetic pathways involved, which may lead to reduced rates of application. Attempts at prolonging the herbicidal effect are likely to result in damage to nontarget plant species. The pathogenic capability of *Trichoderma* spp. is limited in most instances to artificial conditions. As a result of the work with potential herbicidal strains of *Trichoderma*, we have a good understanding of how to limit their detrimental effects on plants thereby allowing greater use of *Trichoderma* spp. for their beneficial effects.

Scientists working with *Trichoderma* spp. have been inventive in exploring new ways of using *Trichoderma* in plant production systems. It is not sufficient to simply evaluate *Trichoderma* in terms of the direct effects on plant pathogens. In doing so, the potential usefulness of many *Trichoderma* strains may be underestimated. *Trichoderma* has real potential for plant growth promotion and induced resistance much like that which has been observed with other microorganisms. A great deal of effort should be put into understanding the mechanisms involved in these effects so that their potentials can be fully realized.

References

AGRIOS, G. N. 1978. *Plant Pathology*, 2nd edn. Academic Press, New York, 703 pp.

AHMAD, J. S. and BAKER, R. 1987. Rhizosphere competence of *Trichoderma harzianum*. *Phytopathology* **77**: 182–189.

AHMAD, J. S. and BAKER, R. 1988a. Implications of rhizosphere competence of *Trichoderma harzianum*. *Can. J. Microbiol.* **34**: 229–234.

AHMAD, J. S. and BAKER, R. 1988b. Rhizosphere competence of benomyl-tolerant mutants of *Trichoderma* spp. *Can. J. Microbiol.* **34**: 694–696.

AHMAD, J. S. and BAKER, R. 1988c. Growth of rhizosphere-competent mutants of *Trichoderma harzianum* on carbon substrates. *Can. J. Microbiol.* **34**: 807–814.

ANDERSON, J. D., BAILEY, B. A., TAYLOR, R., SHARON, A., AVNI, A., MATTOO, A. K., and FUCHS, Y. 1993. Fungal xylanase elicits ethylene biosynthesis and other defense responses in tobacco. In J. C. Pech, A. Latche', and C. Balague (eds), *Cellular and Molecular Aspects of the Plant Hormone Ethylene*, Vol. 16. Kluwer, Dordrecht, pp. 197–204.

AUBE, C. and GAGNON, C. 1970. Influence of certain soil fungi on alfalfa. *Can. J. Plant Sci.* **50**: 159–162.

AVNI, A., AVIDAN, N., ESHED, Y., ZAMIR, D., BAILEY, B. A., STOMMEL, J. R., and ANDERSON, J. D. 1994. The response of *Lycopersicon esculentum* to a fungal xylanase is controlled by a single dominant gene. *Plant Physiol.* (Suppl.) **105**: 872.

BAILEY, B. A., KORCAK, R. F., and ANDERSON, J. D. 1992. Alterations in *Nicotiana tabacum* L. cv Xanthi cell membrane function following treatment with an ethylene biosynthesis-inducing endoxylanase. *Plant Physiol.* **100**: 749–755.

BAILEY, B. A., KORCAK, R. F., and ANDERSON, J. D. 1993. Sensitivity to an ethylene biosynthesis-inducing endoxylanase in *Nicotiana tabacum* L. cv Xanthi is controlled by a single dominant gene. *Plant Physiol.* **101**: 1081–1088.

BAILEY, B. A., TAYLOR, R., DEAN, J. F. D., and ANDERSON, J. D. 1991. Ethylene biosynthesis-inducing endoxylanase is translocated through the xylem of *Nicotiana tabacum* cv Xanthi plants. *Plant Physiol.* **97**: 1181–1186.

BAKER, C. J., WHALEN, C. H., and BATEMAN, D. F. 1977. Xylanase from *Trichoderma pseudokoningii*: purification, characterization, and effects on isolated plant cell walls. *Phytopathology* **67**: 1250–1258.

BAKER, R., ELAD, Y., and CHET, I. 1984. The controlled experiment in the scientific method with special emphasis on biological control. *Phytopathology* **74**: 1019–1021.

BEAGLE-RISTAINO, J. E. and PAPAVIZAS, G. C. 1985. Survival and proliferation of propagules of *Trichoderma* spp. and *Gliocladium virens* in soil and in plant rhizospheres. *Phytopathology* **75**: 729–732.

BJÖRKMAN, T., BLANCHARD, L. M., and HARMAN, G. E. 1998. Growth enhancement of shrunken-2 sweet corn by *Trichoderma harzianum* 1295 22: effect of environmental stress. *J. Am. Soc. Hort.* **123**: 35–40.

BOLLER, T. 1989. Primary signals and secondary messengers in the reaction of plants to pathogens. In W. F. Boss and D. J. Morré (eds), *Second Messengers in Plant Growth and Development*. Alan Liss, New York, pp. 227–255.

CALDERON, A. A., ZAPATA, J. M., MUNOZ, R., PEDRENO, M. A., and BARCELO, A. R. 1993. Resveratrol production as a part of the hypersensitive-like response of grapevine cells to an elicitor from *Trichoderma viride*. *New Phytol.* **124**: 455–463.

CALVET, C., PERA, J., and BAREA, J. M. 1993. Growth response of marigold (*Tagetes erecta* L.) to inoculation with *Glomus mosseae, Trichoderma aureoviride* and *Pythium ultimum* in a peat–perlite mixture. *Plant Soil* **148**: 1–6.

CHANG, Y.-C., CHANG, Y.-C., and BAKER, R. 1986. Increased growth of plants in the presence of the biological control agent *Trichoderma harzianum. Plant Dis.* **70**: 145–148.

CHAO, W. L., NELSON, E. B., HARMAN, G. E., and HOCH, H. C. 1986. Colonization of the rhizosphere by biological control agents applied to seeds. *Phytopathology* **76**: 60–65.

CHEN, Z., MALAMY, J., HENNING, J., CONRATH, U., SANCHEZ-COSAS, P., SILVA, H., RICIGLIANO, J., and KLESSIG, D. F. 1995. Induction, modification, and transduction of salicylic acid signal in plant defense responses. *Proc. Natl. Acad. Sci.* **92**: 4134–4137.

CHET, I. and BAKER, R. 1981. Isolation and biocontrol potential of *Trichoderma hamatum* from soil naturally suppressive to Rhizoctonia solani. *Phytopathology* **71**: 286–290.

CONWAY, W. S. 1983. *Trichoderma harzianum*: a possible cause of apple decay in storage. *Plant Dis.* **67**: 916–917.

CUTLER, H. G. and JACYNO, J. M. 1991. Biological activity of (−)-harzianopyridone isolated from *Trichoderma harzianum. Agric. Biol. Chem.* **55**: 2629–2631.

CUTLER, H. G., COX, R. H., CRUMLEY, F. G., and COLE, P. D. 1986. 6-Pentyl-α-pyrone from *Trichoderma harzianum*: its plant growth inhibitory and antimicrobial properties. *Agric. Biol. Chem.* **50**: 2943–2945.

CUTLER, H. G., HIMMELSBACH, D. S., YAGEN, B., ARRENDALE, R. F., COLE, P. D., and COX, R. H. 1989. Koninginin A: a novel plant growth regulator from *Trichoderma koningii. Agric. Biol. Chem.* **53**: 2605–2611.

CUTLER, H. G., HIMMELSBACH, D. S., YAGEN, B., ARRENDALE, R. F., JACYNO, J. M., COLE, P. D., and COX, R. H. 1991. Koninginin B: a biologically active congener of Koninginin A from *Trichoderma koningii. J. Agric. Food Chem.* **39**: 977–980.

DEAN, J. F. D., GAMBLE, H. R., and ANDERSON, J. D. 1989. The ethylene biosynthesis-inducing xylanase: its induction in *Trichoderma viride* and certain plant pathogens. *Phytopathology* **79**: 1071–1078.

DE CLERCQ, E., BILLIAU, A., OTTENHEIJM, H. C. J., and HERSCHEID, J. D. M. 1978. Antireverse transcriptase activity of gliotoxin analogs. *Biochem. Pharmacol.* **27**: 635–639.

DEWAN, M. M. and SIVASITHAMPARAM, K. 1988. Identity and frequency of occurrence of *Trichoderma* spp. in roots of wheat and rye-grass in western Australia and their effect on root rot caused by *Gaeumannomyces graminis* var. *tritici. Plant Soil* **109**: 93–101.

ELAD, Y., CHET, I., and HENIS, Y. 1982. Degradation of plant pathogenic fungi by *Trichoderma harzianum. Can. J. Microbiol.* **28**: 719–725.

ENGLISH, H. 1944. Notes on apple rots in Washington. *Plant Dis. Rep.* **28**: 610 622.

FALLOON, R. E. 1985. Fungi pathogenic to ryegrass seedlings. *Plant Soil* **86**: 79–86.

FUCHS, Y., SAXENA, A., GAMBLE, H. R., and ANDERSON, J. D. 1989. Ethylene biosynthesis-inducing protein from cellulysin is an endoxylanase. *Plant Physiol.* **89**: 138–143.

HARAGUCHI, H., HAMATANI, Y., HAMADA, M., and FUJII-TACHINO, A. 1996. Effect of gliotoxin on growth and branched-chain amino acid biosynthesis in plants. *Phytochemistry* **42**: 645–648.

HARMAN, G. E. 1992. Development and benefits of rhizosphere competent fungi for biological control of plant pathogens. *J. Plant Nutr.* **15**: 835–843.

HARMAN, G. E., TAYLOR, A. G., and STASZ, T. E. 1989. Combining effective strains of *Trichoderma harzianum* and solid matrix priming to improve biological seed treatments. *Plant Dis.* **73**: 631–637.

HOWELL, C. R. 1991. Biological control of Pythium damping-off of cotton with seed-coating preparations of *Gliocladium virens. Phytopathology* **81**: 738–741.

HOWELL, C. R. and STIPANOVIC, R. D. 1984. Phytotoxicity to crop plants and herbicidal effects on weeds of viridiol produced by *Gliocladium virens*. *Phytopathology* **74**: 1346–1349.

HOWELL, C. R. and STIPANOVIC, R. D. 1994. Effect of sterol biosynthesis inhibitors on phytotoxin (viridiol) production by *Gliocladium virens* in culture. *Phytopathology* **84**: 969–972.

HOWELL, C. R., STIPANOVIC, R. D., and LUMSDEN, R. D. 1993. Antibiotic production by strains of *Gliocladium virens* and its relation to the biocontrol of cotton seedling disease. *Biocontrol Sci. and Technol.* **3**: 435–441.

INBAR, J., ABRAMSKY, M., COHEN, D., and CHET, I. 1994. Plant growth enhancement and disease control by *Trichoderma harzianum* in vegetable seedlings grown under commercial conditions. *Eur. J. Plant Pathol.* **100**: 337–346.

ISHII, S. 1987. Generation of active oxygen species during enzymic isolation of protoplasts from oat leaves. *In Vitro Cell. Devel. Biol.* **23**: 653–658.

JONES, R. W. and HANCOCK, J. G. 1987. Conversion of viridin to viridiol by viridin-producing fungi. *Can. J. Microbiol.* **33**: 963–966.

JONES, R. W., LANINI, W. T., and HANCOCK, J. G. 1988. Plant growth response to the phytotoxin viridiol produced by the fungus *Gliocladium virens*. *Weed Sci.* **36**: 683–687.

KEISTER, D. L. and CREGAN, P. B. 1991. *The Rhizosphere and Plant Growth.* Proc. Beltsville Symp. XVII. Kluwer, Dordrecht, 386 pp.

KLEIFELD, O. and CHET, I. 1992. *Trichoderma harzianum* – interaction with plants and effect on growth response. *Plant Soil* **144**: 267–272.

KLOEPPER, J. W., ZABLOTOWICZ, R. M., TIPPING, E. M., and LIFSHITZ, R. 1991. Plant growth promotion mediated by bacterial rhizosphere colonizers. In D. L. Keister and P. B. Cregan (eds), *The Rhizosphere and Plant Growth.* Kluwer, Dordrecht, pp. 315–326.

KNOSEL, D. and SCHICKEDANZ, F. 1976. Temperatureanspruche und extracellulare enzymatische activitat einiger aus citrus-importen isolierter plize. *Phytopathol. Z.* **85**: 217–226.

LINDERMAN, R. G. 1988. Mycorrhizal interactions with the rhizosphere microflora: the mycorrhizosphere effect. *Phytopathology* **78**: 366–371.

LINDSEY, D. L. and BAKER, R. 1967. Effect of certain fungi on dwarf tomatoes grown under gnotobiotic conditions. *Phytopathology* **57**: 1262–1263.

LJUNDHAL, L. G. and ERRIKSSON, K.-E. 1985. Ecology of microbial cellulose degradation. *Adv. Microb. Ecol.* **8**: 237–299.

LO, C.-T., NELSON, E. B., and HARMAN, G. E. 1996. Biological control of turfgrass diseases with a rhizosphere competent strain of *Trichoderma harzianum*. *Plant Dis.* **80**: 736–741.

LOTAN, T. and FLUHR, R. 1990. Xylanase, a novel elicitor of pathogenesis-related proteins in tobacco, uses a non-ethylene pathway for induction. *Plant Physiol.* **93**: 811–817.

LUMSDEN, R. D. and RIDOUT, C. J. 1991. Regulation of the production of the metabolites gliotoxin and viridin by the biocontrol fungus, *Gliocladium virens*. *Phytopathology* **81**: 703.

LUMSDEN, R. D. and WALTER, J. F. 1996. Development of *Gliocladium virens* for damping-off disease control. *Can. J. Plant Pathol.* **18**: 463–468.

LUMSDEN, R. D., RIDOUT, C. J., VENDEMIA, M. E., HARRISON, D. J., WATERS, R. M., and WALTER, J. F. 1992. Characterization of major secondary metabolites produced in soilless mix by a formulated strain of the biocontrol fungus *Gliocladium virens*. *Can. J. Microbiol.* **38**: 1274–1280.

LYNCH, J. M., WILSON, K. L., OUSLEY, M. A., and WHIPPS, J. M. 1991. Response of lettuce to *Trichoderma* treatment. *Lett. Appl. Microbiol.* **12**: 59–61.

MCINTOSH, D. L. and SNEH, B. 1972. *Trichoderma koningii* as a parasite of maize. *Proc. Can. Phytopathol. Soc.* **39**: 34.

MEERA, M. S., SHIVANNA, M. B., KAGEYAMA, K., and HYAKUMACHI, M. 1994. Plant growth promoting fungi from zoysiagrass rhizosphere as potential inducers of systemic resistance in cucumbers. *Phytopathology* **84**: 1399–1406.

MENZIES, J. G. 1993. A strain of *Trichoderma viride* pathogenic to germinating seedlings of cucumber, pepper and tomato. *Plant Pathology* **42**: 784–791.

MOFFATT, J. S., BU'LOCK, J. D., and YUEN, T. H. 1969. Viridiol, a steroid-like product from *Trichoderma viride*. *Chem. Commun.* **139**: 839.

MOREAU, R. A., POWELL, M. J., WHITAKER, B. D., BAILEY, B. A., and ANDERSON, J. D. 1994. Xylanase treatment of plant cells induces glycosylation and fatty acylation of phytosterols. *Physiol. Plant.* **91**: 575–580.

OUCHI, S. 1983. Induction of resistance or susceptibility. *Annu. Rev. Phytopathol.* **21**: 289–315.

OUSLEY, M. A., LYNCH, J. M., and WHIPPS, J. M. 1993. Effect of *Trichoderma* on plant growth: a balance between inhibition and growth promotion. *Microb. Ecol.* **26**: 277–285.

PAPAVIZAS, G. C. 1982. Survival of *Trichoderma harzianum* in soil and in pea and bean rhizospheres. *Phytopathology* **71**: 121–125.

PETERBAUER, C. K., HEIDENREICH, E., BAKER, R. T., and KUBICEK, C. P. 1992. Effect of benomyl and benomyl resistance on cellulase formation by *Trichoderma reesei* and *Trichoderma harzianum*. *Can. J. Microbiol.* **38**: 1292–1297.

REID, C. P. P. 1990. Mycorrhizas. In J. M. Lynch (ed.), *The Rhizosphere*. Wiley, Chichester, England, pp. 281–315.

RYALS, J., LAWTON, K. A., DELANEY, T. P., FRIEDRICH, L., KESSMANN, H., NEUENSCHWANDER, U., UKNES, S., VERNOOIJ, B., and WEYMANN, K. 1995. Signal transduction in systemic acquired resistance. *Proc. Natl. Acad. Sci. USA* **92**: 4202–4205.

SALT, G. A. 1979. The increasing interest in "minor pathogens." In B. Schippers and W. Gams (eds), *Soil-Borne Plant Pathogens*. Academic Press, San Francisco, pp. 289–312.

SCHRAUFSTAETTER, I. U., BROWNE, K., HARRIS, A., HYSLOP, P. A., JACKSON, J. H., QUEHENBERGER, O., and COCHRANE, C. G. 1990. Mechanisms of hypochlorite injury of target cells. *J. Clin. Invest.* **85**: 554–562.

SHARON, A., FUCHS, Y., and ANDERSON, J. D. 1993. The elicitation of ethylene biosynthesis by a *Trichoderma* xylanase is not related to the cell wall degradation activity of the enzyme. *Plant Physiol.* **102**: 1325–1329.

SHIVANNA, M. B., MEERA, M. S., and HYAKUMACHI, M. 1994. Sterile fungi from zoysiagrass rhizosphere as plant growth promoters in spring wheat. *Can. J. Microbiol.* **40**: 637–644.

SHIVANNA, M. B., MEERA, M. S., KAGEYAMA, K., and HYAKUMACHI, M. 1996. Growth promotion ability of zoysiagrass rhizosphere fungi in consecutive plantings of wheat and soybean. *Mycoscience* **37**: 163–168.

SIVAN, A. and CHET, I. 1989. The possible role of competition between *Trichoderma harzianum* and *Fusarium oxysporum* in rhizosphere colonization. *Phytopathology* **79**: 198–203.

SIVAN, A. and HARMAN, G. E. 1991. Improved rhizosphere competence in a protoplast fusion progeny of *Trichoderma harzianum*. *J. Gen. Microbiol.* **137**: 23–29.

STASZ, T. E. and HARMAN, G. E. 1990. Nonparental progeny resulting from protoplast fusion in *Trichoderma* in the absence of parasexuality. *Exp. Mycology* **14**: 145–159.

STIPANOVIC, R. D. and HOWELL, C. R. 1994. Inhibition of phytotoxin (Viridiol) biosynthesis in the biocontrol agent *Gliocladium virens*. In P. A. Hedin (ed.), *Bioregulators for Crop Protection and Pest Control*. American Chemical Society, Washington, DC, pp. 136–143.

THUY, L. B. 1991. Rhizosphere competence of two selected *Trichoderma* strains. *Acta Phytopath. Entomol. Hung.* **26**: 327–331.

Van Der Pyl, D., Inokoshi, J., Shiomi, K., Yang, H., Takeshima, H., and Omura, S. 1992. Inhibition of farnesyl-protein transferase by gliotoxin and acetylgliotoxin. *J. Antibiot.* **45**: 1802–1805.

Windham, M. T., Elad, Y., and Baker, R. 1986. A mechanism for increased plant growth induced by *Trichoderma* spp. *Phytopathology* **76**: 518–521.

Wright, J. M. 1951. Phytotoxic effects of some antibiotics. *Ann. Bot.* **15**: 493–499.

Wright, J. M. 1956. Biological control of a soil-borne *Pythium* infection by seed inoculation. *Plant Soil* **8**: 132–140.

Wyss, P., Boller, T., and Wiemken, A. 1992. Testing the effect of biological control agents on the formation of vesicular arbuscular mycorrhiza. *Plant Soil* **147**: 159–162.

Zimand, G., Elad, Y., and Chet, I. 1996. Effect of *Trichoderma harzianum* on *Botrytis cinerea* pathogenicity. *Phytopathology* **86**: 1255–1260.

10

Industrial production of active propagules of *Trichoderma* for agricultural uses

E. AGOSIN and J. M. AGUILERA

Department of Chemical and Bioprocess Engineering, Pontificia Universidad Católica de Chile, Santiago, Chile

10.1 Introduction

In the previous chapters of this book, priority has been given to the biology, ecology, biochemistry and molecular biology of *Trichoderma* and *Gliocladium* spp. In this chapter we will focus on what is necessary to achieve large-scale, cost-effective production of active biomass of these biocontrol agents.

Translation of scientific discoveries in biotechnology into tangible commercial products requires the joint effort of molecular biologists and bioprocess engineers. The latter are crucial in moving newly discovered bioproducts into the hands of consumers. Major requirements to market *Trichoderma*-based products as biopesticides are the following:

- abundant and cost-effective production of microbial propagules;
- high propagule/vegetative biomass ratio;
- ability to survive downstream processing, particularly drying, which is required to avoid spoilage and contamination of the bioprotectant;
- stability and adequate shelf life of the final product upon storage, preferably without refrigeration;
- ability to withstand environmental variations in temperature, desiccation, radiation and relative humidity in order to survive and establish active populations both in the soil and on the phylloplane;
- consistent efficacy under varying field conditions at commercially feasible rates.

10.2 *Trichoderma* propagules for agricultural purposes

Trichoderma produces three types of propagules: hyphae, chlamydospores, and conidia (Papavizas, 1985). Since hyphae cannot survive after rapid drying, hyphal biomass is not useful (Jin *et al.*, 1991). Chlamydospores have been suggested as promising propagules for biocontrol programs since they are able to survive in soil for extended periods of time (Lewis and Papavizas, 1983; Papavizas *et al.*, 1984).

However, fermentation times of 2–3 weeks are necessary to achieve production of mature chlamydospores, and at most, only 8% of air-dried chlamydospores were able to germinate. Conidia have been the most widely employed propagules in biocontrol programs (Elad *et al.*, 1993c) and the rest of this chapter will consider only this type of propagule. Conidial biomass can be obtained by either submerged- (Elad and Kirshner, 1993) or solid-substrate (Lewis and Papavizas, 1983) cultivation techniques.

Industrial producers of *Trichoderma* conidia must consider the following stages:

1 optimization of culture conditions at bench scale to obtain high yields of active biomass;
2 experimentation at the pilot-plant level to determine and solve critical process engineering variables such as mass and energy transfer limitations in bioreactors, selection of the most efficient bioseparation techniques, selection of scale-up methods to maintain laboratory yields and quality of resulting biomass during scale-up; the resulting products can provide samples for field experiments;
3 integration of selected unit operations from fermentation, bioseparation, and formulation into a single process;
4 industrial plant-scale production of conidiospores of *Trichoderma* spp.

10.3 Producing high-quality conidial biomass of *Trichoderma*

10.3.1 Process abilities of available Trichoderma *strains*

Several strains of *T. harzianum* have been patented for their ability to control phytopathogenic fungi (Elad *et al.*, 1993b,c; Harman, 1988; Harman *et al.*, 1994; Kubota, 1995; McBeath, 1995; Paau *et al.*, 1993, among others). However, their capacity to sporulate abundantly in culture is rarely reported, even though industrial production of *Trichoderma* spp. is sensitive to this parameter; indeed, significant differences in sporulation yields have been found in both native and culture collection strains of *Trichoderma* (Latorre and Agosin, unpublished).

10.3.2 Culture conditions for the production of abundant, high-quality conidiospores

Chemically defined media have been employed to determine optimal cultural conditions that allow for the rapid production of high levels of active and vigorous conidiospores.

Conidiogenesis in *Trichoderma* can be achieved by manipulating the levels of nutrients, particularly using carbon starvation (Agosin *et al.*, 1997b). However, increases in biomass concentration, such as those observed when the carbon concentration was raised, did not support higher conidia production yields (Agosin *et al.*, 1997b). This suggests that some inhibitory compound(s) could be produced together with sporulation (Bodo *et al.*, 1985). Furthermore, the C:N ratio, but not the carbon concentration, appeared to play a key role in spore longevity. At pH 7.0, a medium containing a C:N ratio of 14:1 consistently produced conidia with the longest shelf life. In this medium, both carbon and nitrogen sources were consumed simultaneously. When these nutrients are depleted in the culture medium, derepression of several metabolic systems at the onset of sporulation occurs (Dahlberg and van Etten, 1982). As a consequence, further metabolites are synthesized which may play an active role in conidia survival.

Another way to obtain higher conidial productivities is through the reduction of hyphal production. When fungi form spores almost immediately after germination with limited vegetative growth, this is called microcycle conidiation (Smith *et al.*, 1981). Induction of premature phialoconidiogenesis in *Trichoderma* spp. has been achieved by applying a nutritional shock with glucose to germinated conidia (Zuber and Turian, 1981). Interestingly, the carbon source could be effectively replaced by a non-assimilable sugar, D (−) arabinose. Hence, osmotic pressure, which is dependent upon sugar concentration, appears to have a major role in the onset of microcycle conidiation in *Trichoderma.*

Harman *et al.* (1991) further developed this concept of inducing microcycle conidiation in *Trichoderma* by appropriate manipulation of the fermentation medium composition. A modified Richard's medium supplemented with V-8 juice and amended with an osmoticant, such as polyethylene glycol (PEG) or glycerol, was developed. The addition of the latter lowers the medium water potential to approximately −2 to −4 MPa. Production of high levels of desiccation-tolerant conidia of *Trichoderma harzianum* was achieved (Jin *et al.*, 1991, 1996).

Other environmental stresses, and their effects on spore quality, have also been evaluated. For example, a heat shock applied at the end of the sporulation phase resulted in the production of heat-resistant conidial biomass with high shelf life (Pedreschi *et al.*, 1997).

Temperature, pH, and culture age are key operating parameters to manipulate for better quantity and quality of resulting spores (Agosin *et al.*, 1997b; Jin *et al.*, 1996). Since most of the results reported above have been obtained in shake flasks, the use of larger scale bioreactors should result at least in similar productivities due to the adequate control of important parameters such as dissolved oxygen concentration, temperature, and pH. Finally, most but not all strains of *Trichoderma* require light for maximum conidiation.

10.3.3 ***Conidia production by submerged fermentation***

Defining optimal industrial media

In the production of fermentation products on a large scale, economics plays a key role in medium development. With spore production, chemically defined media will be prohibitive. Inexpensive, abundant, and readily available raw materials are required because low-cost *Trichoderma*-based products will have to be produced. Miller and Churchill (1986) published a complete list of these materials.

Several raw materials, mostly derived from agricultural commodity products, could be employed for *Trichoderma* fermentation. Molasses and potato dextrose broth combined with glucose, starch, malt extract, etc., have been evaluated as carbon sources. A variety of nitrogen sources, including the preferred ammonium nitrate, also could be used. Minerals and some trace elements are also required (Harman *et al.*, 1994; Tabachnik and Ziona, 1989).

The main fermentation parameter that characterizes the efficiency of a production configuration is product volumetric productivity, expressed as grams of product per liter per hour (g product L^{-1} h^{-1}) or number of propagules L^{-1} h^{-1}. In early attempts, Lewis and Papavizas (1983) compared different media in shake flasks. This was followed by the use of larger fermentation vessels (20 liters) to produce both conidia and chlamydospores (Papavizas *et al.*, 1984). As shown in Table 10.1, the volumetric productivities of these fermentation trials were only about 2–3 × 10^8 propagules L^{-1} h^{-1}, and included conidia, chlamydospores, and immature chlamydospores.

These results were improved to 10^{10} conidia L^{-1} h^{-1} by Tabachnik and Ziona (1989), mainly through medium modifications. However, the viability percentages of propagules after drying and final product formulation were not investigated.

Comparable volumetric productivities were achieved by Harman *et al.* (1994) by adding an osmoticant to the culture medium, though most of these resulting conidia were able to germinate even after rapid drying. Moreover, by adequately timing the addition of the osmoticum, hyphal production decreased while conidia production was enhanced. The nature of the osmoticum, however, has a significant effect on process economics and scale-up. For example, PEG 8000 had to be replaced with PEG 200 to reduce medium viscosity problems and increase oxygen transfer rates in large-scale fermentations (see below). In any case, the quantities of PEG necessary for appropriate water potentials are too high to be economically feasible. More recently, glycerol at 9% has been suggested as the most appropriate and cost-effective osmoticant (Jin *et al.*, 1996).

Scaling-up the results obtained in the laboratory

When the optimum production conditions are found at the laboratory scale, translation of these findings is required for use in large bioreactors. Scale-up means reproducing in plant-scale equipment the results from a successful fermentation made in laboratory- or pilot-scale equipment. Several scale-up methods, such as fixed input of the agitator power, fixed impeller tip speed, fixed oxygen transfer coefficient, etc., have often been proposed for bioconversion processes.

In aerobic systems, including those used with filamentous fungi, a constant oxygen transfer coefficient, $K_L a$, is frequently used as a scale-up parameter (Moo-

Table 10.1 Production of propagules of *Trichoderma* spp. in submerged fermentation

Reference	Propagules L^{-1}	Time (h)	Propagules L^{-1} h^{-1}
Papavizas *et al.* (1984)[a]	3 × 10^{10}	144	2.1 × 10^8
	2 × 10^{10}	72	2.8 × 10^8
Tabachnik and Ziona (1989)[b]	5 × 10^{11}	50	1.0 × 10^{10}
Harman *et al.* (1994)[c]	5 × 10^{11}	66	7.6 × 10^{10}
Jin *et al.* (1996)			

[a] Conidia, chlamydospores and immature chlamydospores; [b] conidia or chlamydospores; [c] conidia.

Young and Blanch, 1987). An often-quoted $K_L a$ correlation, valid for non-Newtonian filamentous fermentations, is given by the following equation:

$$K_L a = \text{constant} \times (P_g/V)^{0.33} V_s^{0.56}$$

where P_g/V is the power input per unit volume and V_s is the superficial gas velocity through the vessel. Hence, if $K_L a$ is to be kept constant during scale-up, P_g and V_s – as well as their corresponding dependent variables – will have to vary (McDonough, 1992).

Scale-up of conidial biomass production has shown to be a relatively simple task, mainly because moderate biomass concentrations – generally lower than 10 g L^{-1} – are sufficient for high conidia production. As a consequence, the rheological properties of the medium remain quite stable throughout the fermentation; viscosity also remains low and therefore high oxygen transfer rates should not be limiting during scaling-up. Indeed, similar productivities of conidia were achieved by our group when we scaled-up conidial biomass production from 100 mL culture shake flasks to 200 L liquid fermentors.

Medium contamination problems are of secondary importance here considering the relatively short duration of the fermentation runs, provided that asepsis was maintained throughout. If necessary, antibiotics can be added (Harman *et al.*, 1994; Tabachnik and Ziona, 1989). Finally, foam production was found to be closely related to spore production levels; medium release of spore wall hydrophobins might be responsible for this phenomenon (see below). Efficient foam control could be achieved by constant addition of antifoaming agents, particularly silicone-derived compounds, throughout the fermentation (G. Muñoz, personal communication).

To our knowledge, more sophisticated fermentation strategies for greater production of active conidia, such as a fed-batch culture condition that can induce continuous microcycle conidiation, have not been evaluated. Basically, the fed-batch technique consists of operating the bioreactor as a batch fermentor, until the biomass concentration has reached a plateau. At this time, a controlled continuous input of a limiting nutrient, i.e., carbon or nitrogen, is added to the bioreactor. Under these conditions, it is possible to maintain the existing biomass with a minimum of intracellular flow. Actually, the biomass will be under maintenance energy conditions, which will result in a minimum accumulation of intracellular metabolic intermediates. This will probably trigger the differential expression of genes involved in microcycle conidiation. It is worth noting that the fed-batch system has already been successfully employed to induce microcycle conidiation in cells of the biocontrol agent *Colletotrichum gloeosporioides* (Cascino *et al.*, 1990).

The scaling-down approach

The question to answer at this stage is "How can we mirror the conditions prevailing in large production vessels in a small fermenter?" (Moo-Young and Blanch, 1987). Experiments in full-scale bioreactors are difficult to pursue, time consuming and expensive. Therefore, it is suitable to optimize plant-scale operating conditions in small fermenters. For example, a successful scale-up based on fixed $K_L a$ is achieved when high spore productivities are obtained at the value of $K_L a$ which is found in the large vessel – generally not the optimum one!

10.3.4 *Conidia production by solid-substrate fermentation (SSF)*

Solid-substrate fermentation processes (SSF) can be defined as a four-phase system. The first phase is air, a continuous system which usually flows through a solid bed. The solid second phase is composed of a water-insoluble support that contains the third phase, an aqueous solution of nutrients. This solution is tightly absorbed within the matrix of the insoluble support; however, some water may exist in a free state in capillary regions of the material, although never exceeding the water-holding capacity of the matrix (Mudgett, 1986). The fourth phase is the micro-organism itself, which grows inside the support and/or on its surface and/or in the interparticular free space.

It is recognized that SSF processes have the advantages of superior productivity, simpler techniques, reduced energy requirements, low wastewater output and improved product recovery (Tengerdy, 1985). For production of biocontrol agents, SSF does have a major advantage in allowing the fungus to sporulate naturally, i.e., via aerial mycelium. The following features have been identified for conidiogenesis using SSF.

- Conidia produced in aerial mycelium persist longer under harsher environmental conditions than those produced under submerged culture conditions (Muñoz *et al.*, 1995). For example, viability of aerial spores was fully maintained after 60 days of storage at 75% relative humidity, while viability decreased to less than 15% for submerged spores over the same time period.
- Wall thickness of aerial conidia is nearly twice that of submerged ones (Muñoz *et al.*, 1995). As the spore wall is the first barrier against external conditions, it may be involved with spore viability, shelf life, and establishment of the agent to enable successful biocontrol (see Figure 10.1).
- Only aerial spores exhibit hydrophobic properties. Hydrophobicity has been identified as a key parameter in the adherence and establishment of micro-organisms in the phylloplane (Doss *et al.*, 1992; St Leger *et al.*, 1992); therefore, this critical property will allow rapid and effective adhesion of aerial spores to leaves and other host surfaces. On analysis, hydrophobicity was associated with several low molecular weight compounds, such as trichorzianines (Bodo *et al.*, 1985) and hydrophobin-like proteins (Muñoz *et al.*, 1995). The latter were nearly absent from walls of spores produced by submerged fermentation, since they are released to the culture medium (Agosin *et al.*, 1997b; Schirmböck *et al.*, 1994). The gene encoding a major hydrophobin-like protein of *T. harzianum* has recently been cloned and partially sequenced (Muñoz *et al.*, 1997).

Optimizing cultural conditions at the bench scale

Laboratory "bioreactors" Most experiments in SSF have been conducted in naturally ventilated conical flasks containing a thin layer of wet fermenting support (generally less than 5 cm deep). Limited control of cultural conditions can be achieved in this system. Raimbault and Alazard (1980) devised a more suitable experimental system for bench-scale optimization of culture conditions. This basically comprises a small packed bed bioreactor containing a static substrate supported on a perforated plate through which forced aeration is applied. In these thin cylindrical columns, the

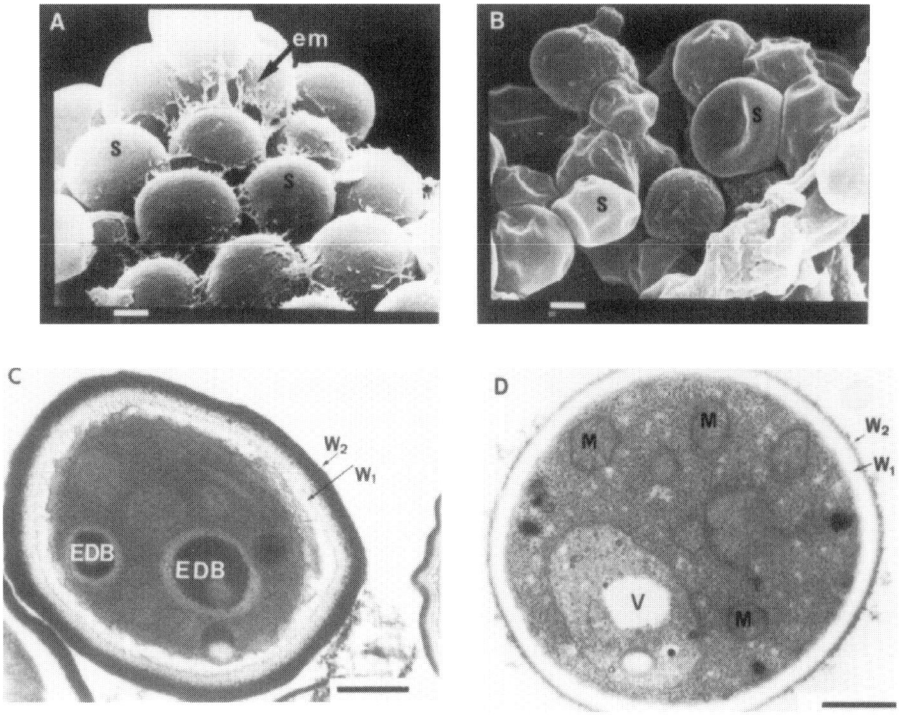

Figure 10.1 Scanning elecgtron micrographs (A,B) and transmission electron micrographs (C,D) of aerial (A,C) and submerged (B,D) conidia of *Trichoderma harzianum*. Symbols: em, mucilage; W1, inner wall layer; W2, outer wall layer; V, vesicles; M, mitochondria; EDB, electrondense bodies. Bars represent 0.5 mm. From Munõz *et al.*, (1995).

humidity of the fermenting bed, the bed temperature, as well as the humidity, temperature, and flow of the incoming air could be manipulated, allowing more rigorous control of operating conditions.

More recently, dynamic laboratory-scale systems, such as rotating drum reactors (de Reu *et al.*, 1993; Laroche and Gros, 1986) and rocking vessels (Sargantanis *et al.*, 1993), have been developed. These systems are equipped with advanced measurement and control features; thus, fermentation can take place in these devices under controlled conditions of temperature, gas composition, and relative humidity.

Medium optimization SSF kinetic parameters reported in the literature have been expressed in several units. Thus, comparisons are difficult to make and conclusions must be expressed with care. As most optimization studies reported results in terms of mass productivity, i.e., number of propagules kg^{-1} initial dry weight h^{-1}, these units will be used (see Table 10.2). However, volumetric productivity might be more suitable for the comparison of both fermentation systems.

SSF has traditionally used agricultural products, including wheat, rice, corn, and their related by-products. These organic materials play a dual role, as both nutrition and support, for the microorganism. As shown in Table 10.2, high conidiospore yields are achieved with organic support cultures, particularly wheat bran-based media (Agosin *et al.*, 1997b; Kubota, 1995; Mitchell and Lonsane, 1992).

Table 10.2 Production of propagules of *Trichoderma* spp. by solid-substrate fermentation[a]

Reference	Propagules kg^{-1}	Time (h)	Propagules kg^{-1} h^{-1}
Roussos *et al.* (1991)	5×10^{13}	144	3.5×10^{11}
Kubota (1995)	1×10^{13}	168	6.0×10^{10}
Agosin *et al.* (1997)	1×10^{13}	72	1.4×10^{11}

[a] Results are expressed per kg of initial dry weight.

The inert support culture (ISC), an alternative SSF system, is based on the sorption of a defined nutrient solution in an inert surface matrix. The latter could have an inorganic (diatomaceous earth, clay granules, vermiculite, etc.) or an organic (corn cobs, sugar beet pulp, sugar cane bagasse, sawdust, etc.) nature. This technique has been evaluated for the production of several metabolites (Auria *et al.*, 1992; Barrios-González *et al.*, 1988; Lakshminarayana *et al.*, 1975). It enables higher product purity, constant substrate geometry throughout the cultivation, better control of microbial metabolism through culture medium manipulation, and higher sorption capacity of the culture medium (Auria *et al.*, 1990). However, it has some drawbacks, including higher substrate costs, the requisite of concentrated media, which could lead to catabolite repression problems and low water potential, and a laborious operating system.

Fungal spores for biocontrol purposes have been successfully produced by ISC. Early ISC studies were conducted on diatomaceous earth impregnated with a 10% molasses solution for production and delivery of conidial biomass of *Trichoderma harzianum* (Backman and Rodriguez-Kabana, 1975). Unfortunately, productivity was not reported. Table 10.2 shows that the highest spore productivities were reached in an ISC system containing bagasse impregnated with a starch-based medium and supplemented with feather meal (Roussos *et al.*, 1991). As expected, resulting spores remained viable even after long periods of storage.

In conclusion, high-quality spores are easily and abundantly produced by SSC at bench scale. Their major advantage is that they are durable. Their greater resistance to harsh environmental conditions probably is related to the apparent physiological resting state of spores produced in an aerial environment. In this respect, the hydrophobicity of phialoconidia of *T. harzianum* may contribute to the maintenance of the dormant state of aerial spores. Conversely, spores produced under submerged liquid conditions could react rapidly with the external medium. Quick water or nutrient uptake may allow shorter germination times than for aerial spores. However, they also will be more susceptible to deterioration reactions and would necessarily need further downstream processing (see section 10.4)

Scaling-up the SSF process

The engineering aspects related to SSF have not been developed as extensively as the research on probable applications (Auria *et al.*, 1993). As a consequence, commercial exploitation of SSF in Western and European countries is scarce.

The most important limitations to SSF are process monitoring and control. The multiphase system of SSF, as already defined, makes non-destructive on-line monitoring more difficult than in liquid fermentations. Nevertheless, some parameters such as incoming air and substrate temperatures, pH, the oxygen uptake rate (OUR), and the carbon dioxide production rate (CDPR), as well as the pressure drop, can be followed on-line (Auria *et al.*, 1993; Saucedo-Castañeda *et al.*, 1994). The last three measurements have been used extensively for growth estimation. Temperature and pressure drop variations are employed to define the agitation policy in industrial Japanese SSF bioreactors (Tomomori, Fujiwara Company, personal communication).

Critical operating conditions The most critical problem with the SSF process is the difficulty of heat removal, arising primarily from the low thermal conductivity of the fermented matter. Among the conventional methods tested to remove excess heat, evaporative cooling with insertion of partially saturated air at a lower temperature was the most promising, provided that an effective system of uniform aeration existed (Fernández *et al.*, 1996). However, the airflow caused variation in the water content profile and, consequently, had an effect on biomass growth and product formation. Periodic addition of fresh sterile water was therefore necessary.

Thus, it is not surprising that maintenance of constant heat and water balances have been recently proposed as a feasible scale-up criterion in SSF systems and even suggested as the key for extensive commercialization of SSF processes (Lonsane *et al.*, 1992).

Furthermore, the heterogeneous and essentially distributed nature of the system demands the necessity of periodic agitation of the fermenting bed. This critical parameter ensures homogeneity with respect to temperature, water, and gaseous environment and provides gas–liquid interfaces for gas to liquid and liquid to gas transfers (Lonsane *et al.*, 1992). However, the mixing of solid substrates can also have deleterious effects, including adverse effects on substrate porosity, disruption of the attachment of microorganisms to substrates, and damage to fungal mycelium due to shear forces caused by abrasion between particles (Mitchell *et al.*, 1992). Shear forces can rupture both surface and interparticle hyphae. Disruption of interparticle hyphae is clearly advantageous, since the fungus is prevented from binding the substrate particles into a single mass. Conversely, if aerial hyphae are crushed onto the surface of the substrate, sporulation will be inhibited. Therefore, static cultures seem necessary for the production of spore inocula (Cannel and Moo-Young, 1980; Laroche and Gros, 1986). However, agitation systems in SSF reactors can be designed to minimize the deleterious effects of mixing (Tengerdy, 1985). In any case, mixing should only be provided as necessary, and periodic agitation is usually sufficient (Mitchell *et al.*, 1992). Clearly, research to understand and optimize mixing systems in SSF should be done.

SSF reactor types for production of conidial biomass Unlike reactors for submerged fermentations, which have roughly the same design, different reactor types for solid-state cultivation have been reported. A change in reactor configuration can effectively improve the process productivity. Several reactor devices have been evaluated for industrial production of *Trichoderma conidia* by SSF, and most of the developed systems have been placed under patent protection.

Tray bioreactors. Tray bioreactors have been successfully used at laboratory, pilot, and commercial scale (Backman and Rodriguez-Kabana, 1975; Kubota, 1995; Toet and Sommers, 1994). In this system, a thin layer of substrate is spread over a shallow pan or aluminum/stainless steel tray. There is no forced aeration, although the base of the tray could be perforated. Temperature/humidity may vary with the ambient, or the whole system may be placed in a temperature/humidity-controlled room. An alternative system with the same principles is conidia production in perforated polyethylene bags instead of trays (Paau *et al.*, 1993). A major disadvantage of tray systems is that large-scale operation is difficult to automate and therefore tends to be labor intensive (Mitchell *et al.*, 1992). Maintaining sterile conditions is difficult as well. Moreover, a large space is required.

Packed-bed bioreactors. High conidia productivity was also achieved in the *Zymotis*, a novel 50 kg packed bed fermenter (Roussos *et al.*, 1991, 1993). In this unit, the whole mass to be fermented is divided into several compartments separated by heat exchanger plates to facilitate efficient heat removal. The unit can be operated at different capacities simply by adding or removing compartments. Interestingly, the superficial area in *Zymotis* is lower than in tray bioreactors, suggesting that high yields could also be achieved deep inside the fermenting mass and not exclusively on its surface.

Stirred bioreactors. Even though agitation has been reported to inhibit sporulation (see above), Durand (1995) developed a cost-effective industrial process with agitated bioreactors for conidiospore production of the fungal bioinsecticide *Beauveria bassiana.* A clay-granulated product is currently commercialized by Calliope Inc., France. The process is conducted in a non-aseptic prototype bioreactor developed by INRA-Dijon (Durand and Chereau, 1988). This one-ton vertical stirred bioreactor is a rectangular device based on barley malting equipment. A one-meter thick layer of substrate is supported on a wire mesh. Conditioned air is forced through the substrate mass. Agitation is provided by three vertical screws situated at the top of the reactor that hang down into the substrate mass. These screws, which rotate at 22 rpm, are mounted on a conveyer that moves them back and forth across the reactor. The bed temperature is automatically controlled. Water content is estimated by water balance. No reports on production yields of conidial biomass of *Trichoderma* spp. are available.

A different pilot prototype bioreactor is available at the Fermentation Facility Plant of our Department of Chemical and Bioprocess Engineering (Agosin *et al.*, 1997b; Fernández *et al.*, 1996). This 50 kg horizontal stirred bioreactor has the advantage of being run aseptically. As shown in Figure 10.2, the following parts of the fermenter can be identified:

- a lid that can be raised and can be hermetically sealed; the agitation and feeding systems are mounted on the lid;
- a 1.2 m rotating basket with a base containing 2 mm perforations;
- an air chamber, located below the rotating basket, for homogenization of incoming air.

The air-conditioning system, situated upstream of the fermenter, consists of a prefilter, a variable-speed fan, a cooler to cool and dry the inlet air, a sterilizing filter able to retain particles larger than 0.25 μm, a heater for reactor sterilization and process control needs, and a vapor generator to regulate the relative humidity of the incoming air.

Basket movement is provided by a connection between the motor and basket shafts. A static, double arm bed agitation system containing blades and curved paddles ensures adequate mixing. The basket is supported on roller bearings to get

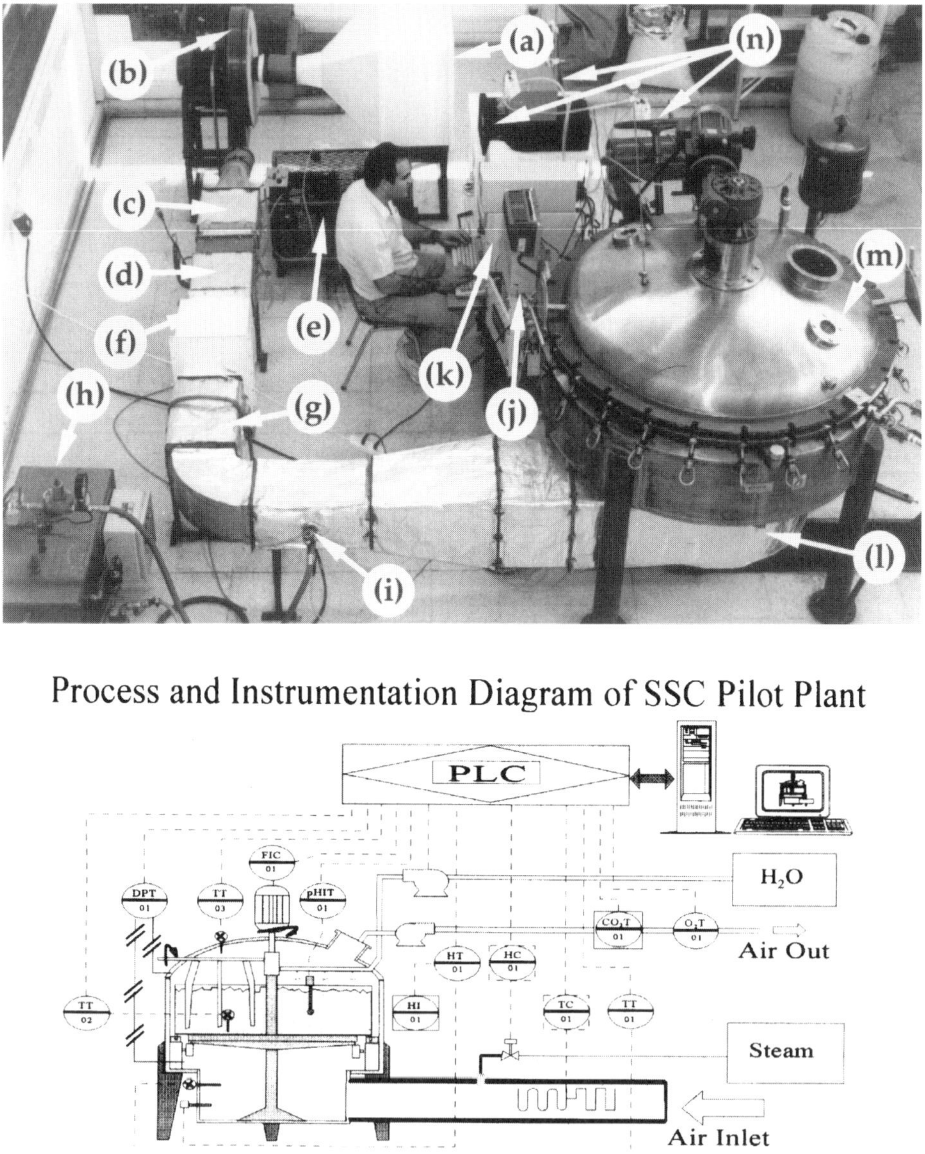

Figure 10.2 General view and process diagram of a pilot plant dynamic solid substrate fermentor: (a) prefilter; (b) fan; (c) air purge; (d) fins cooler; (e) refrigerator; (f) absolute filter; (g) electric heaters; (h) steam boiler; (i) solenoid valve; (j) programmable logic controller (PLC); (k) control system; (l) air chamber; (m) SSC reactor; (n) instruments (CO_2, O_2, flowmeter, pH, differential pressure, and relative humidity).

a uniform and gentle motion. Water and nutrients can be added periodically, according to processing needs, through a set of three sprinklers.

A Programmable Logic Controller (PLC), provided with modules for input/output analog and digital signals and an IBM-PC interface, is used for on-line data acquisition. Process parameters are monitored through temperature (TT), inlet air relative humidity (HT), CO_2 concentration in outlet air (CO_2T), air pressure drop (DPT) and airflow rate. The O_2 concentration and pH can also be measured.

An automated system for controlling temperature and water activity of the substrate mass was achieved. For this purpose, novel and specific control software was developed in our department. It allows, through a graphic interface, direct process intervention and visualization of the trends of process variables. Using this interface, the reactor can be operated manually or automatically. The control algorithms are executed in the PC, according to options and parameters selected from the keyboard. Different ON/OFF and PID algorithms have been implemented. The detailed control strategy has recently been published (Fernández *et al.*, 1996).

Productivities of 10^9 to 10^{10} conidia kg^{-1} h^{-1} were obtained in preliminary trials in the prototype. Vermiculite impregnated with a malt extract-based medium was used as a support. Even though these results were lower than those obtained in the lab, conidial productivity could be improved by optimization of the culture medium and reactor operating conditions, particularly mixing.

Downstream processing in SSF processes In SSF, the need for a separate step for recovery of conidial biomass can be avoided and the spore-containing fermented mass can be used directly, as it is with a number of food fermentations (Lonsane and Krishnaiah, 1991). Alternatively, it can be dried *in situ* by passing dry hot air through the loop that supplied humid air during fermentation. The operation is continued until appropriate moisture content is reached (generally lower than 10% on a dry weight basis; see below) without any appreciable loss of spore viability (Roussos *et al.*, 1991). This formulation will be particularly suitable when soil applications are pursued.

10.4 Preserving the quality of a dry commercial biopesticide

10.4.1 Background

In order to become an efficient commercial product, a biological agent must withstand post-fermentation processing, be stable during storage, and retain its activity under field conditions. In Table 10.3, we have summarized the main unit operations involved as well as the critical parameters affecting viability.

Good results in the laboratory do not guarantee a successful productive process at the commercial level. Reagent quality, manipulation techniques, and control of

Table 10.3 Main post-fermentation processing steps for a commercial bioproduct

Product goal	Unit operations involved	Critical viability bioproduct
Separation	Centrifugation, filtration, pumping, milling	Shear, retention time, heat
Stabilization	Concentration, granulation, drying, encapsulation	Heat, shear, mixing, osmotic stress
Shelf-life	Formulation, packaging, storage	Temperature, time, moisture
Survival	Dispersion, adhesion	Nutrients, environmental stresses (UV, temperature, etc.)

the propagules, drying method and level of key parameters (including the rate of drying), residual moisture content, and storage conditions. At the beginning of the century, Rogers (1914) reported that lactic acid bacteria stored in sealed containers for 157 days at 1.39% and 0.90% moisture were more viable than those maintained at 5.8%. In a thorough review on convective air drying of bacteria published recently by Lievense and van't Riet (1993), survival rates varied widely between <1% and 91% depending on the aforementioned factors. Interestingly, from 26 references cited, most RMCs reported were between 0.04 and 0.07 g water per g dry weight, which may be regarded as the lowest MCs achieved in practice by this drying method. Water activity (a_w) is defined as the ratio between the partial pressure of water in a material and the vapor pressure of pure water at the same temperature under equilibrium conditions. It is a preferred concept to describe the physicochemical state of water in hygroscopic materials than moisture content, to which it is related through a sorption isotherm (Karel, 1975). The viability of spray-dried *Rhizobium* after 45 weeks of storage at 4°C and $a_w < 0.23$ was three orders of magnitude higher than that of samples kept at $a_w = 0.44$. Increasing the temperature to 20°C drastically affected the survival rates of samples stored at a_w values of 0.44, 0.75, and 0.87, while samples having a_w of 0.03 (silica gel) and 0.11 showed survival rates insensitive to temperature increases (Mary *et al.*, 1993). However, very low moisture materials may present handling problems such as dusting and physical breakage. In summary, careful control of low RMC (even through post-drying operations) and moisture during storage are likely to have a positive influence on viability.

10.4.3 Assessment of shelf life of a commercial product

Shelf life is considered a pivotal factor that determines the commercial success of a biocontrol agent as well as its field efficacy (Feng *et al.*, 1994). An 18-month shelf life with minimal losses in viability is recommended for the agricultural market (Couch and Ignoffo, 1981) – although it is seldom attained, because of the strong effect of moisture and temperature. Testing product stability at high temperatures and/or high moisture contents is widely used to rapidly assess the storage life of foods and pharmaccuticals (Labuza and Schmidl, 1985). In this way it is possible to gather data in a few days or weeks instead of the prohibitive time (e.g., several months) that would be needed under real storage conditions. Data from accelerated testing is translated into real-time storage conditions using a temperature-shifting model such as Q_{10} or the energy of activation (E_a) of an Arrhenius-type equation (Saguy and Karel, 1980). However, there are a number of practical and thcorctical crrors that should be avoided before drawing incorrect conclusions from accelerated tests, such as the existence of coupled reactions, changes in mechanism at higher temperatures, insufficient data points to accurately determine E_a, or passage through a glass transition (Franks, 1994).

Mathematical models

A basic assumption is that there is a set of parameters affecting the quality (Q) of the product (in our case viability) which can be measured. Quantitative analysis further requires a relationship between quality changes during time (t) and these factors (F_i), which is usually expressed by an equation of the following form:

$$\frac{dQ}{dt} = f(F_1, F_2, \ldots, F_n) \quad (1)$$

Furthermore, it is also necessary to know how each of these factors changes with time. For example, temperature is usually a key factor for stability of biological materials and, although it is often considered a constant in controlled experiments, it may fluctuate through the day or season. The influence of moisture, the other important parameter that affects the rate of reaction, has been explained in terms of water activity. This approach is now well established in predicting the stability of foods and pharmaceuticals (Labuza and Kamman, 1983). Since most quality changes during storage of a product derive ultimately from chemical reactions, the simplest and most widely used empirical model has the following form:

$$\frac{dQ}{dt} = \pm kQ^n \quad (2)$$

where k is the rate constant and n is a reaction order (normally 0 or 1 for foods and drugs). This model completely disregards the actual mechanism involved and is suitable to the extent in which data fits one of the possible integrated equations derived by varying n in equation (2). In practice, parameter Q is plotted against time in an adequate coordinate system (e.g., semi-log plot for first-order reactions) and good fitting (high r^2) is evaluated by common statistical means. Usually, over 50% completion of the reaction is required to generate data that give a well-defined n value (Labuza and Kamman, 1983).

Moisture and temperature are assumed to be effected through the rate constant k. The Arrhenius equation is the most commonly used mathematical model to describe the dependence of k on T:

$$k = k_0 \exp\left(\frac{-E_a}{RT}\right) \quad (3)$$

where k_0 is the pre-exponential factor, R is the ideal gas constant, and E_a is the activation energy. If a plot of ln(k) versus T^{-1} is a straight line, the Arrhenius model is applicable and the activation energy is constant over the range of temperatures. Data for at least four temperatures are needed to be confident on the linearity of the relationship. Deviations from the straight line may indicate that a change in reaction mechanism operates above a certain temperature (or simply that the data are poor!). In the case of low-moisture products where the existence of a glassy state is suspected, it may be more appropriate to use the Williams–Landel–Ferry (WLF) equation to depict variations with temperature:

$$\log \frac{k_{ref}}{k} = \frac{-C_1(T - T_{ref})}{C_2 + (T - T_{ref})} \quad (4)$$

where C_1 and C_2 are constants that depend on the material and T_{ref} is a reference temperature, usually the glass transition temperature. The influence of moisture on product stability has been well established in terms of the water activity concept (Karel, 1975). A first approach is to test a linear relationship between the rate constant and water activity (defined in section 10.4.2):

$$\ln k = \alpha a_w \tag{5}$$

This requires a further relationship between moisture content, *m* (as g water per g dry spores), and water activity (at constant *T*), the so-called sorption isotherm. Again, a widely used model for this isotherm is the following:

$$m = \frac{m_0 K C a_w}{(1 - K a_w)[1 + K(C - 1)a_w]} \tag{6}$$

We have combined equations (3) and (5) to give a combined dependence of the rate constant on temperature and water activity:

$$\ln k = \alpha_1 + \frac{\beta}{T} + \gamma a_w + \delta \frac{a_w}{T} \tag{7}$$

where α, β, γ, and δ are constants that need to be determined by statistical methods, e.g., non-linear regression.

10.4.4 *Example: Viability of dried* Trichoderma *under storage*

To illustrate the use of the aforementioned mathematical expressions in a practical case, we will use our results for the viability of dry *T. harzianum* spores (Box 10.1) (Pedreschi and Aguilera, 1997).

Experimental data and kinetics of change in relative viability (*V*) at 42, 33, and 8°C for samples stored at different a_w are shown in Figure 10.3. Fitting of a first-order reaction seems satisfactory in predicting variations in relative viability with time and suggests that no change in death mechanism occurs. No kinetic expression was fitted for 8°C as almost no change in *V* occurred at the two lowest a_w (Figure 10.3C). Loss in viability of *T. harzianum* was faster at higher temperatures and a_w. Figure 10.3C shows that the highest viability of stored spores was found at the control temperature of 8°C and a low a_w. At this temperature (close to refrigeration)

Box 10.1 Main mathematical expressions used to predict the loss of viability of *T. harzianum* spores during storage

Integrated kinetic expression for loss of viability with time (*V* is residual viability in %):

$$\ln V = \ln(100) - kt$$

Dependence of rate constant (days^{-1}) on *T*(*K*) and a_w:

$$\ln k = 70.2 - \frac{2.3 \times 10^4}{T} - 42.7 a_w + \frac{1.46 \times 10^4}{T} a_w$$

Sorption isotherm:

$$\frac{a_w}{m} = -37.8 a_w^2 + 32.3 a_w + 2.43$$

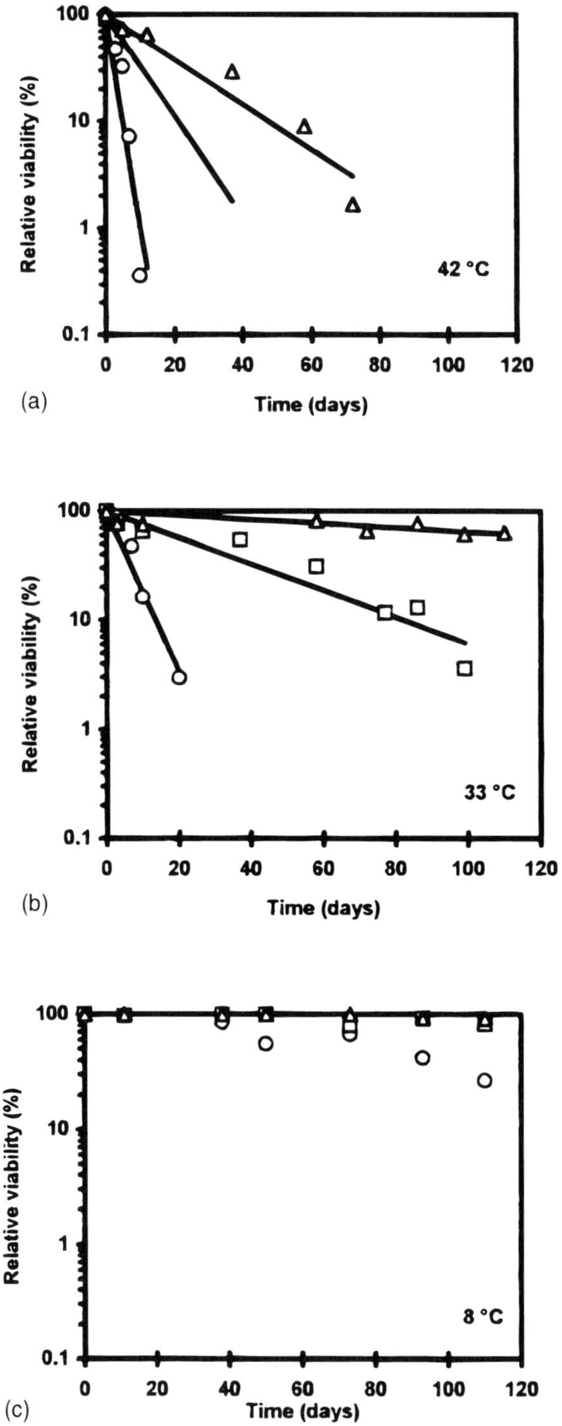

Figure 10.3 Kinetics of change on relative viability of *T. harzianum* spores under storage at 42, 33 and 8°C. Symbols represent experimental data at constant a_w(○ = 0.75, □ = 0.33 and △ = 0.03) and lines represent the model. From Pedreshi and Aguilera (1997).

the effect of a_w on viability was negligible for 0.03 and 0.33 but significant for 0.75. At 33°C over 70% of spores stored at $a_w = 0.03$ remained viable after 110 days. We also verified the applicability of the model for data gathered at 33°C/$a_w = 0.89$ and 42°C/$a_w = 0.40$ for 18 days (after this time relative viability was <1%). These conditions were selected to see whether they could be used for an accelerated test that represented the same mechanism as that of the previous storage assay. These results suggest that slow drying and storage under low moisture conditions (e.g., $m < 4$ g/g) would provide the dry spores with an extended shelf life of up to several months in cool places (e.g. <20°C) inside a hermetic package.

Acknowledgments

This study was supported by grants from FONDEF AI-21 and FONDECYT No. 1960360 and 196389. Collaboration of Mr F. Pedreschi and Dr G. Muñoz is greatly appreciated.

References

AGOSIN, E., COTORAS, M., MUÑOZ, G., SAN MARTÍN, R., and VOLPE, D. 1997a. Comparative properties of the spores of *Trichoderma harzianum* produced by solid state and submerged cultivations. In S. Roussos, B. Lonsane, M. Raimbault, and G. Viniegra-Gonzalez (eds), *Advances in Solid State Fermentation*, Vol. 38. Kluwer, Dordrecht, pp. 461–471.

AGOSIN, E., VOLPE, D., MUÑOZ, G., SAN MARTÍN, R., and CRAWFORD, A. 1997b. Effect of culture conditions on spore shelf life of the biocontrol agent *Trichoderma harzianum*. *World J. Microbiol. Biotechnol.* **13**: 225–232.

AGUILERA, J. M. and KAREL, M. 1997. Preservation of biological materials under desiccation. *CRC Crit. Rev. Food Sci. Nutr.* **37**: 287–309.

AURIA, R., PALACIOS, J., and REVAH, S. 1992. Determination of the interparticle diffusion coefficient for CO_2 and O_2 in solid state fermentation. *Biotechnol. Bioeng.* **39**: 898–902.

AURIA, R., HERNANDEZ, S., RAIMBAULT, M., and REVAH, S. 1990. Ion exchange resin: a model support for solid state growth fermentation of *Aspergillus niger*. *Biotechnol. Tech.* **4**: 391–396.

AURIA, R., MORALES, M., VILLEGAS, E., and REVAH, S. 1993. Influence of mold growth on the pressure drop in aerated solid state fermentors. *Biotechnol. Bioeng.* **41**: 1007–1013.

BACKMAN, P. and RODRIGUEZ-KABANA, R. 1975. A system for the growth and delivery of biological control agents to the soil. *Phytopathology* **65**: 819–821.

BARRIOS-GONZÁLEZ, A., TOMASINI, A. G., VINIEGRA-GONZÁLEZ, G., and LOPEZ, J. 1988. Penicillin production by solid state fermentation. *Biotechnol. Lett.* **10**: 793–798.

BEKER, M. J. and RAPOPORT, A. I. 1987. Conservation of yeast by dehydration. *Adv. Biochem. Engin.* **35**: 127–171.

BODO, B., REBUFFAT, S., EL HAJJI, M., and DAVOUST, D. 1985. Structure of trichorzianine A IIIc, an antifungal peptide from *Trichoderma harzianum*. *J. Am. Chem. Soc.* **107**: 6011–6017.

CANNEL, E. and MOO-YOUNG, M. 1980. Solid state fermentation systems. *Process Biochem.* **15**: 2–7.

Cascino, J. J., Harris, R. F., Smith, C. S., and Andrews, J. H. 1990. Spore yield and microcycle conidiation of *Colletotrichum gloeosporioides* in liquid culture. *Appl. Environ. Microbiol.* **56**: 2303–2310.

Couch, T. L. and Ignoffo, C. M. 1981. Formulation of insect pathogens. In H. D. Burges (ed.), *Microbial Control of Pests and Plant Diseases* 1970–1980. Academic Press, London, pp. 621–634.

Dahlberg, K. R. and van Etten, J. L. 1982. Physiology and biochemistry of fungal sporulation. *Annu. Rev. Phytopath.* **20**: 281–301.

De Reu, J., Zwietering, M., Rombouts, F., and Nout, M. 1993. Temperature control in solid substrate fermentation through discontinuous rotation. *Appl. Microbiol. Biotechnol.* **40**: 261–265.

Doss, R. P., Potter, S., Chastagner, A., and Christian, J. 1992. Adhesion of nongerminated conidia of *Botrytis cinerea* to several substrata. *Appl. Environ. Microbiol.* **59**: 1786–1791.

Durand, A. 1995. The INRA-Dijon reactors: design and applications. In Second International Symposium on Solid-state Fermentation. FMS 95, ORSTOM, Montpellier, France, p. 6.

Durand, A. and Chereau, D. 1988. A new pilot reactor of solid state fermentation: application to the protein enrichment of sugar beet pulp. *Biotechnol. Bioeng.* **31**: 476–486.

Elad, Y. and Kirshner, B. 1993. Survival in the phylloplane of an introduced biological control agent (*Trichoderma harzianum*) and populations of the plant pathogen *Botrytis cinerea* as modified by abiotic conditions. *Phytoparasitica* **21**: 303–313.

Elad, Y., Zimand, G., Zaqs, Y., Zuriel, S., and Chet, I. 1993a. Use of *Trichoderma harzianum* in combination or alternation with fungicides to control cucumber grey mould (*Botrytis cinerea*) under commercial greenhouse conditions. *Plant Pathology (Oxford)* **42**: 324–332.

Elad, Y., Shmuel, G., Zimand, G., Ruth, K., Chet, I., and Ezrahi, S. 1993b. Isolate of *Trichoderma harzianum* I-952, fungicidal compositions containing said isolate and use against *B. cinerea* and *S. sclerotiorum.* United States Patent 5,266,316.

Elad, Y., Shmuel, G., Zimand, G., Ruth, K., Chet, I., and Ziona, N. 1993c. Isolate of *Trichoderma* fungicidal compositions containing said isolate and use against *B. cinerea* and *S. sclerotiorum.* United States Patent 5,238,690.

Feng, M. G., Poprawski, T. J., and Khachatourians, G. G. 1994. Production, formulation and application of the entomopathogenic fungus *Beauveria bassiana* for insect control: current status. *Biocontrol Sci. Technol.* **4**: 30–34.

Fernández, M., Pérez-Correa, J. R., Solar, I., and Agosin, E. 1996. Automation of a solid substrate cultivation pilot reactor. *Bioprocess Engin.* **16**: 1–4.

Franks, F. 1994. Accelerated stability testing of bioproducts: attractions and pitfalls. *TIBTECH* **12**: 114–117.

Fu, W.-Y. and Etzel, M. R. 1995. Spray drying of *Lactococcus lactis* ssp. *lactis* C2 and cellular injury. *J. Food Sci.* **60**: 195–200.

Gee, O. H., Probert, R. J., and Coomber, S. A. 1994. "Dehydrin-like" proteins and desiccation tolerance in seeds. *Seed Sci. Res.* **4**: 135–141.

Harkonen, H., Koskinen, M., Linko, P., Siika-aho, M., and Poutanen, K. 1993. Granulation of enzyme powders in a fluidized bed spray granulator. *Food Sci. Tech. (Zurich)* **26**: 235–241.

Harman, G. E. 1988. Fused biocontrol agents. European Patent 0,285,987.

Harman, G. E. 1991. Seed treatments for biological control of plant disease. *Crop Protection* **10**: 166–170.

Harman, G. E., Jin, X., Stasz, T. E., Peruzzotti, G., Leopold, A. C., and Taylor, A. G. 1991. Production of conidial biomass of *Trichoderma harzianum* for biological control. *Biol. Control* **1**: 23–28.

HARMAN, G. E., JIN, X., STASZ, T. E., PERUZZOTTI, G., LEOPOLD, C., and TAYLOR, A. G. 1994. Method of increasing the percentage of viable dried spores of a fungus. United States Patent 5,288,634.

HUBER, S. and MENNER, M. 1996. A new laboratory dryer for analyzing the deterioration kinetics of biomaterials. *Drying Technol.* **14**: 1947–1966.

JIN, X., HARMAN, G. E., and TAYLOR, A. G. 1991. Conidial biomass and desiccation tolerance of *Trichoderma harzianum* produced at different medium water potentials. *Biol. Control* **1**: 237–243.

JIN, X., TAYLOR, A. G., and HARMAN, G. E. 1996. Development of media and automated liquid fermentation methods to produce desiccation-tolerant propagules of *Trichoderma harzianum. Biol. Control* **7**: 267–274.

JONES, J., ASHER, W., BOMBEN, J., and BOMBERGER, D. 1993. Keep pilot plants on the fast track. *Chem. Engng.* **100**: 98–106.

KAREL, M. 1975. *Physical Principles of Food Preservation.* Marcel Dekker, New York, pp. 237–263.

KUBOTA, T. 1995. *Trichoderma harzianum* SK-55 fungus, fungicide containing it, method of manufacture of the same and its use. United States Patent 5,422,107.

LABUZA, T. P. and KAMMAN, J. F. 1983. Reaction kinetics and accelerated test simulation as a function of temperature. In I. Saguy (ed.), *Computer-Aided Techniques in Food Technology.* Marcel Dekker, New York, pp. 71–115.

LABUZA, T. P. and SCHMIDL, M. K. 1985. Accelerated shelf-life testing of foods. *Food Technol.* **39**: 57–64.

LAKSHMINARAYANA, K., CHAUDHARY, K., ETHIRAJ, S., and TAURO, P. 1975. A solid-state fermentation method for citric acid production using sugar cane bagasse. *Biotechnol. Bioeng.* **17**: 291–298.

LARROCHE, C. and GROS, J. B. 1986. Spore production of *Penicillium roqueforti* in fermenters filled with buckwheat seeds: batch and semi-continuous cultivation. *Appl. Microbiol. Biotechnol.* **24**: 134–139.

LEWIS, J. A. and PAPAVIZAS, G. C. 1983. Chlamydospore formation by *Trichoderma* spp. in natural substrates. *Can. J. Microbiol.* **30**: 1–6.

LEWIS, J. A. and PAPAVIZAS, G. C. 1991. Biocontrol of plant diseases: the approach for tomorrow. *Crop Protection* **10**: 95–105.

LIEVENSE, L. C. and VAN'T RIET, K. 1993. Convective drying of bacteria. I: The drying process. *Adv. Biochem. Engineer./Biotechnol.* **50**: 45–63.

LONSANE, B. K. and KRISHNAIAH, M. 1992. Product leaching and downstream processing. In H. W. Doelle, D. A. Mitchell, and C. E. Rolz (eds), *Solid Substrate Cultivation.* Elsevier Applied Science, London, pp. 147–171.

LONSANE, B. K., SAUCEDO-CASTAÑEDA, G., RAIMBAULT, M., ROUSSOS, S., VINIEGRA-GONZÁLEZ, G., GHILDYAL, N., RAMAKRISHNA, M., and KRISHNAIAH, M. 1992. Scale-up strategies for solid state fermentation systems. *Process Biochem.* **27**: 259–273.

MARY, P., MOSCHETTO, N., and TAILLIEZ, R. 1993. Production and survival during storage of spray-dried *Bradyrhizobium japonicum* cell concentrates. *J. Appl. Bacteriol.* **74**: 340–344.

MCBEATH, J. 1995. Cold tolerant *Trichoderma.* United States Patent 5,418,165.

MCDONOUGH, R. 1992. *Mixing for the Process Industries.* van Nostrand Reinhold, New York, 241 pp.

MILLER, T. L. and CHURCHILL, B. W. 1986. Substrates for large scale fermentations. In A. L. Demain and N. A. Solomon (eds), *Manual of Industrial Microbiology and Biotechnology.* American Society for Microbiology, Washington, DC, pp. 122–137.

MITCHELL, D. A. and LONSANE, B. K. 1992. Definition, characteristics and potential. In H. W. Doelle, D. A. Mitchell, and C. E. Rolz (eds), *Solid Substrate Cultivation.* Elsevier Applied Science, London, pp. 1–13.

Mitchell, D. A., Lonsane, B. K., Durand, A., Renaud, R., Almanza, S., Maratray, J., Desgranges, C., Crooke, P. S., Hong, K., Tanner, R. D., and Malaney, G. W. 1992. General principles of reactor design and operation in SSC. In H. W. Doelle, D. A. Mitchell, and C. E. Rolz (eds), *Solid Substrate Cultivation*. Elsevier Applied Science, London, pp. 115–136.

Moo-Young, M. and Blanch, H. 1987. Transport phenomena and bioreactor design. In J. Bu'Lock and B. Christiansen (eds), *Basic Biotechnology*. Academic Press, New York, pp. 133–172.

Mudgett, R. E. 1986. Solid-state fermentations. In A. L. Demain and N. A. Solomon (eds), *Manual of Industrial Microbiology and Biotechnology*. American Society for Microbiology, Washington, DC, pp. 66–83.

Muñoz, G., Nakari-Setala, T., Agosin, E., and Penttilä, M. 1997. Hydrophobin gene, *srh1*, expressed during sporulation of the biocontrol agent *Trichoderma harzianum*. Current Genetics, **32**: 225–230.

Muñoz, G., Agosin, E., Cotoras, M., San Martín, R., and Volpe, D. 1995. Comparison of aerial and submerged spore properties of *Trichoderma harzianum*. *FEMS Microbiol. Letters* **125**: 63–70.

Nwaka, S., Kopp, M., Burgert, M., Deuchler, I., Kienle, I., and Holzer, H. 1994. Is thermotolerance of yeast dependent on trehalose accumulation? *FEBS Letters* **344**: 225–228.

Paau, A., Bennett, M., and Graham, L. 1993. Production of enhanced biocontrol agents. United States Patent 5,194,258.

Papavizas, G. C. 1985. *Trichoderma* and *Gliocladium*: biology, ecology, and potential for biocontrol. *Annu. Rev. Phytopathol.* **23**: 23–54.

Papavizas, G. C., Dunn, M., Lewis, J. A., and Beagle-Ristaino, J. 1984. Liquid fermentation technology for experimental production of biocontrol fungi. *Phytopathology* **74**: 1171–1175.

Pedreschi, F. and Aguilera, J. M. 1997. Viability of dry *Trichoderma harzianum* spores under storage. *Bioprocess Eng.* **17**: 177–183.

Pedreschi, F., Aguilera, J. M., Agosin, E., and San Martín, R. 1997. Induction of trehalose in spores of the biocontrol agent *Trichoderma harzianum*. *Bioprocess Eng.* **17**: 317–322.

Raimbault, M. and Alazard, D. 1980. Culture method to study fungal growth in solid fermentation. *Eur. J. Appl. Microbiol. Biotechnol.* **9**: 199–209.

Rogers, L. A. 1914. The preparation of dried cultures. *J. Infect. Dis.* **14**: 100–123.

Roussos, S., Raimbault, M., Prebois, J., and Lonsane, B. 1993. Zymotis, a large scale solid state fermenter. *Appl. Biochem. Biotechnol.* **42**: 37–51.

Roussos, S., Olmos, A., Raimbault, M., Saucedo-Castañeda, G., and Lonsane, B. 1991. Strategies for large-scale inoculum development for solid-state fermentation system: conidiospores of *Trichoderma harzianum*. *Biotechnol. Techniques* **5**: 415–420.

Saguy, I. and Karel, M. 1980. Modeling of quality deterioration during food processing and storage. *Food Technol.* **34**: 78–85.

Sargantanis, J., Karim, M., Murphy, V., Ryoo, D., and Tengerdy, R. 1993. Effect of operating conditions on solid substrate fermentation. *Biotechnol. Bioeng.* **42**: 149–158.

Saucedo-Castañeda, G., Trejo-Hernández, M. R., Lonsane, B. K., Navarro, J. M., Roussos, S., Dufour, D., and Raimbault, M. 1994. On-line automated monitoring and control systems for CO_2 and O_2 in aerobic and anaerobic solid-state fermentations. *Process Biochem.* **29**: 13–24.

Schirmböck, M., Lorito, M., Wang, Y.-L., Hayes, C. K., Arisan-Atac, I., Scala, F., Harman, G. E., and Kubicek, C. P. 1994. Parallel formation and synergism of hydrolytic enzymes and peptaibol antibiotics, molecular mechanism involved in the antagonistic action of *Trichoderma harzianum* against phytopathogenic fungi. *Appl. Environ. Microbiol.* **60**: 4364–4370.

St Leger, R. J., Staples, R. C., and Roberts, D. W. 1992. Cloning and regulatory analysis of starvation-stress gene, *ssgA*, encoding a hydrophobin-like protein from the entomopathogenic fungus *Metharizium anisopliae*. *Gene (Amsterdam)* **120**: 119–124.

Tabachnik, M. and Ziona, N. 1989. Method of growing *Trichoderma*. United States Patent 4,837,155.

Tengerdy, R. P. 1985. Solid substrate fermentations. *TIBTECH* **3**: 96–99.

Thevelein, J. M. 1984. Regulation of trehalose mobilization in fungi. *Microbiol. Rev.* **48**: 42–59.

Toet, M. and Somers, A. 1994. Production of *Trichoderma harzianum* rifai CMI CC No. 333646. United States Patent 5,330,912.

van Dijck, P., Colavizza, D., Smet, P., and Thevelein, J. M. 1995. Differential importance of trehalose in stress resistance in fermenting and nonfermenting *Saccharomyces cerevisiae* cells. *Appl. Environ. Microbiol.* **61**: 109–115.

Wechsberg, G. E., Probert, R. J., and Bray, C. M. 1994. The relationship between "dehydrin-like" proteins and seed longevity in *Ranunculus sceleratus* L. *J. Exp. Botany* **45**: 1027–1030.

Wiemken, A. 1990. Trehalose in yeast: stress protectant rather than reserve carbohydrate. *Antonie Leeuwenhoek* **58**: 209–217.

Zuber, J. and Turian, G. 1981. Induction of premature phialoconidiogenesis on germinated conidia of *Trichoderma harzianum*. *Trans. Br. Mycol. Soc.* **76**: 433–440.

11

Potential and existing uses of *Trichoderma* and *Gliocladium* for plant disease control and plant growth enhancement

G. E. HARMAN and T. BJÖRKMAN

Departments of Horticultural Sciences and Plant Pathology, Cornell University, Geneva, New York, USA

11.1 Historical perspective

The ability of *Trichoderma* and *Gliocladium* species to control plant disease and enhance plant growth has been known for many years. The first major discoveries were made by Richard Weindling in the 1930s and 1940s. He made a number of discoveries that have provided much of the impetus for basic and applied research on the biocontrol abilities of these fungi ever since. Among the contributions that Weindling made are the following:

1 *T. lignorum* (this taxon was changed to *G. virens*, and now is considered to be *T. virens* (Bissett, 1991)) parasitizes hyphae of plant pathogenic fungi in two ways, i.e., by coiling about them and by killing at a short distance away (Weindling, 1932) (see Chapter 7 for the current status of this topic).

2 A lethal principle is produced by *T. lignorum*, and this principle was strongly affected by the medium in which it was used, especially by pH (Weindling, 1934). This substance was crystallized and identified subsequently as gliotoxin (Weindling, 1941). Other strains, which were characterized by a coconut-like odor, were also toxic to fungi (Weindling, 1934) (Chapter 8 of Volume 1 and Chapter 8 of this volume describe our current understanding of these compounds).

3 These fungi were effective in control of diseases such as damping-off of citrus. The conditions under which the biocontrol agent was applied made a great difference; control was effective in acidic, but not in neutral or alkaline, soils. Weindling attributed this difference to the stability of the toxic factor, which was stable in acidic conditions but degraded rapidly at higher pH levels (Weindling and Fawcett, 1936).

Evidence suggests that *Trichoderma* spp. can enhance plant growth (see Chapter 9 and this chapter). Lindsey and Baker (1967) demonstrated that gnotobiotic tomato plants grown in the presence of *T. viride* were taller than microbe-free plants, although dry weight of roots or shoots was not significantly increased. In the 1980s, Baker at Colorado State University and Chet in Israel, and their colleagues, demonstrated that *T. harzianum* strains had the ability to enhance plant growth and that this effect apparently was separate from the biocontrol ability of these fungi. For example, pepper seeds treated with *T. harzianum* germinated 2 days earlier than untreated seeds; also, the addition of the organism to potting medium hastened flowering of periwinkle, increased numbers of blooms on chrysanthemum, and increased growth of cucumber, radish, tomato, and pepper (Baker *et al.*, 1984; Chang *et al.*, 1986).

In spite of these promising results, biocontrol products and systems based on these fungi are only just becoming available for general use in commercial agriculture. In the meantime, researchers around the world have published thousands of papers describing the beneficial effects of *Trichoderma* and *Gliocladium* spp. on agricultural and ornamental plants. However, not until the 1990s has it been possible to successfully identify, manufacture, and market products based on these fungi. Only now has the understanding of the organisms and development of the necessary components been sufficiently advanced for commercialization to be feasible.

It is the purpose of this review to consider (a) the requirements for commercially successful biocontrol, (b) some of the efforts to develop commercial biocontrol products and methods, and (c) some commercially available biocontrol products, including their concepts and uses.

11.2 Requirements for development of successful biocontrol systems and products

11.2.1 Biological components

Jin *et al.* (1991) and Elad and Chet (1995) have both considered the necessary biological components of successful biocontrol *systems*. Biocontrol systems are indeed necessary and are composed of three general components: (1) a superior biocontrol strain, (2) a production system that produces appropriate materials, and (3) delivery systems that provide propagules to the appropriate location in a milieu permitting the organism to grow and proliferate. A summary of these components is provided below.

Superior biocontrol strains

Clearly, any biocontrol system must start with a highly effective strain. In most cases, these strains have been obtained through selection from nature (Chet and Baker, 1981; Elad *et al.*, 1993; Hadar *et al.*, 1984; Sivan and Chet, 1986; Smith *et al.*, 1990), in some cases after mutation or selection for enhanced resistance to fungicides (Ahmad and Baker, 1987; Tronsmo, 1989). Further, one strain used commercially was prepared by fusion of two strains followed by extensive selection to obtain a strain that grew more rapidly than either parent and that had a high level of ability to colonize roots (rhizosphere competence) (Harman *et al.*, 1989; Stasz *et al.*, 1988). Transgenic strains with improved properties will undoubtedly soon be available, and so commercially useful strains may be discovered or created using a variety of methods.

However, regardless of the attributes of the strain, it cannot be used successfully and reliably until production methodology and appropriate delivery systems are identified. In our opinion, identification of an effective biocontrol strain is the easiest part of development of a biocontrol system. Once an effective strain is identified, years of work remain before biocontrol products can be produced, tested, formulated, and used commercially. However, as commercialization proceeds and investigators and companies gain greater experience, the techniques and methods developed with one biocontrol strain will probably be applicable to succeeding strains. Therefore, in the future, the time from the discovery of an effective strain to the use of the strain in commercial agriculture should decrease substantially.

Production methodology

Biological control agents must be produced in large quantities for use in commercial agriculture. Not only must large quantities be prepared, but the production method must provide material with the following characteristics (modified from Jin *et al.*, 1991): (1) the preparation must contain high levels of the organism of interest, and contaminating organisms and off-types of the organism being produced must be absent or at very low levels; (2) the product must have good shelf life, for most applications about a year without refrigeration or other special storage; (3) the material must be uniform from batch to batch; (4) products must be inexpensive; and (5) products must be formulated in a way that permits application with existing agricultural equipment and practices.

Biomass with such properties can be produced in either liquid or semi-solid fermentation, but the physiology and type of propagule (hyphae, conidia, or chlamydospores) must be studied and optimized for either general method (Jin *et al.*, 1991). Detailed information on fermentation systems will not be considered further here since Chapter 10 considers this topic in detail.

Delivery systems

The method of application of a superior strain that has been produced and formulated into a useful product is critical. Successful use of biocontrol strains will require application of a biologically and economically effective dose, may require the use of adjuvants, may be most successfully used in combination with fungicides, may need to be formulated or used at particular pH values, or may require other special handling or delivery systems. The best way to use a specific biocontrol agent or preparation frequently differs between crops and application, due to both biological and economic factors. Later sections of this chapter will examine the use of some commercial biocontrol products, and the method of delivery required for successful biocontrol will be stressed.

Product and target concepts

All of the information about a particular biocontrol system, including the strain itself and the agricultural system to which it will be applied, must be formulated into a product concept. The product concept is based on identification of a specific advantage for control of one or more pathogens and/or enhanced growth and yield of the crop in question. It is this product concept that finally results in a formulation that is marketed as a product with specific use instructions for the user. It is important to understand exactly what objectives and economic advantages are conferred by the product in question. In most cases, it is best to identify uses and advantages not conferred by chemical fungicides or to target uses where chemical fungicides have clearly identified disadvantages of cost, toxicity, or environmental damage that are recognized by the potential user. The specific examples described later in this chapter will be identified as to product and target concepts; interestingly, product concepts differ substantially even for products intended for the same use.

11.2.2 *Legal and economic considerations*

Successful use of biocontrol systems requires a number of other components. For example, legal and economic considerations are essential. First, it is imperative that the biocontrol agent and/or processes used to prepare or use it be protected by patents or proprietary information. Otherwise, successful products can easily be copied. Second, it usually is required that strains and products undergo toxicological testing and registration with the proper state or federal regulatory agencies before sales are legal. Third, appropriate marketing and product validation across diverse geographical areas is necessary. All of this requires an investment of several millions of dollars, and so a corporation must be involved. Since most current biocontrol agents and systems have been identified in publicly funded institutions, the transition from public agency to private company must be made. In most cases, it is valuable to the private company to continue to access the experience and knowledge of the originator of the technology as product development occurs.

11.3 Product examples and concepts of use

Biocontrol methods and products must be directed toward specific uses, which necessitates that concepts of use must be developed. It is the aim of this section to examine some of these in detail and to consider the concepts involved in their development, with the expectation that these examples will provide useful models and concepts.

11.3.1 *Seed treatments*

There are two separate and important kinds of seed treatment objectives possible for at least some strains of biocontrol fungi. The first is *seed protection*. Seed protection can be defined as the ability of a treatment to protect seeds against seed- or soil-borne pathogenic fungi and, for this purpose, needs to be effective for a short time period of only 7 to 14 days. *Trichoderma* and *Gliocladium* species have been tested for many years for this purpose (see Harman, 1991; Howell, 1991; Jin *et al.*, 1991; Taylor and Harman, 1990; Wu, 1982).

The second type of activity is *long-term protection of the subterranean portions of plants*. Because biocontrol fungi can grow on the seed surface, they increase in mass. As considered earlier, some strains are rhizosphere competent, and this ability provides some important advantages to seed treatments with strains that possess this property.

Rhizosphere competence is the ability of a fungus to colonize and proliferate on and near the root surface. Most rhizosphere-inhabiting fungi will grow on food sources from plant roots, but they differ in their abilities to compete for those food sources, to grow on roots where carbohydrates and other nutrients are in short supply, and to grow fast enough to continuously inhabit the root tip. They also vary in their abilities to colonize the various niches on the root system, including the actively elongating tip, active uptake zones (root hairs and lateral roots), and suberized mature roots that have lost their cortex. Individual strains respond differently to temperature, food supply, and competition from other organisms.

Rhizosphere competence is a distinct trait because the rhizosphere is a different ecological niche from the bulk soil, and the attributes conferring competitive advantage in this environment are specific to it (Schmidt, 1979). The complex community of microbes normally found in the rhizosphere reflects the diversity of niches on the root surface. In order to have strong rhizosphere competence, a fungus must be able to compete with these inhabitants in each of several niches.

The types of fungal structures responsible for root colonization have been the subject of substantial speculation and concern (Harman, 1991). Most demonstrations of rhizosphere competence have relied upon plating of roots to determine whether or not colonization occurs and to quantify the fungal population levels on roots. If presumed rhizosphere competent strains were simply capable of producing large numbers of conidia on subterranean plant parts and these conidia were distributed by mass water flow, then plating experiments would provide misleading results. However, Ahmad and Baker (1987) found that the total length of hyphae of rhizosphere competent *Trichoderma* strains roughly paralleled population densities as determined by dilution plating. We have recently utilized a novel technique for visualizing microbial growth on roots (M. McCully, personal communication) and have determined that *T. harzianum* strain 1295-22, which is highly rhizosphere competent, grows on the surface of the root. Chlamydospores and three kinds of hyphae present on treated roots were absent on untreated roots when grown in a sandy soil in the greenhouse. The hyphae fell into three classes based on size. Thin hyphae were 0.3 μm in diameter and stained uniformly with 4′,6′-diamidino-2-phenylindole (DAPI) (Figure 11.1A), medium hyphae were 0.8 μm with ovoid DAPI staining (Figure 11.1B), and thick hyphae were 2 μm with a pearl string of stained spheres (Figure 11.1C). All were present in clusters at various places on the root, with greater diversity of structures on root parts older than 10 days. Thin hyphae formed a net-like structure that probably is critical for protecting roots and increasing plant productivity as described later in this chapter. These thin hyphae probably represent young, vigorous hyphae. In cultures, we have noted that similar thin vigorous hyphae contain many nuclei per cell, so that they appear continuous when viewed in mass following staining with DAPI. Older hyphae frequently were thicker and vacuolated, so groups of nuclei appear separated from one another by the vacuoles (Harman *et al.*, 1993; Chapter 11, Vol. 1).

Kinds of seed treatments

A number of distinctly different kinds of seed treatment methods have been developed. Some of these methods are summarized below.

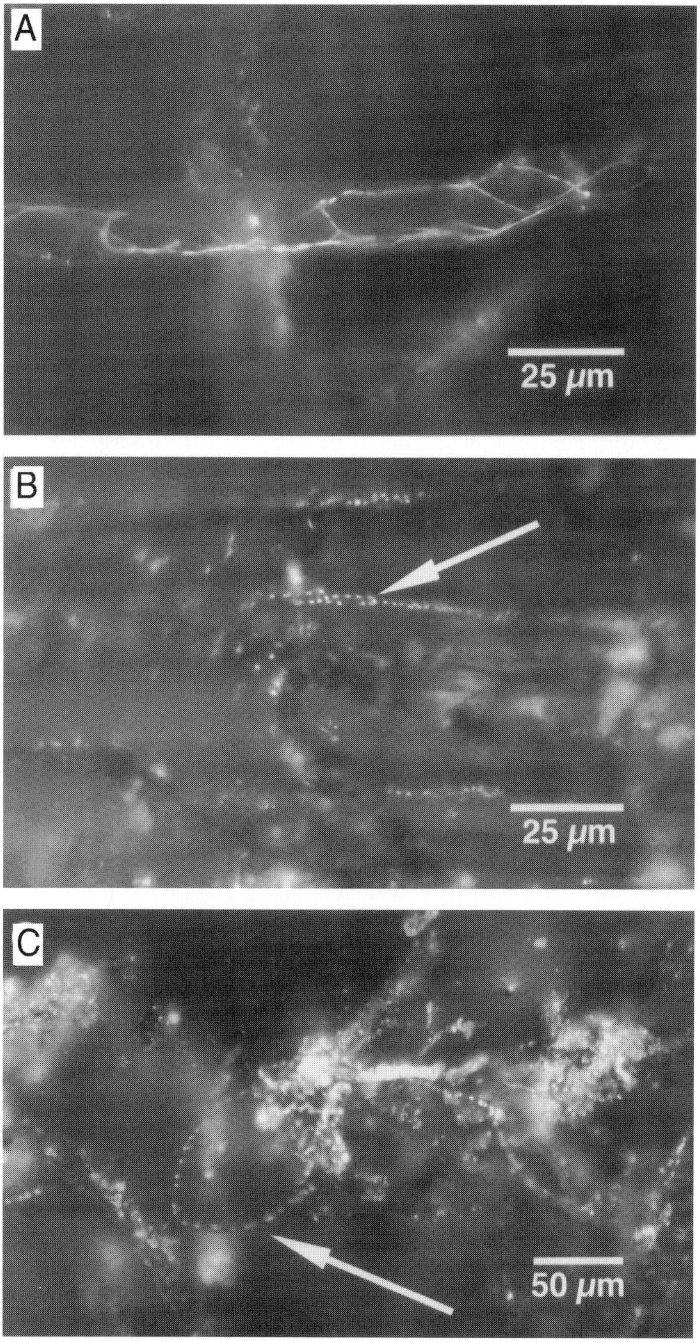

Figure 11.1 Microscopic examination of root colonization (primarily root hairs) of sweet corn plants by *T. harzianum* strain 1295-22 following a seed treatment. Hyphae of the biocontrol fungus were visualized with fluorescence microscopy after staining with DAPI. No hyphae were seen on roots grown from nontreated seeds. Micrograph A shows root hairs with a fibrillar network of hyphae that stain continuously with DAPI. B and C show roots colonized with larger hyphae of the fungus having discontinuous DAPI staining. Similar staining patterns and sizes of hyphae can be seen in cultures of this fungus. Younger, thinner hyphae stain continuously with DAPI, due to a very high density of nuclei. Older or larger hyphae stain discontinuously since vacuoles interrupt clusters of nuclei.

1 *Planter box dust treatments.* Formulations for these treatments are prepared as dry powders and usually contain inert ingredients such as graphite to enhance adherence to seeds. These materials are sold directly to farmers who add them to the seeds just at the time of planting. Coatings provided by this method usually are less uniform than those provided by other seed treatment processes.
2 *Slurry and film coating treatments.* In slurry or film coating treatments, the active ingredient is mixed with various materials that act as stickers for the treatment. The mixture is dissolved or suspended in liquid, which is usually but not always water based, and coated onto seeds in an even layer. Equipment for such applications frequently is quite sophisticated; uniform and blemish-free treatments are usually achieved.
3 *Seed priming treatment.* Seed priming is a process whereby seeds are brought to a moisture level just below that required for germination by immersion in a liquid or solid medium capable of controlling seed hydration precisely. Temperature is carefully controlled and optimized for the seed being treated. During this treatment, either biological or chemical seed treatment materials can be added. This treatment permits activation of the cellular machinery required for seed germination. Even after drying the seed following priming, seeds increase in vigor and ability to produce stands under adverse conditions (Taylor and Harman, 1990).
4 *Coating and pelleting methods.* A layer of varying thicknesses may be added to the seeds by successive additions of adhesives plus particulates of various types. The resulting layer may range from <0.1 to >3 mm thick on the surface of seeds.

Seed protectants

The differing kinds of seed treatment methods can provide very different results when tested as biological seed protectants. Protection against *Pythium* spp. must be provided since they are the primary seed rotting fungi in most soils. These fungi have evolved remarkable abilities to detect and infect seeds just after they are planted. They begin to germinate within 2 h of planting and begin to infect seeds 4–6 h after planting (Harman and Nelson, 1994; Taylor and Harman, 1990) (Figure 11.2). Thus, the response time required of biological seed protectants is very short. *Trichoderma* and *Gliocladium* spp. usually must grow and proliferate on the seed coat if protection is to occur, but spores of these fungi germinate more slowly after planting than do propagules of *Pythium* spp. Therefore, the beneficial fungi are at a temporal disadvantage that must be overcome if seed protection is to be effective. An exception to this generalization may occur if antibiotic-producing strains are used since the biocontrol preparation may contain effective levels of the antibiotic in question. Strains of *G. virens* may be particularly effective in this regard (Howell, 1991; Howell and Stipanovic, 1983) (Chapter 8).

However, even in the absence of preformed antibiotics in the biocontrol preparation, highly effective seed protection can be obtained with *Trichoderma* and *Gliocladium.* The delivery system is critical for highly effective control.

Since timing also is critical, it is logical to develop techniques that slow the arrival of germlings of *Pythium* spp. to the infection court, i.e., the seed surface. One way to accomplish this is to provide a coating on the seed surface; even a layer of <0.1 mm thick can make a difference. Taylor *et al.* (1991) applied *T. harzianum* to

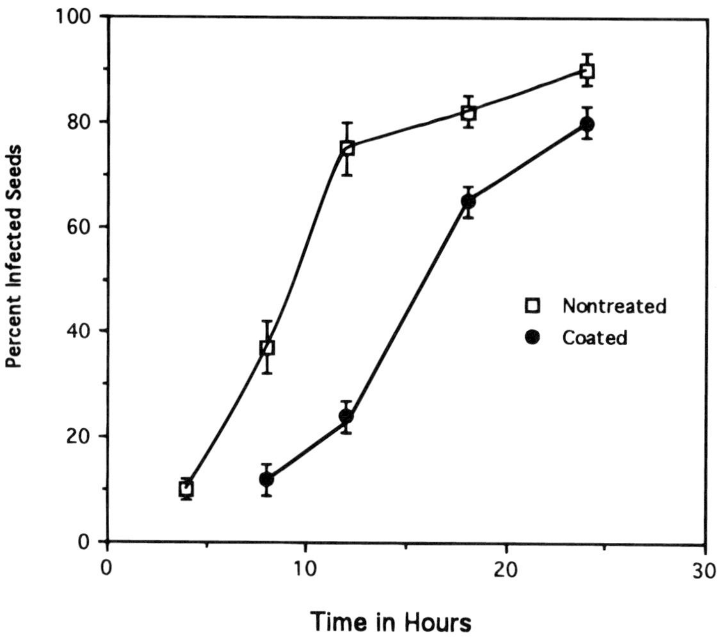

Figure 11.2 The effect over time of no treatment or of application of a thin organic coating onto cucumber on infection of cucumber seeds by *Pythium ultimum*. Data are from Taylor *et al.* (1991).

the surface of cucumber seeds and then applied a layer of Agro-Lig (a highly organic Leonardite shale) together with a binder in a slurry to provide a uniform layer about 0.1 mm thick (liquid coating). This latter treatment delayed seed infection of the cucumber seeds for several hours (Figure 11.2).

These modifications to the seed treatment procedures had a profound effect upon efficacy of seed protection provided by *T. harzianum* (Table 11.1). At the high levels of *P. ultimum* employed in the experiment, a slurry treatment provided no protection to cucumbers. However, changing the time relationships (Figure 11.2) by applying the bioprotectant under a thin coating of organic material that *Pythium* hyphae must traverse before they can infect the seed increased protection by *T. harzianum* dramatically. However, the coating without *T. harzianum* had no effect on numbers of seeds killed by the pathogen (Table 11.1). The protection of seeds by *T. harzianum* could be increased even more if seeds were placed at 100% relative humidity for a few days after treatment. Under these conditions, conidia of *T. harzianum* germinated and colonized seeds (Taylor *et al.*, 1991) prior to planting (Table 11.1). Similar results were obtained when seeds were primed in a matrix of Agro-Lig at a moisture level just below that permitting seeds to germinate (Harman and Taylor, 1988; Harman *et al.*, 1989). This growth of *T. harzianum* (a) permits seeds to be colonized prior to planting in soil and (b) increases the effective dose of the biocontrol agent which resulted in near perfect control of diseases caused by *Pythium* spp. (Table 11.1).

To summarize, seeds treated with *T. harzianum* in a slurry treatment were not protected against high levels of *P. ultimum*. However, if a thin layer of an organic substance was placed over the *T. harzianum* treatment, good seed protection was

Table 11.1 Effects of seed treatment modifications on cucumber seed emergence and survival 10 days after planting in an Arkport sandy loam soil highly infested with *Pythium ultimum*

Treatment	Maximum emergence[a]	Final stand[b]
Slurry alone	2 c[c]	0 a
Liquid coating	72 b	60 b
Liquid coating + high RH	98 a	90 a

In all cases, seeds were treated with 1 mg of *T. harzianum* containing 10^6 cfu/g of seed in a slurry treatment containing 10% Pelgel as an adhesive. In the liquid coating procedure, seeds were then coated with Agro-Lig in Pelgel to provide a thin continuous layer <0.1 mm thick. Seeds were air dried after all treatments prior to planting in soil infested with the pathogen. For the high relative humidity (RH) treatment, liquid-coated seeds were placed at 100% relative humidity for 4 days at 25°C in the dark. Data are from Taylor *et al.* (1991).

[a] Percentage of seeds that produced seedlings 4 days after planting.

[b] Percentage of seeds that produced seedlings that survived for 7 days.

[c] Mean separation within columns by least significant differences at $P = 0.5$.

obtained. Finally, if the layered seeds were placed at high relative humidity, which permitted the biocontrol fungus to colonize the seeds prior to planting, nearly perfect control was obtained. This is an excellent example of the importance of delivery method as a determinant of biocontrol efficacy.

Long-term root protection

As noted earlier, seed treatments with *Trichoderma* or *Gliocladium* spp. may be applied (1) to protect seeds or, if rhizosphere competent strains are used, (2) to provide long-term protection against root pathogens and to enhance root performance. Of course, both results could also be obtained. Rhizosphere competent strains can indeed provide some very significant long-term effects.

The requirements for a seed treatment are quite different for root protection alone than if both seed and root protection are required. For root colonization, the biological seed treatment fungus need only grow on the seed sufficiently well to colonize the root. Of course, a strongly rhizosphere competent strain is required. Further, it needs only to grow well at a time prior to root emergence from the seed, which is usually several days after planting. Therefore, the time relationships for root colonization and seed protection are very different, and the requirements for a root protectant seed treatment are less stringent than for a seed protectant. In fact, simple planter box treatments can provide useful levels of root disease control and yield enhancement.

Benefits of a commercial planter box seed treatment A commercial formulation of *T. harzianum* strain 1295-22 (T-22) is now being manufactured by BioWorks, Inc., Geneva, NY, USA, and sold through a variety of distributors as T-22 Planter Box™. The formulation is designed for application to large-seeded crops such as corn, beans, cotton, and soybeans. In most cases, it is expected to be applied to seeds already treated with chemical fungicides.

The ability of T-22 planter box to result in root colonization has been tested on sweet corn planted in a wide variety of soils and soil temperatures on farms in New York. The soils used included gravel, sand, and muck as well as several different loams. The soil temperature at planting varied from 9°C to 27°C. The range of conditions were expected to favor a range of many competing organisms and should cause the plant to create different conditions in the rhizosphere.

Colonization of sweet corn roots was consistently good, with 10^4 to 10^6 colony forming units (cfu) detected per gram of root when assayed one month after planting (Figure 11.3). This consistent colonization demonstrated that rhizosphere competence of this strain was robust over different rhizosphere environments. The greater rhizosphere competence of T-22 over wild strains of *Trichoderma* was seen in three ways. First, the colonization by T-22 is about 100-fold greater than that on roots from nontreated plants. Second, well-drained soils, such as gravel and sand, supported lower indigenous populations of *Trichoderma*, but populations of T-22 were not lower than in other soils. Third, T-22 populations were high regardless of planting date, whereas indigenous populations were lower early in the season. The ability of a very low level of this strain applied as a seed treatment (only 1–2 g of actual fungal biomass per hectare) to result in long-term root colonization over a range of conditions demonstrates the high level of rhizosphere competence of T-22.

Colonization with the planter box formulation of T-22 was compared with an in-furrow granule application and with a slurry seed treatment to determine the relative effectiveness in producing early colonization. The planter box and slurry treatments both place the *Trichoderma* spores in close contact with seed-applied fungicides. All three methods provided effective and similar root colonization, although the in-furrow granule was somewhat less effective than either seed treatment method (Table 11.2). The in-furrow granular treatment is intrinsically more expensive because at least 30 times more material must be metered into seed

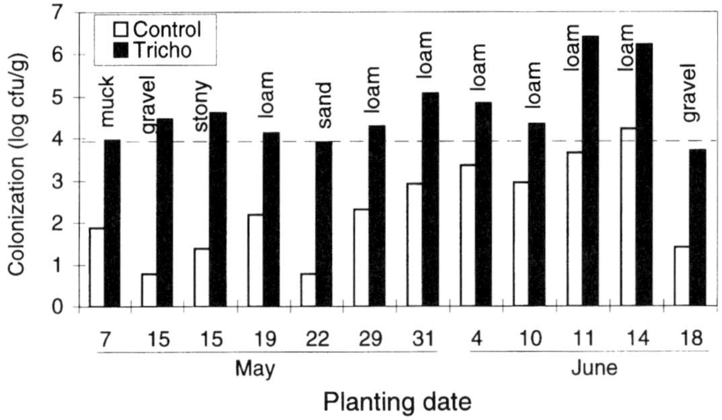

Figure 11.3 Colonization of 4-week-old corn roots on diverse farms in New York by *T. harzianum* strain 1295-22. Roots were sampled from plants from commercial fields sown with inoculated or uninoculated seeds in replicated adjacent blocks. Colonies were enumerated from rhizosphere soil by dilution plating on *Trichoderma*-selective medium. The error of each estimate is 0.4 units.

Table 11.2 Effectiveness of planter box application of T-22 in colonization of 4-week-old roots compared with in-furrow granules and a slurry seed treatment, all on seeds treated with captan and imazilil, both of which are toxic to *Trichoderma*

Treatment	Trichoderma level (CFU/g roots)
Nontreated	$10^{2.5}$ a
Granule	$10^{3.4}$ b
Planter box	$10^{3.9}$ c
Slurry	$10^{4.3}$ c

Numbers followed by different letters are significantly different at 95% with Duncan's means separation test.

furrows than is required for seed treatment, with the additional biocontrol agent being applied to soil rather than to the seed. Interestingly, strain T-22 is able to colonize roots even when applied to seeds treated with chemical fungicides that are highly toxic to it; apparently there is sufficient microsite space separation to avoid killing the majority of the applied biocontrol agent.

On some crops the planter box treatment has not been effective. Peas are sown when the ground is too cold for growth of T-22, so the pea roots can grow without remaining colonized. Furthermore, one of the causal organisms of pea root rot, *Aphanomyces eutiches*, infects plants via production of zoospores, and these structures are not affected by T-22 (Harman *et al.*, unpublished). Spinach and fall-sown small grains also develop much of their root system when it is too cold in the soil for the *Trichoderma* hyphae to grow well. In such a situation the inoculum must be distributed throughout the soil, not just on the seed. There are *Trichoderma* isolates that do have cold tolerance (see Chapter 6), so there is promise that a planter box treatment would be effective if a strain that possessed both cold tolerance and rhizosphere competence were developed.

Crop yields are frequently increased when roots are colonized by *T. harzianum.* Sweet corn responses to T-22 Planter Box were tested by us in replicated trials on commercial farms in New York (Figure 11.4). Sweet corn yields were greater only when yield-reducing stress conditions existed. When root function was limiting, the response could be dramatic. In these trials, low control yields were observed where growers were targeting specific markets, such as early harvest, that involve management that is unfavorable to root growth. Likewise, in a dry year, when root development is more limiting to growth, there was a greater benefit from T-22 use. In 1995 (a dry year), the response to T-22 was seen where control yields were below 12 t/ha, whereas in 1996 (a high rainfall year), the threshold was around 9 t/ha. Apparently, in 1995 the enhanced root development of the crop in the presence of T-22 resulted in greater drought tolerance.

The response of field corn was similar to that of sweet corn, with a positive *Trichoderma* effect when control yields were below 7 t/ha (Figure 11.4). With both

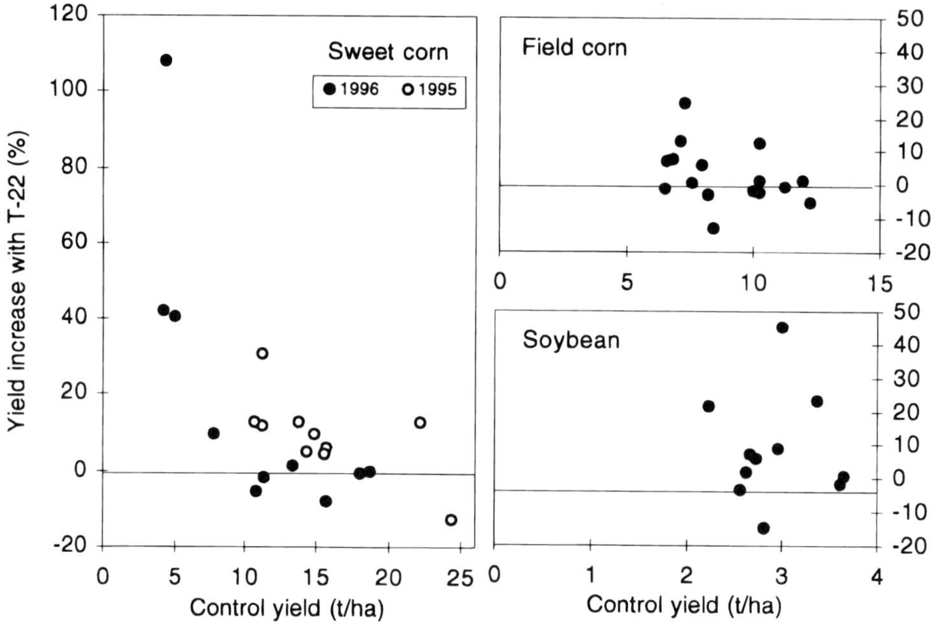

Figure 11.4 Yield increase with T-22 Planter Box™ as a function of control yield in commercial fields. The increase in yield of the plots treated with T-22 over control plots was greatest in lower-yielding sites for sweet and field corn, but such a relationship was not observed with soybeans. In 1995, T-22 was applied to sweet corn in the seed furrow rather than directly on the seed at all but one site. All data are replicated trials under commercial conditions at a variety of locations described in the text. The yields of field corn and soybeans were corrected for moisture content of 15.5% and 13%, respectively.

sweet corn and field corn, substantial increases in plant growth were sometimes noted. In one field corn trial, plants were dramatically larger with T-22.

Soybean yields also responded positively, but the variation was not explained by the simple approach of using control yield as an indicator of stress (Figure 11.4). Data were obtained from a variety of sources and are consistent with results from past years.

Silage corn, beans and potatoes also had higher yields with T-22 treatment. These crops were tested at fewer sites, so relationships like those above could not be determined. Silage corn yield increased in all four trials conducted (7%, 16%, 25%, and 27%), with two trials in New York by Seedway, Inc. and two in Oregon by the Wilbur-Ellis Co. Silage corn yields are a function of dry matter production, whereas yields for other crops tested also reflect partitioning into reproductive structures.

Dry beans were tested in four commercial trials in New York, in which the yield responses were −3%, +6%, 11%, and 21%. Snap beans were tested at one commercial site with a 24% increase. In research plots, root colonization was less consistent with snap beans than with corn or soybeans. Insufficient colonization occurred when the dose was below 10^5 cfu per seed, even though this is ample on corn. Root colonization also was poorer than anticipated when the moderately toxic fungicide captan was present and the *Trichoderma* dose was lower. Again, this combination worked well with corn but was inconsistent on beans. The rhizosphere competence is good, but it is not infallible.

Potatoes also responded to seed-piece treatment. In six trials performed by the Wilbur-Ellis Co. in the Pacific Northwest, T-22 increased the usable yield by −3%, 1%, 11%, 15%, 20%, and 24% over the appropriate control.

The mechanism of yield enhancement is likely to be complex. Control of clinical or sub-clinical root diseases is the usually accepted mechanism for larger roots in the presence of biocontrol organisms. However, there are several other physiological interactions that have been observed, some of which may contribute to increased root vigor and function even when no disease is evident on control roots. Lynch (1996) has demonstrated that *Trichoderma* strains are able to detoxify HCN in soil and that this mechanism may overcome limitations of root growth caused by production of this metabolite by rhizobacteria. Windham *et al.* (1986) showed that a diffusible factor from *Trichoderma* enhanced root growth, and they suggested that this fungus may produce an unidentified metabolite that functions as a plant hormone. We have observed another hormone-like activity: application of T-22 conidia to cuttings of vegetatively propagated *Lycopersicon esculentum X Solanum lycopersicoides* hybrids enhanced rooting about as effectively as commercial growth hormones (Björkman and Harman, unpublished), although *T. harzianum* does not induce callus on the cut surfaces. This same organism reduces insoluble Mn^{4+} to soluble Mn^{2+} in culture, and on roots it reduces Cu^{2+} to Cu^{1+} and Fe^{3+} to Fe^{2+} (Altomare *et al.*, 1996). This mechanism may provide a means of solubilizing insoluble minor nutrients in soil so that they do not limit plant growth. Finally, we also have evidence that this same strain reverses oxidative injury to seedlings (Björkman *et al.*, 1998).

Further, in several cases in 1996 the limiting factor in sweet and field corn yields in the absence of T-22 appeared to be nitrogen. In New York, 1996 was a particularly good season to see nitrogen limitations in corn because heavy rains in the early part of the year leached more soluble nitrogen from soil than expected and so nitrogen was frequently at lower levels than anticipated. This was apparent at several locations. The first one we noted was a trial on organic corn production (T-22 is sufficiently benign that it has been approved for organic growers in several states). In this location, corn was planted in a field where nitrogen was supplied by a previous clover crop and from an additional 30 kg/ha N in the form of chicken manure. At harvest, plants grown in the presence of T-22 were noticeably greener and larger than in its absence. In the presence of T-22, numbers of ears were increased from 1447 to 1995 doz. ears/ha and yields were increased from 4940 to 7184 kg/ha; both increases were significant. Another sweet corn field was located on a sandy loam soil and nitrogen leaching was probably quite large in 1996. Results similar to those obtained on the organic farm site were obtained with yields increasing from 2260 to 2570 doz. ears/ha and 7630 to 8530 kg/ha, which again was a significant yield increase. In a field corn trial, plants grown in the absence of T-22 were small and yellow, while in the presence of the beneficial fungus, plants were greener and more robust (Figure 11.5B). In this trial, yields were increased from 7.6 to 8.7 t/ha by the presence of T-22.

However, other explanations for the enhanced greenness of plants treated with *T. harzianum* are possible. Inbar *et al.* (1994) added *T. harzianum* strain T-203 to greenhouse potting mixes and observed increases in leaf area, plant dry weight, general plant vigor, and increased chlorophyll content in cucumber and pepper seedlings

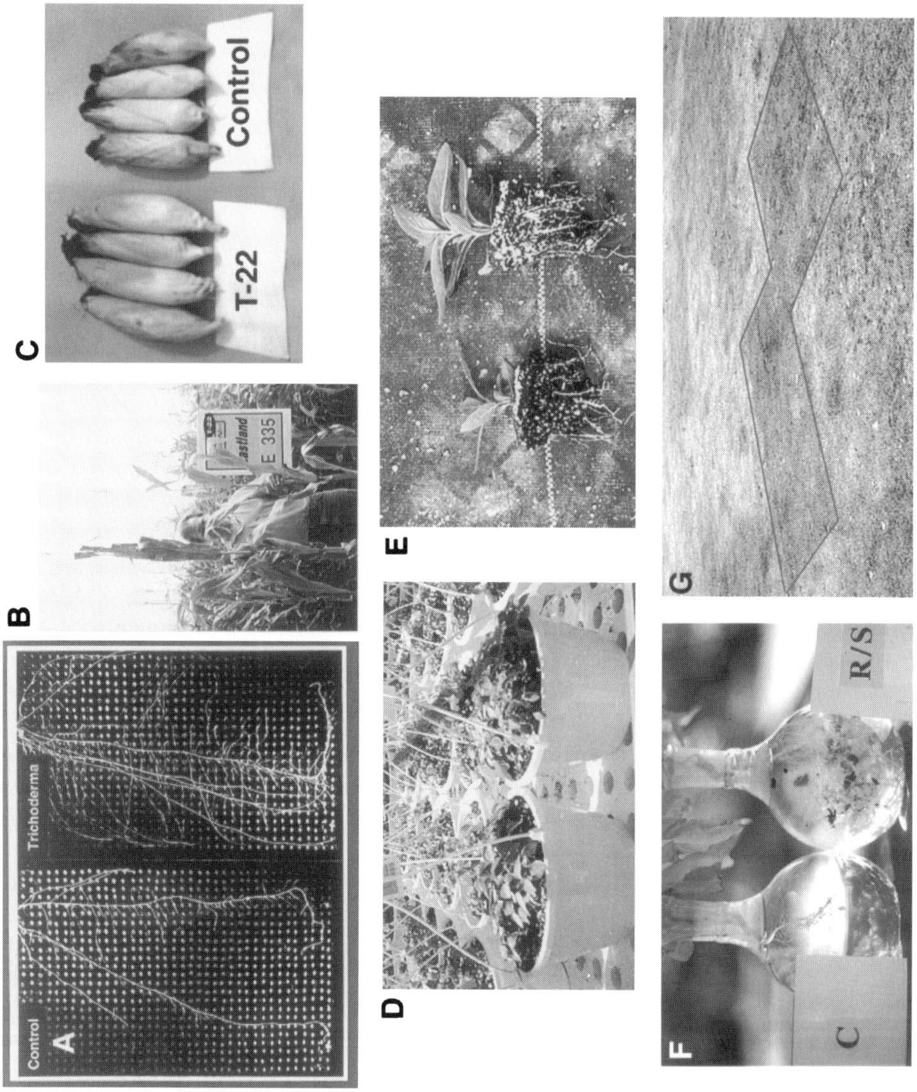
A
Control
Trichoderma
B
E 335
C
T-22
Control
D
E
F
C
R/S
G

relative to plants grown in the absence of the biocontrol fungus. These changes occurred even though plants received optimal fertilizer levels in a commercial greenhouse setting and there were no differences in N, P, or K levels in the plants. In other studies, however, an increase of P and K content in *Trichoderma*-treated tomato seedlings has been noted (Kleifeld and Chet, unpublished, cited in Inbar *et al.*, 1994). The increases in plant vigor and chlorophyll content were associated with a decrease in damping-off caused probably by *Pythium* spp. These data suggest that factors, at present unknown but other than nutrition, contribute to the abilities of *Trichoderma* spp. to increase plant growth and chlorophyll levels.

Therefore, there are a number of possible mechanisms whereby root-colonizing *Trichoderma* spp. may increase yields. Any or all of these may be operating at the same time.

Average yields of all crops were higher with T-22 treatment. However, yield increases were not obtained in all field locations. In general, yield increases were seen where there was a yield-limiting factor in the field that could be eliminated by enhanced root growth. Thus, T-22 improved yields only where conditions limited root function and hindered delivery of nutrients to the above-ground parts of the plants.

Concepts of use of seed treatments

Seed treatments with *Trichoderma* and *Gliocladium* spp. can therefore be used either to protect seeds, to colonize and protect roots, or both. Requirements for effective seed protectants are more difficult to meet than for root protectants.

Chemical fungicides are also used for seed protection. They are relatively inexpensive and readily available and they cause little environmental damage since they are used in small amounts. However, they are effective only for a short time (at most, less than a month) and they do not spread through the soil with the root system. Therefore, they do not provide effective root protection.

Figure 11.5 Examples of enhancement of root or plant growth by commercial products containing *T. harzianum* strain 1295-22 (T-22). Photograph A shows roots of sweet corn produced from nontreated and *Trichoderma*-treated seeds in a mini-rhizotron in greenhouse studies in a sandy loam field soil. Roots from treated seeds contain more small fibrous roots even in the absence of any clinical root disease. The roots shown are a portion of a study that measured root development of a large number of plants, and the roots shown represent mean root development in the presence or absence of T-22. B shows representative plants from seeds treated with T-22 Planter Box™ (left) or without this treatment in a commercial field corn trial near Port Byron, New York. Plants grown in the presence of T-22 were larger and darker green than plants grown without the treatment. The control plants, but not the T-22-treated plants, show symptoms of nitrogen deficiency. C shows ears from the plants from the plots shown in B. D shows impatiens plants in a commercial greenhouse grown in the presence of RootShield™ granules (foreground) and in the absence of this treatment, and E shows *Vinca* plants with similar treatments (nontreated plants are on the left and treated plants on the right), also from a commercial greenhouse. F shows lettuce roots from plants grown in a commercial potting mix treated with RootShield™ drench (right) and in its absence (left). G shows a research trial on creeping bentgrass in November 1994. The areas outlined were treated with Bio-Trek™ 22G granules once in June and again in July 1994. The darker color occurs because of enhanced greenness in the areas treated with the granules.

Biological agents or chemical fungicides applied to seeds must remain active in storage under nearly all conditions where seed germination remains adequate. Unfortunately, viability of *Trichoderma* and *Gliocladium* on treated seeds for the duration of storage (sometimes several years) is unlikely. These shelf life problems frequently are exacerbated by "inert" adhesives and other components of seed treatments. Materials such as polyvinyl alcohols may damage membranes of spores embedded within them. Such materials frequently are used in the preparation of seed treatments, especially in the newer film coating processes.

Therefore, for many applications, especially large-volume, low-cost seeds such as corn, wheat, beans, cotton, and other agronomic crops, pure biological seed treatments for the purpose of seed protection probably are not competitive with chemical pesticides. Chemical pesticides will probably be less expensive and easier to use than effective biological seed protectants.

However, chemical seed treatments cannot provide the long-term protection to roots that biological seed treatments can. Therefore, it seems most reasonable to us to use chemical seed treatments in combination with compatible biological seed treatments, at least for large-volume agronomic crops. *Trichoderma* and *Gliocladium* spp. are excellent candidates for this integrated concept of seed treatment. This combination uses the best aspects of both biological and chemical seed treatments to provide an effective system.

This integrated approach, therefore, provides the following advantages to growers, all at a very modest cost (US \$3–6/ha for the biological and with similar costs for the chemical component that usually is part of the cost of seed). These advantages are as follows:

1 a high level of seed protection primarily by the chemical component;
2. a high level of root colonization by the biological component that provides an increase in root health and function not previously possible;
3 together, these treatments provide an insurance for growers against economically devastating low yields;
4 an excellent overall return on investment of at least 5- to 50-fold based on the average yield increase on various crops.

In addition, pure biological seed treatment systems may be useful for some applications. High-value vegetable and flower seeds do not have the same economic constraints as agronomic seeds, and so treatments that combine the advantages, for example, of seed priming and biological treatments may be highly attractive. They not only will obtain excellent seed protection and root protection, but seed quality and vigor also can be enhanced. Further, there is a growing market in organic crop production, and *Trichoderma* and *Gliocladium* spp. can usually qualify as components of such methods. In this application, pure biological treatments are likely to be both valuable and attractive.

11.3.2 Greenhouse / plant nursery planting mix amendments

Greenhouse crops receive heavy inputs of chemical pesticides for control of root diseases, and there have been substantial efforts to find biological replacements for these (Hoitink *et al*., 1991; Lumsden and Locke, 1989). Greenhouse soil mixes usually are synthetic and contain less microflora that compete with biocontrol agents than natural field soil. Also, extremes of temperature and moisture stress usually are avoided in the greenhouse so, consequently, biocontrol in the greenhouse may be less complex and variable than in the field (Harman and Lumsden, 1989). Several distinctly different approaches to biocontrol for greenhouse and nursery crops are now in commercial production.

Composts and fortified composts

Many organic materials may be composted, i.e., allowed to degrade by the action of microorganisms, whereby the properties and the microbial communities of such materials are substantially altered. Composting essentially yields a complex ecosystem in a matrix of partially decomposed organic material. The physical properties of properly composted materials make them highly useful as planting mix components for plant culture in greenhouses and other locations or as a soil dressing in ornamental or other planting situations. Depending on the composting process and the microbial community that becomes established, such composts may be highly conducive or highly suppressive to such root and seed rotting pathogens as *Pythium*, *Phytophthora*, *Rhizoctonia*, and *Fusarium* spp. Composts prepared from similar materials, which should give comparable results, may be either conducive or suppressive to disease. To add to the complexity, some composts may be conducive to one pathogen and suppressive to another. Examples of materials used to prepare composts include tree barks, sewage sludge, plant wastes, animal manures, and other organic wastes (Hadar *et al.*, 1992; Hoitink *et al.*, 1991).

Over the past two decades, much has been learned about the nature of suppressiveness and conduciveness of various composts, including the complex interactions of such factors as microbe populations, temperature, moisture, and the nature of the organic matter. Processes to control the fermentation must be employed if consistently suppressive materials are to be obtained (Hoitink *et al.*, 1997).

Composting consists of three distinct phases. The first phase is the first 24–48 h when temperatures begin to rise due to microbial degradation of sugars and other similar materials. During the second phase, temperatures rise to 40–65°C; at this time cellulose and other components break down while more resistant materials may be partially degraded. This temperature is sufficient to kill plant pathogens, weed seeds, and most beneficial microbes, with the exception of *Bacillus* spp. (Hoitink *et al.*, 1991). If composts are to have highly useful properties, they must be turned, aerated, or otherwise handled in order to expose all parts to the high temperatures and to produce a homogeneous product. Once temperatures decline as readily biodegradable substances are utilized, then the third, or curing, phase begins. The composted material is at that time recolonized by mesophilic organisms from the outer layers of the composts, and humic substances increase in quantity. Mature composts have a dark color and consist largely of lignins, humic substances, and microbial biomass. Among the beneficial biocontrol organisms that recolonize the compost are bacteria in the genera *Bacillus*, *Enterobacter*, *Flavobacterium*, *Pseudomonas*, and *Streptomyces* and fungi in the genus *Trichoderma* (Hoitink *et al.*, 1991).

At least two separate mechanisms of biocontrol exist in suppressive composts. The first is general suppressiveness that operates primarily against *Pythium* and *Phytophthora*, while the second is more specific and is required for pathogens such as *Rhizoctonia solani.* Propagules of *Pythium* and *Phytophthora* require nutrients to germinate (see section 11.1.2). If composts lack readily-available nutrients, these pathogens may be suppressed. This process has been designated as general suppressiveness (Cook and Baker, 1993). The high level of microbial activity that exists in composts is sufficient to limit or prevent germination of propagules of these fungi, and a biochemical test based on the rate of microbial hydrolysis of fluorescein diacetate has been developed to predict the general suppressiveness of composts (Hoitink *et al.*, 1991).

Consequently, most *properly prepared* (see succeeding paragraphs) composts are suppressive to *Pythium* and *Phytophthora.* Even some natural products, such as light-colored (nondecomposed) sphagnum peat, may be suppressive, at least for a short time due to high general microbial activity levels (Hoitink *et al.*, 1991). Such nondegraded organic materials slowly release nutrients, maintain high levels of microbial activity, and hence possess general suppressiveness. Hoitink *et al.* (1991) suggest that similar phenomena, e.g., slow release of organic nutrients to enhance general microbial activity, constitute the essence of biological control associated with organic farming, where relatively high levels of partially degraded organic materials are maintained in soils.

This mechanism, which is based on growth and effectiveness of general microfloral types, is dependent upon the quality of the organic matter in the compost. Boehm *et al.* (1997) characterized the organic matter in peats relative to their suppressive ability and the characteristics of the microbial community inhabiting these substrates, using ^{13}C nuclear magnetic resonance. The energy available in the organic fraction, primarily complex carbohydrates, was critically important. In less-decomposed peats, readily degradable carbohydrates such as cellulose were more abundant than in more-decomposed materials, especially in the fine particulate fractions. This abundance was associated with gram-negative bacteria such as pseudomonads, of which a high percentage (10%) had biocontrol ability. As carbohydrate concentrations decreased, gram-positive bacteria and oligotrophs became prevalent, and only a few of these ($<1\%$) had biocontrol ability. As a consequence, these peats lost their disease-suppressive abilities (Boehm *et al.*, 1997).

The production of generally suppressive composts, especially those based on pine bark, has been developed for large-scale commercial use. The bark is first milled and screened to provide materials of appropriate sizes for aeration during composting. A controlled amount of water and, in some cases, a small amount of nitrogen in the form of ammonia is added. The material is stored in wind-rows on a concrete pad and turned several times during an 8 to 11 week period. Temperature is monitored, and wind-rows are sized relative to the expected ambient temperature to provide the proper level of heating during the composting process. Electrical conductivity, pH, and moisture are also monitored and adjustments are made to maintain optimal microbial activity.

Control of *Rhizoctonia solani*, however, is quite different. *R. solani* produces large resting sclerotia, and only a limited group of organisms is capable of suppressing or controlling this pathogen. Two papers (Kuter *et al.*, 1983; Nelson *et al.*, 1983) examined the time course of colonization of composts by various fungi, related this to the occurrence of suppressiveness to *R. solani*, and identified specific strains with good biocontrol ability. The most effective in all regards were strains of *Trichoderma*, especially *T. hamatum* and *T. harzianum.* Selected strains of these species were capable of inducing suppressiveness when added to composts rendered conducive to disease by pasteurization. Substantial efforts since this time have identified several effective strains, especially *T. hamatum* 382. This strain consistently provides good levels of suppression (Hoitink *et al.*, 1997). However, single organisms are considered by Hoitink's group not to provide adequate levels of suppression of all common diseases. Kwok *et al.* (1987) examined bacteria that were particularly prevalent and effective in suppressive composts. A number of effective bacteria were identified including species of *Enterobacter*, *Flavobacterium*, and *Pseudomonas*. Some were most effective in combination with *T. hamatum* 382 than either the fungal or the bacterial antagonist alone. A mixture of *T. hamatum* 382 and *F. balustinum* 299r was particularly effective.

A goal has been to prepare composts that, when combined with other organisms, can be admixed into commercial potting mixes to provide control of a wide range of plant pathogenic fungi. To this end, fortified composts are now being prepared. After the primary heating period of composting is complete, *T. hamatum* 382 and *F. balustinum* 299r are added to composts. These two organisms increase to high levels in the composts and create reliably-suppressive materials for addition to greenhouse planting media. The composts thus produced are fortified composts. They are as effective as, or in many cases more effective than, chemical fungicides for control of diseases caused by *R. solani* and by species of *Pythium*, *Phytophthora*, and *Fusarium*. In the USA, materials that make claims of disease suppressiveness must be registered with the US Environmental Protection Agency, and fortified composts amended with *T. hamatum* and *F. balustinum* currently are being registered.

It is important to consider mechanisms of action by which suppressive composts operate. As mentioned earlier, general suppressiveness can be induced by a range and mixture of organisms and is effective against *Pythium* and *Phytophthora* spp. Essentially, general suppressiveness relies on competition for nutrients. Propagules of *Pythium* and *Phytophthora* are stimulated to germinate by the presence of simple sugars and other factors such as unsaturated fatty acids (Harman and Nelson, 1994). The presence of large numbers of competitive microflora reduce the levels of such compounds to levels too low for propagules to germinate.

However, suppression of *R. solani* is more specific and appears to be related primarily to mycoparasitism. The propagules of the pathogen are colonized by *Trichoderma* spp. and other organisms, and so only effective mycoparasites can control this pathogen (Hoitink *et al.*, 1997).

As indicated in Chapter 7, mycoparasitism is a highly complex process dependent upon enzymes and other inducible factors, including antibiotics (Schirmböck *et al.*, 1994). If quantities of simple sugars are present or are readily available to microbes in the form of materials such as cellulose, production of suppressive composts is unlikely. Production of enzymes and antibiotics involved in mycoparasitism is prevented by simple sugars through catabolite repression (Tronsmo and Harman, 1992). Consequently, organisms such as *T. hamatum* and *F. balustinum* increase to high levels in fresh agricultural materials such as tree barks but are ineffective in biocontrol since the appropriate enzymes are not expressed (Hoitink *et al.*, 1997) and, probably, the entire mycoparasitic process is blocked. This factor explains the necessity of proper composting processes to obtain suppressive products and is another example of the need to understand the delivery system to provide an effective product.

Recently, suppressive composts have been shown to possess yet another mechanism of biocontrol. Plants grown in suppressive composts acquire resistance to a range of plant pathogens (Zhang *et al.*, 1996). This resistance is similar to the systemic acquired resistance demonstrated by root-associated microflora, especially bacteria, in other systems. Such SAR responses are highly useful and probably account for the ability of suppressive composts to control Fusarium wilts. However, since this plant response is systemic, other diseases such as those causing foliar symptoms also are controlled. Further, diseases caused by bacteria and viruses may be controlled along with fungal diseases (Zhang *et al.*, 1996).

In summary, composts, and especially fortified composts, provide highly effective control of a range of plant pathogens. *Trichoderma* spp., especially *T. hamatum* and *T. harzianum*, have a primary role in disease suppression, especially in combination with bacteria such as *Flavobacterium* spp. Composts must be properly prepared with the simple sugars and celluloses largely degraded through the composting processes. A number of mechanisms are responsible for the efficacy of composts including competition, mycoparasitism, and systemic acquired resistance.

Suppressive composts are in widespread use in greenhouses and may also be useful in other applications such as top dressings for golf courses and for control of tree diseases. Such materials are likely to be more effective and reliable when fortified composts are readily available (Hoitink *et al.*, 1997).

Suppressive planting mixes based on antibiosis

Researchers at the USDA (United States Department of Agriculture) Biocontrol of Plant Diseases Laboratory have also developed a biocontrol system for control of soil-borne diseases of greenhouse and nursery crops. This system is quite different from that employed for the production of composts.

Researchers sought strains of *Trichoderma* and *Gliocladium* that provided high levels of control of the prevalent greenhouse pathogens *P. ultimum* and *R. solani*. They initially screened many strains of fungi and bacteria and found that only strains of *T.* (formerly *Gliocladium*) *virens* significantly improved seedling survival in the presence of either pathogen (Lumsden and Locke, 1989). Therefore, they screened twenty strains of *T. virens* and found that strains differed remarkably in their ability to control disease. As a consequence of this screening effort, they obtained a single effective strain. This strain produces gliotoxin along with low levels of other materials, including dimethylgliotoxin, viridiol, and a mixture of phenolic substances (Lumsden *et al.*, 1992b). The production of this strain for commercial use has been investigated using several different methods such as alginate pellets (Lewis and Papavizas, 1985). The W. R. Grace Company has been successful in developing methods for the large-scale manufacture of this strain of *T. virens*. This effort has led to the manufacture and sale of SoilGard™, which is now sold by the Thermo-Trilogy Company.

These granules, when added to soil, support the growth of *T. virens*. The production of gliotoxin occurs quickly and reaches maximal levels within a few hours after addition (Figure 11.6). The gliotoxin persists for several days in planting mixes, and during this time, high levels of suppression are obtained. Gliotoxin is the primary factor responsible for disease control as was demonstrated when (a) disease control was reduced when the fungus was grown in a medium containing bark ash (charcoal) which absorbed the antibiotic, (b) aqueous extracts of planting mix containing *T. virens* controlled disease when added to fresh medium infested with

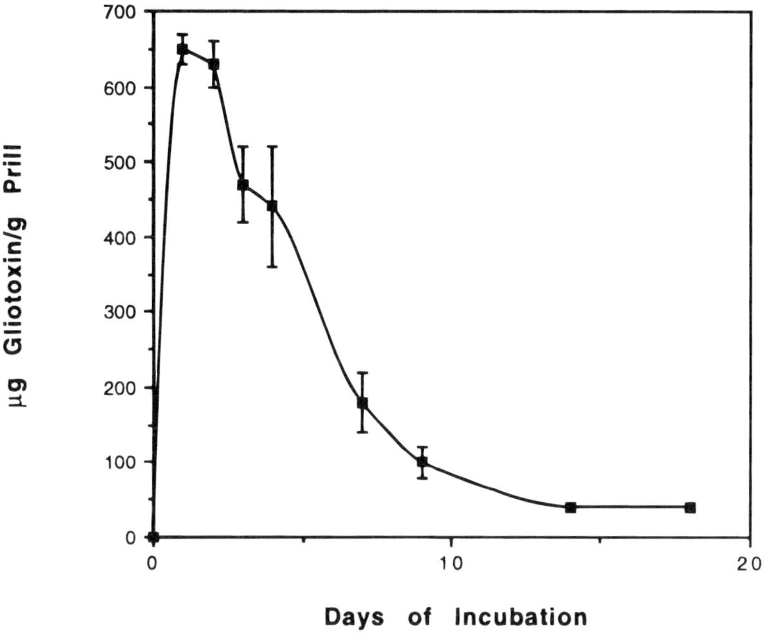

Figure 11.6 Effect of time on the quantity of gliotoxin extracted from a peat-moss–vermiculite planting medium containing *T. virens* applied as an alginate prill. Used with permission from the authors and *Phytopathology* from Lumsden *et al.*, 1992a. Error bars indicate the standard error of each measurement.

pathogens, and (c) disease suppressiveness closely paralleled the level of gliotoxin produced in planting mixes (Lumsden *et al.*, 1992a).

Gliotoxin produced by the biocontrol agent results in high levels of disease control as long as sufficient quantities are present. Fungi that are controlled include *R. solani*, *Pythium* spp., *Phytophthora* spp., *Fusarium* spp., *Sclerotinia* spp., *Sclerotium rolfsii*, and *Thelaviopsis* spp. Control of *Thelaviopsis basicola* is enhanced if sulfur is included with the biocontrol agent. The sulfur may enhance activity by (a) acidification of the soil, as per Weindling and Fawcett (1936), or (b) by enhancing the level of the sulfur-containing antibiotic gliotoxin (Jim Walters, Thermo-Trilogy Corporation, personal communication).

To summarize, *T. virens*, when formulated as SoilGard, provides effective control of a range of plant diseases mainly by production of gliotoxin. Consequently, the length of time that diseases are controlled is primarily determined by the period of time in which gliotoxin remains active, usually a period of one to two weeks (Wilhite and Straney, 1996). Thus, it is dissimilar from composts in at least two ways: it relies on a single biocontrol organism rather than on a range of them, and it relies on antibiosis by a single compound for its activity rather than on the wide range of mechanisms that occur in composts.

Rhizosphere competent T. harzianum

Earlier in this chapter we described the mechanisms and benefits of *T. harzianum* strain 1295-22 (T-22) and its use as a planter box seed treatment. This strain is also useful when formulated and used as a greenhouse soil amendment. The concepts, practices, and benefits of products based on this strain are quite different from either composts or SoilGard™.

There are two commercial products based on this strain, called RootShield™ granules and RootShield drench. The granular product contains the entire thallus of the fungus colonized on clay particles. This material, like SoilGard or composts, is formulated to be mixed into greenhouse potting mixes. The drench, on the other hand, consists primarily of conidia plus inert ingredients and is designed to be suspended in water and used as a drench, i.e., to be added in sufficient water to carry the conidia through the soil volume. Greenhouse operators frequently prefer drenches to granular products since many of them have no soil mixing facilities. In either case, the expected result is that the *T. harzianum* strain will colonize the entire root system.

One limitation of non-rhizosphere competent *Trichoderma* and *Gliocladium* spp. added to soil is that a nutritive substrate is needed to permit growth and, in the case of SoilGard, to produce gliotoxin. Consequently, conidia provided as drenches are unlikely to be effective since they probably will not germinate and proliferate in the absence of a nutrient source (Lewis and Papavizas, 1984). The exceptions to this are strains that provide benefits by growth and development on roots, as opposed to proliferation in the potting soil matrix. Strain 1295-22 of *T. harzianum* described earlier is one such strain. It protects roots of a wide range of ornamentals and vegetables, and so such roots are healthier and more vigorous when grown in the presence of RootShield drench (Figure 11.5A, E, F). For example, lettuce and poinsettia roots grown in the presence of this strain of *T. harzianum* were larger, as was shown with the corn described earlier; in lettuce, the total root length in the presence of this strain was several-fold greater than in its absence (Figure 11.5F). Plants produced in the presence of RootShield are larger and more vigorous, an effect that can frequently be seen even in the absence of clinical root disease (Figure 11.5B, D, E). As a consequence, some greenhouse managers have been able to reduce the total time required for production of crops and to reduce the quantity of fertilizer used because the roots are more vigorous.

However, application of this strain may not result in highly disease-suppressive soils since the bioprotectant proliferates on roots and not in the soil. As a consequence, in cases where severe disease pressure from soil-borne pathogens already exists, such as high levels of damping-off caused by *Pythium*, a compatible fungicide may need to be used. This strain, and products based on it, are not designed to provide high levels of protection against diseases that attack, for example, the hypocotyls of young plants.

Because the strain of *T. harzianum* in RootShield provides excellent root colonization, root disease control usually extends for the life of the plant. For example, transplant tomatoes produced in the presence of RootShield have been shown to have less Fusarium crown and root rots in the production field than those without the biocontrol agent (Datnoff *et al.*, 1995). In commercial nurseries, it has controlled root diseases of azalea and chrysanthemum. It controls root diseases caused by *Fusarium, Rhizoctonia*, and *Pythium* species. It does not control *Phytophthora.*

To summarize, RootShield provides root protection. It also provides more vigorous plants so that greenhouse bench time and fertilizer use in commercial greenhouses may be reduced. Usually, in commercial greenhouse practice, fungicides are not required.

Summary of greenhouse uses of Trichoderma *spp.*

The three examples in this section indicate that widely different approaches can be used for biological control of plant diseases in the greenhouse. The three products described differ in their concepts of use, in their modes of action, and in their expected benefits. These differences are summarized in Table 11.3.

11.3.3 *Control of diseases of fruit, foliage, and flowers*

The sections to this point have considered how *Trichoderma* spp. can be used to control subterranean plant diseases and enhance root function. The remainder of this chapter will examine the control of above-ground plant diseases using these fungi and integrated systems and demonstrate how similar principles apply as in the systems already described. The same components, i.e., superior biocontrol agents, good production methods and formulations, and effective delivery systems, are critical to success. However, as we shall see, the specific approaches to control of above-ground diseases differ from control of root diseases.

Two different general applications of *Trichoderma*-based biocontrol have been extensively studied. These are control of diseases of turf, especially in golf courses (Harman and Lo, 1996; Lo *et al.*, 1996, 1997b), and control of the gray mold fungus *Botrytis cinerea* on a variety of crops (Elad *et al.*, 1993, 1996; Gullino, 1992; Harman *et al.*, 1996; Tronsmo and Dennis, 1977; Tronsmo and Ystaas, 1980). These applications will be examined as examples of the uses of *Trichoderma* and *Gliocladium* that are now becoming available.

These two disease situations need biological alternatives to chemical pesticides. Both of them use large quantities of fungicides: over $100 million are spent annually for control of *B. cinerea* on grape alone, and over 2 million pounds of fungicides are applied to turf each year in the USA alone (Ayers and Gilmore, 1991). Thus non-target impacts of fungicides for these uses can be great, not only on the environment but also on consumers or users of these commodities. For control of *B. cinerea*, only two related dicarboximide fungicides, vinclozolin and iprodione, are primarily used in the USA (captan is used in some locations for some purposes). The pathogen has developed resistance to dicarboximides in many locations (Sanders, 1989). Further, there are increasing regulatory concerns due to their adverse (feminizing) effects on mammals (Anonymous, 1994). In many countries in Europe, there are no effective registered fungicides for the control of *B. cinerea*, since both the dicarboximides and captan have been banned. There are many chemicals available for control of turf diseases, but resistance to some of them has already developed in this intense application (Sanders, 1989). Therefore, alternatives to chemical pesticides for these applications are urgently needed and *Trichoderma* spp. will probably begin to fill these needs in the next few years.

Control of turf diseases

We have been very interested in the possibility of using *T. harzianum* strain 1295-22 for control of turf diseases. Turf pathogens may cause either root or foliar diseases, and some of the pathogens are soil-borne but become foliar diseases. This development may be particularly likely in the short, dense, highly managed grass of golf

Table 11.3 Different *Trichoderma*-based products for use in greenhouses, their benefits and concepts of use

Product	Active organism(s)	Concept of use	Mechanisms of action	Formulations/ uses	Benefits	Pathogens controlled
Fortified and nonfortified composts	Various, including *T. hamatum* and *Flavobacterium balustinum*	Production of highly suppressive soils through activity of many microbes	Competition, mycoparasitism, antibiosis (?), and systemic induced resistance	Composted materials to be mixed into soilless potting mixes	Suppression of a wide range of pathogens in potting mixes	*Phytophthora*, *Fusarium*, *Pythium*, and *Rhizoctonia*
SoilGard	*Trichoderma virens* G20	Production of disease suppressive soils through production of gliotoxin	Antibiosis	Granules mixed into soilless potting mixes	Short-term suppression of a wide range of pathogens	*Phytophthora*, *Fusarium*, *Pythium*, *Thelaviopsis* (with sulfur), *Sclerotinia*, *Sclerotium* and *Rhizoctonia*
RootShield	*Trichoderma harzianum* 1295-22	Protection of roots through root colonization, plant growth enhancement	Mycoparasitism and direct plant growth enhancement	Granules mixed into soilless potting mixes; drenches to incorporate conidia into potting mixes at the time of planting	Enhancement of plant growth, including enhanced root growth, suppression of pathogens	*Fusarium*, *Pythium*, and *Rhizoctonia*

greens. In this location, a premium on cosmetic appearance exists and the highly stressed grass is particularly susceptible to a variety of diseases.

Our first approach to this problem was to utilize the strong rhizosphere competence of strain 1295-22 for control of soil-borne diseases. The organism was formulated into a granule that could be broadcast over the surface of the golf course using standard fertilizer spreaders. This product is now manufactured by BioWorks, Inc. and sold by the Wilbur-Ellis Co. as Bio-Trek™ 22G. Application of this granular material resulted in good colonization of the turf root zone; levels of 10^4 to 10^6 colony forming units (cfu) are usually obtained, and, once established, populations remain stable even over the winter and into the next year (Lo *et al.*, 1996, 1997). In some cases, as with other crops, root colonization results in increased plant vigor, as expressed in greener areas where Bio-Trek was applied (Figure 11.5G). Interestingly, the most numerous fungal colonist of turf roots prior to addition of strain 1295-22 of *T. harzianum* is *T. virens*. This natural population of *T. virens* appears largely ineffective in control of turf diseases, and as *T. harzianum* increases, *T. virens* decreases, perhaps because they colonize similar niches.

Colonization of the turf root-soil zone results in a reduction of disease (Harman and Lo, 1996; Lo *et al.*, 1996, 1997). Early in the season, symptoms of dollar spot caused by *Sclerotinia homoeocarpa* were lower in the presence of Bio-Trek, probably because of reduction of soil-borne inoculum. However, once established, the disease progress curve was identical in the presence and absence of the biocontrol preparation. Similarly, application of this formulation also resulted in a reduction of brown patch caused by *Rhizoctonia solani* and of Pythium root rot caused by *P. graminicola*. In addition, root colonization frequently resulted in an increase in turf quality as indicated by enhanced greenness where the biocontrol agent was applied.

However, application of Bio-Trek 22G is only a partial solution to diseases caused by turf pathogens. It slows the appearance of dollar spot but does not provide long-term control. Further, the level of control of brown patch is less than that provided by chemical fungicides. The surviving fungi become foliar pathogens and spread rapidly by means of foot traffic and mowing. *T. harzianum* applied and established in soil cannot control foliar phases of diseases, which are the most destructive and unsightly phases of turf diseases. Consequently, Bio-Trek must be followed by standard chemical fungicides. Fortunately, most of them are at least partially compatible with *T. harzianum*, which possesses a high level of innate fungicide resistance. Even fungicides that are highly toxic to *T. harzianum*, such as propiconazole, have no detectable effect upon established soil populations of the biocontrol agent. Apparently, soil adhesion to the fungicide, microbial degradation, or other factors prevent toxic fungicides from becoming present in concentrations that are toxic to the biocontrol agent in the soil. Therefore, once the biocontrol agent is established in turf, standard chemical control practices can be used to control foliar diseases.

Therefore, we began to develop formulations for control of foliar phases of turf diseases. We prepared a wettable powder formulation of the biocontrol agent and applied it to turf challenged with any of several pathogens in the greenhouse. Only modest levels of control were obtained. This result characterizes usual first attempts at biocontrol in a new system, and we investigated the use of a variety of substances to enhance biocontrol activity. Certain surfactants such as Triton X-100 proved effective, while other surfactants or other substances such as nutritive adhesives were ineffective (Lo *et al.*, 1997). Trials in 1994 indicated that the level of control of *S. homoeocarpa*, *R. solani*, and *P. graminicola* was as effective with strain 1295-22 plus Triton X-100 (0.1% v/v) as with highly effective fungicides (Lo *et al.*, 1997).

Trials in 1995 attempted to duplicate the results obtained in 1994 but were uniformly unsuccessful. Therefore, trials in 1996 focused on a range of surfactants and it was found that R-11 (an alkyl aryl ethoxylate, compounded silicone and linear alcohol) at 0.2% was highly effective. The other critical component was the level of *T. harzianum* applied to the turf. In 1995, tests used about 7×10^{10} cfu per 100 m^2, while about 1.1×10^{11} were used in 1996. Further, we considered that it might be more effective as a component of an integrated biological–chemical regime (Ondik *et al.*, 1997). We determined that the fungus is resistant to the fungicides iprodione (Harman and Lo, 1996) and also to triadimefon. We therefore tested a tank mix of this strain with 0.5× rates of the fungicides. With iprodione, this mixture was applied every second week, while with the triadimefon combination, the tank mix was applied, then 2 weeks later *T. harzianum* plus the surfactant was applied, and then the tank mix again was applied, and so forth. In trials in New York, dollar spot was well controlled by weekly applications of the biological plus surfactant treatment alone and by the integrated biological–chemical combinations (Table 11.4). However, in trials at the University of Kentucky, the biological–surfactant combination was less effective than in New York, although the iprodione–*T. harzianum* combination worked well (Table 11.4).

These data indicate that spray applications of strain 1295-22 can be used successfully in the demanding turf protection arena. Like nearly all pest control products, it will have limitations. It does not control all diseases; for example, anthracnose (*Colletotrichum graminicola*) is not adequately controlled. Further, the ability of the organism to function effectively in very hot locations (the surface of golf greens in the southern USA can exceed 45°C regularly and this may explain the difference between results in Kentucky and New York) is yet to be determined. However, as a component of an integrated system of biological–chemical control it should have an important place.

The mechanism of action of this strain of 1295-22 has been examined (Lo *et al.*, 1998) using strains of the fungus transformed to be resistant to the toxicant hygromycin B or to express the enzyme β-glucuronidase. This enzyme can be detected in living organisms by addition of an appropriate reagent that provides a blue pigment in transformed strains. In this way, it is possible to accurately determine the presence of the specific organism under microscopic observation and to examine its growth and effects upon other organisms. Spores of the transformed strains were sprayed onto creeping bentgrass foliage being attacked by *R. solani.* Spores of the fungus were found on both foliage and roots following application, and they germinated to form hyphae in both locations. When hyphae of strain 1295-22 were in close proximity (1–3 hyphal diameters) to those of *R. solani*, vacuolation and necrosis of the pathogen's hyphae were noted. Shortly thereafter, structures typical of mycoparasitism were seen. These data suggest that the biocontrol fungus produces metabolites that damage the pathogen when both are in close proximity; enzymes and inducible antibiotics are likely candidates for such an effect. Both may be induced in *T. harzianum* by the presence of fungal cell wall components (Schirmböck *et al.*, 1994). Thus, actively growing hyphae are required for biocontrol activity, and mycoparasitism-like activities probably are responsible for activity of this strain.

Table 11.4 Control of dollar spot caused by *Sclerotinia homoeocarpa* on creeping bentgrass with *T. harzianum* strain 1295–22 (T-22) applied as a foliar spray or as an integrated biological–chemical application in New York and Kentucky

Treatment	*Severity (% of plot area with disease)*
Trial 1: Dollar spot severity in New York	
R-11 + T-22	0.7 A[a]
0.5 × iprodione + T-22	1.0 A
0.5 × triadimefon + T-22	0.9 A
Fungicide check (full rate triadimefon)	2.0 A
Untreated	9.6 B
Trial 2: Dollar spot severity in New York	
R-11 + T-22	5.2 A[a]
0.5 × iprodione + T-22	5.2 A
0.5 × triadimefon + T-22	4.0 A
Fungicide check (full rate triadimefon)	3.0 A
Untreated	13 B
	Incidence
Dollar spot incidence in Kentucky	*(numbers of dollar spot lesions)*
R-11 + T-22	27 B[b]
0.5 × iprodione + T-22	4.3 A
Fungicide check (iprodione)	0 A
Untreated	68 C

Treatments included T-22 applied weekly as a spray with 0.2% of the surfactant R-11 (Wilbur-Ellis Co., Fresno, CA, USA); T-22 applied every 2 weeks in a tank mix with iprodione (Chipco 26019 Flowable) at 0.5 × the manufacturer's rate; and T-22 applied in a tank mix with triadimefon (Bayleton 25W) at 0.5 × the manufacturer's rate, followed two weeks later with T-22 plus R-11, and then with the tank mix of T-22 and Bayleton, and so forth. Disease levels shown are at the middle of the disease-prone time for grass (September in New York and early July in Kentucky). Data from Kentucky are from a trial performed by Vincelli and Doney (1997).

[a] Numbers followed by the same letter are not significantly different at $P = 0.05$ according to Fisher's Protected LSD test.

[b] Numbers followed by the same letter are not significantly different at $P = 0.05$ according to the Waller-Duncan k-ratio test.

Control of diseases caused by Botrytis cinerea

As mentioned at the beginning of this section, control of *B. cinerea* has been the focus of a number of *Trichoderma* and *Gliocladium* researchers. This section will concentrate on potential or existing commercially useful biocontrol products. Product concepts for various uses are different, so we will separately discuss greenhouse applications, control of *B. cinerea* on strawberry, and finally, control of *B. cinerea* on grape.

Control of B. cinerea *on greenhouse crops* *B. cinerea* is an opportunistic organism that can attack above-ground parts of plants by macerating the tissue and producing many spores. Frequently, the infected tissue droops and touches adjacent healthy tissues, thereby increasing the infection rate. High moisture levels facilitate the spread of the disease. Leaves, flowers, and stems may all be attacked; attacks on stems can cause girdling and kill the entire plant. Infection of flowers may make them unmarketable. All of these factors make a variety of greenhouse crops, both ornamentals and vegetables, highly susceptible to this pathogen (Hausbeck and Moorman, 1996). If the pathogen is not controlled, serious losses, including loss of the entire crop, may be incurred.

Elad and his coworkers (1994) obtained strain T39 of *T. harzianum* from cucumber fruit, and it has proven to be highly effective in the control of *B. cinerea.* It is produced by Makhteshim Chemical Works and is marketed as Trichodex™ in Israel and Europe, and Abbott Laboratories expects to produce and market this strain in the USA.

Over the past several years, Makhteshim and Elad have arranged a number of trials around the world on greenhouse crops at rates of 0.2% (w/v) of a commercial preparation containing about 10^{10} cfu/g of *T. harzianum* (Elad *et al.*, 1993). At least 60 greenhouse trials have been conducted, and, in most of these, Trichodex used alone provided control – but frequently at levels below those obtained with standard chemical treatments. Consequently, a strategy has been developed in which standard chemical treatments are alternated with the biological control product. With either chemical or biological control products, multiple applications must be made. Therefore, the strategy employed by Elad *et al.* has been to replace every other chemical treatment with a biological treatment. This strategy has proven effective: control with this integrated approach is about as effective as with a full chemical program (Elad *et al.*, 1994). Such approaches reduce pesticide usage and may reduce the rate at which *B. cinerea* develops resistance to standard fungicides. The frequency of chemical fungicide application is reduced and strains resistant to the fungicide may be controlled by the biocontrol agent.

We also are interested in control of *B. cinerea* in greenhouses. We therefore tested the surfactant–*T. harzianum* 1295-22 combination described earlier. Strain 1295-22 used alone was about as effective as the standard fungicide iprodione, but the combination of biocontrol agent and surfactant was superior to either the fungicide or the biocontrol agent used alone (C. Hayes, BioWorks, Inc., personal communication). This combination is currently being extensively tested in greenhouses on a variety of crops.

Control of B. cinerea *on strawberries* *B. cinerea* is a serious problem on strawberries and large amounts of fungicides frequently are required to provide adequate control in conditions where heavy disease pressure is present. For example, in Florida as many as 25 applications of the fungicide captan are permitted and frequently used. This very high frequency of application is of concern not only because of toxicity but because the pathogen is developing resistance to some of the most effective fungicides.

Control of *B. cinerea* on strawberry can be achieved if the pathogen is prevented from attacking the flower. *B. cinerea* colonizes strawberry flowers and then remains quiescent as the fruit develops. Once the fruit is ripe, the fungus ramifies extensively to form the typical gray mold hyphal mass familiar to most people (Cooley *et al.*, 1996). Therefore, fungicides usually are applied to the flower with the expectation that these treatments also will protect the developing fruit.

Initial studies on biocontrol of *B. cinerea* on strawberry focused on spray applications of organisms to flowers. There were several notable successes to this strategy; for example, Tronsmo and Dennis (1977) applied select *Trichoderma* isolates and obtained good control of the disease. Peng and Sutton (1991) tested a wide range of microorganisms (bacteria, yeasts, and fungi) obtained from strawberry in greenhouse tests. Isolates of *G. roseum*, *Penicillium* sp., *T. viride*, *Colletotrichum gloeosporioides*, *Epicoccum purpurascens*, and *Trichothecium roseum* were among those effective, and the best of these were as effective as standard fungicides. A *G. roseum* strain was the most effective, especially at low temperatures (10–15°C). They further tested the ability of the biocontrol agents and found that they also suppressed the pathogen on green, but not dead or senescing, leaves (Sutton and Peng, 1993).

This group also developed a novel method of delivery of biocontrol agents to strawberries. They considered that spray applications were inefficient because only a small percentage of the total spray material was applied to the flower and because flowers that remain closed at the time of spray application are not protected (Peng *et al.*, 1992). For these reasons, they evaluated the possibility that honey bees contaminated with spores of the biocontrol fungus might provide an effective method of delivery of *G. roseum* to strawberry flowers. When the bees forage in the flowers they would be expected to deliver conidia of the fungus. They therefore devised an apparatus that would hold a powder formulation of *G. roseum* and through which the bees would have to exit on their way from the hive. These experiments were highly successful. Bees became coated with spores of the fungus and effectively delivered them to the flower. In their trials, bee delivery resulted in higher levels of *G. roseum* on strawberry flowers than did spray applications. Both methods were effective in reducing the incidence of *B. cinerea* on various floral parts in field trials (Peng *et al.*, 1992).

Over the past several years, we (J. Kovach, W. Wilcox, and G. E. Harman) have investigated both spray and bee delivery methods further by using strain 1295-22 of *T. harzianum*. Either spray application or bee application consistently resulted in effective control of the pathogen; Table 11.5 provides results of one set of trials. Over a number of trials, the biocontrol agent gave consistently good results, and, in every case, bee delivery was more effective than spray delivery of the same formulation of the organism. However, when flowers or fruits were plated, spray delivery provided heavier colonization of strawberries than did bee delivery. The fungus survived well following either application. Maturing fruits were plated and found to be colonized by the biocontrol fungus even several weeks after application. The reasons for the differences between the efficacy of *T. harzianum* 1295-22 used as bee delivery and as a spray are not clear at present. However, it may be that spray delivery provides levels of spores too high for maximum control. Conidia of *T. harzianum* are self-inhibitory, and excessive levels may prevent good germination, and so protection of flowers may be greater at lower than at higher levels of conidia of this organism.

There are currently no commercially available preparations based on *Trichoderma* or *Gliocladium* spp. for the control of *B. cinerea* on strawberry. Once these results are verified in other locations, however, it is expected that products based on *T. harzianum* strain 1295-22 will be offered for sale.

Control of B. cinerea *on grapes* *B. cinerea* Pers. ex Fr. causes substantial disease losses on grapes. Control is obtained primarily by using the same problematic fungicides considered earlier (Gullino, 1992). There are no adequate replacements. Biological agents could be important components for control of *B. cinerea* if effective and reliable biocontrol preparations were registered and marketed.

Table 11.5 Control of *Botrytis cinerea* on strawberry in grower fields in 1996 using *Trichoderma harzianum* strain 1295-22

Treatment	*% Infected berries*
None	8.5b[a]
Trichoderma spray	4.9b
Bee delivered *T. harzianum*	2.1a
Standard fungicide used by grower	1.2a

Data are means from four grower sites. Berries were harvested from plants and incubated for 4 days at 20°C and 90–95% relative humidity and then the percentage of berries with symptoms of *B. cinerea* was determined. Two spray applications were made at the rate of about 8 lb/acre of a preparation containing 10^9 colony forming units/g in the presence of 0.2% of the surfactant R-11. The same material was used for bee delivery but was packed into a delivery device through which bees must pass in exiting the hive; the bees delivered about 0.33 lb/week. Data from J. Kovach, Integrated Pest Management Program, Cornell University, Geneva, NY, USA.

[a] Data were subjected to arcsine transformations, and values not followed by the same letters are significantly different ($P = 0.05$ according to Fisher's Protected LSD test).

Requirements for biocontrol of *B. cinerea* on grape are quite different than for strawberry. The pathogen may infect grape flowers, developing berries, and the maturing fruit over the entire season. For maximum protection, as many as 5–6 applications may be required each season at various growth stages through the development of the fruit. In fact, if early disease pressure is not great, two late applications may provide adequate control. Further, the grape flower is lacking in succulent tissues, and so there perhaps is less nutritional material available to support growth of the biocontrol agents applied to them. Finally, grapes are not bee pollinated, so this method of delivery cannot be used. These factors indicate that biocontrol of *B. cinerea* on grape is more difficult than on strawberry.

However, there have been substantial efforts to control this pathogen on grape with *Trichoderma* spp. (Dubos, 1987; Elad, 1994; Elad *et al.*, 1994; Gullino, 1992; Harman *et al.*, 1996) because the need for biological alternatives to chemical pesticides is well documented. Two issues that must first be addressed are the quantity of conidia (usually measured in cfus) that must be used and the timing at which they should be applied. Dubos (1987) therefore determined that *Trichoderma* should be applied at a minimal rate of 10^8 spores/ml, which probably translates to about $3\text{–}4 \times 10^7$ cfu/ml. She also provided evidence that the biocontrol fungus colonizes flowers and that this colonization prevents colonization of the same structures by *B. cinerea.* Applications should therefore be made at the time of flowering. Data on fruit rot indicated, however, that four applications ranging from flowering to 3 weeks before harvest were more effective than fewer applications (Dubos, 1987).

In our trials in 1990, two late applications to ripening fruit were as effective as those plus additional treatments made earlier (Harman *et al.*, 1996). However, in other years, early applications increased efficacy over late-only applications. These variables probably are not a consequence of the biocontrol applications but rather are a reflection of the environmental conditions favoring infection by *B. cinerea.* Cool, wet conditions favor infection, while warm, dry conditions are not conducive. The pathogen infects whenever conditions are appropriate, so results of studies to examine spray frequency will vary from year to year, depending on climatic conditions.

We also were concerned with the possible lack of nutritive substances, in grape flowers especially, and considered that biocontrol efficacy would be enhanced if we added a nutritive adhesive to spray formulations for grape. We identified a mixture, PelGel (LiphaTech, Milwaukee, WI), that is composed of carboxymethyl cellulose and gum arabic. This mixture provided better biocontrol of *B. cinerea* on grape than applications in its absence (Harman *et al.*, 1996). Unfortunately, it is too expensive for large-scale spray applications and so cannot be used commercially. We have not discovered an adequate substitute.

Given the difficulty associated with control of gray mold on grape, there have been substantial efforts to integrate chemical and biological control measures. Further, the efficacy of the *T. harzianum* alone has been less than with fungicides alone (Elad, 1994), and our data concurs with this observation (Harman *et al.*, 1996). Elad (1994) has advocated an alternation of sprays with dicarboximide fungicides and has shown that this protocol usually results in control as effective as with the fungicides alone. However, strain 1295-22 of *T. harzianum* is resistant to dicarboximide fungicides. We therefore applied this strain to flowers and developing fruits, and then followed with a tank mix of the fungicide iprodione at 0.5 × the recommended rate plus *T. harzianum* and had excellent results in 1993. Similar treatments in 1994 were less effective.

Elad has taken this concept one step further. He considers that *T. harzianum* is capable of providing adequate protection when conditions are moderately favorable to *B. cinerea* but probably not under severe conditions, a conclusion with which we concur. Forecasting systems can be developed to predict severity of *B. cinerea* on grape and other crops. Three specific situations can exist: (1) no expected disease in hot, dry conditions, (2) moderate disease pressure under humid conditions, and (3) extremely favorable conditions for disease in rainy situations. A decision support system named Botman based on these factors has been developed (Elad *et al.*, 1994). This permits different spray recommendations to be made. Under conditions where no disease is expected, no sprays should be made. Under moderate conditions, the biocontrol spray applications are adequate, but under severe conditions, the maximum control provided by chemical protectants should be employed. This permits good control under all conditions, and minimizes the use of chemical fungicides (Elad *et al.*, 1994).

Mechanisms of control of B. cinerea *by* Trichoderma *and* Gliocladium *spp.* It is useful to briefly consider how *Trichoderma* and *Gliocladium* spp. control *B. cinerea.* Dubos (1987) and Peng and Sutton (1991) have all emphasized that successful biocontrol is associated with colonization of grape or strawberry flowers. Lo *et al.* (1998) demonstrated that *T. harzianum* grows and proliferates on leaf blades of grass, and microscopic observations suggested a mycoparasitism-like interaction with *R. solani.* It is likely that a similar process may occur with *T. harzianum* and *B. cinerea*; certainly all data of which we are aware indicate that the biocontrol agent must be growing and active in the infection court if biocontrol is to be achieved. Species of

Trichoderma or *Gliocladium* active in biocontrol are known to be mycoparasitic and some may also produce antibiotics. *T. harzianum* strains may produce peptaibols (Schirmböck *et al.*, 1994) and other antibiotics; *G. roseum* also may produce antifungal compounds (Volume 1, Chapter 8). In this environment, it seems likely that biocontrol may occur because of competition, mycoparasitism, antibiosis, or some combination of these. However, Zimand *et al.* (1996) recently have described another possibility. They demonstrated that the presence of *T. harzianum* in the site where *B. cinerea* is infecting has an adverse effect upon activities of enzymes involved in pectin degradation. Such enzymes are critical factors involved in the infection process by *B. cinerea* and other pathogens; if the biocontrol agent minimizes their activity, this may also serve to limit disease caused by the pathogen.

To summarize, biocontrol of *B. cinerea* with *T. harzianum* is possible and is becoming commercially available. Control of the pathogen in the three situations examined (greenhouse, strawberries, and grapes) requires different strategies. The pathogen infects different crops at different times, and the environmental conditions at the time of infection determine the degree and severity of infection. The interaction with *T. harzianum* and different crops is different as well, with some sites and crops probably providing a more hospitable environment for the biocontrol agent than others. Adjuvants such as surfactants or nutritive amendments may be helpful, but the same strain on different crops may respond differently. Probably the most effective levels of control of grapes, especially, will be with integrated biological–chemical agents. However, regulations, especially in Europe, may preclude the use of effective chemical fungicides for the control of *B. cinerea.* Success of biocontrol based on *T. harzianum* will be on a crop-by-crop basis and will require fundamental knowledge of the interaction of crop, pathogen, and biocontrol agent. Timing and composition of the delivery system will be critical.

11.4 Summary

Biological control based on *Trichoderma* and *Gliocladium* spp. can be an effective method for control of plant diseases and for enhancing plant performance. The biocontrol agents have abilities that chemical pesticides do not possess. For example, some strains are rhizosphere competent, and this permits season-long protection of roots against pathogens and enhances root function. However, successful commercial use of *Trichoderma* or *Gliocladium* spp. as biocontrol agents is dependent upon development of production methods that provide appropriate formulations that have good activity, have good shelf life, are adapted to local farming practices, and are reasonably priced. Moreover, delivery systems that enhance the activity of the bioprotectant and that maximize its benefits in agricultural systems are critically important. In most cases, it is not a good concept to attempt to simply replace chemical pesticides in situations where they are inexpensive and readily available; instead, it is necessary to determine a specific benefit for the biocontrol agent. Benefits may include attributes not possible for chemical agents, such as season-long root colonization. Other opportunities for biocontrol products may include the necessity of providing an alternative to minimize development of resistance of pathogens, to reduce risk in situations where very high chemical use is highly undesirable for environmental or food residue concerns, or to provide a method of disease control if regulatory concerns remove effective chemical products.

Strategies for employing *Trichoderma* and *Gliocladium* spp. may differ from crop to crop even when the same pathogen is to be controlled. Different conditions and different crops elicit different responses from both the pathogen and the biocontrol agent. Successful control is achieved when the living biocontrol agent is able to control the growth of the pathogen through any of a variety of methods. However, *Trichoderma* and *Gliocladium* spp. can be effective in a wide range of applications if proper attention is paid to achieving proper formulations and delivery systems for the crop system in question. Some strains exhibit very wide adaptation to a variety of different environments and have the ability to control a wide range of pathogens. Therefore, single strains of *T. harzianum*, for example, can be formulated into a number of products for different uses in a wide range of agricultural applications.

References

AHMAD, J. S. and BAKER, R. 1987. Rhizosphere competence of *Trichoderma harzianum*. *Phytopathology* **77**: 182–189.

ALTOMARE, C., BJÖRKMAN, T., NORVELL, W. A., and HARMAN, G. E. 1996. Solubilization of manganese dioxide by the biocontrol fungus *Trichoderma harzianum* strain 1295-22. Abstracts Int. Union Microbiol. Soc., Jerusalem, p. 171.

ANONYMOUS. 1994. Another feminizing hormone found. *Sci. News* **146**: 15.

AYERS, J. E. and GILMORE, B. K. 1991. Fungicide Benefits Assessment: Turf. National Agricultural Pesticide Impact Program, Washington, DC.

BAKER, R., ELAD, Y., and CHET, I. 1984. The controlled experiment in the scientific method with special emphasis on biological control. *Phytopathology* **74**: 1019–1021.

BISSETT, J. 1991. A revision of the genus *Trichoderma*. II: Infrageneric classification. *Can. J. Bot.* **69**: 2357–2372.

BJÖRKMAN, T., BLANCHARD, L. M., and HARMAN, G. E. 1998. Growth enhancement of shrunken-2 sweet corn with *Trichoderma harzianum* 1295–22: effect of environmental stress. *J. Am. Soc. Hort. Sci.*, **123**: 35–40.

BOEHM, M. J., WU, T., STONE, A. G., KRAAKMAN, B., IANNOTTI, D. A., WILSON, G. E., MADDEN, L. V., and HOITINK, H. A. J. 1997. Cross-polarized magic-angle spinning ^{13}C nuclear magnetic resonance spectroscopic characterization of soil organic matter relative to culturable bacterial species composition and sustained biological control of Pythium root rot. *Appl. Environ. Microbiol.* **63**: 162–168.

CHANG, Y.-C., CHANG, Y.-C., BAKER, R., KLEIFELD, O., and CHET, I. 1986. Increased growth of plants in the presence of the biological control agent *Trichoderma harzianum*. *Plant Dis.* **70**: 145–148.

CHET, I. and BAKER, R. 1981. Isolation and biocontrol potential of *Trichoderma hamatum* from soil naturally suppressive to *Rhizoctonia solani*. *Phytopathology* **71**: 286–290.

COOK, R. J., and BAKER, K. F. 1993. *The nature and practice of plant pathogens.* American Phytopathological Society, pp. 539.

COOLEY, D. R., WILCOX, W. F., KOVACH, J., and SCHLOEMANN, S. G. 1996. Integrated pest management programs for strawberries in the northeastern United States. *Plant Dis.* **80**: 228–237.

DATNOFF, L. E., NEMEC, S., and PERNEZNY, K. 1995. Biological control of Fusarium crown and root rot of tomato in Florida using *Trichoderma harzianum* and *Glomus intraradices*. *Biol. Control* **5**: 427–431.

DUBOS, B. 1987. Fungal antagonism in aerial agrobiocenoses. In I. Chet (ed.), *Innovative Approaches to Plant Disease Control.* Wiley, New York, pp. 107–135.

ELAD, Y. 1994. Biological control of grape grey mold by *Trichoderma harzianum*. *Crop Protect.* **13**: 35–38.

Elad, Y. and Chet, I. 1995. Practical approaches for biocontrol implementation. In R. Reuveni (ed.), *Novel Approaches to Integrated Pest Management*. Lewis, Boca Raton, FL, pp. 323–338.

Elad, Y., Malathrakis, N. E., and Dik, A. J. 1996. Biological control of *Botrytis*-incited diseases and powdery mildews in greenhouse crops. *Crop Protect.* **15**: 229–240.

Elad, Y., Shtienberg, D., and Niv, A. 1994. *Trichoderma harzianum* T39 integrated with fungicides: improved biocontrol of grey mold. Brighton Crop Protection Conference, pp. 1109–1114.

Elad, Y., Zimand, G., Zaqs, Y., Zuriel, S., and Chet, I. 1993. Use of *Trichoderma harzianum* in combination or alternation with fungicides to control cucumber grey mould (*Botrytis cinerea*) under commercial greenhouse conditions. *Plant Pathol.* **42**: 324–332.

Gullino, M. L. 1992. Control of Botrytis rot of grapes and vegetables with *Trichoderma* spp. In E. C. Tjamos, G. C. Papavizas, and R. J. Cook (eds), *Biological Control of Plant Diseases: Progress and Challenges for the Future*. Plenum Press, New York, pp. 125–132.

Hadar, Y., Harman, G. E., and Taylor, A. G. 1984. Evaluation of *Trichoderma koningii* and *T. harzianum* from New York soils for biological control of seed rot caused by *Pythium* spp. *Phytopathology* **74**: 106–110.

Hadar, Y., Mandelbaum, R., and Gorodecki, B. 1992. Biological control of soilborne plant pathogens by suppressive composts. In E. C. Tjamos (ed.), *Biological Control of Plant Diseases*, Plenum Press, New York, pp. 79–83.

Harman, G. E. 1991. Seed treatments for biological control of plant disease. Proc. AAAS Symposium. *Crop Protect.* **10**: 166–171.

Harman, G. E. and Lo, C.-T. 1996. The first registered biological control product for turf disease: BioTrek 22G. *TurfGrass Trends* **5**(5): 8–15.

Harman, G. E. and Lumsden, R. D. 1989. Biological disease control. In J. M. Lynch (ed.), *The Rhizosphere*. Wiley, Chichester, UK, pp. 259–280.

Harman, G. E. and Nelson, E. B. 1994. Mechanisms of protection of seed and seedlings by biological seed treatments: implications for practical disease control. In T. Martin (ed.), *Seed Treatment: Progress and Prospects*. British Crop Protection Council, Surrey, UK, pp. 283–292.

Harman, G. E. and Taylor, A. G. 1988. Improved seedling performance by integration of biological control agents at favorable pH levels with solid matrix priming. *Phytopathology* **78**: 520–525.

Harman, G. E., Hayes, C. K., and Lorito, M. 1993. The genome of biocontrol fungi: modification and genetic components for plant disease management strategies. In R. D. Lumsden and J. L. Vaughn (eds), *Pest Management: Biologically Based Technologies*. Proceedings of Beltsville Symposium XVIII. American Chemical Society, Washington, DC, pp. 347–354.

Harman, G. E., Taylor, A. G., and Stasz, T. E. 1989. Combining effective strains of *Trichoderma harzianum* and solid matrix priming to improve biological seed treatments. *Plant Dis.* **73**: 631–637.

Harman, G. E., Lattore, B., Agosin, A., San Martin, R., Riegel, D. G., Nielsen, P. A., Tronsmo, A., and Pearson, R. C. 1996. Biological and integrated control of Botrytis bunch rot of grape using *Trichoderma* spp. *Biol. Control* **7**: 259–266.

Hausbeck, M. K. and Moorman, G. W. 1996. Managing *Botrytis* in greenhouse-grown flower crops. *Plant Dis.* **80**: 1212–1219.

Hoitink, H. A. J., Inbar, Y., and Boehm, M. J. 1991. Status of compost-amended potting mixes naturally suppressive to soilborne diseases of floricultural crops. *Plant Dis.* **75**: 869–873.

Hoitink, H. A. J., Stone, A. G., and Han, D. Y. 1997. Suppression of plant diseases by composts. *HortSci.*, **32**: 184–187.

Howell, C. R. 1991. Biological control of Pythium damping-off of cotton with seed-coating preparation of *Gliocladium virens*. *Phytopathology* **81**: 738–741.

Howell, C. R. and Stipanovic, R. D. 1983. Gliovirin, a new antibiotic from *Gliocladium virens*, and its role in the biological control of *Pythium ultimum*. *Can. J. Microbiol.* **29**: 321–324.

Inbar, J., Abramsky, M., Cohen, D., and Chet, I. 1994. Plant growth enhancement and disease control by *Trichoderma harzianum* in vegetable seedlings grown under commercial conditions. *Eur. J. Plant Pathol.* **100**: 337–346.

Jin, X., Hayes, C. K., and Harman, G. E. 1991. Principles in the development of biological control systems employing *Trichoderma* species against soil-borne plant pathogenic fungi. In G. C. Letham (ed.), *Symposium on Industrial Mycology*. Am. Mycol. Soc., Brock-Springer Series in Contemporary Bioscience.

Kuter, G. A., Nelson, E. B., Hoitink, H. A. J., and Madden, L. V. 1983. Fungal populations in container media amended with composted hardwood bark suppressive and conducive to Rhizoctonia damping-off. *Phytopathology* **73**: 1450–1456.

Kwok, C. H., Fahy, P. C., Hoitink, H. A. J., and Kuter, G. A. 1987. Interactions between bacteria and *Trichoderma hamatum* in suppression of Rhizoctonia damping-off in bark compost media. *Phytopathology* **77**: 1206–1212.

Lewis, J. A. and Papavizas, G. C. 1984. A new approach to stimulate population proliferation of *Trichoderma* species and other potential biocontrol fungi introduced into natural soils. *Phytopathology* **74**: 1240–1244.

Lewis, J. A. and Papavizas, G. C. 1985. Characteristics of alginate pellets formulated with *Trichoderma* and *Gliocladium* and their effect on the proliferation of the fungi in soil. *Plant Pathol.* **34**: 571–577.

Lindsey, D. L. and Baker, R. 1967. Effect of certain fungi on dwarf tomatoes grown under gnotobiotic conditions. *Phytopathology* **57**: 1262–1263.

Lo, C.-T., Nelson, E. B., and Harman, G. E. 1996. Biological control of turfgrass diseases with a rhizosphere competent strain of *Trichoderma harzianum*. *Plant Dis.* **80**: 736–741.

Lo, C.-T., Nelson, E. B., and Harman, G. E. 1997. Improved biocontrol efficacy of *Trichoderma harzianum* 1295-22 for foliar phases of turf diseases by use of spray applications. *Plant Dis.*, **81**: 1132–1138.

Lo, C.-T., Nelson, E. B., and Harman, G. E. 1998. Ecological studies of transformed *Trichoderma harzianum* strain 1295-22 in the rhizosphere and on the phylloplane of creeping bentgrass. *Phytopathology*, **88**: 129–136.

Lumsden, R. D. and Locke, J. C. 1989. Biological control of damping-off caused by *Pythium ultimum* and *Rhizoctonia solani* with *Gliocladium virens* in soilless media. *Phytopathology* **79**: 361–366.

Lumsden, R. D., Locke, J. C., Adkins, S. T., Walter, J. F., and Ridout, C. J. 1992a. Isolation and localization of the antibiotic gliotoxin produced by *Gliocladium virens* from alginate prill in soil and soilless media. *Phytopathology* **82**: 230–235.

Lumsden, R. D., Ridout, C. J., Vendemia, M. E., Harrison, D. J., Waters, R. M., and Walter, J. F. 1992b. Characterization of major secondary metabolites produced in soilless mix by a formulated strain of the biocontrol fungus *Gliocladium virens*. *Can. J. Microbiol.* **38**: 1274–1280.

Lynch, J. M. 1996. *Trichoderma*: potential in regulatory soil function. Abstracts Int. Union Microbiol. Soc., Jerusalem, p. 12.

Nelson, E. B., Kuter, G. A., and Hoitink, H. A. J. 1983. Effects of fungal antagonists and compost age on suppression of Rhizoctonia damping-off in container media amended with composted hardwood bark. *Phytopathology* **73**: 1457–1462.

Ondik, K. L., Nelson, E. B., and Harman, G. E. 1997. Biological and integrated control of turfgrass diseases using *Trichoderma harzianum* (Abstr.). *Phytopathology* **87**: S72.

Peng, G. and Sutton, J. C. 1991. Evaluation of microorganisms for biocontrol of

Botrytis cinerea in strawberry. *Can. J. Plant Pathol.* **13**: 247–257.

PENG, G., SUTTON, J. C., and KEVAN, P. G. 1992. Effectiveness of honeybees for applying the biocontrol agent *Gliocladium roseum* to strawberry flowers to suppress *Botrytis cinerea. Can. J. Plant Pathol.* **14**: 117–129.

SANDERS, P. L. 1989.Fungicide Resistance in the United States. National Agricultural Pesticide Impact Assessment Program, Washington, DC.

SCHIRMBÖCK, M., LORITO, M., WANG, Y.-L., HAYES, C. K., ARISAN-ATAC, I., SCALA, F., HARMAN, G. E., and KUBICEK, C. P. 1994. Parallel formation and synergism of hydrolytic enzymes and peptaibol antibiotics, molecular mechanisms involved in the antagonistic action of *Trichoderma harzianum*: against phytopathogenic fungi. *Appl. Environ. Microbiol.* **60**: 4364–4370.

SCHMIDT, E. L. 1979. Initiation of plant root-microbe interactions. *Annu. Rev. Microbiol.* **33**: 335–376.

SIVAN, A. and CHET, I. 1986. Biological control of *Fusarium* spp. in cotton, wheat and muskmelon by *Trichoderma harzianum. J. Phytopathol.* **116**: 39–47.

SMITH, V. L., WILCOX, W. F., and HARMAN, G. E. 1990. Potential for biological control of Phytophthora root and crown rots of apple by *Trichoderma* and *Gliocladium* spp. *Phytopathology* **80**: 880–885.

STASZ, T. E., HARMAN, G. E., and WEEDEN, N. F. 1988. Protoplast preparation and fusion in two biocontrol strains of *Trichoderma harzianum. Mycologia* **80**: 141–150.

SUTTON, J. C. and PENG, G. 1993. Biocontrol of *Botrytis cinerea* in strawberry leaves. *Phytopathology* **83**: 615–621.

TAYLOR, A. G. and HARMAN, G. E. 1990. Concepts and technologies of selected seed treatments. *Annu. Rev. Phytopathol.* **28**: 321–339.

TAYLOR, A. G., MIN, T.-G., HARMAN, G. E., and JIN, X. 1991. Liquid coating formulation for the application of biological seed treatments of *Trichoderma harzianum. Biol. Control* **1**: 16–22.

TRONSMO, A. 1989. Effect of fungicides and insecticides on growth of *Botrytis cinerea, Trichoderma viride* and *T. harzianum. Norw. J. Agric. Sci.* **3**: 151–156.

TRONSMO, A. and DENNIS, C. 1977. The use of *Trichoderma* species to control strawberry fruit rots. *Neth. J. Plant Pathol.* **83**(Suppl. 1): 449–455.

TRONSMO, A. and HARMAN, G. E. 1992. Coproduction of chitinases and biomass for biological control by *Trichoderma harzianum* on media containing chitin. *Biol. Control* **2**: 272–277.

TRONSMO, A. and YSTAAS, J. 1980. Biological control of *Botrytis cinerea* on apple. *Plant Dis.* **64**: 1009.

VINCELLI, P. and DONEY, J. C. 1997. Evaluation of fungicides for control of dollar spot on creeping bentgrass, 1996. Fungicide and Nematocide Tests, American Phytopathological Society, **52**: 377.

WEINDLING, R. 1932. *Trichoderma lignorum* as a parasite of other soil fungi. *Phytopathology* **22**: 837–845.

WEINDLING, R. 1934. Studies on a lethal principle effective in the parasitic action of *Trichoderma lignorum* on *Rhizoctonia solani* and other soil fungi. *Phytopathology* **24**: 1153–1179.

WEINDLING, R. 1941. Experimental consideration of the mold toxins of *Gliocladium* and *Trichoderma. Phytopathology* **31**: 991–1003.

WEINDLING, R. and FAWCETT, H. S. 1936. Experiments in the control of Rhizoctonia damping-off of citrus seedlings. *J. Agric. Sci., CA Agric. Exp. Station* **10**: 1–16.

WILHITE, S. E. and STRANEY, D. C. 1996. Timing of gliotoxin biosynthesis in the fungal biological control agent *Gliocladium virens* (*Trichoderma virens*). *Appl. Microbiol. Biotechnol.* **45**: 513–518.

WINDHAM, M. T., ELAD, Y., and BAKER, R. 1986. A mechanism for increased plant growth induced by *Trichoderma* spp. *Phytopathology* **76**: 518–521.

WU, W.-S. 1982. Seed treatment by applying *Trichoderma* spp. to increase the emergence of soybeans. *Seed Sci. Technol.* **1**: 557–563.

ZHANG, W., DICK, W. A., and HOITINK, H. A. J. 1996. Compost-induced systemic acquired resistance in cucumber to Pythium root rot and anthracnose. *Phytopathology* **86**: 1066–1070.

ZIMAND, G., ELAD, Y., and CHET, I. 1996. Effect of *Trichoderma harzianum* on *Botrytis cinerea* pathogenicity. *Phytopathology* **86**: 1255–1260.

12

Trichoderma as a weed mould or pathogen in mushroom cultivation

D. SEABY

Applied Plant Science Division, Department of Agriculture for Northern Ireland, Belfast, Newforge Lane, UK

12.1 Introduction

12.1.1 Literature review

There are a variety of situations in which *Trichoderma* species have caused economic losses in mushroom production worldwide. Komatsu (1976) details studies on *Hypocrea, Trichoderma,* and allied fungi that are antagonistic to shiitake (*Lentinus edodes*). Tokimoto (1982) reports that chitinolytic enzymes produced by *T. harzianum* were responsible for shiitake crop failures in Japan. Liao (1985) in China deals with the efficacy of fungicides in controlling *Trichoderma* in sawdust cultivation of shiitake. Yoshida and Takao (1982) report that *T. viride* was the predominant weed mould in the culture of *Pholiota nameko* on rice bran in Japan. In Belgium, several *Trichoderma* species are reported as weed moulds of *Pleurotus* (Lelley, 1987; Poppe *et al.*, 1985). Lelley (1987) records *T. hamatum* as more damaging to *Pleurotus* culture than to *Agaricus*. Shandilya and Guleria (1984) described *Trichoderma* spp. causing losses in *A. bitorquis* crops in India.

12.1.2 Trichoderma *spp. affecting production of* Agaricus bisporus

Several earlier studies, including those of Beach (1937) and Kligman (1950), dealt with blotch caused by *Trichoderma viride*, while Kneebone and Merek (1959) suggested that *T. viride* also secreted a toxin into the casing which caused sunken brown lesions on mushrooms. Sinden and Hauser (1953) described *T. koningii* as causing a mildew that rapidly grew over mushrooms after the third flush, which they distinguished from *Trichoderma* blotch. Kligman (1950) and Harvey *et al.* (1982) observed green mould in compost, listing *T. viride* and *T. koningii*, although their description of large patches of sporulation bursting through the casing would be characteristic of *T. harzianum*. Gandy (1977) found that compost supplemented with sugar rapidly became overgrown with *Trichoderma* spp., and Seaby (1987) reported that small quantities of skim milk applied to compost with the intent to repress *Trichoderma* actually stimulated the growth of the fungus rather than that of inhibitory bacteria. Fergus (1978) isolated a large number of different fungi, including *T. viride*, from black patches in compost. Steane (1978) measured *Trichoderma* contamination in relation to compost batches produced weekly, finding higher spore counts in those with lower than normal nitrogen levels. Most texts on mushroom production (e.g., Fermor *et al.*, 1985; Harvey *et al.*, 1982) state that *Trichoderma* spp. occur mainly in substrates that have not been composted for long enough and are wet or have had faulty peak heating and still contain high levels of readily metabolizable carbohydrates. They also suggest that *Trichoderma* may be found on cereal grains used in mushroom spawn. Supplementation, for example with Soya meal, may stimulate *Trichoderma* growth; however, risk of losses is reported to be less if supplements are applied near the end of spawn run. The first references to an epidemic of green mould in compost were by Staunton (1987) and by Seaby (1987). The latter attributed it to a specific taxon of *T. harzianum* (*Th2*). This epidemic was observed by Seaby (1989, 1996b) to spread from Northern Ireland to compost makers and growers in the Republic of Ireland (ROI), to Scotland and to England, later being unofficially reported in The Netherlands and identified in Germany in 1996 (McKay and Doyle, personal communication). The outbreak was also studied in the ROI by Doyle *et al.* (1990) and Morris *et al.* (1995a,b). The uniqueness of the *Th2* taxon was shown culturally (Seaby, 1987, 1996a) and genetically by DNA techniques (Muthumeenakshi *et al.*, 1994). The description of this taxon as a separate species under a new name is currently in progress (Gams, Muthumeenakshi and Mills, personal communication; see Volume 1, Chapters 1 and 2), but in this review the name *Th2* will be maintained for convenience. A molecular PCR method for specific determination of this fungus has been published by Muthumeenakshi and Mills (1996). Similarly, Seaby (1996a) and Muthumeenakshi (1996) showed that an equally important epidemic in North America was caused by another taxon of *T. harzianum* (*Th4*), also with a distinctive DNA pattern. At the time of writing, this taxon has not yet been renamed. Seaby (1987, 1996a), Doyle *et al.* (1990), and Morris *et al.* (1995a,b) found that of two additional, morphologically distinct *T. harzianum* taxa, one (*Th1*) was very common but virtually harmless, and the other (*Th3*, which is genetically identical to *T. atroviride*) was potentially damaging but was less common and seldom caused economic losses. The importance of red-pepper mites in the spread of *Th2* was stressed by Seaby (1987, 1989, 1996b) and by Terras and Hales (1995) who reported that they possess spore carrying pouches. Grogan and Gaze (1995) found evidence that losses caused by green mould were greater in composts that had undergone an atypical peak-heat regimen. Grogan *et al.* (1996) found that a single *Th2* contaminated grain of spawn in a 45 kg tray of compost reduced yields by 12–46%. In another experiment, they found that dusting spawn with benzimidazole fungicide was an effective control measure, reducing yield-loss from 62% to 16%. The grain treatment method was many times more economical than applying the fungicide to compost, a result which substantiated Fletcher's earlier, unpublished observations.

12.1.3 *Some* Trichoderma *identification anomalies*

Following the work of Bisby (1939), *Trichoderma* was considered to comprise one species. Until recently, despite Rifai's (1969) reclassification, taxonomic difficulties have led to the predominant use of two names for new isolates: *T. viride* for taxa with round, obovoid spores, and *T. koningii* for those with cylindrical ones. Thus some of the earlier literature identified strains as belonging to these taxa exclusively. This situation contrasts with Samuels' (1996) statement that there are currently 75 described species, with an estimate that over 100 probably exist (see also Volume 1, Chapter 1).

Species names in the recent literature, based on Rifai's (1969) reclassification, are currently in a state of flux while his species' aggregates are being divided. The word taxon is frequently used here and indicates a group of isolates that all have several characteristics in common, e.g., each of the four *T. harzianum* taxa. These four fit into Rifai's single *T. harzianum* species aggregate because of their similar spores and phialides. They are distinguished here by initials plus a number, i.e., *Th1–Th4*. This does not imply a close relationship between them; indeed, molecular methods suggest major differences (Muthumeenakshi *et al.*, 1994). Eventually, each may be given species status. Currently, Gams' suggestion (W. Gams, personal communication) is that *Th1* corresponds to the neotype of *T. harzianum*, *Th2* will be designated a new species, and *Th3* is co-specific with *T. atroviride*, leaving only *Th4* to be renamed. However, due to the body of literature that details the work on various aspects of green mould and other *Trichoderma* activities, confusion may arise if *Th1*, *Th2*, etc., are used by future authors to merely denote numbered isolates of *T. harzianum.*

The usefulness of a cultural identification aid

For morphological identification of *Trichoderma*, the characters of size and shape of spores and phialides, plus phialide grouping, have only a moderate usefulness. Furthermore, a character such as spore shape, which was used as a main diagnostic feature for *T. harzianum* (Rifai, 1969), is not sufficiently consistent. For example, *T. viride* as well as *T. citrinoviride* are known to display both obovoid as well as round spores, but the different spore forms are not reflected in different DNA patterns (McKay, personal communication). Thus Rifai (1969), in grouping all smooth, short-obovoid spored taxa together in his *T. harzianum* aggregate, has included *Th1* along with round-spored strains of *T. citrinoviride* (which stains media yellow) and *T. atroviride* (*Th3*). Thus he states, "Similar (yellow, media) discoloration, though less frequent, may also be found in colonies of a few isolates of *T. harzianum*" and also states "a few isolates have tufted and dark green colonies – emit typical coconut odour (like *T. viride*) – but have subglobose, smooth-walled phialospores".

However, although molecular methods of identification can be definitive, they are often not practical with large numbers of isolates. Nevertheless, given occasional DNA backup, Seaby or Morris in Ireland and Rinker in Canada have proved that, with experience, most taxa can be reliably identified by appearance, provided that a strictly controlled cultural method is used (see section 12.2.2). Ideally, some microscopical observations should also be employed.

DNA techniques are essential in differentiating taxa that might otherwise appear morphologically very similar (see Volume 1, Chapter 2). Nevertheless, difficulties could arise if they were used alone to taxonomically combine morphologically different species, simply on the basis that they have DNA patterns that are *almost* identical. For example, there is only a very small difference in DNA between *T. viride* and *Th3* (*T. atroviride*) (Volume 1, Chapter 2). It should be noted that there is no general rule as to which DNA-sequence differences (i.e., how many bases must be different) define a new species, as phylogenetic dynamics and intraspecific sequence variations are strongly different in different genera. The problem of definition of species is more fully addressed in Volume 1, Chapter 2.

12.2 Green mould epidemic in Ireland

12.2.1 Method of mushroom production in Ireland

As compost production methods are relevant to epidemiology, the following gives those in use during the green mould epidemic in Ireland. These remain similar, except that now the aim is for higher dry matter (DM) and nitrogen (N) levels (Seaby, 1995). A satellite system of growing is used, whereby a central compost producer supplies up to 150 or more growers with batches of compost (8–24 tonnes) packed in bags of about 20 kg. Many growers employ a 10–12 week growing cycle, crop 3–5 flushes of mushrooms, and order compost from more than one producer.

Compost production

Compost ingredients are mainly wheat (or occasionally barley) straw and a mixture of wood shavings and chicken droppings (litter) from rearing houses. The straw is wetted in a shallow lagoon containing water recycled from the compost yard, and then mixed with chicken litter (about 10:7, w/w). Rough stacks are formed and occasionally are later mixed with horse bedding plus chicken litter. Urea may also be added to soften straw or enhance the percentage of nitrogen. After 1–3 weeks, stacks are reformed by machinery into long wind-rows, about 2 m × 2 m in section. These are turned three times on alternate days. Gypsum is added to improve aeration and to retain ammonia. Cottonseed meal or dried molasses may also be added if fermentation is not sufficiently active, or water may be applied if the compost is dry.

Compost is then conveyed to a peak-heat tunnel filling machine. This has an oscillating head to evenly build the compost between the vertical internal walls of an insulated semi-circular tunnel. The floor of this is slatted and may be covered with a heavy net to aid emptying. Tunnels vary in size, holding 50–150 tonnes. They are aerated with recirculated air into which fresh air is bled. Compost temperature is allowed to rise rapidly to about 58°C (sometimes aided by steam), maintained for 9 h, and conditioned for approximately 4.5 days at about 45°C. Ammonia released during peak heating is either utilized by thermophilic microorganisms or vented. When ammonia has fallen to below 10 ppm, the compost is cooled by filtered air to about 17°C, unloaded, spawned and bagged by machinery. Phase II compost dry matter (DM) ranges from 24% to 42%, and N%, on a dry matter basis, from 1.6 to 3.2.

Growing methods

Compost is usually delivered on pallets. Rows of 2, 4, 4, 2 bags are set out, i.e., 12 bags wide, with alternate alleyways. A few houses also have central or side wall racks to increase capacity. At levelling, some compost is transferred between bags to even them before tamping down and pressing the surface flat. In 1985 it was customary to sprinkle extra spawn on the surface of the compost to aid colonization. Levelling may be contracted out to groups of men who help a number of growers. Houses are now mainly of double-skinned, insulated plastic over steel hoops, but in 1985 they were equally likely to be made from wood and concrete blocks. After 2–3 weeks of spawn run, with compost temperatures in the range 18–32°C (optimum is 25°C), the compost was "cased" by covering it with 3–6 cm of a mixture of peat and ground limestone (pH $\geqslant$ 7). A small handful of chopped spawn run compost may be mixed into the casing of each bag (CACing), in order to speed mycelium growth. This CACing is now largely replaced by proprietary casing spawn. The casing moisture is then raised almost to saturation. When *Agaricus* mycelium reaches the surface in 5–11 days, the house is "aired" to reduce carbon dioxide to below 900 ppm, relative humidity to below 90% and compost temperature to below 18°C. Initially, "pin heads" form, and after about 3 weeks the first crop of mushrooms is picked. Three or more "flushes" are cropped at intervals of about 7 days.

In Ireland, when bag cultivation is employed, it is rare to use cooking out. In this process, house temperatures are raised to more than 60°C to kill pathogens at the end of cropping. When spent bags were emptied into a garbage trailer inside a house, clouds of dust sometimes contaminated the ventilation system. Furthermore, a minimal amount of disinfectant was used before 1985, that is, until the start of the *Th2* epidemic.

12.2.2 *Epidemiology of* Trichoderma harzianum *group II in Ireland (from Seaby, 1987, 1989, 1996a,b, unless otherwise stated)*

In March 1985, growers obtaining compost from a single Northern Ireland compost maker noticed bags of phase II compost with large white patches. These rapidly turned green with *Trichoderma* spores, usually soon after casing (Figure 12.1). Such bags produced a scant crop and were often covered with red-pepper mites (*Pygmephorus mesembrinae*). Mites sometimes also moved to other bags and congregated on the mushrooms, making them unsellable (Figures 12.2 and 12.3). Each month, a progressively larger number of bags was affected. By 1986, 30% of production was reported lost by growers using compost from this producer. Many bags were brought back by the composter (designated 'A' here) for observation in his own growing houses, which were situated around the compost yard. Initially these bags were tended by workers who also worked in the bagging shed, thus aiding cross-contamination.

In different growers' houses, the proportion of contaminated bags varied from 0% to 85%. Serious green mould problems soon started to affect other producers, and by 1987, five major compost makers in Ireland, plus two in Scotland and one in England, had all requested urgent help. Soon all compost makers in Ireland were affected. *Th2* also affected straw compost of *Pleurotus ostreatus* in the single oyster-mushroom farm in Ireland. Losses due to *Th2* in Ireland in 1986 alone were estimated at one million pounds.

Figure 12.1 Three 20 kg bags of mushroom compost showing green patches of sporulation, indicating contamination with *T. harzianum* 2 (*Th2*). Note the sparse production of mushrooms in these bags. It is now unusual for mushroom houses in Ireland to have racks.

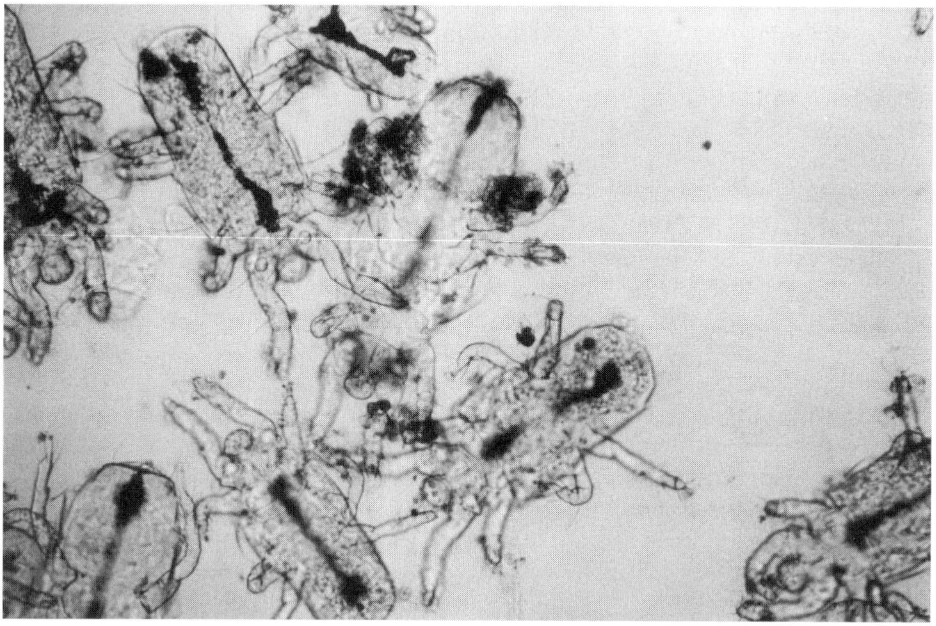

Figure 12.2 Red-pepper mites, *Pygmephorus mesembrinae* (× ~100), heavily contaminated with clusters of spores of *T. harzianum* 2. These are mainly attached to body hairs, although Terras and Hales (1995) suggested that spores are also carried in pouches at the base of each of the fourth pair of legs.

Investigation of peak-heating

According to Fermor *et al.* (1985) or Fletcher *et al.* (1986), green mould was most likely to occur if *Trichoderma* survived peak heating. However, if such a high level of contamination had come through modern, computer-controlled peak-heat tunnels, there were serious implications for the survival of the industry. Two tests indicated otherwise. In each, twelve 450 ml glass jars with perforated lids were half-filled with compost on which *T. harzianum* (*Th2*) was growing and were placed at various levels and positions within the phase I compost during tunnel filling. Jars were retrievable, by virtue of cords tied to them and to wooden stakes. Upon emptying six days later, four samples from each jar were plated out onto antibiotic 2% malt agar (AMA). The jars were then half-filled with AMA. No *Th2* was found. However Staunton (1987), using net-bags in which the compost was compressed, found that *Trichoderma* appeared in this compost following peak heating and spawning. He concluded that spore survival could occur if tunnels were either unevenly filled or over-filled. Nevertheless, this did not explain why only compost maker A was affected at a time when others using similar methods and ingredients were not. Neither was a correlation found between increased green mould and tunnels which yielded an abnormally high number of bags.

Use of compost makers' record books

Record books were invaluable. The data included the date of peak-heat tunnel filling and emptying, its identification number, the maximum and minimum concentrations of ammonia attained, number of bags, total weight delivered, customers'

Figure 12.3 After feeding on *Trichoderma* mycelium in the compost, red-pepper mites sometimes congregate on the mushrooms produced on surrounding bags, making them unsellable.

names (from this, degree of experience and house types were known), brands and rates of spawn used, compost moisture content, N% (undried samples), pH, and results of life tests (for mites, nematodes, fly larvae, etc.). Eventually, most of the yields and numbers of contaminated bags were also obtained along with local records of wind and rainfall of the Northern Ireland Meteorological Office.

For composter A, who was unusual in having nearby growing houses, *Trichoderma* contamination was 60% higher than average during weeks when dry, windy conditions had prevailed at bagging. Furthermore, the first loads out after the weekend averaged 30% infection over the year, compared with 11% for other loads. Similarly, the least sheltered peak-heat tunnels also had the worst records. Closing the bagging shed doorway with heavy plastic strip curtains and covering the compost-carrying conveyors with polythene sheeting greatly reduced contamination. Furthermore, these plastic surfaces when tested by swabbing were found to have *Th2* on them within two days. Thus, although *Trichoderma* spores from culture dishes never became airborne, compost dust that carried spores was a major source of contamination.

Using compost analysis records, 50 severely attacked composts were compared with an equal number that were free of green mould. Data from the analyses showed no differences in moisture levels, N%, or pH. Staunton (1987) made similar observations. However, in an experiment using small bags containing widely different composts, some differences in degree of *Trichoderma* growth were observed (Seaby, 1996b). This agrees with the conclusions of Grogan and Gaze (1995) that compost attributes can affect degree of green moulding but with no single factor identified.

Culture methods used to aid identification

Nine media were tested to determine which produced the most consistent differences in culture appearance between species. Of these, one containing 2% malt-agar and 45 mg/L chloramphenicol gave the best results. Dark malt (Ju-ton™), prepared solely from malted barley by Thornton and Ross, gave the results shown in the culture morphology key (Seaby, 1996a), those below, and those in Figure 12.4.

To aid identification, a precise culture method was vital. Petri dishes were inoculated with a stab of spores, 1 cm from the edge of the dish, and were placed in an incubator free of other cultures, particularly of bacteria. Non-vented petri dishes were distributed, not more than three on top of each other and were not taped. After 48 h, dishes were very briefly removed from the incubator, held up to the light and two fine scratches were made on the back of each dish in order to delimit the colony edge. After another 24 h, this was repeated, thus giving a measure of growth rate. At these examinations, colony edge (smooth, wavy, or toothed) and the first day of sporulation were noted. On day four, dishes were moved to the laboratory bench. The culture smell, the amount of aerial mycelium, and the type and pattern of sporulation were recorded. Phialide size and form were observed using slides from cultures just turning green. For spore size and ornamentation, only mature colonies were used.

Methods of sampling to collect isolates

Lagoon water was examined on seven occasions. A 5 L sample was shaken, subsampled in 25 μL aliquots, and added to 10 mL of sterile 2% malt solution. This

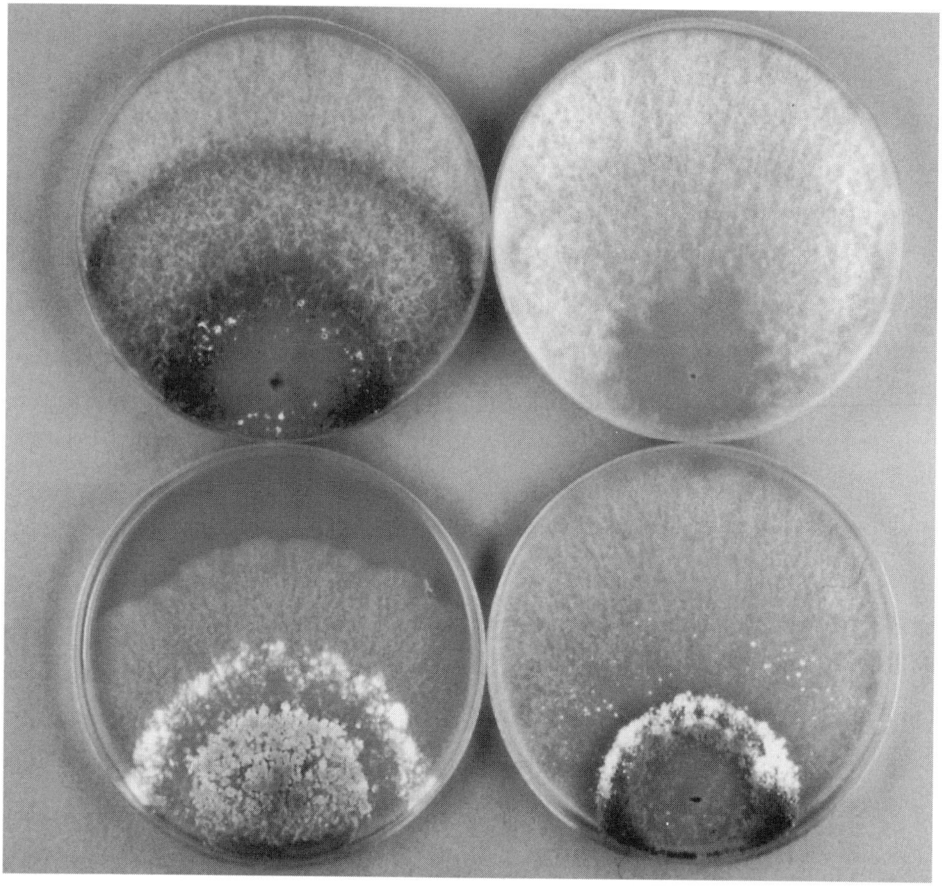

Figure 12.4 The visual appearance of isolates belonging to *Th1–4*. Clockwise from top left: *Th1*, *Th2*, *Th4*, *Th3* (= *T. atroviride*). The cultures were grown using a standard method, including incubation at 27°C for 4 days with two very brief exposures to light on days 2 and 3, which stimulated the characteristic patterns of sporulation. After further incubation of *Th2*, small, white, scattered cushions will develop, which finally become covered with green spores.

was shaken, re-sampled as 0.5 mL, added to 800 mL of liquid AMA, shaken and then poured into 40 sterile plastic Petri dishes. Incubation was for 3 days at 27°C and 2 weeks on the bench. Results indicated that lagoon water contained 5–20 000 *Trichoderma* spores per mL, but no *Th2* was detected.

Sterile, polyether foam cubes of about 3 × 3 cm size were used to swab concrete yards, floors, walls, pallets, load covers, trailers, bags, shrink-wrap, machinery, spawn boxes, growers' houses, and ventilation ducts. Swabs were dabbed onto sterile AMA dishes in thc field and incubated as above. During the epidemic, most of these surfaces were found to be contaminated with *Th2*.

Dust was sampled by drawing air through a polyether foam disc (20 mm diameter × 6 mm in length) retained in a tapered plastic tube (the upper half of a 5 mL plastic pipette tip). Upon return to the laboratory, the disc from each tube was removed and dabbed onto two dishes of AMA and plated out onto a third one followed by incubation as above. The reuseable tubes were fitted with fresh discs and autoclaved in universal bottles. In the field, airflow was provided by a small vacuum cleaner, powered by a car's cigarette-lighter socket. This was a particularly useful method since it indicated which personnel were carriers of *Th2*. Tracking their movements indicated further sources of contamination.

Either plating methods or incorporation in molten agar at 45°C were used to isolate fungal spores from mushroom-flies, mice, and mites. All of these creatures frequented mushroom houses and were found to be *Trichoderma* carriers (see Figure 12.2). Flies carried *T. viride*, *Th2*, and (in a later experiment) *Th4*. Mice carried five species, including *Th2*, as did red-pepper mites. Hussey *et al.* (1969) reported that these mites are frequently found in rodent nests.

Laboratory and field experiments to determine Th2 *virulence*

In vitro experiments compared the virulence of *Th2* with the other compost environment species. This was done in several replicated tests using 100 mL jars. Three spawn grains (one dusted with spores) were placed at the bottom of a jar, covered with a layer of compost, then another 3 grains were added and also covered with compost. Jars were incubated for 14 days in the dark with lids loosely closed, followed by incubation on the bench for 7 days. They were examined at 6 × magnification, and colonization was scored with levels from 1 to 5 (level 1 indicated that growth was restricted to the inoculum vicinity only; level 5 indicated strong growth throughout compost). Results showed *Th2* to be significantly the most virulent taxon (excluding *Th4*, see below). *Th2* scores averaged 3.56 compared with 1.25 for *Th3* and about 1 for *Th1* or any of the other species from the mushroom environment. With *Th2*, colonization was shown to be significantly greater at 27°C than at 18 or 22°C. At 30°C, even *Th1* became potentially damaging.

Th2 was found to grow strongly only in autoclaved or microwaved compost, or in compost that had *Agaricus* growing in it. Fletcher (personal communication) suggested that this was due to the grain in spawned compost providing a food base. He and Grogan *et al.* (1996) found that dusting spawn with a benzimidazole fungicide greatly reduced yield loss. However, Seaby (1996b) found that *Th2* could grow strongly in compost alongside *Agaricus*, even in the absence of grain spawn. Bacteria were isolated from compost that could inhibit *Trichoderma*; compost bacteria are known to be lysed by *Agaricus* (Hayes, 1968). *Agaricus* could thus "aid" *Th2* colonization of compost by eliminating bacterial inhibition.

Cotton bags contaminated inside with *Th2* failed to yield viable spores after 30 min of tumble drying at 60°C. This decontamination method was widely adopted for overnight treatment of overalls.

Results of field experiments using twenty replicate 20 kg bags indicated that one contaminated spawn grain, placed either in freshly spawned compost or after 7 days of spawn run, produced a colony about 30 cm in diameter and that sporulation occurred after casing (at about day 24). However, similar inoculation after 14 days of spawn run only produced four small colonies by day 38. At this stage, the aforementioned treatments had been completely overrun. Addition of micronutrients, such as iron, zinc, copper, or manganese ions, to compost to favour *Agaricus* growth did not reduce green mould.

Other observations on contamination included the following: (a) *Th2* was not found in lagoon water, and addition of lagoon water to phase II compost did not cause green mould. In several isolations from the compost ingredients by Seaby (unpublished observations) or Morris *et al.* (1995a), *Th2* was rarely detected; (b) bags levelled using hands contaminated with *Th2* spores always produced green mould, which usually occurred only in the upper half of the bag; (c) mites taken from contaminated compost that were added either dead or alive to new bags which were then filled with freshly spawned phase II compost often led to green mould. (Seaby, unpublished observations); (d) spawn sprinkled on the surface of the compost was shown to encourage mice (known to carry *Th2* spores). This spawn was also shown to be liable to *Th2* contamination when handled by growers.

Identification of taxa found in the mushroom environment in Ireland

The cultivation method described above, in conjunction with descriptions given by Rifai (1969), identified seven *Trichoderma* species comprising 11 taxa, including four of *T. harzianum.* Three of the latter (*Th1–3*) were found in the UK and Ireland, and *Th4* was found in North America. Figure 12.4 shows isolates of *Th1–4*. Differences in their DNA patterns were shown by Muthumeenakshi *et al.* (1994). A key to cultural characters, after incubating at 27°C, is presented below (Seaby, 1996a).

Th1 (*T. harzianum 1*) produced abundant aerial mycelium, often strongly radially striate, usually sporulating in less than 4 days. Spores formed on tiny pyramidal structures on the aerial mycelium and gave cultures a dusty dull-green appearance. Soon after sporulation, the aerial mycelium collapsed. The growth rate was about 1.1 mm/h. Slow sporing isolates have recently been found in bagging sheds, but these are easily distinguished from *Th2* because later they produce aerial hyphae. *Th1* was very common in the compost ingredients and was occasionally found in compost or on casing, usually not causing problems.

Th2 (*T. harzianum 2*) produced a generous amount of white, sterile, aerial mycelium, apart from a bare central area, and had a smooth colony edge. The mean growth rate at 27°C was 1.1 mm/h and usually took at least 4 days to sporulate. Spores were formed on scattered cushion-like structures. This taxon was found in over 98% of compost advisory cases in Ireland since 1985.

Th3 (*T. harzianum* 3, also co-specific with *T. atroviride*) produced scant aerial mycelium, often radially striate with a wavy colony edge, like that of *T. viride*. However, unlike *T. viride*, it usually produced large central masses of blue-green conidiation within 4 days. *Th3* isolates varied in growth rate at 27°C (like *T. viride*) but had double the mean extension rate (0.85 mm/h) of *T. viride*. Spores of *Th3* were smooth walled and nearly as large as those of *T. viride* (which were rough walled and could also be obovoid). *Th3* phialides were shorter and more compact than *T. viride*. However, both produced cultures that usually smelled of coconut. *Th3* was fairly common in the compost ingredients and was capable of causing extensive green mould in experiments. However, it was only found in 1–2% of advisory cases. *T. viride* was even less common as a cause of green mould and was then usually associated with contaminated spawn.

Th4 (*T. harzianum 4*) produced plentiful aerial mycelium with a wavy colony edge and a faint sour, coconut smell when it was grown as outlined in 12.2.2. The growth rate was 0.9 mm/h. Sporulation occurred within 4 days as cushions and/or small aggregations of phialophores, principally on the surface of the agar, forming a dense yellow-green ring with a white outer edge often around a bare central area. After day 4, aerial mycelium also became dusty in appearance, due to aerial sporulation. Spores were small (2.7–3.1 μm), and small clusters of them rapidly turned from green to yellow in lactophenol mounting fluid, unlike *Th1*, *2*, or *3*. *Th4* appeared to be a better colonizer of unspawned compost and grain spawn than *Th2* (Seaby, 1996a).

The significance of polythene to green mould in the bag method of growing

Opening polythene bags used for compost creates an electrostatic charge. This was shown to attract *Trichoderma*-infected dust (Seaby, 1987, 1996b; Terras and Hales 1995). It also provided a sterile surface, ideal for initial *Th2* growth, as was noted in several experiments when small bags of contaminated compost were observed under the microscope. Carbon dioxide levels greater than 3% develop in compost bags. Experiments have shown that these levels stimulate *Trichoderma* growth when in competition with *Agaricus* (Seaby and Sharma, unpublished).

Contamination routes on tray, block, and shelf farms

In an advisory case on a tray farm, although only a few large green patches of *Th2* were observed, many masses of red-pepper mites appeared on the casing on different trays, thus indicating *Trichoderma* growth. Many mushrooms were also blotched, and yields had declined sharply. On this farm, compost was peak heated in bulk and filled into steamed, wooden trays on a conveyor line. The most convenient sequence of use meant this was not cleaned, in spite of having been used the previous day for casing spawn-run compost. The rollers and guides were found to be heavily contaminated with *Th2*, yet over these ran semi-sterile wooden trays. Compost also became contaminated by contact with workers' overalls and from spray, thrown up by forklift truck wheels. Furthermore, freshly filled boxes of compost were briefly stacked, adjacent to a ramp where spent compost was being dumped! After a complete overhaul of farm hygiene and a large increase in the use of disinfectants, the outbreak was eventually controlled, in spite of cooking-out not being feasible.

On a compost block-producing farm, airborne dust was able to contaminate complex machinery. This had many crevices that retained small amounts of spawned compost where *Trichoderma* was able to proliferate. The plastic wrapping for blocks also electrostatically attracted dust bearing *Th2*.

On an aluminium shelf farm, large patches of *Th2* conidia appeared on the underside of the compost retaining nets. This may have been the cause of extensive blotch on the mushrooms growing beneath. The nets proved difficult to sterilize. Furthermore, because no cooking-out facilities were available, the surrounding area became heavily contaminated when shelves were emptied. At filling this resulted in cross-contamination of the following load. Furthermore, mites escaped house disinfection by getting inside support tubing. The outbreak was eventually controlled by using the shelves for bagged compost only, thus defeating the original labour-saving purpose of the shelves. On tray and shelf farms, especially after infection with *Th2* and *Th4*, efficient cooking-out after cropping is highly recommended for green mould control (see Appendix).

Further control measures

It was found that after *Th2* became established on a compost producer's premises, only unremitting hygiene controlled it. In order to limit cross-contamination, under no circumstances should workers in the bagging shed have contact with growers or growing houses. Trailers entering the bagging shed should be washed free of *Th2*, and load covers (curtain-sided trailers being preferable) should be regularly dipped in a persistent disinfectant. Rolls of new plastic bags and buckets of fly control granules should be treated with the same care as spawn. Moulds such as *Th2* can readily grow on the cardboard of spawn boxes if these are allowed to become damp. Metal pallets are preferable to wooden ones. Spawn should only be handled by one worker using disposable gloves. Mites carrying *Th2* often infest mushrooms sent to market, thus returned pallets should never end up under spawn or bales of bags. In no circumstances should mushrooms and spawn share refrigerators or refrigerated lorries.

Field experience suggests that only strong disinfectants, usually phenol based with residual action, are efficient. Formaldehyde is counter-productive as it appears to leave surfaces suitable for rapid re-colonization, particularly in the bagging shed.

12.3 A *Trichoderma harzianum* group 4 green mould epidemic in North America

An epidemic, similar to the one in Ireland but caused by *Th4*, is outlined by Rinker (1993 and personal communication; see also Appendix). This was first recognized in British Columbia in 1990 and spread to Ontario by autumn 1992. By 1996 it was also found in the Canadian prairie and maritime provinces and losses had totalled an estimated ten million Canadian dollars. In the USA, it reached Pennsylvania in 1993 and has since caused heavy losses (see Appendix). In North America, there are a wide variety of tray, shelf, and bag systems. Wooden (hemlock) shelves are a common container used on older farms and are difficult to sterilize. Furthermore, spawn-run compost is often used for CACing. However, for many years these farms operated successfully, relying mainly on cooking-out with a minimal amount of other hygiene measures prior to the appearance of *Th4*. This epidemic struck some enterprises much more severely than others. *Th4* totally colonized some compost house fills, each valued at thousands of dollars. Many growers were forced out of business. Rinker's recommendations were to reduce inoculum levels by sanitation, especially by effective cooking-out; to prevent vectoring by personnel, insects, or mites; and also to optimize *Agaricus* compost and the subsequent growing conditions.

The origin of *Th4*, like that of *Th2*, is obscure. It is not known if *Th4* is a widespread contaminant of any of the ingredients of compost, which in North America commonly comprises a greater variety of materials, including corn cobs, hay, cocoa husks, or grape pomace.

12.4 Green mould in Australia

In Australia, methods similar to those in the USA are used for growing *Agaricus bisporus*, including trays, shelves, and bags. According to Fletcher (1994), mushroom cap spotting caused by *Trichoderma* appears commonplace and is often associated with distortion. He found green mould in compost on two bag farms and one shelf farm but not on any tray farms. These observations agree with those of Nair and Clift (1993). Among 7 isolates they collected and of the 30 tested using jar experiments similar to those outlined in section 12.2.2, only 10 isolates fractionally colonized compost. They were thus unlike *Th2* or *Th4*.

The latter isolates were not among the 35 Australian isolates sent to Muthumeenakshi for identification by DNA methods. In 9 cases, where a green mould outbreak had occurred, 6 yielded typical *Th3* (*T. atroviride*) and the other 3 isolates were designated *Th1*-Australia (identical to *T. inhamatum* from CBS).

In comparison with Europe and North America, epidemic green mould outbreaks have not been seen. However, two new species have been reported from the mushroom environment, e.g., 7 isolates were identified as *T. inhamatum* and one as *T. parceramosum.*

To control these infections, Fletcher (1994) tested several chemicals, of which phenolics were the most effective.

12.5 The *Trichoderma* spp. found on mushroom casing

Trichoderma spp. are widely found in peat and soil (Danielson and Davey, 1973), and most small patches of sporulation found on casing, particularly at the end of cropping, appear harmless. In Northern Ireland, these are usually *T. viride* or less often *T. hamatum*; *Th1* or *Th3* are only occasionally found. Casing colonization by *Trichoderma* was shown by Visscher (1976) to dramatically increase at a pH below 7. Self-heating in large piles of compost destined to become casing can lead to strong growth of *T. viride*, *Th1*, and several other *Trichoderma* spp. which later appear on the casing.

BINAB-T, a proprietary mix of *T. harzianum and T. polysporum* (Richard, 1981) that is manufactured for control of wood rots, was tested as a means to control *Verticillium* spp. on casing. However, it was found to stimulate red-pepper mites and also to damage mushrooms (Morris, personal communication).

Mushrooms have often been reported in the literature as being killed by *T. koningii*, which causes them to decay, first producing downy white mycelial growth on which large numbers of spores are eventually produced (Fletcher *et al.*, 1986). This symptom may also be found when *Th2* grows over casing and onto mushrooms (Seaby, unpublished). The common factor in the abundant mycelial growth could be that both *T. koningii* and *Th2* are fast growing but slow to sporulate. Occasionally, *T. viride* mycelium, which also sporulates slowly, may also be found growing over living mushrooms.

12.6 Summary and Discussion

12.6.1 Impact of bioengineering on mushroom growing?

Because of the losses they have caused, there has been considerable speculation in the mushroom industry as to the origin of *Th2* and *Th4*. Furthermore, if both taxa were present on the same farm their impact might well be cumulative. Hence there is an urgent need to confine their distribution. However, there has been research interest shown in the use of *Th2* for the biocontrol of pathogens affecting potatoes, tree wounds, linseed, rape, and glass-house seedlings, which, in the last case, were even associated with the proposed use of DNA-mediated transformation (Seaby, unpublished). While the application of *Th2* (or *Th4*) in the rhizosphere of glass-house seedlings may not immediately impact on mushroom compost production, use in the control of soil-borne diseases of cereals could introduce them into straw and hence directly into compost. The same would apply to the use of *Th2* as a biocontrol agent for vine-leaf pathogens, as grape pommace is sometimes an ingredient of mushroom compost.

At the moment, the origin of *Th2* and *Th4* is not known, but their DNA-sequence analysis suggest that they are of natural and not of artificial (genetically modified) origin (Kubicek, personal communication).

However, there appear to be two possible scenarios where bioengineered *Trichoderma* spp., particularly *Th2* and *Th4*, could have a negative impact: one would be if they were made resistant to bacterial competition, since this normally limits their growth in compost, and the other possibility involves inducing resistance to benzimidazole fungicides such as benomyl, since these could provide a useful control measure as outlined in section 12.1.2.

12.6.2 The origin and consequences of green mould epidemics

Green mould of *Agaricus bisporus* compost can be caused by any one of at least five *Trichoderma* taxa that have either "found" or "become adapted" to an important competitor/pathogen role. Losses amounting to millions of UK pounds or tens of millions of US dollars have resulted, particularly from epidemics of *Th2* and *Th4*. These have put some compost makers and many growers out of business. Even when very strict hygiene measures have led to control, there are still high economic, and possibly environmental or even health, costs resulting from the constant liberal use of strong disinfectants, usually phenolics. The cost of adding a sufficient amount of a selective fungicide such as benzimidazole to compost would be prohibitive, except in the short term. The use of this for dusting spawn (as suggested by Fletcher) shows great promise as an additional control measure. However, there is some evidence that resistance may soon develop if this is widely employed (see also section 12.6.1).

12.6.3 Six avenues leading to contamination of compost

- *Faulty peak heating.* Faulty peak heating may lead to serious green mould outbreaks, depending on the amount of inoculum and its virulence. However, with modern peak-heat tunnels and formulations that produce "active" compost, the experience in Ireland is that this is rare. Furthermore, it will probably only involve a taxon with moderate virulence. This is because compost production and mushroom growing are now often well separated, thus reducing the risk of the compost ingredients becoming cross-contaminated by a taxon that has successfully colonized compost in a growing house.

- *Cross-contamination from growing houses to bagging shed machinery.* Even when the build-up of inoculum in growing houses is remote from compost production, it is often accompanied by masses of spore-carrying red-pepper mites and mushroom flies. These, along with humans and inanimate vectors such as pallets, trailers, and load covers, are often sufficient to recontaminate small pockets of spawned compost hidden in the bagging shed machinery. This usually results in outbreaks of green mould that continue until they are eliminated.
- *Cross-contamination from production yard to bagging shed.* After peak heating, compost (or the outer bag) can become contaminated with *Trichoderma*-bearing dust blown from the substrates that are being mixed in the compost yard. Spores on the bag surface may later be transferred to the compost surface during levelling but, as described above, these *Trichoderma* contaminants are usually of low virulence.
- *Contaminated compost used for making CACing.* When a bag of *Th2*-contaminated compost is accidentally used to make CACing, the result can sometimes be a high proportion of bags with a broad green layer of *Trichoderma* sporulation just below the casing, which terminates mushroom production.
- *Accidental contamination of spawn.* Green mould can result by means of accidental colonization either when manufactured or possibly later by mice in the spawn store or by faulty handling. The importance of care in handling spawn to prevent contamination by either *Th2* or *Th4* cannot be over-emphasized.
- *Cross-contamination as a grower's responsibility.* Growers can contaminate compost with *Trichoderma* spores by transferring them from the outside of the bag to the inside during levelling, from hands (which have touched a contaminated object, such as a door knob) or from clothing, if the leveller has not changed this after emptying a contaminated house. Dust created during emptying may contaminate the ventilation system and later be blown onto fresh compost. Unless controlled, mice may carry mites and spores and move from bag to bag in search of grain spawn, resulting in contamination. This also applies particularly to mites and also to Sciarids, Phorids, and other small flies attracted to compost.

12.6.4 *The importance of taxon identification*

Compost ingredients harbour several *Trichoderma* species. Recycled lagoon water, used to wet straw, may have up to 20 000 *Trichoderma* propagules per mL. Furthermore, a wide variety of *Trichoderma* spp. (taxa) are found in the mushroom environment. It might thus appear pointless to identify the one involved in a green mould outbreak. However, experience suggests that compost makers and growers who have *Th2* or *Th4* on their farms must employ a much higher level of hygiene than those who only have *Th1* (*T. harzianum*) or *Th3* (*T. atroviride*). The two former taxa warrant international scheduling in order to limit their movement from country to country.

12.7 Appendix

After completion of this article, the author became aware of the following information, presented at a Mushroom Green Mould Round Table seminar held at the Department of Plant Pathology, Pennsylvania State University, in December 1996 and chaired by Prof. E. L. Stewart.

12.7.1 *Strain identification and geographic distribution*

J. Bissett *et al.* found that a Biolog identification system using 96-well microplates and photometrically measuring growth could be reliably used to distinguish *Th4* from other *Trichoderma* spp. X. Chen *et al.* were unable to detect *Th4* and *Th2/Th4* using RAPD and PCR amplification of defined DNA-regions, respectively, among 17 isolates of *Trichoderma* spp. that were collected from the USA mushroom industry from 1950 to 1990, prior to the current *Th4* epidemic.

D. L. Rinker recorded *Th4* in British Columbia in 1990, in Ontario in late 1992, and in Alberta, Manitoba, Prince Edward Island, and New Brunswick in 1996. Losses were not recorded; however, the number of farms affected increased year after year (e.g., in Ontario, which produces 50% of Canada's mushrooms, 56% of farms are currently having green mould problems). P. Wuest *et al.* reported the arrival of *Th4* epidemics in Pennsylvania in late 1994. Where green mould was observed in the compost, losses ranged from 15 to 40%. In 1995 *Th4* arrived in Chester County, Pennsylvania, and losses ranged from 30% to 100%; this continued until 1996. Currently, growers consider management successful if losses are held to 5%.

F. Geels reported that the first large-scale problem with *Trichoderma* spp. occurred on cut stumps in The Netherlands in 1971–76. This was counteracted by removing stumps after each flush, applying benlate and raising the casing soil pH to above 7.6. In 1978–81, serious *Trichoderma* problems were associated with a virus problem. In 1994, an epidemic at CNC caused losses that were calculated to be 6% of total production, valued at 31 million Dutch guilders. This could be traced back to contamination of spawn during a commercial repacking process. The species found were *Th1* (or *T. inhamatum*, which could not be distinguished by rDNA typing) and *Th3*. *Th2* was not found. Stopping the repacking and increasing cleaning and disinfection have eliminated the problem. The authors also reported experiments with several *T. harzianum* isolates, including *Th3*, that caused only minimal symptoms except for cap spotting. They did not lead to green mould development when compost quality was reduced or wetted, even at an incubation temperature of 28°C. It should be noted that these findings contrast with those of Seaby (1996b) and Grogan and Gaze (1995), who found some *Th3* isolates virulent.

J. J. Lelley investigated 129 *Trichoderma* isolates obtained from 16 mushroom farms and identified them as *T. harzianum* (which were not further analysed for identity as *Th1–4*), *T. viride*, *T. koningii*, *T. pseudokoningii*, and *T. longibrachiatum*. Strong *Trichoderma* contamination was recorded on nine of the farms and very strong contamination on one.

12.7.2 Reasons for green mould development

Standard cooking-out was reported by Rinker not to be sufficient to kill *Th4*; this required 42 h at 68°C. Since this temperature can damage the facilities, heating for even longer times at lower temperatures should be used.

N. Nair found that in Australia (where *Th2* has not yet been recorded) inoculation of compost with 3×10^8 spores of a not further identified "*T. harzianum*" caused green mould symptoms, whereas 3×10^3 spores did not cause significant yield loss. Similarly, Grogan *et al.* (1995) found that yield loss was dependent on the concentration of spores of *Th2* used for inoculation.

A correlation was reported by Rinker between the type of supplement used and *Th4* contamination, feather-meal being unique as it led to an increase in yield, whereas all other supplements reduced yield in the presence of *Th4*.

During pre-casing stages, inoculation of the trays with *Th4* and incubation at different temperatures showed that green mould contamination was optimal between 25 and 30°C but virtually zero at 37°C; however, 24 days after casing, all trays showed green mould.

Green mould colony size was found to be significantly smaller with brown mushrooms than with either white or off-white strains.

References

Beach, W. S. 1937. Control of mushroom diseases and weed fungi. *Pennsylvania State College of Agriculture Bulletin* **351**: 1–32.

Bisby, G. B. 1939. *Trichoderma viride* Pers. & Fries and notes on *Hypocrea. Trans. Br. Mycol. Soc.* **23**: 149–168.

Danielson, R. M. and Davey, C. B. 1973. The abundance of *Trichoderma* propagules and distribution of species in forest soils. *Soil Biol. Biochem.* **5**: 405–454.

Doyle, O., Morris, E., and Clancy, K. 1990. *Trichoderma* green mould: research at University College Dublin. *Irish Mushroom Review* **2**: 6–7.

Fergus, C. L. 1978. The fungus flora of compost during mycelium colonisation by the cultivated mushroom *Agaricus brunnescens. Mycologia* **56**: 267–284.

Fermor, T. R., Randle, P. E., and Smith, J. F. 1985. Compost as a substrate and its preparation. In P. B. Flegg, D. M. Spencer, and D. A. Wood (eds), *The Biology and Technology of the Cultivated Mushroom*, Wiley, New York, pp. 81–110.

Fletcher, J. T. 1994. Preliminary investigations of the pathogens and weed mould problems of mushrooms in Australia. Unpublished report of project Mu/303 Horticultural Research and Development Corporation, Gordon, NSW, 2072, Australia.

Fletcher, J. T., White, P. F., and Gaze, R. H. 1986. *Mushrooms: Pest and Disease Control*, 1st edn. Intercept, Andover, UK, pp. 94–95.

Gandy, D. G. 1977. Report of the Glasshouse Crops Research Institute, Littlehampton, UK, p. 118.

Grogan, H. M. and Gaze, R. H. 1995. Growth of *Trichoderma harzianum* in traditional and experimental composts. In T. J. Elliot (ed.), *Mushroom Science XIV*, Vol. 2. Balkema, Rotterdam, pp. 653–660.

Grogan, H. M., Noble, R., Gaze, R. H., and Fletcher, J. T. 1996. Control of *Trichoderma harzianum* – a weed mould of mushroom cultivation. Proceedings of the Brighton Crop Protection Conference – Pests and Diseases, Vol. 1, pp. 337–342.

Harvey, C. L., Wuest, P. J., and Schisler, L. C. 1982. Diseases, weed moulds, indicator moulds and abnormalities of the commercial mushroom. In P. J. Weust and G. D.

Bengtson (eds), *Penn State Handbook for Commercial Mushroom Growers*. Pennsylvania State University, pp. 19–33.

HAYES, W. A. 1968. Microbiological changes in composting wheat straw/horse manure mixtures. *Mushroom Sci.* **7**: 173–183.

HUSSEY, N. W., READ, W. H., and HESLING, J. J. 1969. In *The Pests of Protected Cultivation*, Edward Arnold, London, pp. 348–350.

KLIGMAN, A. M. 1950. *Handbook of Mushroom Culture*. J. B. Swayne, Kennet Square, Pennsylvania.

KNEEBONE, L. R. and MEREK, E. L. 1959a. Brief outline of and controls for mushroom pathogens, weed moulds and competitors. *Mushroom Growers Assoc. Bull.* **113**: 146–153.

KNEEBONE, L. R. and MEREK, E. L. 1959b. Brief outline of and controls for mushroom pathogens, weed moulds and competitors. *Mushroom Growers Assoc. Bull.* **114**: 190–193.

KOMATSU, M. 1976. Studies on *Hypocrea*, *Trichoderma* and allied fungi antagonistic to shiitake, *Lentinus edodes*. Report of the Tottori Mycological Institute **13**: 1–113.

LELLEY, J. 1987. Disinfection in mushroom farming – possibilities and limits. *Mushroom J.* **174**: 181–187.

LIAO, Y. M. 1985. Efficacy of fungicides on control of *Trichoderma* spp. in sawdust cultivation of shiitake. *J. Agric. Res. China* **34**: 329–340.

MORRIS, E., DOYLE, O., and CLANCY, K. J. 1995a. A profile of *Trichoderma* species I and II. *Mushroom Sci.* **14**: 611–618.

MORRIS, E., DOYLE, O., and CLANCY, K. J. 1995b. A profile of *Trichoderma* species I and II. *Mushroom Sci.* **14**: 619–626.

MUTHUMEENAKSHI, S. 1996. Molecular taxonomy of the genus *Trichoderma*. Ph.D. Thesis, The Queen's University of Belfast.

MUTHUMEENAKSHI, S. and MILLS, P. R. 1995. Detection and differentiation of fungal pathogens of *Agaricus bisporus*. *Mushroom Sci.* **14**: 603–609.

MUTHUMEENAKSHI, S. and MILLS P. R. 1996. Diagnostic PCR for *Trichoderma harzianum* (Group 2), an aggressive coloniser of mushroom compost. In F. M. McKim (ed), British Crop Protection Council Symposium Proceedings 65, Symposium on Diagnostics in Crop Protection. University of Warwick, Coventry, UK, pp. 217–222.

MUTHUMEENAKSHI, S., MILLS, P. R., BROWN, A. E., and SEABY, D. A. 1994. Intraspecific molecular variation among *Trichoderma harzianum* isolates colonising mushroom compost in the British Isles. *Microbiology UK* **140**: 769–777.

NAIR, T. and CLIFT, A. D. 1993. Integrated pest and disease management, Unpublished report of project Mu/002 Horticultural Research and Development Corporation, Gordon, NSW, 2072, Australia.

POPPE, J., WELVAERT, W., and DE BOTH, G. 1985. Diseases and their control-possibilities after ten years of *Pleurotus* culture in Belgium. *Medelingen van de Faculteit Landbouwweterschappen Rijksuniversiteit Gent* **50**: 1097–1108.

RICHARD, J. L. 1981. Commercialisation of a *Trichoderma* mycofungicide – some problems and solutions. *Biocontrol News and Information* **2**: 95–98.

RIFAI, M. A. 1969. A revision of the genus *Trichoderma*. *Mycol. Papers* **116**: 1–56.

RINKER, L. R. 1993. Commercial mushroom production, Publication 350. Ontario Ministry of Agriculture and Food, Canada.

SAMUELS, G. J. 1996. *Trichoderma*: a review of biology and systematics of the genus. *Mycol. Res.* **100**: 923–935.

SEABY, D. A. 1987. Infection of mushroom compost by *Trichoderma* species. *Mushroom J.* **179**: 355–361.

SEABY, D. A. 1989. Further observations on *Trichoderma* species. *Mushroom J.* **197**: 147–151.

SEABY, D. A. 1995. Mushroom (*Agaricus bisporus*) yield modelling for the bag method of mushroom production using commercial yields and from microplots. *Mushroom Sci.* **14**: 409–416.

SEABY, D. A. 1996a. Differentiation of *Trichoderma* taxa associated with mushroom production. *Plant Pathol.* **45**: 905–912.

SEABY, D. A. 1996b. Investigation of the epidemiology of green mould of mushroom (*Agaricus bisporus*) compost caused by *Trichoderma harzianum*. *Plant Pathol.* **45**: 913–923.

SHANDILYA, T. R. and GULERIA, D. S. 1984. Control of green mould (*Trichoderma viride* Pers. ex Gray) during cultivation of *Agaricus bitorquis*. *Indian J. Plant Pathol.* **2**(1): 7–12.

SINDEN, J. W. and HAUSER, E. 1953. Nature and control of three mildew diseases of mushrooms in America. *Mushroom Sci.* **2**: 177–180.

STAUNTON, L. 1987. *Trichoderma* green mould in mushroom compost. *Mushroom J.* **179**: 362–363.

STEANE, R. G. 1978. Monitoring of disease and pest levels in the mushroom crop as a guide to the application of control measures. *Mushroom Sci.* **9**: 281–302.

TERRAS, M. A. and HALES, D. F. 1995. Red-pepper mites are vectors of *Trichoderma*. *Mushroom Sci.* **14**: 485–490.

TOKIMOTO, K. 1982. Lysis of the mycelium of *Lentinus edodes* by mycolytic enzymes of *Trichoderma harzianum* when the two fungi were in antagonistic state. *Trans. Mycol. Japan* **23**: 13–20.

VISSCHER, H. R. 1976. De pH van dekaarde en het optreden van *Trichoderma*, mede in verband met mechanisch afsnijden. *De Champignoncultuur* **20**: 35–43.

YOSHIDA, T. and TAKAO, S. 1982. Distribution of fungi in artificial culture of *Poliota nameko* in Hokkaido. *Memoirs of the Faculty of Agriculture Hokkaido University* **13**: 81–101.

PART THREE

Protein Production and Application of *Trichoderma* Enzymes

13

Industrial mutants and recombinant strains of *Trichoderma reesei*

A. MÄNTYLÄ,* M. PALOHEIMO† and P. SUOMINEN†

*Röhm Enzyme Finland Oy, Rajamäki, Finland

†Roal Oy, Rajamäki, Finland

13.1 Introduction

Industrial use of *Trichoderma reesei* began in the 1980s. *Trichoderma* cellulases were first used in the animal feed industry and in some food applications such as clarifying juices. During the past ten years, the use of cellulolytic enzymes in general, including *Trichoderma* cellulases, has increased tremendously, the main reason being their increasing use in the textile industry. It can be estimated that cellulases and hemicellulases account for about 20% of the world enzyme market today, and a substantial part of this is *Trichoderma* enzymes.

The first industrial production strains were mutants that produced elevated amounts of cellulases. Motivation for the mutagenesis work came from the concept of hydrolyzing cellulosic wastes to glucose and then fermenting it to ethanol. However, this application is still not in commercial use. Instead, cellulases and hemicellulases of *T. reesei* are used in applications of feed, food, textile, and pulp and paper industries. In these applications the purpose is to modify cellulosic fibers; thus many of the current products on the market are produced by strains genetically modified to produce a mix designed for each application.

For the production of technical enzymes and proteins, the yield and secretion into the culture medium are important factors. Classical strain improvement methods such as mutagenesis and screening as well as modification of fermentation conditions have resulted in significant increases in protein yields. Mutant strains of *T. reesei* secrete substantial amounts of proteins into the culture medium, up to 40 g/l, in optimized cultivation conditions (Durand *et al.*, 1988b). The major proteins are cellulases and, of these, cellobiohydrolase I (CBHI), coded by a single gene, forms the major part. A more recent approach for strain improvement has been to use genetic engineering to further increase the yields and to tailor the secretion patterns to produce better mixtures for specific applications. Efficient gene expression systems are required for production of proteins in large quantities. In *T. reesei* the use of the strong inducible *cbh1* promoter has proven useful, yields from several grams to tens of grams per liter of homologous or heterologous fungal proteins have been obtained. Stable recombinant strains have been obtained following integrative recombination of the transforming DNA.

13.2 Industrial mutants of *Trichoderma reesei*

Interest in *T. reesei* was generated during the oil crisis of the 1970s, when saccharification of cellulose to glucose and its subsequent conversion to ethanol as a fuel source became economically attractive. However, the high cost of enzyme production due to the low yields of the enzymes produced by the available microbial strains prohibited economical use. Strain improvement using classical genetic breeding methods such as mutation and screening was initiated: (1) to increase enzyme yield through the production of hypercellulolytic mutants of *T. reesei*, (2) to isolate catabolite repression-resistant mutants, (3) to isolate mutants resistant to end-product inhibition, (4) to isolate constitutive mutants, so that inexpensive nitrogen and carbon sources could be used as medium components without reducing enzyme yield, and (5) to isolate mutants with elevated β-glucosidase production.

Chemical mutagens, UV irradiation, X-rays, or mixed treatments have been used for the generation of high-yielding *T. reesei* mutants. Screening methods have been devised for specific detection of different types of mutants, i.e., either high-producing or regulatory cellulase mutants. In screening for mutants that hyperproduce all of the enzymes in the cellulase complex, phosphoric-acid-swollen Walseth-cellulose (Walseth, 1952) has been most commonly used as a substrate. Selection of hyperproducing endoglucanase mutants is achieved by substituting a modified cellulase such as carboxymethylcellulose (CMC) for the acid-swollen cellulose (Montenecourt and Eveleigh, 1979). β-Glucosidase mutants can be detected by a number of plate assay techniques; e.g., 4-methylumbelliferyl-β-D-glucoside (Kawamori *et al.*, 1986), p-nitrophenyl-β-D-glucoside (pNPG) and 5-bromo-4-chloro-3-indolyl-β-D-glucoside (Durand *et al.*, 1984) have been used as substrates.

Catabolite repression-resistant mutants have been isolated by adding high concentrations (5% or more) of a catabolite repressor such as glucose or glycerol in combination with the specific cellulosic substrate. The anti-metabolite 2-deoxyglucose has been particularly useful in selecting catabolite repression-resistant mutants since at concentrations above 0.5% it can be used as a catabolite repressor as well as an anti-metabolite (Montenecourt and Eveleigh, 1979).

Mutants resistant to end-product inhibition are obtained by adding high concentrations of the end-product to the selective screening system. Constitutive mutants may be isolated by plating the survivors from mutagen treatment on agar-salts medium containing a carbon source (e.g., glucose or sucrose) that is not an inducing substrate. Following growth, the plates are overlayed with the specific selective-screening medium as described above.

The particular *T. reesei* strain QM6a (Mandels and Reese, 1957) is as far as we know the initial parent of practically all industrially relevant cellulase production strains today. In the early 1970s, the mutant *T. reesei* QM9414 was isolated by Mandels (1975) at Natick Laboratories, MA, USA. Further mutagenesis of QM9414 has produced several high-yielding mutant series in a number of laboratories. Strains developed at Rutgers University originated from the wild-type QM6a strain and form a separate line of high-producing *T. reesei* mutants. The different cellulase mutant series are presented in Table 13.1.

In a number of isolated hyperproducing mutants, there has been enhanced release of extracellular proteins in general, including enzymes of the cellulase complex as well as unrelated enzymes. However, specific changes in the reactive amounts of individual enzymes have been observed. In Rut-NG14 and Rut-C30

Table 13.1 *Trichoderma reesei* high-producing mutant series

Mutant series[a]	Objectives	References
US Army Natick Laboratories, USA QM6a → Natick mutants (incl. QM9414) Rut-C30 → MCG 80	Increased and catabolite repression-resistant production of cellulase	Mandels *et al.*, 1971; Mandels, 1975; Gallo *et al.*, 1978; Andreotti *et al.*, 1980
Rutgers University, USA QM6a → Rutgers' mutants	Increased, catabolite repression-resistant and constitutive production of cellulase; increased production of β-glucosidase	Montenecourt and Eveleigh, 1977, 1979; Montenecourt, 1983; Sheir-Neiss and Montenecourt, 1984
Gulbenkian Institute, Portugal QM9414 → MG strains	Increased and derepressed β-glucosidase production	Beja De Costa and van Uden, 1980
Technical Research Center, Finland QM9414 → VTT series	Increased cellulase production	Nevalainen *et al.*, 1980; Bailey and Nevalainen, 1981
Cetus Corporation, USA QM9414 → L series	Constitutive production of cellulase and β-glucosidase	Shoemaker *et al.*, 1981
Slovak Academy of Sciences, Slovakia QM9414 → Slovak series	Increased and catabolite repression-resistant production of cellulase and β-glucosidase	Farkaš *et al.*, 1981; Labudova and Farkaš, 1983
Indian Institute of Technology, India QM9414 → C strains	Constitutive cellulase production	Mishra *et al.*, 1982
Cayla Laboratories, France QM9414 → CL series	Increased, catabolite-repression resistant and constitutive production of cellulase and β-glucosidase	Durand *et al.*, 1984, 1988a
Kyowa Hakko Kogyo, Co., Japan QM9414 → Kyowa series	Increased and catabolite repression-resistant cellulase production; increased β-glucosidase production	Kawamori *et al.*, 1985, 1986

[a] The parental strain precedes the arrow; the mutant strain or series follows it.

mutants, the cellobiohydrolase is specifically hyperproduced relative to the rest of the enzymes in the cellulase complex (Montenecourt and Eveleigh, 1979). Increase of the amount of an individual cellulase component, endoglucanase, has been reported (Sheir-Neiss and Montenecourt, 1984; Shoemaker *et al.*, 1981). Thus, although cellulases tend to be produced together, different strains appear to differ in the control of the ratios produced. In addition, different ratios have been produced on different substrates (Shoemaker *et al.*, 1981).

Some mutants have shown considerable morphological differences when compared with the parent strain, with the most noticeable being a higher degree of branching of the mutant hyphae. The branched mutants produced 2 to 3 times higher levels of β-glucosidase than the parent strain QM9414 (Farkas *et al.*, 1981). In the case of Rut-C30, increased hyperproduction of cellulases has been attributed to the presence of increased endoplasmic reticulum (ER) leading to enhanced synthesis and/or secretion of cellulases (Ghosh *et al.*, 1982; Montenecourt and Eveleigh, 1979).

Comparisons of the enzyme yields and productivity between different cellulolytic mutant strains have been difficult because the published results are from cultivations where different substrates and culture conditions have been used. Different enzyme activity measurements have been used to follow the productivity and, as mentioned earlier, the amounts of individual enzymes in the cellulase complex may vary among the high-producing mutant strains. As reviewed by Persson *et al.* (1991), the relative contributions from strain development, nature of substrate, substrate concentration, and cultivation conditions on the improvements in enzyme production with *T. reesei* strains were found to be of the same order of magnitude.

Successful mutagenesis, screening, and process development work done at both academic and industrial laboratories led to *T. reesei* strains feasible for commercial cellulase production. The classical strains have not lost their importance after the appearance of gene technology but have been further developed, mainly in industrial laboratories. Mutant *T. reesei* strains are used as production organisms for cellulases and hemicellulases for all the main applications. Genetically modified industrial strains originate from these mutants, and thus the high secretion capacity is further exploited.

13.3 Industrial recombinant strains of *Trichoderma reesei*

Use of random mutagenesis often leads to strains in which the production of all enzymes in the cellulase complex is improved (see above). Using cloned *T. reesei* genes and their promoters, novel strains with completely different cellulase profiles, e.g., devoid of one or more cellulase components, have been constructed. Development of rDNA techniques and tools for *T. reesei* presented an opportunity to use this fungus for overproduction of native proteins as well as for production of heterologous fungal or non-fungal proteins.

13.3.1 *Expression vectors*

The expression vectors that have been used are shuttle vectors containing sequences required for replication and selection in *Escherichia coli*, in addition to the expression cassette. A selection marker, necessary to select transformants, may be included on the shuttle vector or on a separate vector since co-transformation works efficiently in *T. reesei*. The efficiency of co-transformation with unselected DNA is high – approximately 80% (Penttilä *et al.*, 1987). Dominant selectable markers are most preferable for industrial strains, since they provide a system which overcomes the requirement of auxotrophic mutants. The *Aspergillus nidulans amdS* (acetamidase) has been used as a dominant selectable marker also in *T. reesei*, which is naturally unable to utilize acetamide as a sole nitrogen source (Penttilä *et al.*, 1987). Markers conferring resistance against antibiotics have also been used successfully, including phleomycin (Durand *et al.*, 1988a; Harkki *et al.*, 1991) and hygromycin B (Mach *et al.*, 1994; unpublished results of Primalco Biotec). In transformation of *T. reesei* auxotrophic mutants to prototrophy, vectors bearing *A. nidulans argB* (Penttilä *et al.*, 1987), *A. nidulans trpC* (Koivula *et al.*, 1996), *Neurospora crassa pyr4* (Gruber *et al.*, 1990a), *A. niger pyrA* (Gruber *et al.*, 1990a), and homologous *T. reesei pyr4* (Berges and Barreau, 1991; Gruber *et al.*, 1990b; Smith *et al.*, 1991) have been used. Transformation methods and selectable markers used in fungal transformations are described in detail in Volume 1, Chapter 10.

Commonly used expression cassettes, as exemplified by an *egl1* expression cassette in Figure 13.1, include, in addition to the selection marker, the promoter and transcriptional regulatory sequences, the DNA sequence encoding a secretory signal peptide, the coding region for the desired protein product, and the sequences required for the termination and processing of the mRNA transcript. Flanking regions to target the expression cassette into a specified locus are also included when appropriate. The expression cassette is preferentially cut from the vector backbone and transformed to *T. reesei* as a linear fragment to avoid the introduction of unnecessary foreign DNA into the fungal genome. In cases where inactivation or deletion of the gene encoding harmful or unnecessary enzyme activity is wanted, a

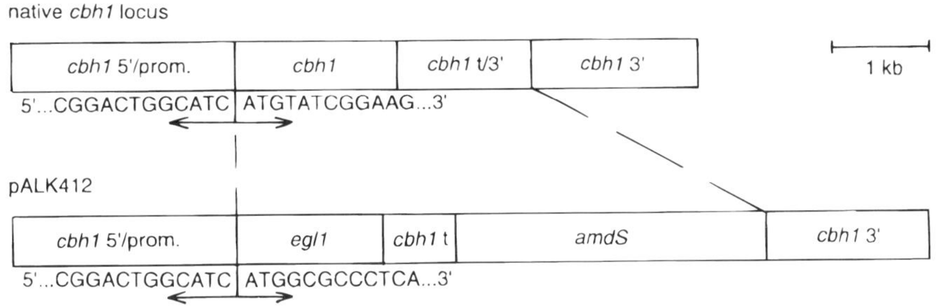

Figure 13.1 Replacement of the *cbh1* locus by the 9.4-kb *egl1* expression cassette isolated from the plasmid pALK412. The *cbh1* 5′-promoter region is precisely fused by PCR to the gene to be expressed. *cbh1 t* indicates the terminator region following the STOP codon (0.7 kb) to ensure the termination of transcription. The *cbh1 t/3′* region (1.4 kb) of the native *cbh1* locus contains the terminator sequence and additional downstream sequence. The transformation cassette contains the gene encoding a selection marker, in this case acetamidase (*amdS*). The *cbh1* 3′ flanking region further downstream of the *cbh1 t/3′* region together with the 5′-promoter region target the expression cassette to the *cbh1* locus. After homologous recombination, the 3 kb region containing the native *cbh1* coding region and the 1.4 kb area downstream of the cbh1 STOP codon is replaced with the 5.5 kb fragment containing the *egl1* gene, *cbh1* terminator, and *amdS* marker gene.

similar construct is used, but only the selection marker is linked between the flanking regions of the gene of interest.

Promoter and terminator sequences

For the production of industrial enzymes, minimal downstream processing is advantageous, and thus it is preferable that the desired protein is secreted to the culture medium as a major component. It is important to be able to control promoter activity easily and efficiently so that the synthesis of the recombinant protein is optimal. For this purpose it is necessary to have an expression vector with a strong, regulated promoter such as the homologous cellobiohydrolase 1 (*cbh1*) promoter. Use of the strong promoter that produces large amounts of mRNA substrate can compensate in part for mRNA instability, poor translation efficiency, or an unstable product (Sawers and Jarsch, 1996). Use of promoters active in glucose-based media may provide another possibility for producing the desired enzyme free of most other hydrolases (Nakari *et al.*, 1993). Several *T. reesei* genes active in the presence of glucose have been isolated, as reviewed by Nyyssönen *et al.* (1995).

The 3′-untranslated region (terminator) of a homologous gene is usually placed after the protein-encoding DNA sequence to ensure termination and processing of the mRNA transcript. Terminator sequences may also have an effect on mRNA stability.

Secretion signal

The secretion signal is a short aminoterminal leader sequence present on secreted proteins that directs the transfer of the newly synthesized protein across the membrane of the endoplasmic reticulum. The signal sequence is proteolytically cleaved from the mature protein during this translocation. Signal sequences from filamentous fungi are similar to those of higher eukaryotes (Salovuori *et al.*, 1987). Although information concerning the effect of different secretion signals on protein expression and secretion in filamentous fungi is limited, it is generally observed that heterologous signal sequences, including that of calf chymosin (Harkki *et al.*, 1989) and barley endopeptidase B (Nykänen *et al.*, 1997), are recognized by *T. reesei.* More glucoamylase activity was produced by a *T. reesei* transformant strain containing the *Hormoconis resinae* glucoamylase (*gamP*) gene with the natural N-terminal extension than by that containing the *gamP* gene with the *cbh1* gene signal sequence and the codon for the first amino acid of mature CBHI (Joutsjoki *et al.*, 1993). The expression conditions were standardized by targeting one copy of the *gamP* expression cassette to the *cbh1* locus of the host. This shows that instead of only signal peptide regions directing the translocation procedure, the entire structure of the exported proteins may have an important effect on the level of secretion in filamentous fungi.

Genomic DNA or cDNA?

To express the desired protein, either cDNA or a genomic gene can be used. The intron sequences had a positive effect on expression of *Hormoconis resinae* glucoamylase in *T. reesei* (Joutsjoki, 1994). A transformant in which the *cbh1* locus was replaced by a single copy of the glucoamylase genomic gene (expressed from the *cbh1* promoter) secreted more active glucoamylase than did a transformant strain having three copies of the cDNA clone in tandem orientation in the *cbh1* locus.

Similar results were obtained with laccase. Saloheimo and Niku-Paavola (1991) described the expression of the cDNA and genomic gene of *Phlebia radiata* laccase enzyme in *T. reesei.* The amount of laccase-specific mRNA in transformants containing a genomic gene clone was reported to be higher than that in transformants containing the cDNA clone.

These results leave the structure of genomic genes as the probable reason for higher enzyme yields. The reason for this is unknown at the moment; however, fungal genes may carry transcriptional regulatory elements within their introns as shown with higher eukaryotes (Banerji *et al.*, 1983). This could result in enhanced transcription of the genomic clone or a more stable and translation-competent mRNA.

Targeted integration

In *T. reesei*, stable transformants have been obtained due to integrative recombination of the transforming DNA (Penttilä *et al.*, 1987). Stability of the industrial production strains is important since loss of vector DNA can result in decreased production yields. It is necessary to purify the individual transformants by successive platings on selective and non-selective media. This approach has been very successful, and mitotic stabilities of up to 100% have been reported for *argB* transformants (Penttilä *et al.*, 1987).

The frequency of homologous integration of the circular vector DNA in *T. reesei* is low (Harkki *et al.*, 1991; Karhunen *et al.*, 1993; Seiboth *et al.*, 1992) and seems to vary according to the target locus. If targeted integration of the transforming DNA is desired, as in the case of gene inactivations and replacements, transformation with a linear DNA fragment is more advantageous. High replacement frequencies, ranging from 32% to 84% were obtained with all major cellulase loci, *cbh1*, *cbh2*, *egl1*, and *egl2*, when over 1 kb 5′ and 3′ flanking regions surrounding the selectable marker gene were used (Paloheimo *et al.*, 1993; Suominen *et al.*, 1993). Considerable decrease in the replacement frequency was observed with DNA fragments harboring shorter flanking regions (Suominen *et al.*, 1993).

13.3.2 *Altering the production profile using gene replacements*

The ability to inactivate or delete genes encoding major secreted proteins is important for construction of suitable host strains for the production of homologous or heterologous proteins (Berka *et al.*, 1991; Harkki *et al.*, 1991; Koivula *et al.*, 1996; Seiboth *et al.*, 1992; Suominen *et al.*, 1993). To prepare suitable enzyme mixtures for different applications, the use of a one-step gene replacement procedure has proven successful (Karhunen *et al.*, 1993). Genes encoding enzymes having harmful or unnecessary activities for the specified application can be efficiently replaced by genes coding for enzymes with desired activities. Use of such strains in production has improved the quality of the products; for example, strains secreting elevated amounts of β-glucanases and reduced amounts of cellobiohydrolases (Karhunen *et al.*, 1993) produce more effective enzyme mixtures for feed improvement than classical strains. Furthermore, production costs are lowered when the desired enzyme, in this case β-glucanase, is the major secreted protein instead of CBHI, which accounts for over 50% of proteins secreted by mutant strains.

Genes for the major cellulase CBHI, CBHII, EGI, and EGII and their promoters have been replaced with a marker gene (Suominen *et al.*, 1993). Using similar vectors, a set of strains lacking genes for any pair of the four major cellulases, triple replacements, and a strain lacking all four genes have been constructed (Koivula *et al.*, 1996; Primalco Biotec, unpublished).

Small effects were noted on the expression of the other cellulase genes when the amounts of secreted cellulases were measured using ELISA techniques. Deletion of the *cbh1* gene (encoding the major cellulase) led to a two-fold increase in the production of cellobiohydrolase II; however, replacement of the *cbh2* gene did not affect the final cellulase levels, and deletion of *egl1* or *egl2* slightly increased production of both cellobiohydrolases (Suominen *et al.*, 1993). Similar results were obtained by Seiboth *et al.* (1992) when the effects of *cbh2* deletion on the production of other cellulases were determined.

Presented in Table 13.2 and by Suominen *et al.* (1993) are the effects on the production of cellulase activity when each of the four cellulase genes (*cbh1*, *cbh2*, *egl1*, and *egl2*) was replaced with a selectable marker. EGII seems to account for the most part of the hydroxyethylcellulose-(HEC)-degrading activity: in the strain lacking EGII the secreted endoglucanase activity was decreased as much as 60% (compared with the host strain) and only 25% in the EGI replacement strain. A double replacement strain not producing either one of the major endoglucanases secreted less than 10% of the HEC-degrading activity. As expected, filter paper-degrading activity (reflecting total hydrolysis of cellulose) of the culture filtrate of the strain lacking both cellobiohydrolases is below the detectable limit. Furthermore, this activity was decreased in the culture filtrate of all the strains except the one lacking EGI, the largest effect being due to lack of CBHI.

Table 13.2 Cellulase activities in different *T. reesei* deletion strains

Strain	Endoglucanase activity[a] (HEC/ml)	Filter paper degrading activity[b] (FPU/ml)
VTT-D-79125	1100	3.6
↓ *cbh1*	1300	0.96
↓ *cbh2*	1000	2.3
↓ *egl1*	830	3.7
↓ *egl2*	470	2.9
↓ *cbh1* ↓ *cbh2*	1300	–
↓ *egl1* ↓ *egl2*	75	2.5

T. reesei strains were cultivated in shake flasks in 50 ml of cellulase-inducing lactose-based medium (Suominen *et al.*, 1993) for 7 days. The cellulase activities were measured as the release of reducing sugars from [a] hydroxyethylcellulose (HEC), as described in Bailey and Nevalainen (1981), or from [b] filter paper (FPU), according to Mandels *et al.* (1976). ↓ indicates the gene which had been replaced.

13.3.3 *Use of* cbh1 *promoter for high-yield production of fungal proteins by* Trichoderma

Effect of promoter change

Gene targeting has provided a means to compare the effect of promoter change on the expression of homologous genes in order to exclude the potential differences in expression caused by distinct locations in the host genome. Over 50% of the secreted cellulase consists of cellobiohydrolase I enzyme, the product of a single *cbh1* gene, and thus *cbh1* is believed to be preceded by strong promoter-regulatory sequences. In addition, the *cbh1* promoter is inducible by several carbon sources including cellulose, and small oligosaccharides such as cellobiose, lactose, and sophorose (see Chapter 3). The strongly inducible *cbh1* promoter has been used in expressing both homologous *Trichoderma* proteins (Harkki *et al.*, 1991; Karhunen *et al.*, 1993; Paloheimo *et al.*, 1993) and heterologous proteins originating from other fungal species (Joutsjoki, 1994; Joutsjoki *et al.*, 1993; Margolles-Clark *et al.*, 1996; Miettinen-Oinonen *et al.*, 1997; Paloheimo *et al.*, 1993; Saloheimo and Niku-Paavola, 1991). Use of the *cbh1* promoter in the production of heterologous non-fungal proteins has also been described (Harkki *et al.*, 1989; Nykänen *et al.*, 1997; Nyyssönen *et al.*, 1993). Expression of heterologous proteins is discussed in more detail in Chapter 16.

High-yield production of several homologous proteins has been achieved simply by exchanging the natural promoter of the corresponding gene with the *cbh1* promoter. Studies on expression of the intact *cbh1-egl1* cassette at the *cbh1* locus revealed that *egl1* cDNA is expressed from the *cbh1* promoter as efficiently as *cbh1* itself (Karhunen *et al.*, 1993). The level of the *egl1* specific mRNA and EGI protein in the one-copy transformant was about ten-fold higher than that found in the non-transformed host strain, indicating that the *cbh1* promoter is about ten times stronger than the *egl1* promoter (Karhunen *et al.*, 1993). Similar expression levels were obtained when the promoters of *xln1*, *xln2* and *man1* were replaced by the *cbh1* promoter and the enzyme production of transformants carrying single copies integrated into the *cbh1* locus were analyzed (Table 13.3; Paloheimo *et al.*, 1993; Primalco Ltd, Biotec, unpublished). In shake-flask cultivations yields of several grams per liter were obtained corresponding to yield increases reaching 20-fold in the case of mannanase (Table 13.3). The differences in these increases reflect the original low amounts of xylanases or mannanase secreted by the host. In each case, the level of production of the enzyme expressed from the *cbh1* promoter was increased roughly to the level of CBHI of the host. Keeping in mind that even the already very high level of CBHI production can be further increased simply by increasing the number of *cbh1* copies (Paloheimo *et al.*, 1993), higher production levels for the above proteins are expected. Before industrial use of the final production strains, further improvements are often carried out, including increasing copy number of the expression cassette, classical mutagenesis, and optimization of the fermentation process. Thus enzyme yields in industrial-scale fermentations using such improved strains are much higher than the yields of the initial recombinant strains in shake-flask cultivations.

In the high-producing recombinant strains, the production capacity of cellobiohydrolases has been switched to the production of the desired protein, which now accounts for 80–90% of the total secreted protein. Only minimal amounts of

Table 13.3 Production of proteins in one-copy/*cbh1* locus transformants

Strain	Xylanase activity[a] (BXU/ml)	Mannanase activity[b] (MNU/ml)
Host strain	3 300	150
Xylanase I, *cbh1* promoter	16 900	ND
Xylanase II, *cbh1* promoter	27 300	ND
xln2 promoter	4 900	ND
Mannanase I, *cbh1* promoter	ND	3300

T. reesei strains were cultivated in shake flasks in 50 ml of cellulase-inducing lactose-based medium (Suominen *et al.*, 1993) for 7 days.

[a] Xylanase activity was measured according to Bailey *et al.* (1992) using birch xylan as substrate.

[b] Mannanase activity was measured according to Stålbrand *et al.* (1993) using locust bean gum as substrate.

other native proteins, in addition to the desired protein product, are secreted by such recombinant strains (Figure 13.2). This kind of switch can be carried out using routine molecular genetic tools and techniques in two or three strain construction steps as illustrated for mannanase production in Figure 13.3. The bars represent the relative amounts of different protein types, i.e., cellobiohydrolases, endoglucanases and mannanases, secreted into the culture medium by each strain. The strains secreting mannanase as the major protein are obtained after successive transformations

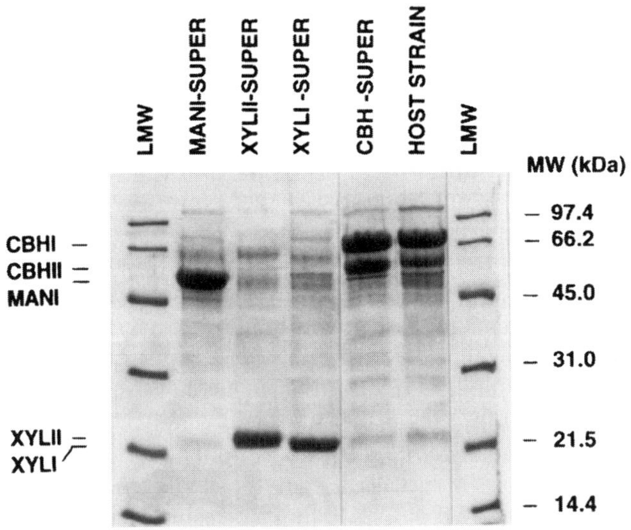

Figure 13.2 SDS-PAGE of the culture media of different high-producing *T. reesei* recombinant strains. The culture medium was centrifuged after cultivation, and equal amounts of the supernatant were used for analysis. CBHI, CBHII, MANI, XYLII, and XYNI indicate the expected position of migration of these enzymes, respectively. CBH-SUPER, MANI-SUPER, XYLII-SUPER, and XYLI-SUPER indicate culture supernatants from recombinant strains designed to overexpress the respective enzyme. Sizes of the low molecular weight markers (LMW, Bio-Rad) are indicated on the right.

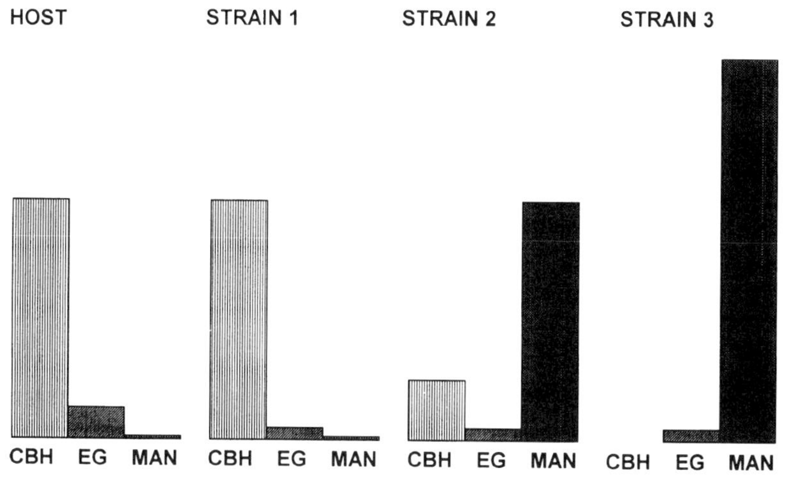

Figure 13.3 Steps involved in construction of strains producing mannanase as major secreted protein. CBH = cellobiohydrolase activity (filter paper degrading activity); EG = endoglucanase activity (hydroxyethylcellulose hydrolyzing activity); MAN = mannanase activity (locust bean gum hydrolyzing activity). The bar height is proportional to the relative volumetric activity (U/ml).

in which chosen cellulase genes are replaced and the mannanase expression is changed to be driven by the *cbh1* promoter.

Effects of gene copy number and integration site

Increasing gene copy number enhances production of target protein levels, although the relationship between copy number and gene expression is not always proportional. This might be explained at least in part by titration of transcription factors. Only three additional copies of the strong *cbh1* promoter may be sufficient to titrate out regulatory proteins or other essential transcription factors. In the work of Karhunen *et al.* (1993), the third copy of the *egl1* expression cassette did not further increase the level of EGI production from that of the two-copy transformant. There seemed to be limiting factors also in the transformants containing three extra copies of *xln1* gene expressed by the *cbh1* promoter (Paloheimo *et al.*, 1993). In *Aspergillus* (Beri *et al.*, 1990; Burger *et al.*, 1991), this phenomenon has been circumvented by simultaneous overexpression of positively acting regulatory factors. However, the possibility of bottlenecks in translation and secretion of the desired protein product cannot be excluded.

There are some indications in filamentous fungi that the genomic site of integration of introduced genes can have profound effects on expression levels (Verdoes *et al.*, 1995). Transforming DNA should therefore be directed to a locus of known high transcriptional activity. The *cbh1* locus of *T. reesei* has been considered to be one of high transcriptional activity and has been used successfully for high level gene expression (Joutsjoki, 1994; Joutsjoki *et al.*, 1993; Karhunen *et al.*, 1993; Nyyssönen *et al.*, 1993; Paloheimo *et al.*, 1993). There are also results showing that secretion levels of homologous proteins obtained from the CBHI+ are as high as those from the CBHI− transformants (Paloheimo *et al.*, 1993). Thus, it seems the *cbh1* locus is not the only locus giving high expression from the *cbh1* promoter.

To circumvent the effect of integration site on expression, the gene of interest could be introduced using a vector that confers high transcription, independent of position. Several studies have shown that matrix attachment regions (MARs) flanking a reporter construct can mitigate position-dependent expression in both homologous and heterologous backgrounds (reviewed by Archer *et al.*, 1994). Three MAR fragments, *trs1*, *2*, and *3*, have been isolated from the *T. reesei* genome, and their presence in transforming plasmids has been shown to enhance the frequency of integrative transformation of *T. reesei* approximately five-fold over plasmids without a *trs* (Belshaw *et al.*, 1994). Their effects on protein production have not yet been evaluated.

13.3.4 Low protease host strains for protein production

The extracellular proteases secreted by many fungi have been reported to be very harmful, especially to heterologous proteins. The amounts of proteases can be diminished by modifying the cultivation procedure and/or preparing protease-deficient strains. Protease-deficient mutants have been constructed, e.g., from *Aspergillus niger* strains (Mattern *et al.*, 1992) and *T. reesei* strains (Mäntylä *et al.*, 1994) using classical mutagenesis and screening. The main protease (an aspartic protease) has been cloned from several fungal species. Deletion of the aspartic proteinase gene from *A. niger* var. *awamori* decreased the degradation of the secreted calf chymosin and increased the yields considerably (Berka *et al.*, 1991).

The occurrence of proteases in *T. reesei* culture fluids has been reported by several authors (Dunne, 1982; Nakayama *et al.*, 1976; Sheir-Ness and Montenecourt, 1984). Haab *et al.* (1990) have purified the main protease component of *T. reesei* QM9414 and shown it to belong to a class of pepstatin-insensitive aspartate proteases (North, 1982).

Table 13.4 Effect of trichodermapepsin gene deletion on acidic protease activities of the *T. reesei* strains VTT-D-79125 and ALKO2221, a low-protease mutant strain derived from VTT-D-79125

Strain	Protein[a] (mg/ml)	Acidic protease[b] (U/ml)	U/mg protein
VTT-D-79125	14.8	640	43.2
ALKO4253 (gene deletion)	11.1	40	3.6
ALKO2221	13.5	50	3.7
ALKO4247 (gene deletion)	15.8	20	1.3

T. reesei strains were cultivated in shake flasks in 50 ml of cellulase-inducing lactose-based medium (Suominen *et al.*, 1993) for 7 days.

[a] Protein in the culture supernatants was determined by the assay of Lowry *et al.* (1951) using bovine serum albumin as standard.

[b] Proteolytic activity in the culture supernatants is expressed as hemoglobin units on the tyrosine basis as described in Food Chemicals Codex (Committee on Codex Specifications, 1981).

A pepstatin A-sensitive aspartic proteinase of *T. reesei* VTT-D-79125 (Bailey and Nevalainen, 1981), trichodermapepsin, has been purified and crystallized (Pitts, 1992; Pitts *et al.*, 1995). Peptide sequences derived from the purified trichodermapepsin were used to isolate the corresponding gene from *T. reesei* QM6a (Mäntylä *et al.*, 1994). Deletion of the trichodermapepsin gene from VTT-D-79125 resulted in 94% reduction in the acid protease activity (Table 13.4), to the same level as in the low protease mutant ALKO2221 derived from the strain VTT-D-79125 by UV-mutagenesis (Mäntylä *et al.*, 1994). Deletion of the trichodermapepsin gene from ALKO2221 had only minor effects on the remaining proteolytic activity, as seen in Table 13.4.

The acid proteases seem to degrade not only heterologous proteins but also native *Trichoderma* enzymes. Proteolysis has been reported to be especially pronounced with CBHI and CBHII, and to a lesser extent with β-glucosidase (Kubicek-Pranz *et al.*, 1991). Degradation of CBHI exclusively starts by removal of the AB-domain (see Chapter 1), whereas CBHII becomes attacked simultaneously from both termini (Hagspiel *et al.*, 1989). Degradation of CBHI protein in the culture medium of a high protease-producing *T. reesei* strain is shown in Figure 13.4. Deletion of the trichodermapepsin-encoding gene decreased the amount of the tailless form of the CBHI protein in the culture medium.

13.3.5 Optimization of growth conditions

Each new production strain requires some medium and fermenter optimization to achieve maximal rates of production. Initial screening of the production levels of transformants is usually performed in relatively small culture volumes in shake flasks. Significant improvements can be obtained by increasing the fungal biomass

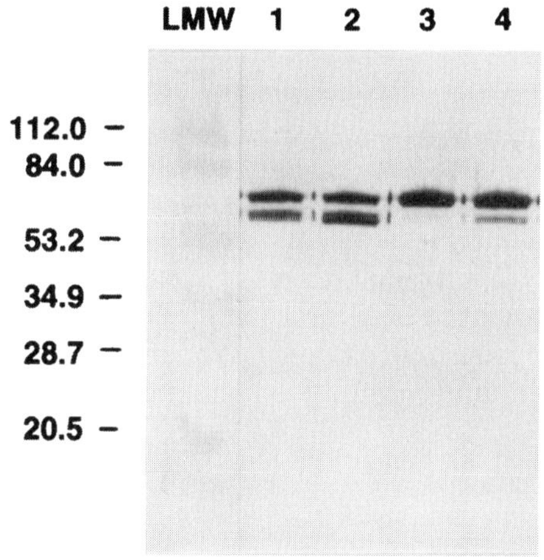

Figure 13.4 Western analysis showing the effect of proteolytic activity on the degradation of CBHI protein. Lanes 1 and 2: non-concentrated and fivefold concentrated media from the high-protease-producing strain, respectively; lanes 3 and 4: non-concentrated and fivefold concentrated culture media from the low-protease-producing strain, respectively. The CBHI-258 monoclonal antibody was used for the detection (Aho *et al.*, 1991). Positions of the low molecular weight markers (Bio-Rad) are shown on the left.

and controlling various critical growth parameters in a bioreactor. Such cultivations have been widely optimized by commercial enzyme producers, but little has been published. The fermentation mode, inoculum conditions, and environmental parameters, including medium composition, temperature, pH, agitation and aeration, influence the enzyme yields from filamentous fungal cultivations (reviewed by Dunn-Coleman *et al.*, 1992 and Jeenes *et al.*, 1991). Altering the composition of the growth medium can relate to specific responses of a given promoter and to more general effects on growth, hyphal morphology, protease production, cell death, etc. However, improvements obtained in small-scale laboratory fermentations using defined substrates may not always lead to similar improvements in larger-scale fermenter cultivations where inexpensive, complex media are used.

13.4 Concluding remarks

Trichoderma reesei has been utilized by the enzyme industry for about 15 years. During this time, its natural cellulases and hemicellulases have found their way into various fiber processing industries. Gene technology has allowed enzyme producers to further develop the high-producing mutant strains, which has led to improvement of both the quality and cost-effectiveness of the products. In addition to the native products, there is a need for efficient production of engineered homologous proteins as well as production of heterologous enzymes with properties required in the specified application. In our laboratory, high yield expression of enzyme genes originating from other filamentous fungi has already been achieved and several products are ready to be commercialized.

Yields of heterologous non-fungal proteins have remained low. The combination of molecular genetics and strain improvement by classical methods such as mutagenesis and screening may be used to increase expression of heterologous non-fungal proteins, as successfully performed in *Aspergillus* (Dunn-Coleman *et al.*, 1991; Ward *et al.*, 1995). More fundamental research for overcoming limitations in gene expression and protein secretion is in progress, and the results may be applicable for increasing the production of non-fungal proteins and for further improving the yields from industrial *T. reesei* strains. Today only one non-fungal protein produced by *T. reesei*, a thermostable xylanase for enzyme-aided bleaching of pulp, has been successfully commercialized (Primalco Ltd, Biotec).

We can say with confidence that *T. reesei* is among the industrially important fungi. We believe that when the bottlenecks in the secretion of heterologous proteins have been solved, the industrial use of *T. reesei* will further grow and expand beyond enzyme production to include production of other useful proteins, possibly even for medical use.

References

Aho, S., Olkkonen, V., Jalava, T., Paloheimo, M., Bühler, R., Niku-Paavola, M.-L., Bamford, D. H., and Korhola, M. 1991. Monoclonal antibodies against core and cellulose-binding domains of *Trichoderma reesei* cellobiohydrolases I and II and endoglucanase I. *Eur. J. Biochem.* **200**: 643–649.

Andreotti, R., Medeiros, J., Roche, C., and Mandels, M. 1980. Effects of strain and substrate on production of cellulases by *Trichoderma reesei* mutants. In T. K. Ghose (ed.), *Proceedings of the Second International Symposium on Bioconversion and Biochemical Engineering.* ITT, Delhi, pp. 353–388.

ARCHER, D. B., JEENES, D. J., and MACKENZIE, D. A. 1994. Strategies for improving heterologous protein production from filamentous fungi. *Antonie van Leeuwenhoek* **65**: 245–250.

BAILEY, M. J. and NEVALAINEN, K. M. H. 1981. Induction, isolation and testing of stable *Trichoderma reesei* mutants with improved production of solubilizing cellulases. *Enzyme Microb. Technol.* **3**: 153–157.

BAILEY, M. J., BIELY, P., and POUTANEN, K. 1992. Interlaboratory testing of methods for assay of xylanase activity. *J. Biotechnol.* **23**: 257–270.

BANERJI, J., OLSON, L., and SCHAFFNER, W. 1983. A lymphocyte-specific cellular enhancer is located downstream of the joining region in immunoglobulin heavy chain genes. *Cell* **33**: 729–740.

BEJA DA COSTA, M. and VAN UDEN, N. 1980. Use of 2-deoxyglucose in the selective isolation of mutants of *Trichoderma reesei* with enhanced β-glucosidase production. *Biotechnol. Bioeng.* **12**: 2429–2432.

BELSHAW, N. J., SUOMINEN, P., and ARCHER, D. B. 1994. The nuclear matrix and matrix attachment regions (MARs) in *Trichoderma reesei.* In Second European Conference on Fungal Genetics, Lunteren, The Netherlands, Abstract C12.

BERGES, T. and BARREAU, C. 1991. Isolation of uridine auxotrophs from *Trichoderma reesei* and efficient transformation with the cloned *ura3* and *ura5* genes. *Curr. Genet.* **19**: 359–365.

BERI, R. K., GRANT, S., ROBERTS, C. F., SMITH, M., and HAWKINS, A. R. 1990. Selective overexpression of the *GUTE* gene encoding catabolic 3-dehydroquinase in multicopy transformants of *Aspergillus nidulans. Biochem. J.* **265**: 337–342.

BERKA, R. M., BAYLISS, F. T., BLOEBAUM, P., CULLEN, D., DUNN-COLEMAN, N. S., KODAMA, K. H., HAYENGA, K. J., HITZEMAN, R. A., LAMSA, M. H., PRZETAK, M. M., REY, M. W., WILSON, L. J., and WARD, M. 1991. *Aspergillus niger* var. *awamori* as a host for the expression of heterologous genes. In J. W. Kelly and T. O. Baldwin (eds), *Applications of Enzyme Biotechnology.* Plenum Press, New York, pp. 273–292.

BURGER, G., TILBURN, J., and SCAZZOCCHIO, C. 1991. Molecular cloning and functional characterization of the pathway-specific regulatory gene *nirA*, which controls nitrate assimilation in *Aspergillus nidulans. Mol. Cell Biol.* **11**: 795–802.

COMMITTEE ON CODEX SPECIFICATIONS. 1981. Proteolytic Activity, Fungal (HUT). In Food Chemicals Codex. National Academy Press, Washington, pp. 496–497.

DUNN-COLEMAN, N. S., BLOEBAUM, P., BERKA, R. M., BODIE, E., ROBINSON, N., ARMSTRONG, G., WARD, M., PRZETAK, M., CARTER, G. L., LACOST, R., WILSON, L. J., KODAMA, K. H., BALIU, E. F., BOWER, B., LAMSA, M., and HEINSOHN, H. 1991. Commercial levels of chymosin production by *Aspergillus. Bio/Technology* **9**: 976–981.

DUNN-COLEMAN, N. S., BODIE, E. A., CARTER, G. L., and ARMSTRONG, G. L. 1992. Stability of recombinant strains under fermentation conditions. In J. R. Kinghorn and G. Turner (eds), *Applied Molecular Genetics of Filamentous Fungi.* Blackie Academic and Professional, Glasgow, UK, pp. 152–174.

DUNNE, C. P. 1982. Relationship between extracellular proteases and the cellulase complex of *Trichoderma reesei.* In I. Chibata, S. Fukui, and L. B. Wingard JR. (eds), *Enzyme Engineering*, Vol. 6. Plenum Press, New York, pp. 355–356.

DURAND, H., CLANET, M., and TIRABY, G. 1984. A genetic approach of the improvement of cellulase production by *Trichoderma reesei.* In H. Egneus and A. Ellegard (eds), *Proceedings of Bioenergy World Conference*, Vol. 3. Elsevier Applied Science, London, pp. 246–253.

DURAND, H., CLANET, M., and TIRABY, G. 1988b. Genetic improvement of *Trichoderma reesei* for large scale cellulase production. *Enzyme Microb. Technol.* **10**: 341–345.

DURAND, H., BARON, M., CALMELS, T., and TIRABY, G. 1988a. Classical and molecular genetics applied to *Trichoderma reesei* for the selection of improved cellulolytic industrial strains. In J. P. Aubert, P. Beguin, and J. Millet (eds), *Biochemistry and Genetics of Cellulose Degradation.* Academic Press, New York, pp. 135–151.

FARKAŠ, V., LABUDOVA, I., BAUER, S., and FERENCZY, L. 1981. Preparation of mutants of *Trichoderma viride* with increased production of cellulase. *Folia Microbiol.* **26**: 129–132.

GALLO, B. J., ANDREOTTI, R., ROCHE, C., RUY, D., and MANDELS, M. 1978. Cellulase production by a new mutant strain of *Trichoderma reesei* MCG 77. *Biotechnol. Bioeng. Symp.* **8**: 89–101.

GHOSH, A., AL-RABIAI, S., GHOSH, B. K., TRIMINO-VAZQUEZ, H., EVELEIGH, D. E., and MONTENECOURT, B. S. 1982. Increased endoplasmic reticulum content of a mutant of *Trichoderma reesei* (Rut-C30) in relation to cellulase synthesis. *Enzyme Microbiol. Technol.* **4**: 110–113.

GRUBER, F., VISSER, J., KUBICEK, C. P., and DE GRAAFF, L. H. 1990a. The development of a heterologous transformation system for the cellulolytic fungus *Trichoderma reesei* based on a *pyrG*-negative mutant strain. *Curr. Genet.* **18**: 71–76.

GRUBER, F., VISSER, J., KUBICEK, C. P., and DE GRAAFF, L. H. 1990b. Cloning of the *Trichoderma reesei pyrG* gene and its use as a homologous marker for high-frequency transformation system. *Curr. Genet.* **18**: 447–451.

HAAB, D., HAGSPIEL, K., SZAKMARY, K., and KUBICEK, C. P. 1990. Formation of the extracellular protease from *Trichoderma reesei* QM9414 involved in cellulase degradation. *J. Biotechnol.* **16**: 187–198.

HAGSPIEL, K., HAAB, D., and KUBICEK, C. P. 1989. Protease activity and proteolytic modification of cellulases from a *Trichoderma reesei* QM 9414 selectant. *Appl. Microbiol. Biotechnol.* **32**: 61–67.

HARKKI, A., UUSITALO, J., BAILEY, M., PENTTILÄ, M., and KNOWLES, J. K. C. 1989. A novel fungal expression system: secretion of active calf chymosin from the filamentous fungus *Trichoderma reesei. Bio/Technology* **7**: 596–603.

HARKKI, A., MÄNTYLÄ, A., PENTTILÄ, M., MUTTILAINEN, S., BÜHLER, R., SUOMINEN, P., KNOWLES, J., and NEVALAINEN, H. 1991. Genetic engineering of *Trichoderma* to produce strains with novel cellulase profiles. *Enzyme Microbiol. Technol.* **13**: 227–233.

JEENES, D. J., MACKENZIE, D. A., ROBERTS, I. N., and ARCHER, D. B. 1991. Heterologous protein production by filamentous fungi. *Biotech. Gen. Eng. News* **9**: 327–367.

JOUTSJOKI, V. 1994. Construction by one-step gene-replacement of *Trichoderma reesei* strains that produce the glucoamylase P of *Hormoconis resinae. Curr. Genet.* **26**: 422–429.

JOUTSJOKI, V., TORKKELI, T., and NEVALAINEN, H. 1993. Transformation of *Trichoderma reesei* with the *Hormoconis resinae* glucoamylase P (*gamP*) gene: production of a heterologous glucoamylase by *Trichoderma reesei. Curr. Genet.* **24**: 223–229.

KARHUNEN, T., MÄNTYLÄ, A., NEVALAINEN, K. M. H., and SUOMINEN, P. L. 1993. High frequency one-step gene replacement in *Trichoderma reesei.* I: Endoglucanase I overproduction. *Mol. Gen. Genet.* **241**: 515–522.

KAWAMORI, M., ADO, Y., and TAKASAWA, S. 1986. Preparation and application of *Trichoderma reesei* mutants with enhanced β-glucosidase. *Agric. Biol. Chem.* **50**: 2477–2482.

KAWAMORI, M., MORIKAWA, Y., SHINSHA, Y., TAKAYAMA, K., and TAKASAWA, S. 1985. Preparation of mutants resistant to catabolite repression of *Trichoderma reesei. Agric. Biol. Chem.* **49**: 2875–2879.

Koivula, A., Lappalainen, A., Virtanen, S., Mäntylä, A. L., Suominen, P., and Teeri, T. T. 1996. Immunoaffinity chromatographic purification of cellobiohydrolase II mutants from recombinant *Trichoderma reesei* strains devoid of major endoglucanase genes. *Prot. Express. Purific.* **8**: 391–400.

Kubicek-Pranz, E. M., Gsur, A., Hayn, M., and Kubicek, C. P. 1991. Characterization of commercial *Trichoderma reesei* cellulase preparations by denaturating electrophoresis (SDS-PAGE) and immunostaining using monoclonal antibodies. *Biotechnol. Appl. Biochem.* **14**: 317–323.

Labudova, I. and Farkaš, V. 1983. Enrichment technique for the selection of catabolite repression-resistant mutants of *Trichoderma* as producers of cellulase. *FEMS Microbiol. Lett.* **20**: 211–215.

Lowry, O. H., Roseborough, N. J., Farr, A. L., and Randall, R. J. 1951. Protein measurement with the folin phenol reagent. *J. Biol. Chem.* **193**: 265–275.

Mach, R. L., Schindler, M., and Kubicek, C. P. 1994. Transformation of *Trichoderma reesei* based on hygromycin B resistance using homologous expression signals. *Curr. Genet.* **25**: 567–570.

Mandels, M. 1975. Microbial sources of cellulase. *Biotechnol. Bioeng. Symp.* No. **5**: 81–105.

Mandels, M. and Reese, E. T. 1957. Induction of cellulase in *Trichoderma viride* as influenced by carbon sources and metals. *J. Bacteriol.* **73**: 269–278.

Mandels, M., Andreotti, R., and Roche, C. 1976. Measurement of saccharifying cellulase. In E. L. Gaden, M. H. Mandels, E. T. Reese, and L. A. Spano (eds), Biotechnology Bioengineering Symposium 6. Wiley, New York, pp. 21–33.

Mandels, M., Weber, J., and Parizek, R. 1971. Enhanced cellulase production by a mutant of *Trichoderma viride*. *Appl. Microbiol.* **21**: 152–154.

Mäntylä, A., Saarelainen, R., Fagerström, R., Suominen, P., and Nevalainen, H. 1994. Cloning of the aspartic protease gene of *Trichoderma reesei*. In Second European Conference on Fungal Genetics, Lunteren, The Netherlands, Abstract B52.

Margolles-Clark, E., Hayes, C. K., Harman, G. E., and Penttilä, M. 1996. Improved production of *Trichoderma harzianum* endochitinase by expression in *Trichoderma reesei*. *Appl. Environ. Microbiol.* **62**: 2145–2151.

Mattern, I. E., van Noort, J. M., van den Berg, P., Archer, D. B., Roberts, I. N., and van den Hondel, C. A. M. J. J. 1992. Isolation and characterization of mutants of *Aspergillus niger* deficient in extracellular proteases. *Mol. Gen. Genet.* **234**: 332–336.

Miettinen-Oinonen, A., Torkkeli, T., Paloheimo, M., and Nevalainen, H. 1997. Overexpression of the *Aspergillus niger* pH 2.5 acid phosphatase gene in a heterologous host, *Trichoderma reesei*. *J. Biotech.* **58**: 13–20.

Mishra, S., Gopalkrishnan, K. S., and Ghose, T. K. 1982. A constitutive cellulase-producing mutant of *Trichoderma reesei*. *Biotechnol. Bioeng.* **24**: 251–254.

Montenecourt, B. S. 1983. *Trichoderma reesei* cellulases. *Trends Biotechnol.* **1**: 156–161.

Montenecourt, B. S. and Eveleigh, D. E. 1977. Preparation of mutants of *Trichoderma reesei* with enhanced cellulase production. *Appl. Environ. Microbiol.* **34**: 777–782.

Montenecourt, B. S. and Eveleigh, D. E. 1979. Selective screening methods for the isolation of high yielding cellulase mutants of *Trichoderma reesei*. *Adv. Chem. Ser.* **181**: 289–301.

Nakari, T., Onnela, M.-L., Ilmén, M., and Penttilä, M. 1993. New *Trichoderma* promoters for production of hydrolytic enzymes on glucose medium. In P. Suominen and T. Reinikainen (eds), Proceedings of the Tricel 93 Symposium on *Trichoderma reesei* Cellulases and Other Hydrolases, Vol. 8. Foundation for Biotechnical and Industrial Fermentation Research, Espoo, Finland, pp. 239–246.

Nakayama, M., Tomita, Y., Suzuki, H., and Nisizawa, K. 1976. Partial proteolysis of some cellulase components from *Trichoderma viride* and the substrate specificity of the modified products. *J. Biochem.* **79**: 955–966.

NEVALAINEN, K. M. H., PALVA, E. T., and BAILEY, M. J. 1980. A high cellulase-producing mutant strain of *Trichoderma reesei. Enzyme Microb. Technol.* **3**: 59–60.

NORTH, M. J. 1982. Comparative biochemistry of the proteases of eukaryotic micro-organisms. *Microbiol. Rev.* **46**: 308–340.

NYKÄNEN, M., SAARELAINEN, R., RAUDASKOSKI, M., NEVALAINEN, K. M. H., and MIKKONEN, A. 1997. Expression and secretion of barley cysteine endopeptidase B and cellobiohydrolase I in *Trichoderma reesei. Appl. Environ. Microbiol.* **63**: 4929–4937.

NYYSSÖNEN, E., DEMOLDER, J., CONTRERAS, R., KERÄNEN, S., and PENTTILÄ, M. 1995. Protein production by the filamentous fungus *Trichoderma reesei*: secretion of active antibody molecules. *Can. J. Bot.* **73**(Suppl. 1): S885–S890.

NYYSSÖNEN, E., PENTTILÄ, M., HARKKI, A., SALOHEIMO, A., KNOWLES, J. K. C., and KERÄNEN, S. 1993. Efficient production of antibody fragments by the filamentous fungus *Trichoderma reesei. Bio/Technology* **11**: 591–595.

PALOHEIMO, M., MIETTINEN-OINONEN, A., TORKKELI, T., NEVALAINEN, H., and SUOMINEN, P. 1993. Enzyme production by *Trichoderma reesei* using the *cbh1* promoter. In P. Suominen and T. Reinikainen (eds), Proceedings of the Tricel 93 Symposium on Trichoderma reesei Cellulases and Other Hydrolases, Vol. 8. Foundation for Biotechnical and Industrial Fermentation Research, Espoo, Finland, pp. 229–238.

PENTTILÄ, M., NEVALAINEN, H., RÄTTÖ, M., SALMINEN, E., and KNOWLES, J. 1987. A versatile transformation system for the cellulolytic filamentous fungus *Trichoderma reesei. Gene* **61**: 155–164.

PERSSON, I., TJERNELD, F., and HAHN-HÄGERDAL, B. 1991. Fungal cellulolytic enzyme production: a review. *Process Biochem.* **26**: 65–74.

PITTS, J. E. 1992. Crystallization by centrifugation. *Nature* **355**: 117.

PITTS, J. E., CRAWFORD, M. D., NUGENT, P. G., WESTER, R. T., COOPER, J. B., MÄNTYLÄ, A., FAGERSTRÖM, R., and NEVALAINEN, H. 1995. The three-dimensional X-ray crystal structure of the aspartic proteinase native to *Trichoderma reesei* complexed with renin inhibitor CP-80794. *Adv. Exp. Med. Biol.* **362**: 543–547.

SALOHEIMO, M. and NIKU-PAAVOLA, M. L. 1991. Heterologous production of the ligninolytic enzyme: expression of the *Phlebia radiata* laccase gene in *Trichoderma reesei. Bio/Technology* **9**: 987–990.

SALOVUORI, I., MAKAROW, M., RAUVALA, H., KNOWLES, J., and KÄÄRIÄINEN, L. 1987. Low molecular weight high-mannose type glycans in a secreted protein of the filamentous fungus *Trichoderma reesei. Bio/Technology* **5**: 152–156.

SAWERS, G. and JARSCH, M. 1996. Alternative regulation principles for the production of recombinant proteins in *Escherichia coli. Appl. Microbiol. Biotechnol.* **46**: 1–9.

SEIBOTH, B., MESSNER, R., GRUBER, F., and KUBICEK, C. P. 1992. Disruption of the *Trichoderma reesei cbh2* gene coding for cellobiohydrolase II leads to the delay in the triggering of cellulase formation by cellulose. *J. Gen. Microbiol.* **138**: 1259–1264.

SHEIR-NEISS, G. and MONTENECOURT, B. 1984. Characterization of the secreted cellulase of *Trichoderma reesei* wild type and mutants during controlled fermentations. *Appl. Microbiol. Biotechnol.* **20**: 46–53.

SHOEMAKER, S. P., RAYMOND, J. C., and BRUNER, R. 1981. Cellulases: diversity amongst improved *Trichoderma strains.* In A. Hollaender (ed.), *Trends in the Biology of Fermentations for Fuels and Chemicals.* Plenum Press, New York, pp. 89–109.

SMITH, J. L., BAYLISS, F. T., and WARD, M. 1991. Sequence of the cloned *pyr4* gene of *Trichoderma reesei* and its use as a homologous selectable marker for transformation. *Curr. Genet.* **19**: 27–33.

STÅLBRAND, H., SIIKA-AHO, M., TENKANEN, M., and VIIKARI, L. 1993. Purification and characterization of two β-mannanases from *Trichoderma reesei. J. Biochem.* **29**: 229–242.

SUOMINEN, P. L., MÄNTYLÄ, A., KARHUNEN, T., HAKOLA, S., and NEVALAINEN, H. 1993. High frequency one-step gene replacement in *Trichoderma reesei*. II: Effects of deletions of individual cellulase genes. *Mol. Gen. Genet.* **241**: 523–530.

VERDOES, J. C., PUNT, P. J., and VAN DEN HONDEL, C. A. M. J. J. 1995. Molecular genetic strain improvement for the overproduction of fungal proteins by filamentous fungi. *Appl. Microbiol. Biotechnol.* **43**: 195–205.

WALSETH, C. S. 1952. Occurrence of cellulases in enzyme preparations from microorganisms. *TAPPI* **35**: 228–233.

WARD, P. P., PIDDINGTON, C. S., CUNNINGHAM, G. A., ZHOU, X., WYATT, R. D., and CONNEELY, O. M. 1995. A system for production of commercial quantities of human lactoferrin: a broad spectrum natural antibiotic. *Bio/Technology* **13**: 498–503.

14

Application of *Trichoderma* enzymes in the textile industry

Y. M. GALANTE, A. DE CONTI and R. MONTEVERDI

Laboratory of Biotechnology, Central R&D, Lamberti s.p.a., Albizzate, Italy

14.1 Introduction and background: Why the textile industry?

Some of the early applications of cellulase enzymes were developed for the agro-food area, including production of alcohol by fermentation following enzymatic saccharification of cellulosic material; improvement of malt wort filtration in brewing; maceration of plant tissues for aroma extraction and fruit juice production; increased machinability and extrusion of high natural fiber-containing food products; and improvement of wine making and olive oil production. Bioconversion of cellulose for energy production purposes turned out to have limited practical feasibility, given the slow kinetics of enzyme degradation of cellulosic material due to the largely crystalline nature of cellulose. Instead, and somewhat unexpectedly, cellulases met their first worldwide success in the textile industry and in industrial laundries, largely because their kinetic properties allow a controlled treatment of cellulosic fibers for product improvement.

Indeed, cellulases for textile applications have become the third largest (after detergent and starch) and one of the fastest growing markets for industrial enzymes, even though they were introduced in laundries, dyehouses and textile mills less than a decade ago. The rapid diffusion and widespread use of these enzymes in the textile field can be explained by considering that most fabrics used for clothing, upholstery, household goods, etc., are at least partially or wholly made of cellulosic material (e.g., cotton, linen, flax, viscose, lyocell).

In the last few years, there has been a wealth of new applications developed for the treatment and modification of cellulosic fabrics, along with the introduction of several new cellulase-based products. More recently, totally new recombinant and bioengineered cellulases have appeared on the market, although most have been described very little, if at all, in the scientific literature, for obvious reasons of industrial confidentiality. For this reason, the authors of the present chapter will refer to a few relevant patents describing new rDNA cellulases and their applications, rather than to scientific journals. While probably the best-known current textile application of cellulases is "biostoning", or the enzymatic stonewashing of denim garments such as jeans, the biofinishing or biopolishing of non-denim cellulosic fabrics and garments is rapidly developing into an industrial technology, beyond the occasional treatments mostly dictated by volatile and changing fashions.

Finally, cellulases are increasingly used in household washing products, because they aid detergent performance and allow the removal of small, fuzzy fibrils from fabric surfaces, thereby improving appearance and color brightness.

14.2 Enzymatic stonewashing of denim garments with fungal cellulases

Blue jeans (the name presumably comes from "bleu de Gènes", after the Italian seaport of Genova and the dark blue apparel worn by dock workers) and other denim garments are more than just clothing. They have become a timeless myth of the modern era and an integral part of cultural heritage as much as of our leisure and work wear. Ever since the middle of the last century when Levi Strauss, a young emigrant from Bavaria, introduced jeanswear to California at the time of the gold rush, the popularity of this peculiar garment has steadily grown. Nowadays, it can be estimated that each year about 800 million pairs of jeans are produced worldwide, which represents a multi-billion dollar industry.

Denim ("de Nimes" – the name comes from Nimes, the town in Southern France where it used to be manufactured) is basically a cotton fabric woven from indigo-dyed warp and white fill yarn. Indigo is a natural dye originally extracted from the swamp plants *Indigofera tinctoria* and *Isatis tinctoria,* known and used since Sumerian times. It has been chemically synthesized on a very large scale for almost a century from N-phenylglycine. More recently, however, a biotechnological process has been developed first by Amgen and later by Genencor International to produce indigo by fermentation technology using a recombinant bacterial strain (Berry, 1996).

In the making of denim fabric, indigo is absorbed on the yarn surface in a reduced, soluble leuco form, followed by precipitation through rapid air re-oxidation to give a blue ring encircling an undyed yarn core. Indigo dye is thus attached to the surface of the yarn and to the most exterior short cotton fibers. This feature can later generate a "wash down" or "aged" effect, and it is the very basis upon which the entire denim finishing industry is built.

At the textile mill, the indigo warp is heavily sized with starch and the denim fabric is woven into a very tight structure (twill weave). This makes for an extremely sturdy and long lasting material but is rather stiff and uncomfortable to wear when new. In "the old days", new jeans had to be worn and washed time and again, to make them soft and "lived-in". Wear and repeated washes finally gave that familiar abraded, faded and worn look reminiscent of tough cowboys and city street fighters seen in countless movies! Fashion makers were quick to sniff out a fashion trend in progress, and "aged" or "faded" blue jeans were definitely "in".

In the late 1970s and early 1980s, worn and aged jeans were obtained by washing the garments in industrial laundries with pumice stones after desizing with α-amylase. This was designated as the "stone-washing" process, which prompted many industrial laundries to enter the denim finishing business. The abrasion effect on denim garments is obtained by locally removing the surface-bound indigo dye by abrasion with pumice stones, thus partially revealing the white interior of the yarn and giving the faded, worn and aged appearance. Typically, 1–2 kg of stones per kg of jeans were used with a 1 h treatment time.

While the final look of the denim garments met market requirements, the massive use of stones caused a long series of problems:

- rapid wear and tear of the washing machine;
- high incidence of second quality garments because of excessive abrasion;
- environmental problems caused by quarrying for pumice and disposal of grit after treatment (the waste is actually a mix of pumice grit and cellulose lint, laced with indigo dye);
- unsafe working conditions due to pumice powder on floors and in sewers of laundries;
- need for manual removal of pumice from pockets and folds of the garments.

In the mid-1980s, biotechnology came around to offer the (almost) perfect alternative to pumice in the stonewashing process: microbial cellulases. Once biostoning was developed, cellulases finally made it big in industry!

14.2.1 *Application of cellulases in denim washing*

The concept of cellulase addition to detergent mixtures was already around in the mid-1970s and early 1980s to improve soil removal and fabric softness (see for example US Pat. No. 4,435,307 issued in 1984 and several other US, UK and European Patents that followed thereafter). It is said that the peculiar fading effect of cellulases on denim was discovered almost by accident in the UK around 1987 (Godfrey, 1996). However, several patents giving a detailed description of the application were filed starting in 1988 and issued between 1990 and 1993 to Ecolab Inc., an American company now part of the Henkel Group. Whatever the true history, the original discovery is widely credited to the work done by Ecolab Inc. with an enzyme produced by Novo of Denmark. Recently, the validity of these patents was questioned, and in 1995 a US Board of Patent Appeals and Interferences ruled to invalidate them.

Regardless of legal issues, in a few years this enzyme technology met with a remarkable success in laundries. The principle is rather simple: the cellulase enzyme complex acts on the cotton fabric by breaking off the small fiber ends on the yarn surface, thereby loosening the indigo, which is easily removed by mechanical abrasion in the wash cycle.

Replacement of pumice stones by a cellulase-based treatment offers several advantages:

- reduced wear and tear of washing machines and shorter treatment times;
- increased productivity of the machines because of higher garment loading;
- substantial decrease of second quality garments;
- less work-intensive and safer working environment;
- an environmentally safer process because substantially less or no pumice powder is generated;
- the possibility to automate the process with computer-controlled dosing devices when using liquid cellulase preparations;
- flexibility to create and consistently reproduce new finished looks.

The nature of the specific cellulase, subunit composition, activity and dosage are all crucial in biostoning, but other important aspects to be considered are equipment type, machine load, liquor ratio, addition of chemical auxiliaries, eventual need for a small pumice stone load (to help move the wet jeans mass uniformily in the drum), washing time, temperature, pH and denim type and quality, as well as the final look required. Only a careful combination of all these parameters will allow the expert laundry operator to achieve optimal results from any enzyme used.

14.2.2 *Indigo dye redeposition:* Humicola *versus* Trichoderma *cellulases*

As well explained in Chapter 1, the *Trichoderma* cellulase complex is composed of several activities: mainly of two exo-cellulases (CBH), four or five endo-cellulases (EG) and one β-glucosidase (CB). Most individual subunits comprise a catalytic domain (CD) and a cellulose binding domain (CBD). It is also recognized that the various activities act synergistically in the degradation of cellulose, although it is not clear which subunit(s), if any, play(s) the most important role in biostoning. The *Trichoderma* cellulase complex has a pH activity profile with an optimum around pH 5.0 and is therefore commonly referred to as "acid cellulase". It is considered that these enzymes act aggressively on cellulosic fibers, causing fast abrasion of the fabric. There is, however, a drawback: significant indigo redeposition (backstaining) occurs during stonewashing with any *Trichoderma reesei* whole cellulase complex.

The phenomenon can be described as follows: abrasion of the indigo yarn causes release of the dye in the liquor bath. The amount of indigo released is a function of abrasion level and, therefore, of enzyme dosage, cycle time, mechanical action, denim type, etc. The dye released has a strong tendency to redeposit on the garments, causing backstaining. The filler yarn and the inner pockets or linings become increasingly bluish in color. As a result, the blue/white contrast of the finished and abraded garment is partially masked by backstaining. Controlling indigo redeposition backstaining has thus become an important goal in enzymatic denim stonewashing, particularly when no post-wash bleaching step is applied and/or when a high level of abrasion and contrast are required. Abrasion and backstaining levels can be quantitatively estimated by measuring reflectance on the outside and inside of the garment with a color matching apparatus at, respectively, wavelengths of 420 nm and 680 nm. In both cases, a higher value measured means more abrasion (at 420 nm) and less backstaining (at 680 nm).

From an industrial standpoint, two types of cellulases can be employed in washing: those with an acid pH optimum (pH 5.0–5.5) and those with a pH optimum near neutrality (pH 6.0–6.5). Although it was first believed that the primary cause of backstaining is the pH of the water bath (i.e., the more acid the pH, the greater the backstaining) regardless of the biochemical nature of the enzyme used, it has become evident that the difference in backstaining between washing garments at pH 5 or 7 is not very significant, as compared with the difference in backstaining when washing at either pH with an "acid" cellulase versus a "neutral" cellulase (Klahorst *et al.*, 1994). In other words, the pH of the bath is much less a cause of backstaining than is the biochemical nature of the cellulase employed in the treatment. The "acid" type cellulase, therefore the *Trichoderma* enzyme complex, consistently causes more extensive backstaining. The neutral cellulases active at pH values from 6 to 8, with an optimum around pH 6.5, are less aggressive and seem to cause little or no backstaining during denim washing.

The best known neutral-type fungal cellulase for stonewashing is produced by the thermophilic soft-rot fungus *Humicola insolens* and is an enzyme commercially available in several liquid and solid formulations (the trade name Denimax™ for all *H. insolens*-derived neutral cellulase preparations is of Novo Nordisk). The *Humicola* cellulase complex is composed of at least seven distinct cellulases and, besides the different pH– and temperature–activity profiles, it has biochemical features very similar to the *T. reesei* complex (Schulein *et al.*, 1993).

Whether its low backstaining property is related to any particular subunit composition or to the presence/absence of a cellulose binding domain (CBD) in any given subunit is not known to us at this time. Figure 14.1 shows a comparison of abrasion level and extent of backstaining of denim fabric following standard washing conditions using a *Humicola* or a *Trichoderma* cellulase at pH 7.0 and 4.8, respectively. The control was done with no enzyme added, but the same washing treatment was applied. The higher reflectance value at 680 nm of the *Humicola* cellulases indicates lower backstaining, while the relatively lower value at 420 nm (i.e., abrasion index) of the acidic cellulase compared with the neutral can be explained by the greater indigo redeposition that partially stains the fiber.

In spite of what can be considered a drawback of *T. reesei* cellulases in this particular industrial application, several means have been devised to offset the backstaining problem and to successfully use these enzymes in stonewashing. Such means include a careful balancing of enzyme dosage, liquor ratio, processing time, non-inhibiting non-ionic detergents and mostly post-wash clean up treatments with

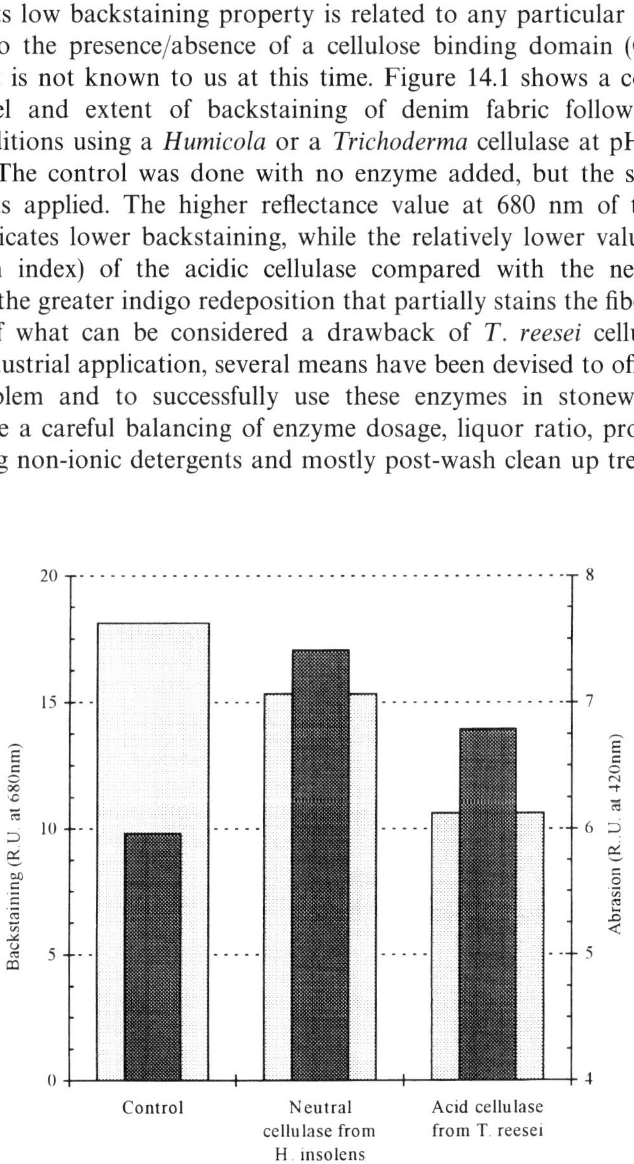

Figure 14.1 Evaluation of abrasion and backstaining of denim fabrics by reflectance measurement comparing neutral cellulase from *H. insolens* to acid cellulase from *T. reesei*. Shaded bars indicate backstaining, full bars abrasion; R.U., reflectance units.

detergents or even chemical bleaching with hypochlorite. Indeed, a few commercial formulations of *Trichoderma* cellulases, in liquid or solid form, have been developed, containing a mixture of enzymes, buffers, antiredeposition chemicals, stabilizing agents, etc., that are ready-to-use, single addition products with excellent cost/performance ratios for the end-user.

The exact mechanism of cellulase-mediated indigo redeposition is not fully explained. There are indications of the existence in nature of low redeposition "acid" cellulases and high redeposition "neutral" cellulases. Therefore, the pH profile of a cellulase complex should not be taken as the sole criterion of its potential performance in stonewashing. *Trichoderma* acid cellulases are more aggressive and thus allow shorter washing times than is possible with the neutral enzymes, but they must be more carefully controlled in order to minimize potentially excessive fabric strength loss (FSL). Also, *Trichoderma* acid cellulases are less costly biochemicals, largely used in laundries in the Americas and the Far East.

It has been reported that the endo-cellulase activity is most important for the removal of indigo dye (Cavaco-Paulo *et al.*, 1996a). These authors also offered an interesting explanation (although experimentally unproven) for the backstaining phenomenon. They suggested that CBH's action produces more soluble sugars, which can in turn reduce indigo and thereby enhance dye redeposition. The relative higher endo-activity of the commercial neutral *Humicola* cellulase could in part account for its lower backstaining feature. In our own experience, no higher redeposition level is observed when supplementing a washing bath of *Humicola* cellulase with reducing sugars. We conclude that the above is an unlikely explanation for the backstaining phenomenon.

An interesting observation has been made and patented by researchers at Genencor International (WO 93/25655 and WO 94/29426). They have found that blue indigo redeposition during stonewashing with acid cellulase can be substantially prevented by adding a microbial protease (e.g., a subtilisin) together with the cellulase in the wash bath or, alternatively, pre-mixing the protease beforehand with the cellulase prior to addition to the washing machine. A tentative explanation given for this curious behavior is that ". . . the added protease effectively removes or prevents the cellulase proteins from binding the colored particles back onto the surface of the denim and yet . . . does not adversely affect the resultant abraded look caused by the action of cellulase". Of course, there is also an optimum ratio of cellulase to protease that achieves the antiredeposition effect without adversely affecting abrasion, presumably because of proteolytic attack on the cellulase itself. In such a system, pH control becomes absolutely critical in order to properly manage the activity of both enzymes. On the other hand, it has not yet been fully established that an endo-enriched or a single component endoglucanase *T. reesei* cellulase preparation performs better in biostoning than the whole complex (although it appears to be the case). The new acid endo-cellulase-enriched preparations and the recombinant, monocomponent acid endo-cellulase that have recently become commercially available will allow the experimental testing of this hypothesis.

The strategy of molecular biology to generate mutants and recombinant strains of *Trichoderma* that produce cellulase complexes with a different subunit composition than the natural complex is illustrated in Chapter 13.

Essentially, three different approaches of genetic engineering are followed in the development of second generation cellulases from *Trichoderma*: either by cloning of a single gene (e.g., of a CBH or EG) to generate a monocomponent enzyme, disruption or overexpression of one or more genes (e.g., of CBHs) to generate a preparation enriched in a wanted group of subunits (e.g., of EGs). For example, the isolation, cloning and expression of the gene coding for *T. reesei* endoglucanase III (EGIII) has been reported, as well as methods for purification of a single, monocomponent EGIII activity (Patents WO 93/20208, WO 92/20209 and WO 94/21801 assigned to Genencor International).

It is suggested that, besides its inclusion in detergent formulations, "... EGIII cellulase has a further use in the stonewashing process of colored fabrics wherein redeposition of a colorant onto the fabric may be reduced by employing purified EGIII" (Patent WO 94/21801). This behavior could also be due to the less acidic pH optimum of EGIII compared with EGI and EGII (from 5.5 to 6.0). The lack of a CBD in EGIII might also have some relevance in relation to its low redeposition properties.

14.3 Biofinishing and enzymatic defibrillation of cellulosic fibers

14.3.1 The aim and mode of cellulase treatments of non-denim fabrics and garments

The main objective of the applications and developmental work on enzymes for textile applications is currently focused on the quality improvement of cellulosic textile fabrics and garments by various treatments with cellulase preparations. Most of the natural material used in fabric manufacturing is composed of cellulosic fibers: cotton, linen, ramie, viscose, lyocell (made by a new extrusion process from wood cellulose with solvent exhaustion) and the various blends with or without synthetic fibers. To various extents, all cellulosic materials have a tendency to fuzz formation. Fuzz is the designation for short fibers protruding from the surface of the yarn and fabrics. Hairiness refers to fuzz levels on the yarn; pilling consists of fluffy agglomerations of loosened fuzz attached to the surface of the fabrics, which are mostly formed during wearing and laundering of the garments (Pedersen *et al.*, 1992). The nature of the cellulosic fiber and the whole fabric construction have a significant impact on the tendency for fuzz and pill formation. Generally, a loose construction fabric will have an increased tendency to create fuzz and pill, whereas a tighter fabric will have an opposite tendency towards lower pilling. In any case, fuzziness and pilling are considered negative features of cellulosic fabrics and garments made thereof, and their prevention or permanent removal substantially increases the commercial value of textiles and garments. Traditionally, chemical agents, such as cationic surface-active compounds, are used to soften fabrics and improve their "hand feel", but these treatments are "natural" neither in feel nor permanence. Alternatively, cellulosic fabrics can be subjected to treatments with cellulolytic enzymes, a process termed "biopolishing" (see also patents WO 93/20278 and WO 94/12578 assigned to Novo Nordisk). Biopolishing is usually carried out in the textile wet processing stage, which includes desizing, scouring, bleaching, dyeing and finishing. Essentially, the action of cellulases removes impurities and small loose fiber ends that protrude from the fabric surface. The currently accepted mechanism is that the enzyme weakens the fibrils but that mechanical action is needed to separate them from the yarn, thus completing the process (Pedersen *et al.*, 1992). Mechanical action can be provided by the process equipment employed and by the abrasion caused by the revolving fabric mass itself, which is directly related to the ratio of liquor bath to weight of goods of the batch, i.e., the lower the ratio the greater the internal attrition obtained.

There are several advantages to the use of cellulases on cellulose-made fabrics and garments:

- removal of dead and immature cotton, mops and surface fuzziness;
- softening and improvement of hand feel, luster and drapability;
- better scouring and surface cleaning;
- improved dyestuff affinity, yield, uniformity and brilliance;
- higher hydrophilicity and moisture absorbancy;
- improved overall quality of fabrics, as piece-dye goods or as meter material, and of finished garments;
- possibility to create new and original finishing and fashionable effects;
- application of a fully ecological process.

An added advantage of enzymatic biofinishing is the long-term effect it confers to treated fabrics: pilling and fuzziness do not re-form (Sarkka and Suominen, 1996). Indeed, the durability of the antipilling properties after several wash cycles is considerably increased. On the contrary, when textile finishing chemicals are applied on fabrics, their effects disappear with time after a few washes or exposure to the open environment.

The cellulase treatment of non-denim fabrics is referred to as "biopolishing" or "biofinishing". In this application, the issue of indigo redeposition is not relevant, as a great variety of other dyes are employed. Instead, the main concerns are to minimize weight and strength loss due to enzyme action, while achieving complete antipilling and defibrillation of fabrics, and to avoid shade changes on dyed goods. Minimizing fabric strength loss is particularly important for linen and viscose rayon fabrics. Linen has a high amorphous cellulose content and low degree of polymerization, therefore it is easily weakened by cellulases. Similarly, viscose rayon is a weak fabric, even more so when wet. The treatment with cellulases can be carried out on woven and knit fabrics or meter pieces and also on garments (e.g., T-shirts, slacks, shirts, sweats, etc.), sheets and towels. It can be applied before or after dyeing or printing. However, biofinishing is often performed after bleaching and before dyeing, but at times (particularly with T-shirts and sweatshirts) also after dyeing. Winches, jiggers, jets, overflows and industrial washing machines for garments are widely available in the textile industry. Thanks to recent developments in equipment engineering for textile finishing, high throughput machines are now available that allow for vigorous mechanical action and are even designed to include an enzyme treatment step. When the treatment is applied on dyed or undyed garments, the mechanical action is vigorous, so a less aggressive or lower enzyme dosage can be used. If piece or meter fabrics have to be treated, the less energetic mechanical action of the equipment is offset by using a more aggressive enzyme or a higher cellulase dosage.

Biofinishing is mostly performed with acid cellulases because of their more aggressive action on cellulose fibers compared with the neutral enzymes. Therefore, cellulases used in this application are composed either of the whole natural complex of *Trichoderma reesei* or of genetically manipulated enzymes lacking one or more subunits or enriched in others. A "typical" biofinishing treatment with cellulase is carried out at 45–55°C, pH 4.5–5.5, for 30–90 min, at a ratio of 5:1 to 20:1 of liquor bath to weight of goods and at enzyme dosages that vary according to the composition and the specific activity of the cellulase preparation used. After treatment, the enzyme must be deactivated either by adding alkali to give a pH of 9–10 or by increasing the temperature to 70-80°C for 10 min, or both. Failure to inactivate the enzyme can result in negative effects on the fabrics. Figure 14.2 shows a 100% cotton knit before and after a 40 min treatment with an acid cellulase (courtesy of Primalco Ltd, Biotec). It can be seen that the impurities and loose fibers are removed from the fuzzy surface during biofinishing so that the surface of the knit appears totally clean.

The effectiveness of cellulase treatment is much influenced by previous treatments applied to the fabric, such as scouring, bleaching, mercerizing, dyeing, etc. For example, the enzyme is more active on mercerized than on non-mercerized cotton, as mercerization of cotton lowers the crystallinity of the substrate. These aspects should be taken into consideration when planning a pilot or an industrial trial with any commercial cellulase. Furthermore, the rate of catalytic hydrolysis of cellulose fabrics by cellulase is inhibited by both direct and reactive dyes (which are anionic) but not by vat dyes (which are non-ionic) (Koo *et al.*, 1994). As mentioned before, enzyme action on textile fibers is also largely determined by mechanical agitation during the process itself (Kumar *et al.*, 1995), all other parameters being equal. Lastly, it is obvious that the textile and dyehouse finisher faces a difficult challenge when dealing with a diverse range of textile fibers and fabric constructions, which, from a biochemical point of view, represent many different substrates for the enzyme. It is therefore difficult, if not impossible, to standardize all possible textile applications. Small-scale experiments are always recommended in order to find the optimal treatment conditions with any cellulase preparation under evaluation, although industrial-scale trials are always necessary to assess real performance, particularly with the new monocomponent cellulases being developed.

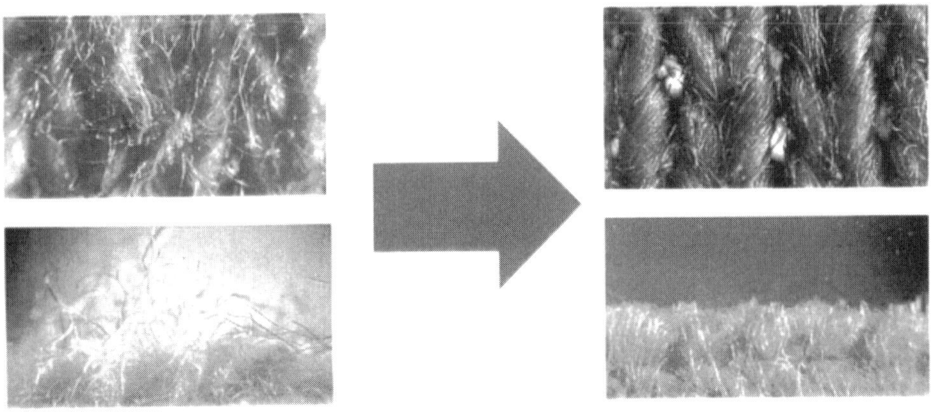

Figure 14.2 Biofinishing of 100% cotton knit with *Trichoderma* cellulase: before (left) and after (right) treatment (courtesy of Primalco Ltd, Biotec).

14.3.2 *Defibrillation of man-made cellulosic fibers (lyocell)*

Lyocell is the generic name for solvent spun, 100% cellulosic fibers of wood pulp from managed plantations in a closed amine oxide solvent system. In the manufacturing process, there is essentially total recycling of the solvent, while the solution is filtered and extruded to form the filaments. The resulting cellulose fiber should have roughly the same ratio of amorphous and crystalline cellulose as the starting material, even though the crystallinity of lyocell is also determined by the extrusion process employed. This is the first new fiber to be developed for the textile industry in over 30 years. The lyocell fiber combines the comfort properties of natural fibers with the high strength typical of synthetics. Tencel® is the registered mark for the new generic class lyocell of Courtaulds Fibres (Holdings) Ltd.

There are several positive attributes of lyocell fabrics:

- good wash stability resulting in low shrinkage;
- high resistance to tearing and good tensile properties, even when wet;
- excellent absorbancy and low static typical of natural cellulosic fibers;
- ease of blending with other fibers, luxurious drape and silk-like hand feel.

However, woven or knitted lyocell fabrics show tangles of primary fibrils on the surface, commonly referred to as "fibrillation". This characteristic is enhanced by high temperature, high pH and mechanical attrition. While this feature can be viewed as a negative aspect of lyocell, it nevertheless offers the opportunity to create a whole range of novel finishing effects. Indeed, fibrillation of lyocell fabrics can be best controlled by cellulase treatment; acid cellulases have proven to be most effective, although mixed lyocell garments can also be treated with neutral cellulases. The benefits of enzyme treatments are overall similar to those of the biofinishing described above for natural cellulosics. They include remarkably enhanced appearance and feel of lyocell fabrics and blends, associated with long-lasting effects such as softening, depilling and pill prevention, improved drapability and surface appearance even after multiple launderings (Kumar *et al.*, 1994).

The final results of Tencel® finishing will be determined by the degree to which mechanical, chemical and enzymatic processes can be combined. Usually, maximum fibrillation of lyocell is intentionally created by a combination of high pH, high temperature and mechanical attrition. This so-called primary fibrillation is mainly composed of long fibrils lying across the fabric surface. Cellulase treatments attack the primary fibrillation, giving a clean fiber surface. If so required, further processing can generate a shorter, stable secondary fibrillation under wet conditions to give a range of new interesting and commercially valuable finishing effects. Figure 14.3 is a scanning electron micrograph of untreated and cellulase-treated Tencel® (courtesy of Genencor International).

The versatility of lyocell is best exploited at the finishing stage of the textile process where cellulase treatments offer the greatest potential. It is no wonder, therefore, that substantial R&D efforts have been devoted by the major enzyme producers to develop new cellulases tailor-made for lyocell. These new cellulases must achieve multiple goals: a wider range of finishing effects, particularly in blends with other fibers; lower losses of strength and weight; better adaptability to high-speed jet machines; possible combination with other chemical treatment steps; "natural" and pleasant hand feel; etc.

Figure 14.3 Scanning electron micrograph of untreated (top) and cellulase-treated (bottom) Tencel®.

As far as we are aware, two endo-enriched acid cellulases obtained from genetic engineering and specifically useful for the treatment of lyocell fabrics have recently become commercially available: one from Novo Nordisk (1995) and the other from Genencor International (1996).

Finally, it should be emphasized that the commercial availability of new man-made fibers, such as lyocell, represents a powerful driving force for the textile application of *T. reesei* cellulases and for the development of optimized, bioengineered components of the natural cellulase complex.

14.3.3 ***New endo-enriched cellulase preparations for biofinishing***

The recent developments in cellulase research have generated new enzymes with which even lightweight cotton and linen can be biofinished without compromising tensile strength and elasticity to any significant extent. It is still not clear which subunit of the *T. reesei* cellulase complex ought to be either deleted or amplified in order to get better practical performance. As in the case of biostoning, it appears that endo-activities are also better suited for biopolishing and biofinishing. The deletion of CBHs from the whole cellulase complex seems to reduce fabric strength and weight loss. Some recent reports, however, appear to offer conflicting views and conclusions about this issue. Miettinen-Oinonen *et al.* (1996) have described the performance of a cellulase preparation produced with a genetically modified strain

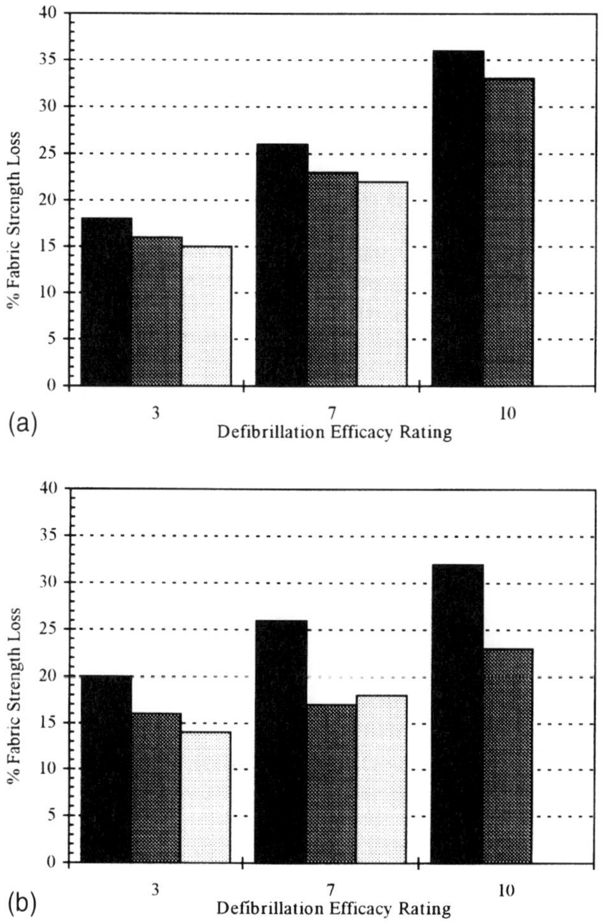

Figure 14.4 (a) Fabric strength loss on 100% lyocell at equal defibrillation efficacy (adapted from Kumar *et al.*, 1996). (b) Fabric strength loss on lyocell/cotton (65/35) at equal defibrillation efficacy (adapted from Kumar *et al.*, 1996). Full bars indicate whole acid cellulase composition #1, dark shaded bars endo-enriched cellulase composition #2, light shaded bars endo-enriched cellulase composition #3.

of *T. reesei.* Deletion of the EGII-encoding gene (*egl2*) produced a cellulase complex with a much reduced endoglucanase activity.

In denim stonewashing using the preparation with low endo-activity, there was up to 15% less strength loss than when using the complete cellulase mixture at an equal abrasion level. Similarly, in biofinishing experiments on cotton the cellulase preparation with low endo-activity gave less strength loss as a function of enzyme dosage. Weight loss was comparable with low endoactivity and complete mixture preparations. They also claimed that when treating the cotton fibres the low endo-activity preparation gave smoother fibers and was almost devoid of fibrils. The authors concluded that a modified *T. reesei* cellulase preparation lacking EGII gives better performance in both biostoning and biofinishing of cotton by causing less strength loss and fewer fiber-damaging effects. A somewhat different claim was made by Kumar *et al.* (1996), who compared the performances of three different acid cellulase enzyme compositions (all derived from *T. reesei*) in controlling fibrillation of several mixed fabrics containing lyocell. The exact subunit composition of each preparation was not reported, but it was stated that one was a whole acid cellulase (while the other two were enriched for different endo-cellulase activities and had exo-cellulases removed). No further biochemical details or exact subunit compositions were given. After treatment with these three different cellulase preparations under the same conditions, the fabric samples were tested for extent of defibrillation, fabric strength and weight loss. Efficacy of the enzyme preparation was measured as a function of the degree of defibrillation achieved. No additional chemical auxiliaries, other than buffer and wetting agents, were used in order to determine only the effect of the enzyme used in each trial. Figure 14.4 shows that a particular endo-enriched *Trichoderma* cellulase can give the highest degree of defibrillation on 100% lyocell or on a 65/35 blend of lyocell/cotton with significantly less fabric strength loss when compared with a whole acid cellulase. Another type of endo-enriched cellulase is more effective when mild surface polishing is required and fabric strength loss is a major concern, as with lyocell/linen blends. The conclusion was that the performance characteristics of whole acid cellulases are quite different from endo-enriched preparations, but that the latter offer better performance in all applications in which losses in fabric strength and weight have to be minimized.

The effect of agitation rates on the mechanism of attack by pure endoglucanases and/or total cellulase complex has been recently studied by Cavaco-Paulo *et al.* (1996b). They demonstrated that mechanical agitation influences endo- and exo-cellulase action on cotton fabric in different ways, presumably due to uneven physical absorption of the enzyme components on the fiber. They concluded that higher agitation levels increase the activity of endoglucanase to a greater extent than that of exo-cellulase in a total cellulase complex, offering the chance to exploit useful synergies of action. It also implies that such phenomena should be considered when developing practical textile applications with the new endo-enriched preparations now available. In a later paper, the same group studied the kinetic parameters measured during cellulase processing of cotton (Cavaco-Paulo and Almeida, 1996) and concluded that the synergism between increasing mechanical friction and cellulase action facilitates the dissociation of the enzyme from the fiber. From our own practical experience in industry, we can state that as attractive as a laboratory hypothesis might be, one should never extrapolate textile performance of any recombinant or bioengineered cellulase preparation purely from its biochemical composition or properties. On the contrary, rigorous and quantitative industrial tests need to be developed and trials must be performed for this purpose.

14.4 Current perspectives and future developments for cellulase applications in textiles

Different textile fibers have different properties that are closely related to their morphology and polymeric structure. Therefore, an enzyme treatment will probably give varying results. While whole acid cellulase complexes were used as a general purpose enzyme in the early days of biofinishing, genetically modified cellulase preparations now allow for selection of the best enzymes for any given fiber composition and treatment type. For example, linen and viscose are highly sensitive to cellulase hydrolysis and should be treated with less aggressive preparations, while 100% cotton or lyocell and their blends are inherently sturdier and can afford a more aggressive defibrillation treatment. In either case, it is no longer just a matter of enzyme dosage but mostly a matter of cellulase composition (with the optimal ratio of endo- to exo-activities and of dissociation constants under actual process conditions) and of treatment conditions.

In the field of biostoning, the first step was to validate the usefulness of cellulases in industrial laundries in order to eliminate or decrease the use of pumice stones. Later, the main concern became the minimization of indigo redeposition by using neutral modified or recombinant cellulases from *Trichoderma.* Currently, good abrasion and low redeposition on denim garments are taken for granted with any cellulase-containing product. The emphasis has now shifted to the reduction of fabric strength loss, which becomes the next target for the enzyme producers. In biopolishing and biofinishing, good defibrillation with low fabric strength loss are the minimal requirements expected from any commercial cellulase. The new *Trichoderma* and other acid cellulases appear to fully meet this challenge. The next objectives appear to be (a) obtaining softer fabric hands with no loss of strength properties and (b) adapting cellulase use in continuous fabric processes rather than only batch, which would represent a major breakthrough in textile technology. Selection of the correct cellulase is therefore of paramount importance to obtain the best results in any textile application. In addition, process conditions should be optimized to eliminate or reduce unwanted side-effects.

In the future, the range of modified cellulases in terms of subunit composition and biochemical features will surely expand, along with our understanding of their modes of action. At that point, one will be able to choose the optimal biofinishing treatment for each type of fabric and blends. Finally, enzyme applications will have to be made even more user-friendly, reproducible, sturdy and economically competitive. Enzyme biotechnology must not only be understandable to the final user, but it must also be affordable, in order to become a (bio)tool for everyday work!

Acknowledgments

We wish to thank the following persons for kindly providing literature and other written material in the preparation of this chapter: Pirkko Suominen and Arja Miettinen-Oinonen of Primalco Ltd Biotech, Kathleen Clarkson, Hugh McDonald and Akhil Kumar of Genencor International; Henrik Lund of Novo Nordisk. A. Kumar and A. Cavaco-Paulo also contributed precious comments and suggestions. Expert collaboration of Lamberti's Textile Division and the continuous support and encouragement of Lamberti's management are gratefully acknowledged.

References

BERRY, A. 1996. Improving production of aromatic compounds in *Escherichia coli* by metabolic engineering. *Trends Biotech.* **14**: 250–256.

CAVACO-PAULO, A. and ALMEIDA, L. 1996. Kinetic parameters measured during cellulase processing of cotton. *J. Textile Inst.* **87**: 227–233.

CAVACO-PAULO, A., ALMEIDA, L., and BISHOP, D. 1996a. Cellulase activities and finishing effects. *Textile Chemist and Colorist* **28(6)**: 28–32.

CAVACO-PAULO, A., ALMEIDA, L., and BISHOP, D. 1996b. Effects of agitation and endoglucanase pretreatment on the hydrolysis of cotton fabrics by a total cellulase. *Text. Res. J.* **66(5)**: 287–294.

GODFREY, T. 1996. Textiles. In T. Godfrey and S. West (eds), *Industrial Enzymology*, 2nd edn. Macmillan, Basingstoke, UK, pp. 361–371.

KLAHORST, S., KUMAR, A., and MULLINS, M. 1994. Optimizing the use of cellulase enzymes. *Textile Chemist and Colorist* **26(2)**: 13–18.

KOO, H., UEDA, M., WAKIDA, T., YOSHIMURA, Y., and IGARASHI, T. 1994. Cellulase treatment of cotton fabrics. *Textile Res. J.* **64(2)**: 70–74.

KUMAR, A., LEPOLA, M., and PURTELL, C. 1994. Enzyme finishing of man-made cellulosic fabrics. *Textile Chemist and Colorist* **26(10)**: 25–28.

KUMAR, A., PURTELL, C., and YOON, M. Y. 1996. Performance characterization of endo-enriched cellulase enzymes in the treatment of 100% lyocell and lyocell-blended fabrics. In Proceedings of the Textile Institute's 77th World Conference, May 21–24, Tampere, Finland, pp. 177–189.

KUMAR, A., YOON, M. Y., and PURTELL, C. 1995. Optimizing the use of cellulase enzymes in finishing cellulose fabrics. In Proceedings of the 1995 AATCC International Conference, Atlanta, GA, pp. 238–247.

MIETTINEN-OINONEN, A., ELOVAINIO, M., PALOHEIMO, M., SUOMINEN, P., PERE, J., and OSTMAN, A. 1996. Effects of cellulases on cotton fibers and fabric. In Proceedings of the Textile Institute's 77th World Conference, May 21–24, Tampere, Finland, pp. 197–209.

PEDERSEN, G. P., SCREWS, G. A., and CEDRONI, D. A. 1992. Biopolishing of cellulosic fabrics. *Can. Text. J.* **Dec**: 31–35.

SARKKA, P. and SUOMINEN, P. 1996. Cellulolytic enzymes in biofinishing of cellulosic fabrics. VTT Symposium 163 on Chemical Reaction Mechanisms, Jan 29–30, Espoo, Finland, pp. 29–35.

SCHULEIN, M., TIKHOMIROV, D. F., and SCHOU, C. 1993. *Humicola insolens* alkaline cellulases. In Proceedings of the Second TRICEL Symposium on *Trichoderma reesei* Cellulases and Other Hydrolases, Espoo, Finland, pp. 109–116.

UEDA, M., KOO, H., WAKIDA, T., and YOSHIMURA, Y. 1994. Cellulase treatment of cotton fabrics (II): inhibitory effect of surfactants on cellulase catalytic reaction. *Textile Res. J.* **64(10)**: 615–618.

15

Application of *Trichoderma* enzymes in the food and feed industries

YVES M. GALANTE, ALBERTO DE CONTI and RICCARDO MONTEVERDI

Laboratory of Biotechnology, Central R&D, LAMBERTI s.p.a., Albizzate, Italy

15.1 Brewing

The production of alcoholic beverages by fermentation of cereal extracts is probably as old as civilization itself. Traces of brewing activities can be dated back to Babylonian times. However, unlike wines made from crushed grapes or other fruits, the fermentation process does not spontaneously take place with grains, unless their starch and protein contents are partially modified by enzymes in order to provide the necessary fermentable substrates and nutrients to the yeasts.

Beer is traditionally made of malt, hops, yeast and water, but brewing technology is largely based on the action of enzymes activated during malting and later during fermentation itself. Malting of barley relies upon seed germination to initiate the biosynthesis and activation of specific endogenous hydrolytic enzymes that modify the seed reserves until germination is stopped by heating and water removal. Four main categories of enzymes are involved: α- and β-amylases, (carboxy)peptidases and β-glucanases; all have to act in synergy and to an optimal extent. The quality of the brewer's mash will therefore be determined by the ingredients and the skill put into the malting stage. In many breweries around the world, alternative sources of fermentable carbohydrates are used in brewing when malt is of poor quality, is not readily available or happens to be too expensive. For example, unmalted barley and other cereals (e.g., corn, rice) can be used in combination with malt at the mashing step. However, there is a problem with this process since barley contains, on a dry-weight basis, about 6–10% of non-starch polysaccharides (NSP), of which the most important fraction are the β-glucans, the main component of the so-called "barley gums".

Typically, barley β-glucans contain 30% of 1,3- and 70% of 1,4-linked glucose units, structures which, unlike that of cellulose, make this polysaccharide more or less soluble in water, depending upon the molecular size. Soluble barley β-glucans may form gels during the brewing process and cause serious problems, including poor filtration of the wort, slow run-off times, low extract yields, or development of a haze in the final product. Pentosans (e.g., arabinoxylans) and proteins can also add to the problem. In addition, the quality of barley and malt used in brewing depends very much on seasonal exposure to weather conditions. Unfavorable conditions, especially high humidity or rainfall, may result in poor quality barley and malt.

Unmalted barley contains only low levels of β-glucanase activity, while the activity levels of endogenous β-glucanases in malt are variable on account of several factors (e.g., barley cultivar, seasonal variations, storage conditions), and the enzymes produced during germination are rather heat sensitive. From all of the above, it is obvious that excess viscosity in the wort or fermentation broth due to the presence of β-glucans is unlikely to be eliminated by endogenous β-glucanases. To alleviate or offset most filtration difficulties in the brewing process, microbial β-glucanases are added either during mashing or during primary fermentation. The enzyme splits 1,4-glycosidic bonds of β-glucans, thereby generating shorter polymers with few 1,3-linkages and causing a drop in medium viscosity. β-glucanases commonly used in brewing are derived from *Penicillium emersonii, Aspergillus niger, Bacillus subtilis* or *Trichoderma reesei.*

In a comparative study of different β-glucanase performances in beer wort production, Pajunen (1986) concluded that the most economically advantageous enzyme preparation was the one from *Trichoderma*, as judged by its cost/performance ratio. In another study, it was suggested that of all activities of the *Trichoderma reesei* cellulase complex, reduction of the degree of polymerization and wort viscosity were caused primarily by EGII and CBHII acting on barley β-glucans (Oksanen *et al.*, 1985). Typically, addition of a commercial *Trichoderma reesei* cellulase preparation (in the range of 0.05 to 0.1 ml per kg of grists) causes a 90% decrease in β-glucan content and up to 30% shorter wort filtration times. Filterability (expressed as V_{max} in ml) is consistently enhanced with increasing doses when tested in pilot-scale conditions similar to brewery processes. The same enzyme can also be used as an efficient filtration aid when added to the primary fermentation.

Whether the enzyme is added to the mash or during primary fermentation, the final beer taste is comparable to that of a control lager with no enzyme addition. Three commercially available β-glucanases derived from different microorganisms were also compared in pilot- and industrial-brewing plant trials by Canales *et al.* (1988) on a variety of grist bills (65% malt/35% barley, 65% malt/35% rice, 50% malt/15% barley/35% rice). The preparation derived from *Trichoderma* outperformed those from *Bacillus subtilis* and from *Aspergillus niger* in terms of higher filtration rates and overall brewhouse efficiency. As in other published studies, the beers obtained could not be distinguished from normal production by a taste panel.

Therefore, *Trichoderma* β-glucanases represent an efficient processing aid for the brewery. A different approach can be taken by using cellulase enzymes during the malting process itself (Home *et al.*, 1983). The use of cellulases in malting improves the quality of malt, regardless of seasonal differences between barley crops. Filterability of wort and of final beer are markedly enhanced, as well as brewhouse yield and fermentation rate. Presumably, the enzyme degrades to a large extent the cell wall structure, as indicated by electron microscopy analysis, improving penetration of added gibberellic acid into the grain and thus accelerating the synthesis of endogenous hydrolytic enzymes.

More recently, glucanolytic brewer's yeast strains have been constructed by transferring and integrating a β-glucanase gene from *Trichoderma reesei* into industrial brewing yeast strains (Penttilä *et al.*, 1987). The new strains expressing and secreting β-glucanase were reported to efficiently hydrolyze β-glucans and to decrease viscosity during primary fermentation, which resulted in a markedly improved filterability of beer with unaltered quality (Suihko *et al.*, 1991).

In the future, these glucanolytic recombinant strains may be precious for those breweries where addition of exogenous microbial enzymes is prevented by traditional processes or is prohibited by local legislations that limit the use of additives.

15.2 Wine making and fruit juices

Wine making is, by its very own nature, a "biotechnological" process, in which both yeasts and enzymes play a fundamental role. Efforts to improve yeast strains used in fermentation of grape juice have been under way for decades; on the other hand, the use of exogenous microbial enzymes has developed rather slowly in the last 30 years. Their use still is perceived with a certain degree of suspicion by wine makers and consumers alike. Nevertheless, our knowledge of the macromolecular composition of grapes, must and wine has advanced considerably, as has our understanding of how to use enzymes to improve the process and the final product that we all enjoy: the wine, of course!

Three main exogenous enzyme activities are exploited in wine making: pectinases, β-glucanases and hemicellulases. These enzymes are also used in the maceration of olives and in the production of fruit and berry juices. Recently, a fourth use has been attracting considerable interest: β-glycosidases for aroma development from naturally present, glycosylated precursors (Caldini *et al.*, 1994; Gunata *et al.*, 1990; Williams *et al.*, 1982; Wilson *et al.*, 1984). The benefits that are mainly sought by the winemaker using enzymes are better skin maceration, improved color extraction, easier must clarification and filtration, and increased wine quality and stability.

Pectinases from *Aspergillus* spp. were the first microbial enzymes to be introduced into wineries in the 1960s and 1970s. Pectins form a heterogeneous group of polysaccharides located in the intermediate lamella and the primary membrane of the grape skin cells. They have an essential function in the ripening of the fruits, but also strongly influence pressing efficiency, juice yield, color extraction, clarification and filtration of the must.

Commercial pectinases are all enzyme mixtures containing different amounts of pectin esterase and depolymerizing activities (polygalacturonases and pectin lyases). The enzymes from *Aspergillus* spp. also contain minor amounts of hemicellulases, which degrade neutral pectins associated with hemicelluloses, and cellulases, which attack the cell wall structure. The exact nature of these so-called side activities and their ratios are not necessarily established or constant yet have proved crucial in obtaining optimal performance from enzyme technology in wine making.

The use of pectinase preparations, added during the crushing of the grapes or directly to the must, enhances the extraction yield of first juice, shortens the clarification step and increases the content of terpenes in wine. Preparations high in pectin lyase and low in pectin methyl esterase are preferred so that methanol liberation from methylated polygalacturonic acid is minimized.

Only a phenomenological description of these effects is given here; it is virtually impossible to quantify them generically, as grape cultivars, seasonal varieties and mode of plant operation all contribute to give widely different quantitative results. Overall, better clarification and must filtration rates are obtained with pectinases.

However, these enzymes have a somewhat limited effect in the improvement of "trouble" musts or wines and of wine quality. It was first suggested by Dubourdieu in the early 1980s (Dubourdieu *et al.*, 1981; Villetaz *et al.*, 1984) that a *Trichoderma* β-glucanase could be successfully used in the processing of wine made from *Botrytis cinerea*-infested grapes. This mold usually attacks nearly-ripe grapes under certain conditions of temperature and humidity, thus producing a high molecular weight colloidal polysaccharide (once wrongly assumed to be a dextran) that causes great difficulties in wine filtration. It turned out that the soluble polysaccharide produced by *Botrytis* is a β-(1,3)-glucan with short side chains linked through β-(1,6)-bonds (Dubourdieu *et al.*, 1981). This glucan can be specifically hydrolyzed by a purified β-glucanase produced by a selected strain of *Trichoderma harzianum.* A patent was granted covering this discovery, and the commercial preparation of the enzyme was given approval by the European Commission around 1995 for use in wine making after alcoholic fermentation (Novo Nordisk). It was subsequently noticed that treatment of wine with *Trichoderma* β-glucanase helps to hydrolyze other glucans, such as those from yeast, that can also cause serious clarification and filtration problems.

Developments since the 1980s in the characterization and applications of macerating enzymes (i.e., blends with various activity ratios of cellulases, pectinases and hemicellulases) brought significant improvements in grape pressability, settling rate and lees compaction, juice free-run and total yields, when compared with a pectinase added at equal dosage, without negatively affecting juice clarity or quality of the final wine (Harbord *et al.*, 1990). Significant and reproducible improvements in white and red wine production can be obtained only with a correct balance of exogenous pectinolytic, cellulolytic and hemicellulolytic activities added to complement the relatively poor endogenous enzyme activities of the grape. This balance is found by experimenting with various enzyme mixtures in production trials on different grape varieties. Under optimal conditions, significant improvements were obtained in yield of free-run and pressed juice, juice and wine quality (including aroma enhancement) and overall process efficiency.

Using a macerating enzyme preparation (a blend of activities from *Trichoderma* and *Aspergillus* with relative activities of pectinase to cellulase to hemicellulase of 80 : 70 : 55, commercially available as Cytolase 219^{TM}), our group obtained positive results in trials carried out over four successive vintage seasons (1989–92) in Northern Italy on white grape varieties (Soave, Chardonnay, Sauvignon). The enzyme was added to the crusher at dosages of 40–100 ppm (see Figure 15.1) and given a residence time of 1–4 h at a temperature of 0–35°C.

It is difficult to specify the exact composition of macerating enzyme blends used by other groups or that are commercially available, as these products contain several primary and secondary activities.

The benefits obtained with Cytolase 219 can be generally summarized as given in Table 15.1. In industrial trials carried out on other vintages and in different countries, quality improvements were always noticeable, such as: improved color extraction in red grapes; increased aroma and flavor extraction in both red and white grapes; better body and mouthfeel. Particularly with red grape varieties, the amount of color released from the skins was increased by 20–150%, depending on skin contact time, processing temperature, enzyme dosage and grape variety, and was associated with decreased browning during processing and aging (Zent and Inama, 1992).

In conclusion, there are considerable benefits offered to the wine industry by

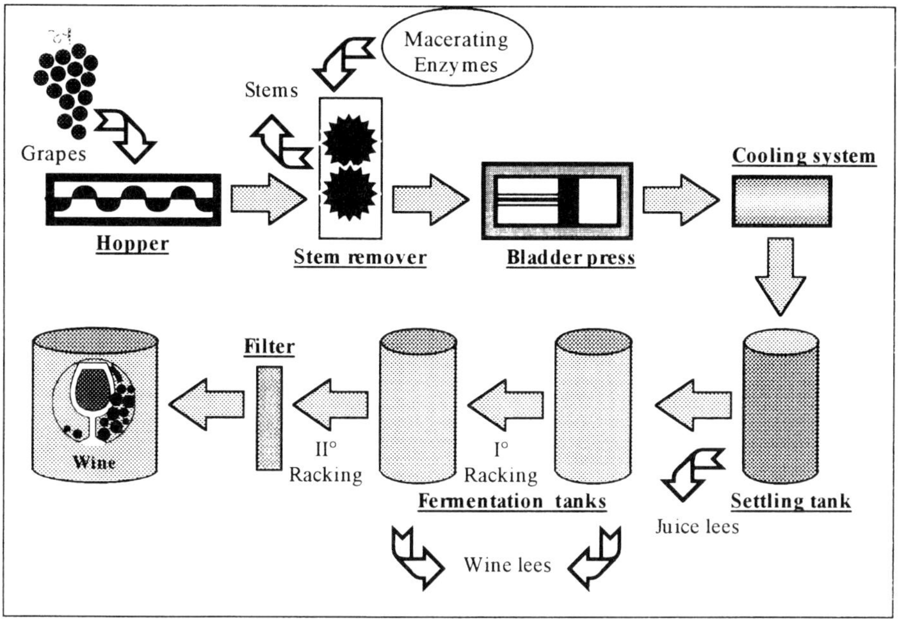

Figure 15.1 Wine making process scheme. Macerating enzymes are added at the stem remover or at the bladder press step.

enzyme technology in general and by *Trichoderma* enzymes in particular. However, further developments in this important field will be made possible by further advances and by a better "biochemical" education of producers and consumers.

The use of macerating enzymes in wine making is actually an outgrowth of developments in enzyme technology for fruit juice production. In the latter, the advantages of using a combination of pectinases, cellulases and hemicellulases (macerating enzymes) are essentially those described before.

In apple and pear fruit juice production, the whole fruits are crushed to obtain a pulp mash that, after mechanical processing (pressing, centrifuging and filtering), gives a clear liquid (the juice) and a solid phase (the pomace). Total yield and overall process performance are increased by addition of enzymes without capital investment in new equipment. Macerating enzymes are used in two steps of the process.

Table 15.1 Benefits of Cytolase 219 in wine making

Extraction of first must ("mosto fiore")	10–35% increase
Pressing time	50–120 min decrease
Must viscosity	30–70% reduction
Main analytical features of must	no difference from control
Must filtration rate	70–180% increase
Energy saving in fermentation cooling	by 20–40%
Settled lees volume at first racking	200–400% decrease
Use of clarifying adjuvants	none in must; 20–100% reduction in wine
Wine shelf life (stability)	much improved

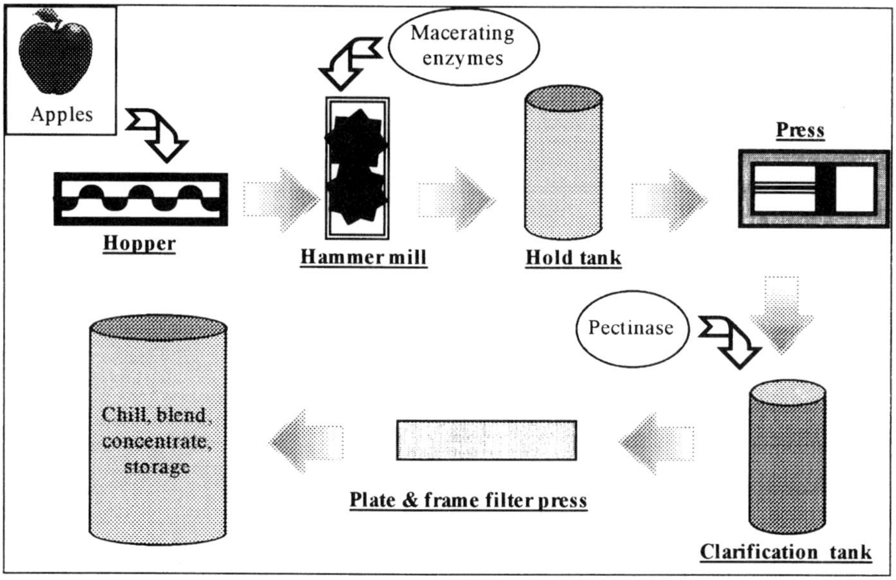

Figure 15.2 Scheme of apple juice production with the aid of macerating enzymes.

(1) After crushing, they are used in the treatment of the pulp mash to macerate the fruit pulp (partial liquefaction) or for complete fruit pulp fluidification (total liquefaction). This increases the yield of juice and reduces processing time while improving the extraction of valuable fruit components. (2) In the treatment of the juice, they are used to clarify it and lower its viscosity prior to concentration, thereby increasing filtering rate and improving final product stability.

A typical scheme of apple juice production with the aid of macerating enzymes is shown in Figure 15.2. The use of macerating enzymes offers appreciable advantages also in the treatment of by-products of the citrus fruit industry. For example, in pulp- and peel-washing (i.e., recovery of juice and soluble solids from exhausted citrus pulp or peel by counter-current washing), a combination of pectinases and cellulases is used to improve the extraction yield and the overall process efficiency. Lastly, in the production of cloudy peel and comminuted products obtained by the crushing/mashing of citrus peel and whole fruit, respectively, enzymatic mash treatment with *Trichoderma*-derived enzymes is one of the most commonly used and economical processes.

15.3 Olive oil production

The olive tree, along with the grape vine and the laurel, is one of the oldest and most fascinating symbols of Mediterranean civilization and lifestyle. Production and use of olive oil are indeed as old as Greek mythology. Olive presently is an important commercial crop in Italy, Spain, Greece, Tunisia and Turkey, where olive oil is a fundamental diet component. These five countries account for 90% of world production, with Spain being the major producing and exporting country. Furthermore, olive oil consumption is on the increase in other countries, such as the USA and Northern Europe, where it is becoming important as a replacement for animal and dairy fats. Indeed, several positive health factors have been attributed to olive oil, including antioxidative properties, lowering of serum cholesterol, and other benefits to the cardiovascular system. In Italy alone, over two hundred thousand metric tons of olive oil are produced per year and have a wholesale market value of about US $1.2 billion. Consumers generally prefer extra-virgin oil.

It is beyond the scope of this chapter to describe the biology of the olive tree and fruit or of the process for oil production. Only the latter will be briefly sketched, in order to appreciate the advantages provided by enzyme technology.

The olive tree is a slow growing plant that takes over seven years after the planting of a young tree for the first olive crop to be harvested. Furthermore, crop yield follows a cyclical high and low pattern on alternating years, with additional significant variations due to seasonal conditions. Olives are picked and processed from October through February–March, according to latitude, cultivar and local tradition. Poor winter weather can adversely affect both olive yield and quality. Severe winter frost almost completely wiped out olive tree groves in certain areas in Italy in January 1985.

There are several cultivars of *Olea europea*, each with different attributes, including ripening features, resistance to pests and environmental conditions, oil content and composition, characteristic aroma and flavor, etc. Regardless of the cultivar, olive oil production is a rather simple process, with the average oil yield given in Figure 15.3. Two different processes are in use: traditional pressing and the continuous centrifugal process. Existing plants range from large, modern, stainless steel factories to small, family-run mills.

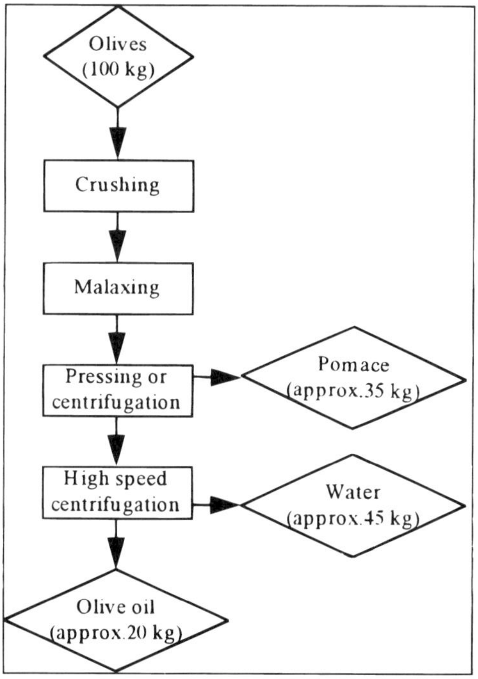

Figure 15.3 Scheme of olive oil production.

Both processes begin with the crushing and grinding of the olives in a stone mill (as has been done for centuries, but electric engines have now replaced human and animal labor) or hammer mill to produce a coarse, minced paste. The paste is further mixed and minced in a series of malaxeurs, through the mechanical action of rotating blades in horizontal cylinders. In the traditional process, the olive paste is layered over round canvas supports, stacked one on top of the other in a hydraulic press, which is then operated at increasing pressure up to 350–400 kg/cm^2 for about 2 h per batch. The oily must is collected by gravity and the oil is recovered by centrifugation or spontaneous decanting. In the continuous centrifugal process, powerful horizontal decanters follow the malaxing step to bring about a three-phase separation of water/oily must/solids (the last referred to as pomace). The oily must phase is an oil/water emulsion, from which the oil fraction is recovered by high-speed centrifugation. The Sinolea System (patented by Rapanelli, Foligno, Italy) is a continuous process in which part of the olive oil (of the highest quality) is recovered during malaxing by a percolating malaxeur.

On the average, 100 kg of olives yield 16–20 kg of oil. The "acidity" of the oil (as oleic acid) determines its classification as extra-virgin (which contains less than 1% free fatty acid), virgin (1–3% FFA) or third choice. Given this simple production scheme, it is easily understood why conditions that favor high quality oil give low yields and vice versa: fruit ripeness and malaxing temperature play an important role.

In the EEC, the quality of extra-virgin oil is mainly established by test panel, although chemical analysis is routinely used to determine various "quality parameters", such as acidity, peroxide number, polyphenol and tocopherol content (vitamin E), etc., and resistance to rancidity. To assure high quality oil, a production plant should process somewhat immature olives that are freshly picked and thoroughly washed of all dirt. Industrial processing must be under cold milling, mixing and pressing conditions. On the other hand, higher yields are obtained with fully ripe fruits, processed at higher than ambient temperature to favor quantitative oil extraction. Such conditions tend to cause the oil to be more acidic, give poorer aroma, and be more prone to develop rancidity.

The pattern of expression and activation of endogenous free and cell-bound hydrolytic enzymes in the olive was thoroughly investigated by Fernandez-Bolanos *et al.* (1995). They concluded that several enzymatic activities besides pectinases are involved in fruit softening. Endoglucanases appear to be active mainly during the late ripening phase, when no major textural changes take place, as at the onset of the senescence process. It is therefore unlikely that these endogenous cellulases play a major role in the maceration of olives and oil extraction during processing.

Just like in wine making, pectinases were the enzymes initially used in olive processing to improve yield (Leone *et al.*, 1977; Fantozzi *et al.*, 1977). The first commercial preparation for this purpose was Olivex®, derived from an *Aspergillus aculeatus* strain that contains several types of enzyme activities. Pectinolytic enzymes predominate in this mixture, but there are also various side activities of hemicellulases and cellulases (Olsen, 1995). This preparation allows the production of an additional 10–20 kg of oil from 1 tonne of olives, on average, but there is considerable variation depending on the cultivar and ripeness of the olives, process temperature, pH and enzyme dosage.

When more systematic studies were conducted in the 1980s, it became apparent that no single activity was sufficient to macerate the olives and to fully increase oil extraction at ambient temperature. Instead, mixtures containing a range of activities performed better. Accordingly, various enzymes were tested in the laboratory and in the mills for their effectiveness in improving olive oil yield and quality profile on different cultivars (Montedoro, 1987).

The usual three types of activities needed to obtain optimal fruit maceration during crushing, mincing and mixing were pectinases, cellulases and hemicellulases. Fungal pectinases alone were not completely effective, while in our experience a combination of *Aspergillus* pectinases with *Trichoderma* glucanases and hemicellulases gave the best performance (Galante *et al.*, 1993). During successive years, at the beginning of each crop season, we carried out field trials in oil mills on freshly picked olives, testing various blends of enzyme activities in which the *Trichoderma* enzymes were the prevalent components, but pectinases were always added. Conventional pressing processes benefit less from enzyme addition than continuous centrifugal ones. The latter are significantly and more consistently improved by enzyme technology.

The enzyme mixture was added either to the crusher or in the first malaxeur, as shown schematically in Figure 15.4, and allowed to macerate during crushing and mixing up to the final centrifugation step, where it separated and was eliminated in the water fraction. Of the various blends tested, the one with the activity profile reported in Table 15.2 consistently gave better yield improvement under a variety of conditions and was chosen for further studies. It contained no lipolytic activity. The enzyme product used was available from Genencor Inc. under the tradename of Cytolase O®. Representative results from industrial trials carried out in central and

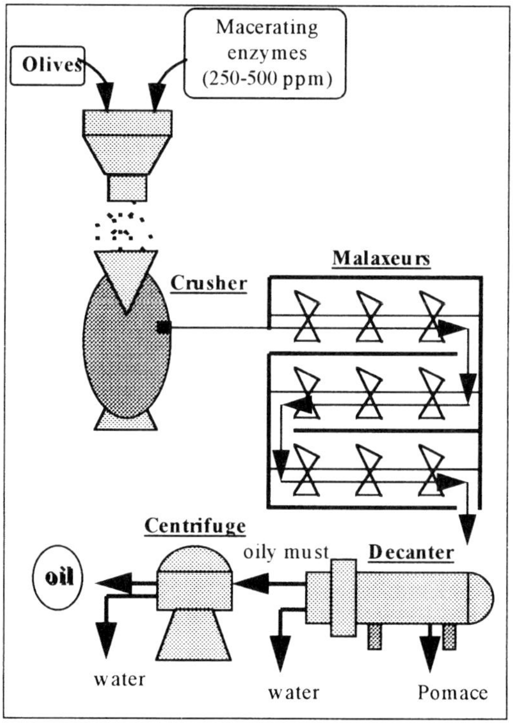

Figure 15.4 Scheme of olive oil production in a continuous centrifugal process.

Table 15.2 Activity profile of the macerating enzyme blend tested in industrial trials during the 1991 season[a]

Enzyme activity	Method	Units
Endo-cellulase	viscosimetric	2400 min/ml
Pectin lyase	UV	24 μmol/min/ml
Endo-pectinase	viscosimetric	170 min/ml
Endo-mannanase	viscosimetric	47 min/ml
Endo-arabinase	colorimetric	0.78 OD/min/ml

[a] Adapted from Caldini (1992).

southern Italy during the 1991 crop season are reported in Tables 15.3a and 15.3b. Controls refer to olive batches matching as closely as possible the experimental batches with respect to weight, fruit ripeness and cultivar, process conditions and time of harvest.

Obviously, there is a significant variation of results and, due to the extreme difficulty of conducting precisely controlled and matched industrial trials during a full

Table 15.3a Use of macerating enzymes in olive oil production with a continuous decanter system[a]

Trial number	Treatment	Olive cultivar	Batch (kg)	Oil yield (kg oil/ 100 kg olives)	Oil increase (kg oil/ 100 kg olives)
A.1.c	Control	*Coratina*	1800	15.45	
A.1.e	0.025% Cytolase O	*Coratina*	1570	17.76	+2.31
A.2.c	Control	*Ogliarola*	1000	12.70	
A.2.e	0.050% Cytolase O	*Ogliarola*	982	14.66	+1.96
A.3.c	Control	*mixed*	1770	18.11	
A.3.e	0.025% Cytolase O	*mixed*	1720	18.89	+0.78

[a] From Galante *et al.* (1993).

Table 15.3b Use of macerating enzymes in olive oil production with a Sinolea system[a]

Trial number	Treatment	Olive cultivar	Batch (kg)	Sinolea step (%)	Oil yield (kg oil/ 100 kg olives)	Oil increase (kg oil/ 100 kg olives)
B.1.c	Control	*Moraiolo*	357	60	14.87	
B.1.e	0.025% Cytolase O	*Moraiolo*	358	73	16.79	+1.92
B.2.c	Control	*Moraiolo*	312	25	16.57	
B.2.e	0.025% Cytolase O	*Moraiolo*	312	23	20.90	+4.33
B.3.c	Control	*mixed*	630	56	21.08	
B.3.e	0.025% Cytolase O	*mixed*	630	72	22.14	+1.06

[a] From Galante *et al.* (1993).

production season, statistical examination of data is not possible. Nevertheless, addition at the milling stage of the enzyme preparation in the range of 25–50 ml per 100 kg of olives (250–500 p.p.m.) gave a yield improvement from 1 to 4 kg oil per 100 kg olives. An increase of 1–2 kg is to be considered "average" under good process conditions.

Also the quality profile of the extra-virgin oil appears to be improved, as summarized in Table 15.4. From Table 15.4, it can be seen that enzyme addition doesn't significantly alter oil acidity and peroxide number, while the content in total anti-oxidants (tocopherols and polyphenols) is often increased. Shelf life of the oil is mostly improved, as indicated by the longer Rancimat Induction Time (which expresses the time it takes for injected oxygen to induce oil rancidity under standard laboratory conditions).

Taken together, these results show that use of *Trichoderma* cellulases, containing hemicellulases side activities and blended with pectinases, can improve both yield and quality of extra-virgin olive oil.

An interesting increase in yield and quality of olive oil using an experimental endopolygalacturonase from *Cryptococcus albidus* has also been reported by Servili *et al.* (1989). However, no industrial development of this enzyme has followed.

The main advantages of using "macerating" enzymes in olive oil production can be summarized as follows:

- increased extraction yield under cold processing conditions by up to 2 kg oil per 100 kg of olives processed;
- better centrifugal fractionation of the oily must;
- improvement of oil quality with higher levels of natural anti-oxidants and vitamin E;
- retarded induction of oil rancidity;
- improved overall plant efficiency;
- lower oil content in the "vegetable" waste water.

Table 15.4 Use of macerating enzyme in olive oil production. Summary of analytical results (*from Galante et al., 1993*).

Trial number	F.F.A. % oleic acid	Peroxide value	Total tocopherols mg/kg	Total poliphenols mg/kg	Rancimat Induction Time (hours)
A.1.c	0.26	9.8	252	456	8.5
A.1.e	0.36	9.3	260	522	10.1
A.2.c	0.83	22.9	266	108	3.0
A.2.e	0.44	17.6	262	171	4.5
A.3.c	0.33	8.5	n.d	n.d	n.d
A.3.e	0.36	11.6	n.d	n.d	n.d
B.1.c	0.22	12.5	115	307	11.9
B.1.e	0.20	9.8	118	265	11.2
B.2.c	0.39	9.5	2	141	6.2
B.2.e	0.47	11.5	2	133	5.7
B.3.c	0.28	14.2	4	540	12.3
B.3.e	0.28	13.3	1	581	12.9

15.4 Animal feed

Animal feed production is an extremely important component of agrobusiness, with an annual output in excess of 600 million tonnes, worth around US $50 billion per year. Feed for poultry, pigs and cows each account for one-third of the total production, and a small percentage goes to fish farming and pet food. The use of enzymes in animal feed is a rather recent development of applied enzymology, although studies in this field have been ongoing for some 40 years (Rexen, 1981; Walsh *et al.*, 1993). Currently, it represents the fastest growing area for industrial enzymes, estimated by some to be around 30% per year. If accurate, this impressive growth rate is largely due to the contribution made by *Trichoderma* enzymes.

While the first generation of "feed enzymes" were experimental or commercial products originally developed for other applications, new specific enzymes are currently available for this field. Essentially all enzymes employed in the feed industry are hydrolases used as additives to achieve one or more of the following objectives: (a) elimination of anti-nutritional factors (ANF) naturally present in grain or vegetable ingredients; (b) degradation of certain cereal components in order to decrease their viscosity and improve their nutritional value; (c) supplementation of the animals' own digestive enzymes (e.g., with exogeneous proteases and amylases), whenever these are insufficient in the post-weaning period, as is often the case with broilers, turkeys and piglets.

Fungal phytase is the best example of an enzyme used to eliminate anti-nutritional compounds (myoinositolhexaphosphates, or phytates) present in certain ingredients of pig feeds (e.g., soya), giving appreciable benefits to animal nutrition and decreasing the phosphorus content in swine waste. Commercial phytases are available from at least two major enzyme producers. However, by far the main enzyme feed application is in the degradation of non-starch polysaccharides (NSP), such as β-glucans and pentosans, to improve feed conversion rates (FCR, expressed as kg feed per kg body weight gain) of monogastric animals and to decrease the environmental load in densely farmed areas. The enzymes in question are β-glucanases and xylanases used in different relative proportions in barley- and wheat-based diets.

Pentosans (mostly arabinoxylans) and β-glucans, the components of NSP, are present in various amounts and ratios in wheat, barley, oats, triticale and rye, and are mainly associated with the endosperm cell walls of grain seeds. These NSP are highly viscous and create difficulties in the absorption and digestion of nutrients. As a consequence, the presence of high levels of these materials in feed causes poor FCR, slow weight gain, and wet droppings, particularly in poultry houses. Indeed, the digestive system of monogastric animals lacks the appropriate enzyme activities to degrade NSP, which explains their poor nutritional value. Addition of NSP-hydrolyzing enzymes during feed production causes the breakdown of poorly digestible feed ingredients and helps the absorption of *all* feed components (noticeably of fats), thereby improving feed conversion and weight gain. Since the early 1980s it has been demonstrated that consumption of barley-containing feed and live weight of broiler chickens is improved by the inclusion of β-glucanases (Hesselman *et al.*, 1982). In the case of laying hens and broilers, supplementation of NSP-degrading enzymes to daily rations can achieve a 10% increase in energy and nitrogen utilization and an increased growth rate and feed efficiency.

It is currently assumed that the beneficial effects of enzyme addition are mainly two-fold: reduction of intestinal viscosity and release of nutrients from grain endosperm and aleurone layers. Viscosity is considered to be an important constraint to animal digestion by interfering with the diffusion of pancreatic enzymes, substrates and reaction products (Morgan *et al.*, 1995). Therefore, with addition to an NSP-rich diet of enzymes that are highly effective in reducing viscosity *in vivo*, namely β-glucanases and xylanases, digestion and assimilation of all nutrients are greatly facilitated.

For example, some of the benefits claimed to be offered by the commercial Econase™ range of *T. reesei* enzymes added to poultry feed include the following:

- greater flexibility in diet formulation;
- use of cheaper raw materials;
- better digestibility of feed ingredients;
- increased energy value of cereals;
- improved growth and feed conversion;
- more uniform animals;
- cleaner eggs (e.g., reduction in number of dirty eggs);
- increased egg yolk colour;
- drier droppings;
- less environmental waste.

Econase™ is a registered tradename of Primalco Ltd and is a family of enzymes, containing different ratios of *Trichoderma* xylanase and β-glucanase activities. Each mixture is indicated for diets containing different percentages and ratios of barley, wheat or oats, although the age of the birds appears to determine the optimal ratio of xylanase to β-glucanase activities, with older broilers requiring more xylanase activity than laying hens. Figure 15.5 reports the effect of varying xylanase/β-glucanase ratios on weight gain of broilers on a diet containing 60% wheat

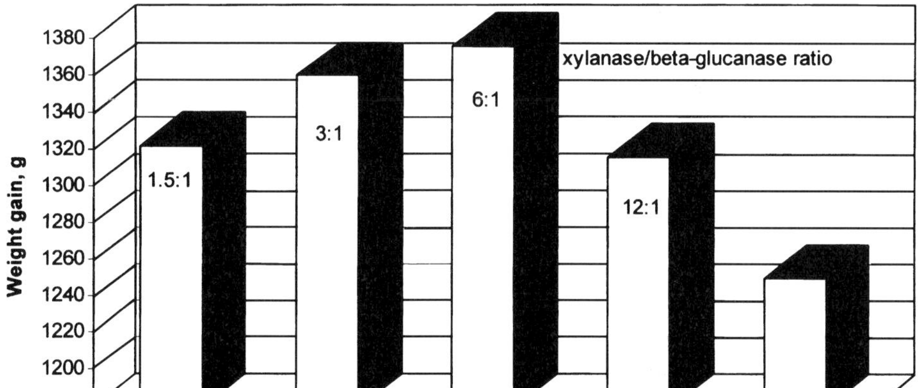

Figure 15.5 The effect of xylanase/glucanase ratio on weight gain of broilers on a diet containing 60% wheat. Weight gain increased by up to 13% at 35 days of age (*courtesy of Primalco Ltd, Biotec*).

(courtesy of Primalco Ltd, Biotec). Similarily, the Avizyme™ (a tradename of Finnfeeds International) range of enzyme products is designed for broilers and laying and breeding hens fed on diets based on barley, wheat, and a wheat and barley mixture. Each product has a ratio of β-glucanase to xylanase activities optimized for each given diet composition. The claimed benefits are similar to those of Econase™.

On the other hand, the presence of β-glucans in ruminant diets causes few problems, given the degrading activity of the rumen microbial population. Also, adult pigs are less susceptible to glucans because the longer retention time in their digestive system and the greater dilution of the viscous NSP that takes place reduces their negative effects on digestion and FCR as compared with chickens (Walsh *et al.*, 1993). Nevertheless, NSP-hydrolyzing enzymes, added with the feed or the liquid feeding system, appear to offer some benefits to these animals as well. For example, when a xylanase-containing preparation is added to wheat-based diets for grower/finisher pigs, there is a clear dose response in average daily gain as a function of the amount of xylanase added (Finnfeeds Intl).

Post-weaning young piglets can benefit most from enzyme supplementation to their diet. Cellulases and hemicellulases are often used in combination with proteases and amylases, but the cereal diet composition will determine the optimal enzyme mixture to use. Finnfeeds International has developed a complete range of enzyme preparations for pig feed, under the Porzyme™ commercial name.

Also in these feed applications, all *Trichoderma*-derived enzymes work better under acidic conditions (between pH 5 and 6), which makes them more efficient in the upper areas of the small intestine, closer to the stomach. Some enzyme action takes place partly during feed production especially at and after the pelleting stage, but mostly it occurs in the animals' own digestive tract. *Trichoderma* enzymes have relatively good thermal stability (with an optimum temperature around 55–60°C) and so can be safely added during feed production. Given the current trend toward higher pelleting temperatures to eliminate *Salmonella* contamination of feeds, β-glucanases are suitable because they maintain their activity through average conditioning temperatures, as proven by their positive effect on performance digestibility of nutrients and gut morphology of broilers (Viveros *et al.*, 1994). The potential negative effect of heat during pelleting on activity is overcome by using granulated enzymes or by spraying liquid enzyme preparations on the pellets at the cooling phase. Studies on the stability of β-glucanase and xylanase activities of *T. reesei* during feed pelleting were reported by Piironen (1996).

A new and challenging industrial application of *Trichoderma* enzymes is in the production of pet foods. Pet foods are becoming sophisticated systems both in nutritional content and in their manufacturing processes and contain several ingredients that are excellent substrates for enzyme action. Fibers are increasingly added to the commercial formulations and their controlled degradation by NSP-hydrolyzing enzymes achieve the positive effects described above (viscosity reduction, improved emulsification of substrates, better cohesiveness and flexibility of food material, increased FCR). They also can be exploited to generate reducing sugars that enhance the Maillard browning reaction, improving food appeal and palatability.

High fiber-containing foods for older dogs can particularly benefit from enzyme addition. Finally, it should be mentioned that, in the continuous search to improve the nutrition of ruminant livestock fed on diets high in plant structural polysaccharides, model transgenic mice expressing a microbial endoglucanase gene in their exocrine pancreas have been created (Hall *et al.*, 1993). A non-glycosylated active enzyme was secreted in the small intestine. This enzyme was resistant to intestinal proteases and demonstrated the feasibility of generating non-ruminant animals with an endogenous capacity to metabolize plant non-starch polysaccharides.

While the transgenic approach offers a fascinating alternative to the addition of enzymes to animal feed, it still appears to be a long-term goal in animal husbandry.

References

CALDINI, C. 1992. Studi cinetici ed immobilizzazione di glicosidasi fungine per il rilascio di aromi. Graduation thesis. University of Milan.

CALDINI, C., BONOMI, F., PIFFERI, P. G., LANZARINI, G., and GALANTE, Y. M. 1994. Kinetic and immobilization studies on fungal glycosidases for aroma enhancement in wine. *Enzyme Microb. Technol.* **16**: 286–291.

CANALES, A. M., GARZA, R., SIERRA, J. A., and ARNOLD, R. 1988. The application of a β-glucanase with additional side activities in brewing. *MBAA Tech. Q.* **25**: 27–31.

DUBOURDIEU, D., RIBEREAU-GAYON, P., and FOURNET, B. 1981. Structure of the exocellular β-D-glucan from *Botrytis cinerea. Carbohydr. Res.* **93**: 294–299.

FANTOZZI, P., PETRUCCIOLI, G., and MONTEDORO, G. 1977. Trattamenti con additivi enzimatici alle paste di oliva sottoposte ad estrazione per pressione unica: Influenze delle cultivars, dell'epoca di raccolta e della conservazione. *Rivista Italiana delle Sostanze Grasse* **54**: 381–388.

FERNANDEZ-BOLANOS, J., RODRIGUEZ, R., GUILLEN, R., JIMENEZ, A., and HEREDIA, A. 1995. Activity of cell wall-associated enzymes in ripening olive fruit. *Physiol. Plant.* **93**: 651–658.

GALANTE, Y. M., MONTEVERDI, R., INAMA, S., CALDINI, C., DE CONTI, A., LAVELLI, V., and BONOMI, F. 1993. New applications of enzymes in wine making and olive oil production. *Italian Biochem. Soc. Trans.* (IBST) **4**: 34.

GUNATA, Y. Z., BAYONOVE, C. L., CORDONNIER, R. E., ARNAUD, A., and GALZY, P. 1990. Hydrolysis of grape monoterpenyl glycosides by *Candida molischiana* and *Candida wickerhamii* β-glucosidases. *J. Sci. Food. Agric.* **50**: 499–506.

HALL, J., SIMI, A., SURANI, M. A., HAZLEWOOD, G. P., CLARK, A. J., SIMONS, J. P., HIRST, B. H., and GILBERT, H. J. 1993. Manipulation of the repertoire of digestive enzymes secreted into the gastrointestinal tract of transgenic mice. *Bio/Technology* **11**: 376–379.

HARBORD, R., SIMPSON, C., and WEGSTEIN, J. 1990. Winery scale evaluation of macerating enzymes in grape processing. *Wine Industry J.* (**May**): 134–137.

HESSELMAN, K., ELWINGER, K., and THOMKE, S. 1982. Influence of increasing levels of β-glucanase on the productive value of barley diets for broiler chickens. *Animal Feed Sci. and Technol.* **7**: 351–358.

HOME, S., MAUNULA, H., and LINKO, M. 1983. Proc. Eur. Brew. Conv., London, pp. 385–392.

LANZARINI, G. and PIFFERI, P. G. 1989. Enzymes in the fruit juice industry. In C. Cantarelli and G. Lanzarini (eds), *Biotechnology Applications in Beverage Production.* Elsevier Science, London, pp. 189–222.

LEONE, A. M., LAMPARELLI, F., LA NOTTE, E., LIUZZI, V. A., and PADULA, M. 1977. L'impiego elaiotecnico di un sistema enzimatico pectocellulosolitico: rendimento in olio e qualità del prodotto. *La Rivista Italiana delle Sostanze Grasse* **54**: 514–530.

MONTEDORO, G. F. 1987. Impiego di preparati enzimatici e drenanti nell'estrazione meccanica di olii di oliva. *La Rivista Italiana delle Sostanze Grasse* **64**: 415–421.

MORGAN, A., BEDFORD, M., TERVILA-WILO, A., HOPEAKOSKI-NURMINEN, M., AUTIO, K., POUTANEN, K., and PARKKONEN, T. 1995. How enzymes improve the nutritional value of wheat. *Zootecnica Int.* (**April**): 44–48.

OKSANEN, J., AHVENAINEN, J., and HOME, S. 1985. Microbal cellulase for improving filtrability of wort and beer. Proc. Eur. Brew. Chem., Helsinki, pp. 419–425.

OLSEN, H. S. 1995. Enzymes in food processing. In G. Reed and T. W. Nagodawithana (eds), *Biotechnology*, Vol. 9. VCH Publisher, Wenheim, Germany, pp. 663–736.

PAJUNEN, E. 1986. EBC-Symposium on Wort Production, Monograph XI, Maffliers, pp. 137–147.

PENTTILÄ, M. E., SUIHKO, M.-L., LEHTINEN, U., NIKKOLA, M., and KNOWLES, J. K. C. 1987. Construction of brewer's yeast secreting fungal endo-β-glucanases. *Curr. Genet.* **12**: 413–420.

PIIRONEN, J. T. 1996. Studies on the stability of β-glucanase and xylanase activities of *Trichoderma reesei* during pelleting. *Proc. Aust. Poultry Sci. Symp.* **8**: 146–148.

REXEN, B. 1981. Use of enzymes for the improvement of feed. *Animal Feed Sci. Technol.* **6**: 105–114.

SERVILI, M., MONTEDORO, G. F., BEGLIOMINI, A. L., PETRUCCIOLI, M., and FEDERICI, F. 1989. Impiego di una endopoligalatturonasi prodotta da *Cryptococcus albidus* var. *albibus* nell'estrazione dell'olio d'oliva. *Ind. Alim.* **28**: 1075–1078.

SUIHKO, M.-L., LEHTINEN, U., ZURBRIGGEN, B., VILPOLA, A., KNOWLES, J., and PENTTILÄ, M. 1991. Construction and analysis of recombinant glucanolytic brewer's yeast strains. *Appl. Microbiol. Biotechnol.* **35**: 781–787.

VILLETAZ, J. C., STEINER, D., and TROGUS, H. 1984. The use of a β-glucanase as an enzyme in wine clarification and filtration. *Am. J. Enol. Vitic.* **35(14)**: 253–256.

VIVEROS, A., BRENES, A., PIZARRO, M., and CASTANO, M. 1994. Effect of enzyme supplementation of a diet based on barley, and autoclave treatment, on apparent digestibility, growth performance and gut morphology of broilers. *Animal Feed Sci. Technol.* **48**: 237–251.

WALSH, G. A., POWER, R. F., and HEADON, D. R. 1993. Enzymes in the animal feed industry. *Trends Biotechnol.* **11**: 424–430.

WILLIAMS, P. J., STRAUSS, C. R., WILSON, B., and MASSY-WESTROPP, R. A. 1982. Studies on the hydrolysis of *Vitis vinifera* monoterpene precursor compounds and model monoterpene β-D-glucosides rationalizing the monoterpene composition of grapes. *J. Agric. Food Chem.* **30**: 1219–1223.

WILSON, B., STRAUSS, C. R., and WILLIAMS, P. J. 1984. Changes in free and glycosidically bound monoterpenes in developing muscat grapes. *J. Agric. Food Chem.* **32**: 919–924.

ZENT, J. B. and INAMA, S. 1992. Influence of macerating enzymes on the quality and composition of wines obtained from red Valpollicella wine grapes. *Am. J. Enol. Vitic.* **43**: 311.

16

Applications of *Trichoderma reesei* enzymes in the pulp and paper industry

J. BUCHERT, T. OKSANEN, J. PERE, M. SIIKA-AHO, A. SUURNÄKKI and L. VIIKARI

VTT Biotechnology and Food Research, Espoo, Finland

16.1 Introduction

The use of enzymes in the pulp and paper industry has grown rapidly since the mid-1980s. Increased understanding of the enzymatic reaction mechanisms on the fibre substrates has been the basis for development of powerful new enzyme preparations. The establishment of cost-effective production technologies of relevant enzymes has led to decreased enzyme prices. The enzymatic techniques are usually easily adoptable by the industry, and represent environmentally benign technologies. However, enzymatic methods must be able to compete with existing and other new technologies. Therefore, enzymatic methods are most applicable when performing reactions with higher specificity or with lower costs or environmental impacts than the competing technologies.

Trichoderma strains are the source of several commercial enzyme preparations developed for applications in various industrial areas, including the pulp and paper industry. This is due to the efficiency of *Trichoderma* as a producer of proteins and also to a wide number of relevant enzymes naturally produced by these strains. *Trichoderma* spp. can extensively degrade wood-derived carbohydrates, especially cellulose and hemicelluloses, but lack the ability to degrade or extensively modify lignin. For this reason, the native enzymes of *Trichoderma* are used for the hydrolysis and specific modification of fibre carbohydrates for different purposes in the pulp and paper industry. As a result of extensive research, the enzymatic degradation mechanisms of the major carbohydrates in fibres, i.e., cellulose and hemicelluloses, by *Trichoderma* enzymes are well established (Poutanen, 1988a; Teeri *et al.*, 1992). Genetic methodologies developed for this organism have enabled the production of tailored commercial enzyme mixtures. The molecular structures of several hydrolases originating from *Trichoderma* have been revealed. The basic knowledge also enables the structural modification of proteins to carry out targeted conversions of pulp carbohydrates in the future. This is especially relevant in the case of cellulases which, while detrimental to the fibres, have found useful applications in tailored mixtures.

Presently, the underlying mechanisms of only a few applications have been studied in detail with purified enzymes. Monocomponent hemicellulases can be applied for the removal of certain fibre components, whereas the mechanisms of cellulases are more complex because of the structural organization of the amorphous and crystalline regions in different types of fibres. Furthermore, the type of fibres, including the chemical composition, coarseness and outer fibrillation, seem to affect the site of the enzymatic attack. The accessibility and interlinkage of different carbohydrates in different types of pulps may also limit the applicability of enzymes. Purified enzymes used for research, however, help us to understand and optimize the enzymatic treatments and to develop useful commercial enzyme preparations.

Several enzymatic applications in the pulp and paper industry have been developed using enzymes produced by *Trichoderma*. This review describes the principles of the known applications, with special emphasis to enzymes derived from *Trichoderma*.

16.2 Reasons for modifying the fibre substrates

The pulp and paper industry processes wood and other cellulose-containing raw materials. Pulp is produced from the raw material by chemical or mechanical means, while paper and board are manufactured by combining one or more types of pulps with various additives designed to achieve the desired properties, such as strength, smoothness, or optical properties. The virgin fibre sources for paper and board products are mechanical and chemical pulp. In addition, recycled fibres, consisting of mechanical or chemical pulps or their mixtures, are becoming more prominent. The compositions of the major fibre sources are presented in Table 16.1. In chemical pulping, the chips are cooked in the presence of chemicals that dissolve particularly lignin. Today, most of the chemical pulp is produced by the alkaline kraft process, in which wood chips are cooked at 160–190°C for about 3 h in a concentrated solution of sodium hydroxide and sodium sulphide. The cooking liquor is recycled through an extensive recovery process, also producing excess energy. Removal of lignin significantly increases the relative cellulose content and also modifies the structure and composition of hemicelluloses. The other major chemical pulping process is sulphite pulping, particularily prevalent in central Europe. In sulphite cooking, the chips are cooked in alkaline, neutral or acid cooking conditions, resulting in weaker fibres than in the kraft process. The kraft

Table 16.1 Chemical composition of fibres from different origins (Sjöström, 1977)

	Average composition (% of d.w.)				
Fibre source	Cellulose	Hemicellulose	Lignin	Extractives	Others
Mechanical pulp (softwood)	39	25	27	4	5
Chemical pulp (softwood)	74	19	6	1	0
Chemical pulp (hardwood)	64	33	4	1	0

and sulphite fibres differ from each other with respect to chemical composition, morphology and technical properties (Rydholm, 1965).

In mechanical pulping, wood logs or chips are fiberized by mechanical means. Refining comprises a series of compressions and decompressions that lead to separation and fibrillation of individual fibres with simultaneous generation of fines. In contrast to chemical pulping, where the yield is usually less than 50% due to solubilization of lignin and hemicellulose, the yield in mechanical pulping processes is as high as 95%. Because of the limited dissolution of wood components during mechanical pulping, the chemical composition of the pulp is rather similar to the woody raw material used. However, depending on the pulping process selected, variations in the fibre composition (Bauer-McNett classification) as well as in chemical composition of fibre surfaces can be detected (Pettersson *et al.*, 1988; Yang *et al.*, 1988).

After pulping, the fibres are usually bleached in order to increase their brightness. During kraft pulping, lignin is almost completely removed, but the residual lignin is highly condensed and is removed during bleaching, in which different chemicals are used stepwise. In conventional bleaching, both chlorine gas and chlorine dioxide together with alkali have been used as the bleaching chemicals. Today, most pulp in Europe is bleached in elementary chlorine free (ECF) sequences and to a lesser extent by totally chlorine free (TCF) sequences. In most cases, mechanical pulp is only mildly bleached with hydrogen peroxide or dithionite.

In both major types of pulping processes, as well as in bleaching, the conditions are too harsh for simultaneous application of enzymes. Enzymes have, however, been found to facilitate the fibre processing in many ways. They are used to decrease energy and chemical consumption by modifying the fibre structure. In addition, enzymes have been studied for improving properties of the fibre products. The reasons for using enzymes based on the chemical modification of fibres are described in Table 16.2. The action of enzymes in pulp is affected by the accessibility of substrates in the fibre matrix. The main factors limiting the access of enzymes in woody materials are the specific surface area and the porosity, i.e., the median pore size of fibres (Grethlein, 1985; Stone *et al.*, 1969). Other factors such as the molecular organization of the fibre components and the linkages between lignin and carbo-

Table 16.2 Reasons for using *Trichoderma* enzymes in the pulp and paper industry

Raw materials preparation	Technical target	Substrate
Chemical	Beatability	Cellulose
	Handsheet properties	Cellulose, hemicellulose
	(air resistance, density)	Hemicellulose
	Bleachability	Hemicellulose
	Charge modification	Hemicellulose
Mechanical	Energy saving	Cellulose
	DCS control	Hemicellulose
	Increase in fines	Cellulose
	Flexibility	Cellulose
Recycled	De-inking	Cellulose + (hemicellulose)
	Drainage	Cellulose

hydrates may also have a significant role in the accessibility of fibre-bound substrates, especially in the case of hemicelluloses.

In general, the median size of pores in wood fibres is about 1 nm (Stone and Scallan, 1968). In chemical pulping the pore size value increases as a function of the decreasing pulp yield, being 5–6 nm at the yield level of 50% in the kraft and sulphite processes. The molecular size and structure of an enzyme are important factors when considering the limited accessibility of the substrates in fibre matrices. The molecular sizes of enzymes vary considerably depending on the molecular weight and sterical configuration of the protein molecules. The molecular sizes of the catalytic domains of xylanase II and CBHI from *T. reesei* are 20 and 65 kDa, with catalytic core dimensions of 3.2 × 3.4 × 4.2 and 4.0 × 5.0 × 6.0 nm, respectively (Divne *et al.*, 1994; Törrönen *et al.*, 1994). According to these figures, the proteins could be expected to penetrate most of the pores in chemical pulps. Thus, while the native wood is highly inaccessible to enzymes, chemical pulps can be structurally modified by enzymes.

16.3 Potential enzymes for commercial applications

16.3.1 Identification of useful enzymes

Commercial preparations intended for industrial use are mixtures of various enzymes. They usually contain, in addition to the desired active component(s), a multitude of other proteins, that may or may not affect the process. The action of multi-component mixtures is often difficult to trace to a single protein because it results from the combined action of various enzymes on a complex mixture of carbohydrate substrates. On the other hand, the effect of a single isolated protein may be different when used alone relative to its effect when acting together with other enzymes. This is due to the synergistic action of carbohydrate depolymerizing enzymes, i.e., the effect of the enzymes acting together is remarkably higher than the sum of the effects of the individual enzymes acting separately.

When novel applications are developed, it is important to be able to distinguish and identify the effects of various enzyme components. Therefore, the characterization of enzymes using both isolated and fibre-bound substrates is extremely important. On a large scale it is usually not possible to use purified proteins or even their defined mixtures. At laboratory scale this can, however, be realized. Care is needed when results obtained with monocomponent preparations are evaluated. The purity of isolated enzymes should be confirmed with appropriate hydrolysis studies because in some cases a minor contamination of other hydrolases may have a notable effect on the hydrolytic properties of the preparation (see, for example, Reinikainen *et al.*, 1995). These kinds of problems can be avoided by using adequate multi-step purification procedures, by purifying the proteins from strains genetically modified to be devoid of unwanted activities, or by using selected host organisms.

Trichoderma, like several other fungi, secretes products derived from the same gene in several iso-forms, which may differ in pI or in MW. Proteins are often present in culture filtrates in a pattern of various pI forms, as in the case of CBHII (Tomme *et al.*, 1988a) or mannanase (Stålbrand *et al.*, 1993). Proteolytic degradation may also lead to truncated but still active forms of the proteins. Typically, the cellulose binding domain (CBD) is cleaved off and the catalytic core proteins can be found in large quantities in culture filtrates. Kubicek-Pranz *et al.* (1991) showed that cellulolytic enzymes are degraded in aged preparations. The multiplicity of enzymes, especially in the case of cellulases, has caused significant confusion. Therefore, to indisputably identify a protein, the original gene coding for the protein studied should be confirmed on the basis of amino acid sequences determined from the protein. This should immediately and directly reveal the identity of the enzyme. For the same reason, the enzyme nomenclatures based on gene sequences should be favoured.

When developing applications for the pulp and paper industry, the substrate specificities of enzymes, especially on the actual fibre, are of utmost importance. The most reliable primary method to study the substrate specificities in mixed-fibre-bound substrates is to identify the type and amount of released sugars or oligomers. Relationships between the rather small chemical changes and the altered technical characteristics of pulps are complex and in many cases poorly understood. The secondary analysis of the effects on the pulp substrate are thus often more complex and require the use of chemical or mechanical treatments or analysis of paper technical properties. Only a few potential applications have been studied using purified enzymes or well-characterized mixtures of them.

The enzymatic modification of cellulose is an interesting application with potential for the modification of properties of fibres. Cellulases have traditionally been considered detrimental for pulp and paper applications, and the applications are still hindered by the well-characterized synergistic action of the cellulolytic system of *Trichoderma*, leading to yield losses and decreased viscosity. The degree of synergy obtained, however, depends on the nature of the substrate and the concentration and ratios of the enzymes used. The highest degree of synergy is observed between specific pairs of cellulases: EGI pretreats cellulose more efficiently for CBHI than for CBHII (Nidetzky *et al.*, 1994). Thus, it seems that the different cellulolytic enzymes have specificities which are not immediately apparent from traditional product analysis. Even the primary product analysis reveals, however, that the *T. reesei* endoglucanase I is an unspecific endo-enzyme, being able to attack also xylan,and the ratio of cellulose and xylan hydrolysis depends on the type of pulp (Buchert *et al.*, 1992; Pere *et al.*, 1995, 1996; Suurnäkki *et al.*, 1997).

16.3.2 **Enzymes acting on pulp carbohydrates**

The relevant enzymes from *Trichoderma* for pulp and paper applications are cellulases and hemicellulases (Table 16.3). Endo-wise acting hemicellulases have been used successfully, whereas the side group cleaving accessory enzymes, needed for the total hydrolysis of substituted hemicellulose substrates, have not been shown to be successful in pulp and paper applications. This is obviously due to the substrate specificities of the accessory enzymes from *Trichoderma* which generally prefer oligomers rather than native substrates. In fibre modification, accessory enzymes capable of acting on polymeric rather than oligomeric substrates would be most desirable. It may be speculated that the excretion of efficient endo-enzymes in large quantities by *Trichoderma* is dependent on the specificity of its accessory enzymes, which act on oligomers rather than on polymers.

Table 16.3 *Trichoderma reesei* enzymes acting on pulp carbohydrates

Activity	Enzyme	References
Endoglucanase (EC 3.2.1.4)	EGI	Penttilä *et al.*, 1986
	EGII	Saloheimo *et al.*, 1988
	EGIII	Ward *et al.*, 1993
	EGIV	Saloheimo *et al.*, 1997
	EGV	Saloheimo *et al.*, 1994
Cellobiohydrolase (EC 3.2.1.91)	CBHI	Shoemaker *et al.*, 1983
	CBHII	Chen *et al.*, 1987
β-Glucosidase (EC 3.2.1.21)	β-Glucosidase I	Barnett *et al.*, 1991; Chen *et al.*, 1992
	β-Glucosidase II	Chen *et al.*, 1992
Xylanase (EC 3.2.1.8)	XYLI (pI 5–5.5)	Törrönen *et al.*, 1992; Tenkanen *et al.*, 1992b
	XYLII (pI ~ 9)	Törrönen *et al.*, 1992; Tenkanen *et al.*, 1992b
α-Arabinosidase	ABFI	Margolles-Clark *et al.*, 1996b; Poutanen, 1988b
α-Glucuronidase	GLRI	Margolles-Clark *et al.*, 1996c; Siika-aho *et al.*, 1994
Acetyl xylan esterase	AXEI	Margolles-Clark *et al.*, 1996a; Sunderg and Poutanen, 1991
β-Xylosidase	BXLI	Margolles-Clark *et al.*, 1996b; Poutanen and Puls, 1988
Mannanase	(pI 5.4 and pI 4.6)	Stålbrand *et al.*, 1995; Stålbrand *et al.*, 1993
α-Galactosidase (EC 3.2.1.22)	AGLI	Margolles-Clark *et al.*, 1996d; Zeilinger *et al.*, 1993
	AGLII	Margolles-Clark *et al.*, 1996d
	AGLIII	Margolles-Clark *et al.*, 1996d

The genes for the *T. reesei* endoglucanases EGI, EGII, EGIII and EGV and for the cellobiohydrolases CBHI and CBHIII, β-glucosidase (see review by Nevalainen and Penttilä, 1995), and EGIV (Saloheimo *et al.*, 1997) have been characterized. All of these cellulases except the endoglucanase III and β-glucosidase have basically similar structures with cellulose binding domains (CBD). The genes for the major xylanases of *T. reesei* (Törrönen *et al.*, 1992) and the major mannanase have also been described (Stålbrand *et al.*, 1993, 1995). In addition to the endoenzymes, the genes of the accessory enzymes required for hemicellulose hydrolysis, i.e., the α-arabinosidase, α-glucuronidase, acetyl xylan esterase, β-xylosidase and α-galactosidase are also known (Margolles-Clark *et al.*, 1996a–d). Of the hemicellulases, mannanase and acetyl xylan esterase also contain a cellulose binding domain.

It is obvious that completely different enzyme combinations are needed for various new applications to bring about the required effects. Development of genetic engineering methods has enabled the specific modification of the enzyme profiles produced by *Trichoderma*. Complete removal on the one hand, and overproduction on the other, of either individual or multiple enzyme components have been reported (Fowler *et al.*, 1993; Harkki *et al.*, 1991; Kubicek-Pranz *et al.*, 1991; Nevalainen *et al.*, 1991; Suominen *et al.*, 1993; Uusitalo *et al.*, 1991). Thus, enzyme mixtures devoid of the major endoglucanases (EGI and EGII) can be produced (Koivula *et al.*, 1996). The cultivation conditions may also dramatically affect the activity profiles obtained, e.g., high cultivation pH during fermentation leads to increased xylanase and lowered cellulase production (Bailey *et al.*, 1993).

It can be expected that engineering of the *Trichoderma* cellulases will produce enzymes with interesting new properties. Today, the CBDs and cores can already be produced separately. It has been observed that the removal of the CBD of CBHII had no influence on the activity of the enzyme on soluble substrates, but both its binding and activity on crystalline substrates was severely impaired (Srisodsuk *et al.*, 1993). The disrupting or swelling of the cellulose structure as a result of the action of the binding domain has been claimed as a reason for the higher activity of the intact enzyme toward crystalline cellulose. However, somewhat contradictory results have been presented concerning *Trichoderma* cellulases as well as bacterial cellulases (Ståhlberg *et al.*, 1991; Woodward *et al.*, 1992).

The *Trichoderma* enzymes usually exhibit highest activity at pH values slightly below neutral and at temperatures around 50°C. Interestingly, the *Trichoderma* xylanases were found to exert higher pH optima on pulp substrates than expected (Buchert *et al.*, 1992). This has, however, later been shown to be an artefact, due to the Donnan effect (Buchert *et al.*, 1997a). It remains to be shown whether attempts to change the pH or temperature optima by protein engineering will prove to be successful. If so, some drawbacks could be eliminated, such as the relatively narrow temperature and pH optima of the *Trichoderma* enzymes which limit the industrial applicability under the harsh conditions that prevail in the pulp and paper industry.

16.4 Applications of *T. reesei* enzymes in the pulp and paper industry

16.4.1 *Mechanical pulping*

In mechanical pulping processes such as refining and grinding, the woody raw material is fiberized by mechanical action. Characteristics of mechanical pulps are high content of fines, high bulk and stiffness, while chemical pulp fibres have flexibility, compressibility and good bonding ability. Chemical and mechanical pulps have different market niches and many paper products contain both pulp types in variable proportions depending on the required properties. The efficient light-scattering ability of mechanical pulps, based on abundant fines and stiff fibres, is advantageous in many paper grades, such as in printing papers, where good optical properties are essential.

The main disadvantage of mechanical pulping is its high energy consumption. Thermomechanical pulping (TMP) especially requires high specific energy input. One way to reduce the energy consumption in TMP is to modify the raw material by biotechnical means prior to refining. The main focus so far has been on pretreatment of wood chips with ligninolytic fungi. Biomechanical pulping with chosen white-rot fungi has resulted in substantial energy savings and improvements in handsheet strength properties (Akhtar, 1994; Leatham *et al.*, 1990). Despite encouraging results on a laboratory scale, the method still awaits commercialization. From the chemical point of view, no correlation between lignin degradation and energy savings has been found, and evidently slight modification of both lignin and carbohydrates was needed for the observed positive effects.

Because of the low accessibility of wood chips for enzymatic modification (Grethlein, 1985), incorporation of an enzymatic step in the mechanical pulping process can be expected to be successful only after the primary refining. Therefore, the effects of enzymatic modification of coarse mechanical pulp with monocomponent *Trichoderma* cellulases and hemicellulases were studied prior to secondary refining (Pere *et al.*, 1996). Slight modification of cellulose or hemicellulose gave rise to energy savings in laboratory-scale secondary refinings; energy savings with an atmospheric disk refiner ranged from 20% for CBHI to 5% for the hemicellulases (mannanase and xylanase). Treatment with EGI decreased the energy consumption slightly but at the expense of pulp quality by sensitizing the fibres for breakage. Interestingly, no positive effect on energy consumption was detected with a cellulase mixture. When the refining was performed with a low-intensity refiner (wingdefibrator), the energy consumption was reduced by 30–40% with CBHI as compared with the untreated reference (Pere *et al.*, 1996). The results obtained with the laboratory refiners could be further verified in a pilot-scale experiment where 900 kg of TMP reject was treated with CBHI prior to the secondary refining. In a two-stage secondary refining, an energy saving of 10–15% with CBHI was obtained (Figure 16.1). The cellulase treatment with CBHI did not have any detrimental effects on pulp quality. In fact, the tensile index was even higher for the CBHI-treated pulp than for the reference. The increase in tensile index could be explained by the intensive fibrillation induced by the CBHI treatment (unpublished data). The good optical properties were also maintained after the CBHI treatment.

16.4.2 *Bleaching of kraft pulps*

Enzyme-aided bleaching of kraft pulps is the first large-scale application of enzymes in the pulp and paper industry. The idea of using hemicellulolytic enzymes to increase the bleachability of chemical pulps was introduced in the 1980s (Viikari *et*

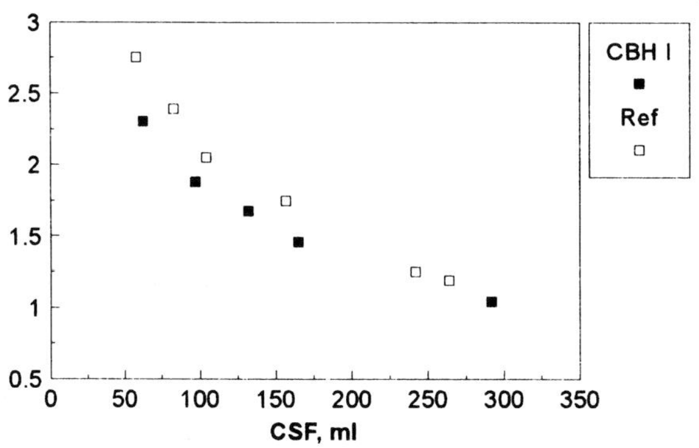

Figure 16.1 Specific energy consumption (SEC) of the CBHI-treated and the reference pulps in the pilot experiment (Pere *et al.*, 1996). CSF = Canadian Standard Freeness.

al., 1986, 1987). The concept of enzyme-aided bleaching was based on the observation that limited hydrolysis of hemicellulose in pulps by hemicellulases, mainly xylanases, increased the extractability of lignin from the kraft pulps in the following bleaching sequences. The xylanase pretreatment permits the use of lower chlorine charges during the bleaching of kraft pulps and as a consequence the chloro-organic materials in the discharge are decreased. The first mill trials were conducted in Finland in 1988 (Lavielle *et al.*, 1992), which is a remarkably short time after the first reports. This was due to the ease of producing xylanases on an industrial scale using strains of *Trichoderma*, for example. As effective and easily applicable process aids, xylanases are nowadays used in several mills prior to chlorine and non-chlorine bleaching sequences of kraft pulps (reviewed by Viikari *et al.*, 1994).

The effect of hemicellulases in bleaching is based on the modification of pulp hemicelluloses which results in increased removal of lignin in chemical bleaching. The main enzyme enhancing pulp bleachability is endoxylanase (Kantelinen *et al.*, 1988; Paice *et al.*, 1988; Viikari *et al.*, 1991). The effect achieved by xylanases is generally independent of the origin of the enzyme, and both fungal and bacterial xylanases have been shown to increase pulp bleachability. The *T. reesei* mannanase either alone or in combination with xylanase has also been reported to increase the pulp bleachability (Buchert *et al.*, 1993). The ability of mannanases to enhance the bleachability is, however, generally limited to certain pulps and highly dependent on the origin of the mannanase used (Cuevas *et al.*, 1996; Suurnäkki *et al.*, 1996a).

The xylanase treatment has been used to enhance pulp bleachability in traditional chlorine gas-based sequences and in ECF and oxygen-chemical-based TCF bleaching sequences. The reduction of bleaching chemical consumption has important implications not only for the environment but also for the economy as a result of cost savings or increased pulp production capacity.

Mechanisms involved

It has been proposed that the action of xylanases is due to the partial hydrolysis of reprecipitated xylan (Kantelinen *et al.*, 1993b) or to the removal of xylan from the lignin-carbohydrate (LCC) complexes (Viikari *et al.*, 1986; Yang and Eriksson, 1992). Both of these hypotheses would allow the enhanced leaching of entrapped lignin from the fibre cell wall. The suggested direct removal of chromophoric groups from the pulp by xylanase (Patel *et al.*, 1993; Wong *et al.*, 1996) is, however, probably mainly an artefact due to removal of the recently discovered hexenuronic acid, containing a double bond, along with xylan (Buchert *et al.*, 1995; Teleman *et al.*, 1995). The enhanced bleachability resulting from the action of xylanase on both reprecipitated and LC-xylan suggests that it is probably not only the type but also the location of the xylan that is important in the mechanism of xylanase-aided bleaching. The xylanase of *T. reesei* has been observed to act rather uniformly in all accessible surfaces of kraft pulps (Saake *et al.*, 1995; Suurnäkki *et al.*, 1996b), indicating that the effect of xylanase on bleachability is not only an outer surface phenomenon.

Compared with xylanase-aided bleaching, the mechanism of mannanase-aided bleaching has attracted only minor interest, probably because of its rather limited effect on most pulp types. However, the work done with purified *T. reesei* mannanase indicates that the mechanism of mannanase-aided bleaching differs from that of xylanase-aided bleaching, because of the different distribution of glucomannan and xylan in pulp fibres (Buchert *et al.*, 1992; Suurnäkki *et al.*, 1996c,d). The composition and configuration of the outer surfaces of pulp fibres seem to be important in the mannanase-aided bleaching (Suurnäkki *et al.*, 1996d).

Trichoderma *hemicellulases in bleaching*

The role of xylanase activity in the delignification of kraft pine pulp has been extensively studied with purified xylanases of *T. reesei* (Buchert *et al.*, 1992; Tenkanen *et al.*, 1992a). The *T. reesei* xylanases have been shown to have different pI values, pH optima and substrate specificities (Tenkanen *et al.*, 1992b). However, in the limited hydrolysis of xylans of pine kraft pulp, the two *T. reesei* xylanases (pI 5.5 and pI 9.0) resulted in approximately the same kappa number reduction and brightness increase in subsequent chemical delignification (Buchert *et al.*, 1992).

Purified *Trichoderma* hemicellulases have been used in studies of bleach-boosting of different types of pulps. The *T. reesei* xylanase was most effective when used in conventionally cooked pulps, and the effect was more pronounced in pulp produced from northern pine than from radiata pine (Suurnäkki *et al.*, 1996a). Similar results regarding the role of the pulp origin and production method on the xylanase effect have been reported with purified and commercial xylanases from various origins (Allison *et al.*, 1995; Nelson *et al.*, 1995; Pedersen *et al.*, 1992; Tolan, 1992; Wong *et al.*, 1996). The mannanases of *T. harzianum* (Clark *et al.*, 1991) and *T. reesei* (Buchert *et al.*, 1993; Suurnäkki *et al.*, 1994, 1996a) have been shown to be effective when used before either peroxide delignification or chlorine chemical based bleaching sequence. The most beneficial use of mannanase in pulp bleaching has been proposed to be in combination with xylanase (Buchert *et al.*, 1992; Suurnäkki *et al.*, 1996a). The accessory enzymes, such as β-xylosidase and α-arabinosidase, have only a minor role in the enzymatic bleach-boosting of pulp (Kantelinen *et al.*, 1993a; Luonteri *et al.*, 1996). Of the individual cellulases, only the endoglucanase I from *T. reesei* was shown to increase bleachability, due to its catalytic activity on xylan (Buchert *et al.*, 1994). The effects of xylanases originating from *Trichoderma* spp. in various bleaching sequences are presented in Table 16.4. Most of the commercial hemicellulase preparations intended for pulp bleaching originate from *T. reesei.*

16.4.3 Modification of fibre properties

The modification of fibre properties by enzymes acting on fibre carbohydrates is a challenge, and several reports have been published on the use of enzyme mixtures with the aim of improving drainage, beatability or runnability of the paper machine (Jokinen *et al.*, 1991; Pommier *et al.*, 1989, 1990; Noé *et al.*, 1986). In these applications, the enzymatic treatment was carried out either prior to or after beating of the pulps. By carrying out the enzymatic stage prior to the refining process the aim has been to improve the beating response or to otherwise modify the fibre properties. Beating is a mechanical process that is essential to enhance fibrillation and interfibre bonding of pulps. As discussed by Kirk and Jeffries (1996) the principal challenge in using enzymes to enhance fibre bonding is to increase fibrillation without reducing pulp viscosity. However, when applying enzymes after beating, the main focus has been on the improvement of the dewatering, i.e., drainage properties of the

Table 16.4 Chlorine-based, ECF and TCF bleaching sequences with *Trichoderma* spp. xylanase pretreatment resulting in +80% ISO brightness

Xylanase origin	Scale	Sequence after xylanase treatment[a]	Pulp	Brightness increase (% ISO)	Chem consumption decrease	Ref.
purified *T. reesei*	lab.	(CD)EDED	pine kraft		6.6%	Buchert *et al.*, 1992
partially purified *T. harzianum*			radiata pine kraft	4.0		Clark *et al.*, 1991
T. longibrachiatum (Genencor)	mill		softwood kraft		21% TAC	Koponen, 1991
Ecopulp	lab.	(CD)EDED	softwood kraft			Lahtinen *et al.*, 1992
			softwood kraft + O_2	2.0–3.0	16% TAC	Latinen *et al.*, 1993
	mill	DEoDEpD	black spruce		16% TAC	Jean *et al.*, 1994
		D_0(EO)DED	softwood kraft + O_2		10–15% TAC	Lahtinen *et al.*, 1993
		D_0EE_0DED	hardwood kraft		15–20% TAC	Suurnäkki *et al.*, 1996e
		QZ(OP/P)ZP	hardwood kraft		>15% H_2O_2	Suurnäkki *et al.*, 1996e
Irgazyme	lab.	ZED	softwood kraft + O_2	3.2–6.0	40% O_3	Brown *et al.*, 1994
	mill	(CD)EoDED	radiata pine kraft	1.1	18.4% TAC	Werthemann *et al.*, 1993
Albazyme	mill	(CD)EopDED	softwood kraft		17% TAC	Lavielle *et al.*, 1992
		DEopDED	softwood kraft		15% TAC	Scott *et al.*, 1993

Commercial enzyme preparations Ecopulp (produced by Primalco Biotec), Irgazyme (Genencor Int.) and Albazyme (Genencor Int.) are from *T. reesei*.
[a] C = elemental chlorine (Cl_2), D = chlorine dioxide (ClO_2), E = alkaline extraction (NaOH), E_0/E_p = oxygen/hydrogen peroxide-reinforced alkaline extraction, H = hypochlorite (HClO), AC = active chlorine, TAC = total active chlorine.

pulps, which determines the speed of paper machine operation. A commercial cellulase/hemicellulase enzyme preparation (Pergalase A-40) based on *Trichoderma* enzymes has been developed and is currently used in several paper mills in the production of release papers and wood-containing printing papers (Freiermuth *et al.*, 1994; Jokinen, 1994; Pommier *et al.*, 1990).

The potential for improving the drainage rates of recycled fibres by cellulase mixtures was discovered in the late 1980s (Fuentes and Robert, 1986; Pommier *et al.*, 1989, 1990). According to Stork *et al.* (1995), endoglucanase activity is a prerequisite for drainage improvement of recycled pulps. This has been confirmed by Oksanen *et al.* (1996) and Kamaya (1996) using purified or partially purified *Trichoderma* cellulases. According to Oksanen *et al.* (1996), endoglucanases I and II were equally effective in decreasing the Schopper-Riegler (SR) value of recycled softwood kraft pulp, indicating improved drainage, whereas cellobiohydrolases had no effect. Xylanase and mannanase treatment resulted in only limited improvement of the SR value (Oksanen *et al.*, 1996). Depending on the pulp furnish, however, the efficiency of the different hydrolases may vary. According to Pere *et al.* (1996), a simultaneous solubilization of xylan and cellulose is required for the drainage improvement of reed canary grass kraft pulp. By using EGI acting on cellulose and xylan, the drainage was improved by 30%, whereas with EGII, being specific to cellulose, only a limited effect was observed (Pere *et al.*, 1996).

The development of enzymatic fibre modification processes requires a profound understanding of the action of different enzymes on different types of pulps. Recently, many reports on this area have been published. Mansfield *et al.* (1996) have investigated the impact of a commercial cellulase mixture (Novozyme SP 342 from *Humicola insolens*) on the different fibre fractions obtained from Douglas fir kraft pulp. The cellulase treatment was found to decrease the freeness of the fibres, indicating defibrillation, which in turn led to reduced fibre coarseness. The strength properties (tensile index, burst index and tear resistance) were found to decrease with increasing cellulase dosage. The impacts of the major *T. reesei* cellulases on the fibre properties of unbleached softwood kraft and dissolving pulps have been investigated by Pere *et al.* (1995) and Rahkamo *et al.* (1996). The four cellulases of *T. reesei* exhibited significant differences in their modes of action on the pulps. The cellobiohydrolases (CBH) were reported to have only a very modest effect on pulp viscosity, whereas the endoglucanases (EG) dramatically decreased viscosity, even at low enzyme dosages. Similar results have also been obtained when unbleached kraft pulp was treated with monocomponent *T. reesei* cellulases prior to bleaching (Buchert *et al.*, 1994). Of the endoglucanases, EGII was shown to decrease the viscosity most drastically (Pere *et al.*, 1995; Rahkamo *et al.*, 1996). Even after PFI-refining, the CBHI treatment had no effect on the handsheet properties, indicating that CBHI caused no structural damage to the fibres. On the other hand, EGII treatment was reported to damage the strength properties, suggesting that EGII attacks cellulose at sites where even low levels of hydrolysis result in large decreases in viscosity and, consequently, a dramatic deterioration of tensile index (Pere *et al.*, 1995).

The effect of the purified *Trichoderma* cellulases and hemicellulases on the beatability and on the technical properties of paper from bleached kraft pulps has been further investigated by Oksanen *et al.* (1997b) and Kamaya (1996). Pretreatment of the pulp with CBHI or CBHII had practically no effect on the development of pulp properties, whereas endoglucanases, especially EGII, were found to improve the beatability of the pulp as measured by SR value, sheet density and Gurley air resistance. Xylanase and mannanase pretreatments did not seem to significantly modify the pulp properties when less than 10% of the respective hemicellulose was hydrolysed (Oksanen *et al.*, 1997b).

16.4.4 *De-inking of recycled fibres*

The application of enzymes in de-inking has been intensely studied at both the laboratory and pilot scales in recent years, and numerous patents exist, as reviewed by Welt and Dinus (1995). However, the application of enzymatic de-inking in commercial installations has not yet been reported. There are two principal approaches to using enzymes in waste paper de-inking: one employs lipases to hydrolyse soy-based ink carriers, and the other uses specific carbohydrate hydrolysing enzymes, such as cellulases, xylanases or pectinases, to release ink from fibre surfaces (Prasad *et al.*, 1992). Hydrolysis of the ink carrier releases the individual ink (carbon black) particles that are too small in size to be effectively flotated. Most applications proposed so far use cellulases and hemicellulases where the detachment of ink results from a partial enzymatic hydrolysis of carbohydrate molecules on the fibre surface (Prasad *et al.*, 1992, 1993; Jeffries *et al.*, 1994).

One of the advantages offered by enzymatic de-inking is the avoidance of alkaline de-inking chemicals. De-inking at low pH prevents alkaline yellowing of pulp and simplifies the de-inking chemistry. In an industrial operation, the use of enzymes as de-inking aids could thus lower the chemical costs and decrease environmental impacts. Offset and letterpress newsprint waste have been enzymatically de-inked at low pH in several laboratory studies (Kim *et al.*, 1991; Prasad *et al.*, 1992, 1993; Putz *et al.*, 1994). In general, these studies revealed that cellulases and hemicellulases increase brightness and pulp cleanliness compared with conventional de-inking. The enzymatic de-inking also changes the ink particle size distribution, apparently reducing the particle size. Besides ink removal, enzymatic de-inking may contribute to improved strength properties and freeness and reduced fines content. Strength improvement has been observed especially with xylanase treatment while cellulases are more efficient in increasing brightness and freeness (Prasad *et al.*, 1993). The laboratory-scale results have also been confirmed in pilot trials (Heise *et al.*, 1996). In all reports on enzyme-aided de-inking thus far, enzyme mixtures have been used, and no reports on the effect of isolated *Trichoderma* enzymes in de-inking have been published.

16.4.5 *Treatment of dissolved and colloidal material*

Various wood components are dissolved and dispersed into the process water during mechanical pulp production. These include lipophilic extractives such as resin or pitch, hydrophilic extractives (lignans) and hydrophilic carbohydrates, mainly hemicelluloses (Ekman *et al.*, 1990; Thornton *et al.*, 1994). These components are generally defined as dissolved and colloidal substances (DCS). During peroxide bleaching of mechanical pulps, other wood-derived components were also shown to be released from the fibres, including pectins (Thornton, 1994). The DCS can cause runnability problems on paper machines, such as pitch depositions, specks in the paper and decreased dewatering.

Because of the chemical composition of DCS, enzymes acting on either polysaccharides or extractives can be expected to affect the properties of DCS. A commercial enzyme preparation (Pergalase A 40), a product of *Trichoderma*, gave a remarkable decrease in the turbidity of TMP filtrates (Kantelinen *et al.*, 1995). As a result of the enzymatic treatment, the lipophilic extractives in the filtrates were destabilized and attached to the TMP fibres. The purified endoglucanase I of *T. reesei* was also tested and found useful in disturbing the steric stability of colloidal pitch, whereas the *T. reesei* xylanase was only effective at very high concentrations (Kantelinen *et al.*, 1995).

16.4.6 Analysis of pulp fibres

Because of their specificity, hydrolases acting on pulp components are potential tools for fibre characterization. Xylanase and mannanase purified from *T. reesei* have been successfully used for selective solubilization of xylan or glucomannan from different types of chemical pulps, as reviewed by Buchert *et al.* (1996a). In some cases non-specific enzymes, such as EGI, can also be exploited for fibre characterization. In analytical applications, high hydrolysis levels are obtained using high enzyme dosages and long hydrolysis times. After the enzymatic solubilization, the chemical characterization of the solubilized oligosaccharide fractions can be carried out directly by NMR or HPLC after a secondary enzymatic hydrolysis (Buchert *et al.*, 1995; Teleman *et al.*, 1995). For the complete hydrolysis of chemical pulps, an enzyme mixture containing all the required activities has been developed (Tenkanen *et al.*, 1995). The selective enzymatic solubilization of the pulp xylan or glucomannan also enables the analysis of impact of the respective hemicellulosic components on the fibre properties, such as pore size distribution (Suurnäkki *et al.*, 1997) or location of lignin (Buchert *et al.*, 1996b), or on the technical properties of pulps, such as brightness reversion (Buchert *et al.*, 1997b) or hornification (Oksanen *et al.*, 1997a). Especially in the analysis of acid-labile pulp components, the enzymatic solubilization of pulp carbohydrates under mild and non-destructive conditions is beneficial. The suitability of this approach has been verified in the structural analysis of kraft xylan using *T. reesei* xylanase, leading to the discovery of hexenuronic acid, a hitherto unknown component in kraft pulps (Buchert *et al.*, 1995; Teleman *et al.*, 1995).

16.5 Conclusions

Trichoderma strains are efficient producers of a variety of hydrolytic enzymes and the source of a number of commercial enzyme preparations designed also for the pulp and paper industry. The cellulolytic and hemicellulolytic system of *Trichoderma* has been widely used as a model to understand the basic enzymatic degradation reactions of the major carbohydrates in fibre materials. This knowledge has been utilized when developing new enzymatic applications for the pulp and paper industry. The present applications where *Trichoderma* enzymes have shown potential are refining of mechanical pulps, bleaching of kraft pulps, de-inking of recycled fibres and modifying fibres for various improvements, e.g., for better drainage. Of these, only the bleach- and drainage-boosting enzymes are so far commercially used. Hemicellulases, especially purified xylanases and mannanases originating from *Trichoderma*, have been used in studies of the mechanisms involved in the enzyme-aided bleaching. The other potential applications are based on the use of cellulases which, as compared with hemicellulases, are more complex due to the potential detrimental effects on the fibre properties. In the future, the monocomponent cellulases as well as the structurally engineered enzymes of *Trichoderma* will, however, offer new tools for tailored modifications of fibre carbohydrates.

References

AKHTAR, M. 1994. Biochemical pulping of aspen wood chips with three strains of *Ceriporiopsis subvermispora. Holzforschung* **48**: 199–202.

ALLISON, R. W., CLARK, T. A., and ELLIS, M. J. 1995. Process effects on the response of softwood kraft pulp to enzyme assisted bleaching. *Appita* **48(3)**: 201–206.

BAILEY, M., BUCHERT, J., and VIIKARI, L. 1993. Application of *Trichoderma* enzymes in the pulp and paper industry. In P. Suominen and P. Reinikainen (eds), *Proc. 2nd* TRICEL *Symposium*, Espoo, Finland, 1993. Foundation for Biotechnical and Industrial Fermentation Research **8**: 255–262.

BARNETT, C. C., BERKA, R. M., and FOWLER, T. 1991. Cloning and amplification of the gene encoding an extracellular β-glucosidase from *Trichoderma reesei*: evidence for improved saccharification of cellulosic substrates. *Bio/Technology* **9**: 562–567.

BROWN, J., CHEEK, M. C., JAMEEL, H., and JOYCE, T. W. 1994. Medium-consistency ozone bleaching with enzyme-pretreatment. *Tappi J.* **77(11)**: 105–109.

BUCHERT, J., TAMMINEN, T., and VIIKARI, L. 1997a. Impact of the Donnan effect on the action of xylanases on fibre substrates. *J. Biotechnol.*, **57**: 217–222.

BUCHERT, J., CARLSSON, G., VIIKARI, L., and STRÖM, G. 1996b. Surface characterization of unbleached kraft pulps by enzymatic peeling and ESCA. *Holzforschung* **50**: 69–74.

BUCHERT, J., RANUA, M., KANTELINEN, A., and VIIKARI, L. 1992. The role of two *Trichoderma reesei* xylanases in the bleaching of pine kraft pulp. *Appl. Microbiol. Biotechnol.* **37**: 825–829.

BUCHERT, J., SALMINEN, J., SIIKA-AHO, M., and VIIKARI, L. 1993. The role of *Trichoderma reesei* xylanase and mannanase in the treatment of softwood kraft pulp prior to bleaching. *Holzforschung* **47**: 473–478.

BUCHERT, J., SUURNÄKKI, A., TENKANEN, M., and VIIKARI, L. 1996a. Enzymatic characterization of pulps. In T. W. Jeffries and L. Viikari (eds), *Enzymes for Pulp and Paper Processing*. ACS Symp Ser 655. American Chemical Society, Washington, DC, pp. 38–43.

BUCHERT, J., BERGNOR, E., LINDBLAD, G., VIIKARI, L., and EK, M. 1997b. Significance of xylan and glucomannan in the brightness reversion of kraft pulps. *Tappi J.* **80**: 165–175.

BUCHERT, J., RANUA, M., SIIKA-AHO, M., PERE, J., and VIIKARI, L. 1994. *Trichoderma reesei* cellulases in the bleaching of kraft pulp. *Appl. Microbiol. Biotechnol.* **40**: 941–945.

BUCHERT, J., TELEMAN, A., HARJUNPÄÄ, V., TENKANEN, M., VIIKARI, L., and VUORINEN, T. 1995. Effect of cooking and bleaching on the structure of xylan in conventional pine kraft pulp. *Tappi J.* **78(11)**: 125–130.

CHEN, C. M., GRIZALI, M., and STAFFORD, D. W. 1987. Nucleotide sequence and deduced primary structure of cellobiohydrolase II of *Trichoderma reesei. Bio/Technology* **5**: 274–278.

CHEN, H., HAYN, M., and ESTERBAUER, H. 1992. Purification and characterization of two extracellular β-glucosidases from *Trichoderma reesei*. *Biochim. Biophys. Acta* **1121**: 54–60.

CLARK, T. A., STEWARD, D., BRUCE, M., MCDONALD, A., SINGH, A., and SENIOR, D. 1991. Improved bleachability of radiata pine kraft pulps following treatment with hemicellulolytic enzymes. *Appita* **44**: 389–404.

CUEVAS, W. A., KANTELINEN, A., TANNER, P., BODIE, B., and LESKINEN, S. 1996. Purification and characterization of novel mannanases used in pulp bleaching. In E. Srebotnik and K. Messner (eds), *Biotechnology in the Pulp and Paper Industry*, Recent Advances in Applied and Fundamental Research, Proc. 6th Int. Conf. Biotechnology in the Pulp and Paper Industry. Facultas-Universitätsverlag, Vienna, p. 123.

DIVNE, C., STÅHLBERG, J., REINIKAINEN, T., RUOHONEN, L., PETTERSSON, G., KNOWLES, J. K. C., TEERI, T., and JONES, A. 1994. The three-dimensional crystal of the catalytic core of cellobiohydrolase I from *Trichoderma reesei*. *Science* **265**: 524–528.

EKMAN, R., ECKERMAN, C., and HOLMBOM, B. 1990. Studies on the behaviour of extractives in mechanical pulp suspensions. *Nord. Pulp Paper Res. J.* **5**: 96–102.

FOWLER, T., GRIZALI, M., and BROWN, R. D. JR. 1993. Regulation of the cellulase genes of *Trichoderma reesei*. In P. Suominen and T. Reinikainen (eds), *Trichoderma Reesei* Cellulases and Other Hydrolases. *Foundation for Biotechnical and Industrial Fermentation Research* **8**: 199–210.

FREIERMUTH, B., GARRETT, M., and JOKINEN, O. 1994. The use of enzymes in the production of release papers. *Paper Technol.* **35**: 21–23.

FUENTES, J. L. and ROBERT, M. 1986. French patent 2604198.

GRETHLEIN, H. E. 1985. The effect of pore size distribution on the rate of enzymatic hydrolysis of cellulosic substrates. *Bio/Technology* **(2)**: 155–159.

HARKKI, A., MÄNTYLÄ, A., PENTTILÄ, M., MUTTILAINEN, S., BÜHLER, R., SUOMINEN, P., KNOWLES, J., and NEVALAINEN, H. 1991. Genetic engineering of *Trichoderma* to produce strains with novel cellulase profiles. *Enzyme Microb. Technol.* **13**: 227–233.

HEISE, O. U., UNWIN, J. P., KLUNGNESS, J. H., FINERAN, W. G. JR., SYKES, M., and ABUBAKR, S. 1996. Industrial scale up of enzyme-enhanced deinking of nonimpact printed toners. *Tappi J.* **79(3)**: 207–212.

JEAN, P., HAMILTON, J., and SENIOR, D. J. 1994. Mill trial experiences with xylanase: AOX and chemical reductions. Proc. 80th CPPA Annu. Meeting, Montreal, Canada, pp. 229–233.

JEFFRIES, T. W., KLUNGNESS, J. H., SYKES, M. S., and RUTLEDGE-CROPSEY, K. R. 1994. Comparison of enzyme-enhanced with conventional deinking of xerographic and laser-printed paper. *Tappi J.* **77**(4): 173–179.

JOKINEN, O. 1994. Entsyymien mahdollisuudet märkäosan hallinnassa. Paperi ja Puu [Paper and Timber] **76**: 491–493.

JOKINEN, O., KETTUNEN, J., LEPO, J., NIEMI, T., and LAINE, J. E. 1991. Unites States Pat. 5,068,009.

KAMAYA, Y. 1996. Role of endoglucanase in enzymatic modification of bleached kraft pulp. *J. Ferm. Bioeng.* **82**: 549–553.

KANTELINEN, A., RANTANEN, T., BUCHERT, J., and VIIKARI, L. 1993a. Enzymatic solubilization of fibre-bound and isolated birch xylans. *J. Biotechnol.* **28**: 219–228.

KANTELINEN, A., HORTLING, B., SUNDQUIST, J., LINKO, M., and VIIKARI, L. 1993b. Proposed mechanism of the enzymatic bleaching of kraft pulp with xylanases. *Holzforschung* **47**: 318–324.

KANTELINEN, A., RÄTTÖ, M., SUNDQUIST, J., RANUA, M., VIIKARI, L., and LINKO, M. 1988. Hemicellulases and their potential role in bleaching. Proc. 1988 Int. Pulp Bleaching Conf., Tappi, Orlando, FL, USA, pp. 1–5.

KANTELINEN, A., JOKINEN, O., SARKKI, M.-L., PETTERSSON, C., SUNDBERG, K., ECKERMAN, C., EKMAN, R., and HOLMBOM, B. 1995. Effects of enzymes on the stability of colloidal pitch. Proc. 8th Int. Symp. Wood and Pulping Chemistry, Helsinki, 6–9 June 1995, Vol. I, pp. 605–612. Gummerus Kirjapaino Oy, Jyväskylä 1995.

KIM, T.-J., OW, S., and BOM, T.-J. 1991. Enzymatic deinking method of wastepaper. Tappi Pulping Conference, Orlando, FL, 3–7 November, p. 1023.

KIRK, T. K. and JEFFRIES, T. W. 1996. Roles of microbial enzymes in pulp and paper processing. In T. W. Jeffries and L. Viikari (eds), *Enzymes for Pulp and Paper Processing.* ACS Symp. Ser. 655, American Chemical Society, Washington, DC, pp. 2–14.

KOIVULA, A., LAPPALAINEN, A., VIRTANEN, S., MÄNTYLÄ, A. L., SUOMINEN, P., and TEERI, T. T. 1996. Immunoaffinity chromatographic purification of cellobiohydrolase II mutants from recombinant *Trichoderma reesei* strains devoid of major endoglucanase genes. *Protein Expression Purif.* **8**: 391–400.

KOPONEN, R. 1991. Enzyme systems prove their potential. *Pulp Paper Int.* **33(11)**: 20–25.

KUBICEK-PRANZ, E. M., GSUR, A., HAYN, M., and KUBICEK, C. P. 1991. Characterization of commercial *Trichoderma reesei* cellulase preparations by denaturing electrophoresis (SDS-PAGE) and immunostaining using monoclonal antibodies. *Biotechnol. Appl. Biochem.* **14**: 317–323.

LAHTINEN, T., OJAPALO, P., LAUKKANEN, A., and SENIOR, D. 1993. Enzymes in TCF-bleaching. 1993 CPPA Spring Conf., Whistler, Canada, pp. 3.

LAHTINEN, T., SUOMINEN, P., OJAPALO, P., PEHU-LEHTONEN, K.,and LASSENIUS, I. 1992. Using selective *Trichoderma* enzyme preparations in kraft pulp bleaching. Proc. 5th Int. Conf. Biotechnology in Pulp and Paper Industry, Kyoto, Japan, pp. 129–137.

LAVIELLE, P., KOLJONEN, M., PIIROINEN, P., KOPONEN, R., REID, D., and FREDRIKSSON, R. 1992. Three large scale uses of xylanases in kraft pulp bleaching. Proc. 4th Int. Conf. New Available Techniques and Current Trends, Eur. Pulp and Paper Week, Bologna, Italy, p. 203.

LEATHAM, G., MYERS, G., and WEGNER, T. 1990. Biomechanical pulping of aspen chips: energy savings resulting from different fungal treatments. *Tappi J.* **73**: 197–200.

LUONTERI, E., TENKANEN, M., SIIKA-AHO, M., BUCHERT, J., and VIIKARI, L. 1996. α-Arabinosidases of *Aspergillus terreus* and their potentials in pulp and paper applications. In E. Srebotnik and K. Messner (eds), *Biotechnology in the Pulp and Paper Industry,* Recent Advances in Applied and Fundamental Research, Proc. 6th Int. Conf. Biotechnology in the Pulp and Paper Industry. Facultas-Universitätsverlag, Vienna, pp. 119–122.

MANSFIELD, S. D., WONG, K. K. Y., DE JONG, E., and SADDLER, J. N. 1996. Modification of Douglas fir mechanical and kraft pulps by enzyme treatment. *Tappi J.* **79**: 125–132.

MARGOLLES-CLARK, E., SALOHEIMO, M., SIIKA-AHO, M., and PENTTILÄ, M. 1996a. The α-glucuronidase-encoding gene of *Trichoderma reesei. Gene* **172**: 171–172.

MARGOLLES-CLARK, E., TENKANEN, M., LUONTERI, E., and PENTTILÄ, M. 1996b. Three α-galactosidase genes of *Trichoderma reesei* cloned by expression in yeast. *Eur. J. Biochem.* **240**: 104–111.

MARGOLLES-CLARK, E., TENKANEN, M., NAKARI-SETÄLÄ, T., and PENTTILÄ, M. 1996c. Cloning of genes encoding α-L-arabinofuranosidase and β-xylosidase from *Trichoderma reesei* by expression in *Saccharomyces cerevisiae. Appl. Environ. Microbiol.* **62**: 3840–3846.

MARGOLLES-CLARK, E., TENKANEN, M., SÖDERLUND, H., and PENTTILÄ, M. 1996d. Acetyl xylan esterase from *Trichoderma reesei* contains an active-site serine residue and a cellulose-binding domain. *Eur. J. Biochem.* **237**: 553–560.

NELSON, S. L., WONG, K. K. Y., SADDLER, J. N., and BEATSON, R. P. 1995. The use of xylanases for peroxide bleaching of kraft pulps derived from different wood species. *Pulp Paper Canada* **96**: T258–261.

NEVALAINEN, H. and PENTTILÄ, M. 1995. Molecular biology of cellulolytic fungi. In U. Kück (ed.), *The Mycota II Genetics and Biotechnology*. Springer-Verlag, Berlin, pp. 303–319.

NEVALAINEN, H., PENTTILÄ, M., HARKKI, A., TEERI, T., and KNOWLES, J. 1991. The molecular biology of *Trichoderma* and its application to the expression of both homologous and heterologous genes. In S. A. Leong and R. Berka (eds), *Molecular Industrial Mycology*. Marcel Dekker, New York, pp. 129–148.

NIDETZKY, B., STEINER, W., HAYN, M., and CLAEYSSENS, M. 1994. Cellulose hydrolysis by the cellulases from *Trichoderma reesei*; a new model for synergistic interaction. *Biochem. J.* **298**: 705–710.

NOÉ, P., CHEVALIER, J., MORA, F., and COMTAT, J. 1986. Action of enzymes in chemical pulp fibres. Part II: Enzymatic beating. *J. Wood Chem. Technol.* **6**: 167–184.

OKSANEN, T., BUCHERT, J., and VIIKARI, L. 1997a. The role of hemicelluloses in the hornification of bleached kraft pulp. *Holzforschung*. **51**: 355–360.

OKSANEN, T., BUCHERT, J., PERE, J., and VIIKARI, L. 1996. Treatment of recycled kraft pulps with hemicellulases and cellulases. In E. Srebotnik and K. Messner (eds), *Biotechnology in the Pulp and Paper Industry*, Recent Advances in Applied and Fundamental Research, Proc. 6th Int. Conf. Biotechnology in the Pulp and Paper Industry. Facultas-Universitätsverlag, Vienna, pp. 177–180.

OKSANEN, T., PERE, J., BUCHERT, J., and VIIKARI, L. 1997b. The effect of *Trichoderma reesei* cellulases and hemicellulases on the paper technical properties of never-dried bleached kraft pulp. Cellulose **4**: 329–339.

PAICE, M., BERNIER, M., and JURASEK, L. 1988. Viscosity enhancing bleaching of hardwood kraft pulp with xylanase from a cloned gene. *Biotechnol. Bioeng.* **32**: 235–239.

PATEL, R. N., GRABSKI, A. C., and JEFFRIES, T. W. 1993. Chromophore release from kraft pulp by purified *Streptomyces roseiscleroticus* xylanases. *Appl. Microbiol. Biotechnol.* **39**: 405–412.

PEDERSEN, L. S., KIHLGREN, P., NIELSEN, A. M., MUNK, N., HOLM, H. C., and CHOMA, P. P. 1992. Enzymatic bleach boosting of kraft pulp: laboratory and mill scale experiences. Proc. Tappi 1992 Pulping Conf., Book 1, Tappi Press, Atlanta, GA, pp. 31–37.

PENTTILÄ, M. E., LEHTOVAARA, P., NEVALAINEN, H., BHIKHABHAI, R., and KNOWLES, J. K. C. 1986. Homology between cellulase genes of *Trichoderma reesei*: complete nucleotide sequence of the endoglucanase I gene. *Gene* **45**: 253–263.

PERE, J., SIIKA-AHO, M., BUCHERT, J., and VIIKARI, L. 1995. Effects of purified *Trichoderma reesei* cellulases on the fiber properties of kraft pulp. *Tappi J.* **78(6)**: 71–78.

PERE, J., PAAVILAINEN, L., SIIKA-AHO, M., CHENG, Z., and VIIKARI, L. 1996. Potential use of enzymes in drainage control of nonwood pulps. In Proc. 3rd Int. Nonwood Fiber Pulping and Papermaking Conf., Beijing, 15–18 Oct. 1996, Vol. 2, pp. 421–430.

PETTERSSON, B., YANG, J., ERIKSSON, K.-E., and NIKU-PAAVOLA, M.-L. 1988. Characterization of pulp fiber surfaces by lignin specific antibodies. *Nordic Pulp Paper Res. J.* **3(3)**: 152–155.

POMMIER, J. C., FUENTES, J. L., and GOMA, G. 1989. Using enzymes to improve the process and the product quality in the recycled paper industry. Part 1: The basic laboratory work. *Tappi J.* **72(6)**: 187–191.

POMMIER, J. C., GOMA, G., FUENTES, J. L., ROUSSET, C., and JOKINEN, O. 1990. Using enzymes to improve the process and the product quality in the recycled paper industry. Part 2: Industrial applications. *Tappi J.* **73(12)**: 197–202.

POUTANEN, K. 1988b. An α-L-arabinofuranosidase of *Trichoderma reesei*. *J. Biotechnol.* **7**: 271–282.

POUTANEN, K. 1988a. Characterization of xylanolytic enzymes for potential applications. Technical Research Centre of Finland Publications 47, Espoo. 59 pp. + app. 50 pp.

POUTANEN, K. and PULS, J. 1988. Characteristics of *Trichoderma reesei*, β-xylosidase and its use in the hydrolysis of solubilized xylans. *Appl. Microbiol. Biotechnol.* **28**: 425–432.
PRASAD, D. Y., HEITMANN, J. A., and JOYCE, T. W. 1992. Enzyme deinking of black and white letterpress printed newsprint waste. *Progress in Paper Recycyling.* **1(3)**: 21–30.
PRASAD, D. Y., HEITMANN, J. A., and JOYCE, T. W. 1993. Enzymatic deinking of colored offset newsprint. *Nord. Pulp Pap. Res. J.* **8(2)**: 284.
PUTZ, H.-J., RENNER, K., GÖTTSCHING, L., and JOKINEN, O. 1994. Enzymatic deinking in comparison with conventional deinking of offset news. Tappi Pulping Conference, San Diego, CA, USA, p. 877.
RAHKAMO, L., SIIKA-AHO, M., VEHVILÄINEN, M., DOLK, M., VIIKARI, L., NOUSIAINEN, P., and BUCHERT, J. 1996. Modification of hardwood dissolving pulp with purified *Trichoderma reesei* cellulases. *Cellulose* **3**: 153–163.
REINIKAINEN, T., HENRIKSSON, K., SIIKA-AHO, M., TELEMAN, O., and POUTANEN, K. 1995. Low level endoglucanase contamination in a *Trichoderma reesei* cellobiohydrolase preparation affects its enzymatic activity on β-glucan. *Enzyme Microb. Technol.* **17**: 888–892.
RYDHOLM, S. A. 1965. *Pulping Processes.* Interscience Publishers, New York, pp. 596–609.
SAAKE, B., CLARK, T., and PULS, J. 1995. Investigations on the reaction mechanism of xylanases and mannanases on sprucewood chemical pulps. *Holzforschung* **49**: 60–68.
SALOHEIMO, M., HENRISSAT, B., HOFFRÉN, A.-M., TELEMAN, O., and PENTTILÄ, M. 1994. A novel, small endoglucanase gene, *egl5*, from *Trichoderma reesei* isolated by expression in yeast. *Molec. Microbiol.* **13**: 219–228.
SALOHEIMO, M., LEHTOVAARA, P., PENTTILÄ, M., TEERI, T., STÅHLBERG, J., JOHANSSON, G., PETTERSSON, G., CLAEYSSENS, M., TOMME, P., and KNOWLES, J. 1988. EGIII, a new endoglucanase from *Trichoderma reesei*; the characterization of both gene and enzyme. *Gene* **63**: 11–21.
SALOHEIMO, M., NAKARI-SETÄLÄ, T., TENKANEN, M., and PENTTILÄ, M. (1997). CDNA cloning of a *Trichoderma reesei* cellulose and demonstration of endoclucanase activity by expression in yeast. *Eur. J. Biochem.* **249**: 584–591.
SCOTT, B. P., YOUNG, F., and PAICE, M. G. 1993. Mill-scale enzyme treatment of a softwood kraft pulp prior to bleaching. *Pulp Paper Can.* **94(3)**: T-75–79.
SHOEMAKER, S., SCHWEICKART, V., LADNER, M., GELFAND, D., KWOK, S., MYAMBO, K., and INNIS, M. 1983. Molecular cloning of exo-cellobiohydrolase I derived from *Trichoderma reesei* strain L27. *Bio/Technology* **1**: 691–696.
SIIKA-AHO, M., TENKANEN, M., BUCHERT, J., PULS, J., and VIIKARI, L. 1994. An α-glucuronidase from *Trichoderma reesei* Rut C-30. *Enzyme Microb. Technol.* **16**: 813–819.
SJÖSTRÖM, E. 1977. The behaviour of wood polysaccharides during alkaline pulping processes. *Tappi J.* **60**(9): 151–154.
SRISODSUK, M., REINIKAINEN, T., and TEERI, T. T. 1993. Role of the interdomain linker peptide of *Trichoderma reesei* cellobiohydrolase I in its interaction with crystalline cellulose. *J. Biol. Chem.* **268**: 20756–20761.
STÅHLBERG, J., JOHANSSON, G., and PETTERSSON, G. 1991. A new model for enzymatic hydrolysis of cellulose based on the two-domain structure of cellobiohydrolase I. *Bio/Technology* **9**: 286–290.
STÅLBRAND, H. SALOHEIMO, A., VEHMAANPERÄ, J., HENRISSANT, B., and PENTTILÄ, M. 1995. Cloning and expression in *Saccharomyces cerevisiae* of a *Trichoderma reesei* β-mannanase gene containing a cellulose binding domain. *Appl. Environ. Microbiol.* **61**: 1090–1097.
STÅLBRAND, H., SIIKA-AHO, M., TENKANEN, M., and VIIKARI, L. 1993. Purification and characterization of two β-mannanases from *Trichoderma reesei*. *J. Biotechnol.* **29**: 229–242.

Stone, J. E. and Scallan, A. M. 1968. The effect of component removal upon the porous structure of the cell wall of wood. Part III: A comparison between sulphite and kraft processes. *Pulp Paper Mag. Can.* **69(6)**: T288–T293.

Stone, J. E., Scallan, A. M., Donefer, E., and Ahlgren, E. 1969. Digestibility as a simple function of a molecule of similar size to an enzyme. *Adv. Chem. Ser.* **96**: 219–241.

Stork, G., Pereira, H., Wood, T., Dusterhöft, E. M., Toft, A., and Puls, J. 1995. Upgrading recycled pulps using enzymatic treatment. *Tappi J.* **78(2)**: 79–88.

Sundberg, M. and Poutanen, K. 1991. Purification and properties of two acetylxylan esterases of *Trichoderma reesei. Biotechnol. Appl. Biochem.* **13**: 1–11.

Suominen, P. L., Mäntylä, A. L., Karhunen, T., Hakola, S., and Nevalainen, H. 1993. High frequency one-step gene replacement in *Trichoderma reesei*. II: Effect of deletions of individual genes. *Mol. Gen. Genet.* **241**: 523–530.

Suurnäkki, A., Kantelinen, A., Buchert, J., and Viikari, L. 1994. Enzyme-aided bleaching of industrial softwood kraft pulps. *Tappi J.* **77(11)**: 111–116.

Suurnäkki, A., Clark, T., Allison, R., Viikari, L., and Buchert, J. 1996a. Xylanase- and mannanase-aided ECF and TCF bleaching. *Tappi J.* **79(7)**: 111.

Suurnäkki, A., Heijnesson, A., Buchert, J., Viikari, L., and Westermark, U. 1996c. Chemical characterization of the surface layers of unbleached pine and birch kraft pulp fibers. *J. Pulp Paper Sci.* **22(2)**: J43–J47.

Suurnäkki, A., Heijnesson, A., Buchert, J., Westermark, U., and Viikari, L. 1996d. Effect of pulp surfaces on enzyme-aided bleaching of kraft pulps. *J. Pulp Paper Sci.* **22**(3): J91–J96.

Suurnäkki, A., Lahtinen, T., Lundell, R., Buchert, J., and Viikari, L. 1996e. Bleaching of kraft pulps with xylanases and mannanases. Proc. Int. Non-Chlorine Bleaching Conf., Orlando, FL, March 24–28, section 6-2. 5 pp.

Suurnäkki, A., Heijnesson, A., Buchert, J., Tenkanen, M., Viikari, L., and Westermark, U. 1996b. Location of xylanase and mannanase action in kraft fibres. *J. Pulp Paper Sci.* **22(3)**: J78–J83.

Suurnäkki, A., Li, T.-Q., Buchert, J., Tenkanen, M., Viikari, L., Vuorinen, T., and Ödberg, L. 1997. Effects of enzymatic removal of xylan and glucomannan on the pore size distribution of kraft fibres. *Holzforschung* **51**: 27–33.

Teeri, T. T., Penttilä, M., Keränen, S., Nevalainen, H., and Knowles, J. K. C. 1992. In D. B. Finkelstein and C. Ball (eds), *Biotechnology of Filamentous Fungi: Technology and Products.* Butterworth-Heinemann, Boston, MA, pp. 417–445.

Teleman, A., Harjunpää, V., Tenkanen, M., Buchert, J., Hausalo, T., Drakenberg, T., and Vuorinen, T. 1995. Characterisation of 4-deoxy-β-L-*threo*-hex-4-enopyranosyluronic acid attached to xylan in pine kraft pulp and pulping liquor by ^{1}H and ^{13}C NMR spectroscopy. *Carbohydr. Res.* **272**: 55–71.

Tenkanen, M., Puls, J., and Poutanen, K. 1992b. Two major xylanases of *Trichoderma reesei. Enzyme Microb. Technol.* **14**: 566–574.

Tenkanen, M., Buchert, J., Puls, J., Poutanen, K., and Viikari, L. 1992a. Two main xylanases of *Trichoderma reesei* and their use in pulp processing. In J. Visser, A. G. J. Voragen, M. A. Kusters-van Someren and G. Beldman (eds), *Xylans and Xylanases.* Elsevier Science, Amsterdam, pp. 547–550.

Tenkanen, M., Hausalo, T., Siika-aho, M., Buchert, J., and Viikari, L. 1995. Use of enzymes in combination with anion exchange chromatography in the analysis of carbohydrate composition of kraft pulps. Proc. 8th Int. Symp. Wood and Pulping Chemistry, Helsinki, 6–9 June 1995. Vol. III, pp. 189–194. Gummerus Kirjapaino, Jyväskylä 1995.

Thornton, J. W. 1994. Enzymatic degradation of polygalacturonic acids released from mechanical pulp in peroxide bleaching. *Tappi J.* **77(3)**: 161–167.

THORNTON, J., EKMAN, R., HOLMBOM, B., and ÖRSÅ, F. 1994. Polysaccharides dissolved from Norway spruce in thermomechanical pulp and bleaching. *J. Wood Chem. Technol.* **14**: 159–175.

TOLAN, J. S. 1992. The use of enzymes to enhance pulp bleaching. Proc. Tappi Pulping Conf., Boston, MA, Book 1, pp. 13–17.

TOMME, P., MCCRAE, S., WOOD, T., and CLAEYSSENS, M. 1988a. Chromatographic separation of cellulolytic enzymes. *Methods Enzymol.* **160**: 187–193.

TÖRRÖNEN, M., HARKKI, A., and ROUVINEN, J. 1994. Three-dimensional structure of endo-1,4-β-xylanase II from *Trichoderma reesei*: two conformational states in the active site. *EMBO J.* **13**: 2493–2501.

TÖRRÖNEN, A., MACH, R., MESSNER, R., GONZALEZ, R., KALKKINEN, N., HARKKI, A., and KUBICEK, C. P. 1992. The two major xylanases from *Trichoderma reesei*: characterization of both enzymes and genes. *Bio/Technology* **10**: 1461–1465.

UUSITALO, J. M., NEVALAINEN, K. M. H., HARKKI, A. M., KNOWLES, J. K. C., and PENTTILÄ, M. E. 1991. Enzyme production by recombinant *Trichoderma reesei* strains. *J. Biotechnol.* **17**: 35–50.

VIIKARI, L., KANTELINEN, A., RÄTTÖ, M., and SUNDQUIST, J. 1991. Enzymes in pulp and paper processing. In G. F. Leatham and M. E. Himmel (eds), *Enzymes in Biomass Conversion.* ACS Symp. Ser. 460, American Chemical Society, Washington, DC, pp. 12–21.

VIIKARI, L., KANTELINEN, A., SUNDQUIST, J., and LINKO, M. 1994. Xylanases in bleaching: from an idea to the industry. *FEMS Microb. Rev.* **13**: 335–350.

VIIKARI, L., RANUA, M., KANTELINEN, A., LINKO, M., and SUNDQUIST, J. 1987. Application of enzymes in bleaching. Proc. 4th Int. Symp. Wood and Pulping Chemistry, Paris, Vol. 1, pp. 151–154.

VIIKARI, L., RANUA, M., KANTELINEN, A., SUNDQUIST, J., and LINKO, M. 1986. Bleaching with enzymes. Proc. 3rd Int. Conf. Biotechnology in the Pulp and Paper Industry, STFI, Stockholm, pp. 67–69.

WARD, M., WU, S., DAUBERMAN, J., WEISS, G., LARENAS, E., BOWER, B., REY, M., CLARKSON, K., and BOTT, R. 1993. Cloning, sequence and preliminary analysis of a small, high pI endoglucanase EGIII from *Trichoderma reesei.* In P. Suominen and T. Reinkainen (eds), *Trichoderma Reesei Cellulases and Other Hydrolases.* Foundation for Biotechnical and Industrial Fermentation Research, Helsinki. Vol 8, pp. 153–158.

WELT, T. and DINUS, R. J. 1995. Enzymatic deinking – a review. Progress in Paper Recycling, **4(2)**: 36–47.

WERTHERMANN, D. P., TANNAR, D., and KOLJONEN, M. 1993. Enzymatic prebleaching of *Pinus radiata* pulp. Proc. 47th Appita Ann. Gen. Conf., Rotorua, NZ, Vol. 1, pp. 249–255.

WONG, K. K. Y., CLARKE, P., and NELSON, S. L. 1996. Possible roles of xylan-derived chromophores in xylanase prebleaching of softwood kraft pulp. ACS Symp. Ser. 618, p. 352.

WOODWARD, J., AFFHOLTER, K. A., NOLES, K. K., TROY, N. T., and GASLIGHTWALA, S. F. 1992. Does cellobiohydrolase II core protein from *Trichoderma reesei* disperse cellulose macrofibrils? *Enzyme Microb. Technol.* **14**: 625–630.

YANG, J., PETTERSSON, B., and ERIKSSON, K.-E. 1988. Development of bioassays for the characterization of pulp fiber surfaces. I: Characterization of various mechanical pulp fiber surfaces by specific cellulolytic enzymes. *Nordic Pulp Paper Res. J.* **3**: 19–25.

YANG, J. L. and ERIKSSON, K.-E. L. 1992. Use of hemicellulolytic enzymes as one stage in bleaching of kraft pulps. *Holzforschung* **46**: 481–488.

ZEILINGER, S., KRISTUFEK, D., ARISAN-ATAC, I., HODITS, R., and KUBICEK, C. P. 1993. Conditions of formation, purification and characterization of an α-galactosidase of *Trichoderma reesei* RUT C-30. *Appl. Environ. Microbiol.* **59**: 1347–1353.

17

Heterologous protein production in *Trichoderma*

M. PENTTILÄ

VTT Biotechnology and Food Research, Espoo, Finland

17.1 General features of *Trichoderma* in protein production

Trichoderma reesei has a long history in industrial enzyme production, and there has been considerable interest in developing *T. reesei* as a more versatile host for production of its own proteins for various industrial applications (see Chapters 13–16). Many of the features that make *Trichoderma* an excellent industrial enzyme producer are also beneficial in foreign protein production. The advantages of prokaryotic and eukaryotic production hosts can be considered to a large extent to be combined in filamentous fungi, *Trichoderma* and *Aspergillus* species being the ones mainly used so far. Enzymes produced by *T. reesei* have also been approved for food and feed applications, and no toxin production has been observed in *T. reesei* (see Volume 1, Chapter 8)

Being a microbe, *T. reesei* is easy and inexpensive to cultivate. It is currently grown in fermenters up to 230 m^3, which shows that its fermenter technical properties are good and it is not susceptible to contamination. Unlike mammalian or insect cell lines, there is no requirement for expensive nutrient additions or special propagation; *Trichoderma* can grow on simple and inexpensive media and on materials such as whey and various plant wastes. Being a eukaryote, it has the advantage of possessing a secretory machinery with protein modifications typical of eukaryotes. Compared with the yeast *Saccharomyces cerevisiae*, filamentous fungi generally secrete higher levels of protein and with shorter high mannose type N-glycosylation (Maras *et al.*, 1997; Salovuori *et al.*, 1987; see Volume 1, Chapter 6). There is, for instance, a clear difference in glycosylation of *T. reesei* cellulases (Penttilä *et al.*, 1987, 1988) and a mannanase (Stålbrand *et al.*, 1995) when expressed in *S. cerevisiae* compared with the native host; production in yeast gives the enzymes a higher molecular weight in SDS-PAGE and a greater heterogeneity. The disadvantages of *Trichoderma*, and other filamentous fungi, are the slower growth rate and more time-consuming and tedious genetic engineering techniques. Thus it may not always be competitive with, for instance, *E. coli* in production of high-value products. The production economy of *E. coli* expression systems is often feasible even when the proteins are produced in small scale and even if they might end up in intracellular inclusion bodies from which they need to be denatured and renatured to obtain the active, correctly folded product. However, *Trichoderma* should have a special advantage in production of foreign proteins that are needed in large amounts.

All essential tools needed for foreign protein production have been developed for *Trichoderma*. There are several different mutant strains of *T. reesei*, such as many hypercellulolytic or cellulase-negative production strains or strains that are partially protease deficient, that have been obtained through conventional mutagenesis or by genetic engineering (reviewed by Nevalainen *et al.*, 1990; Nevalainen and Penttilä, 1995; see Chapter 13). Some of the strains have high synthesis and secretion capacities; yields of cellulases of as much as 40 g per liter of culture medium have been reported (Durand *et al.*, 1988). In addition, Cayla Ltd has selected strains with improved secretion properties. *Trichoderma* can be transformed with a variety of selection markers, including several dominant ones, examples being the utilization of acetamidase or invertase as carbon sources, or resistance conferred towards hygromycin, fleomycin or benomyl (see Volume 1, Chapter 10). In principle, any strain may be transformed and further retransformed. Since transformation occurs through integration into the genome and results in varying copy numbers and places of integration, the transformants differ in production levels and the best producers can be screened from the original transformants obtained. A benefit of integrative transformation is that multicopy transformants can be obtained and these are usually stable without any selection pressure in subsequent cultivations. Certain chromosomal loci appear more beneficial for expression, among which is the locus of the major cellulase, *cbh1*. Targeted integration is thus advantageous and can be achieved with frequencies as high as 80% (Karhunen *et al.*, 1993; Paloheimo *et al.*, 1993; Suominen *et al.*, 1993). Targeted integration also allows gene inactivations and replacements to be carried out, and inactivation of some genes encoding highly expressed native extracellular enzymes may be useful since it is possible that their production overloads the expression, synthesis and secretion machinery and reduces yields of the heterologous product.

17.2 Tools and strategies for heterologous protein production

CBHI represents approximately 50% of all protein secreted by wild-type *T. reesei*. Since this high amount of enzyme is produced from a single copy gene, the *cbh1* promoter is strong and is the most frequently used promoter in protein production by *T. reesei*. The promoter is strongly induced in the presence of cellulose, plant materials, or small oligosaccharides such as cellobiose and especially sophorose. There is tight glucose repression which is at least partly mediated by the regulatory gene *cre1* (Ilmén *et al.*, 1996a,b; see Chapter 3). Mutagenesis of the binding sites for the CREI protein in the *cbh1* promoter partially removes glucose repression. The promoter of the *A. niger glaA* gene encoding glucoamylase also gives reasonable yields in *T. reesei* as demonstrated by Cayla Ltd (M. Parriche, personal communication).

The *cbh1* promoter is in many respects a good promoter for protein production since it is tightly regulated and strongly expressed. However, expression of most of the hydrolases produced by the fungus in large amounts is co-regulated with CBHI, which can in some cases interfere with the purification of the desired product from the fungal culture medium. Inexpensive media promoting *cbh1* expression are complex and undefined and some also induce extracellular proteases. Because of these drawbacks, several genes active in the presence of glucose have been isolated, such as *pgk1*, which encodes phosphoglycerate kinase (Vanhanen *et al.*, 1989, 1991) and *pki1*, which encodes pyruvate kinase (Schindler *et al.*, 1993). Unlike the *S. cerevisiae PGK* promoter, the *T. reesei pgk1* promoter seems to be unsuited for high level protein production (Vanhanen, 1991), and only modest levels have been obtained with the *pki1* promoter (Kurzatkowski *et al.*, 1996). For this reason, a strategy to isolate highly expressed genes and their promoters was developed that did not rely on deciding which genes to isolate (Nakari *et al.*, 1993a). This approach led to the isolation of the *tef1* gene encoding translation elongation factor 1α, *hfb1* gene encoding a hydrophobin (Nakari-Setälä *et al.*, 1996), and several highly expressed but yet unidentified genes (Nakari *et al.*, 1993a). Of the several glucose promoters tested *(pgk1, pki, tef1*, "cDNA1" and the glucose derepressed *cbh1* promoter), the promoter of one of the unidentified genes ("cDNA1") is at present the most efficient, giving product yields of 50% of the total secreted protein on glucose-containing medium (Nakari *et al.*, 1993b; Nakari-Setälä and Penttilä, 1995). These levels are nevertheless clearly lower when compared with those obtainable with the *cbh1* promoter in cellulase-inducing conditions. A strong promoter for *T. reesei* has also been isolated by cloning random fragments in front of a selectable marker (Calmels *et al.*, 1991b). In addition, the promoter of the *Aspergillus gpdA* gene encoding glyceraldehyde phosphate dehydrogenase can be used successfully for protein production on glucose-containing medium (VTT, unpublished).

The high production levels of CBHI would also suggest that the properties of the protein would be optimal for translation, translocation into the endoplasmic reticulum (ER), folding and subsequent glycosylation and secretion into the medium. Fusion protein strategies have proven useful in foreign protein production in many organisms, resulting in increased yields. CBHI has a very useful structure in this respect. Like most other cellulases, it consists of a catalytic core domain and a separate cellulose binding domain (CBD) separated from the core by an extended and flexible O-glycosylated linker (see Chapter 1). Replacing the C-terminal CBD to produce foreign proteins fused to the core and linker region of CBHI has proven beneficial. The linker provides a natural region for separation of the two protein domains, which can fold independently. The linker also provides an exposed region for addition of a cleavage site for *in vitro* or *in vivo* release of the product from the fusion. In this kind of a fusion strategy, the translation initiation site, signal sequence and possible regions suited for efficient secretion are retained from the endogenous fusion partner (Figure 17.1). The glucoamylase of *A. niger* has a similar tripartite structure and, as one might expect, it can also be used as a fusion partner in *T. reesei* (Demolder *et al.*, 1994; Parriche *et al.*, 1996). The work carried out at Cayla Ltd also demonstrates that good yields of a product can be obtained with a heterologous fusion partner, a bacterial phleomycin resistance protein (Parriche *et al.*, 1996).

17.3 Examples of heterologous protein production in *T. reesei*

As is the case with other filamentous fungi such as *Aspergillus*, there are not very many examples of foreign proteins produced in *Trichoderma*. Nevertheless, the pro-

A

B

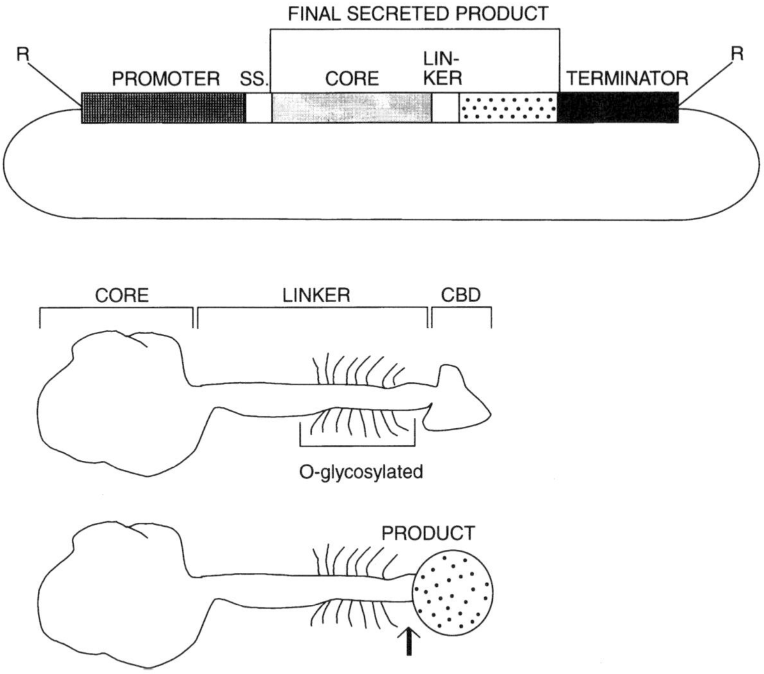

Figure 17.1 A: Schematic presentation of an expression cassette for production of the foreign protein as a fusion to the CBHI core-linker. The boxed areas represent *cbh1* sequences, including the signal sequence (SS), except the dotted area which represents sequences encoding the foreign protein product. The expression cassette can be released from the plasmid at the unique restriction enzyme cleavage sites (R), for instance for targeted integration and replacement of the *cbh1* locus in the fungal genome. B: Schematic presentation of the native CBHI protein and the fusion protein. The arrow marks a possible proteolytic cleavage site introduced for a release of the product from the fusion.

duction of two proteins, calf chymosin and murine antibody Fab fragments, have been studied in detail, and the results illustrate that *Trichoderma* is a potent host for heterologous protein production. As expected, overproduction of homologous proteins or heterologous proteins of fungal origin is usually successful in yielding grams per liter levels of protein and reflecting the strength of the *cbh1* promoter (summarized in Paloheimo *et al.*, 1993), but production of heterologous proteins originating from more distant species remains lower (Harkki *et al.*, 1989; Keränen and Penttilä, 1995; Nyyssönen *et al.*, 1993; Pitts *et al.*, 1993; Saloheimo and Niku-Paavola, 1991; Uusitalo *et al.*, 1991) (Table 17.1). However, the levels obtained thus far using the fusion strategies and without carrying out any further optimization of production conditions are promising.

17.3.1 *Studies on calf chymosin*

Calf chymosin was the first foreign protein expressed in *T. reesei* and also the first heterologous product of mammalian origin on the market produced by filamentous fungi, e.g., *Aspergillus* (Dunn-Coleman *et al.*, 1991; Ward *et al.*, 1990). Chymosin is naturally produced with a propeptide which is autocatalytically cleaved at low pH,

Table 17.1 Secreted heterologous proteins produced in *T. reesei*[a]

Protein	Origin	Fused to CBHI core-linker	Secreted protein	Ref.
lignin peroxidase	*Phlebia radiata*	−	not detectable	1
laccase	*Phlebia radiata*	−	3 mg (20 mg)	2
glucoamylase P	*Hormoconis resinae*	−	0.7 g (several g)	3
phytase	*Aspergillus niger*	−	2 g	4
acid phosphatase	*Aspergillus niger*	−	0.5 g	4
endochitinase	*Trichoderma harzianum*	−	150 mg (several g)	5
chymosin	calf	−	20–40 mg	6, 7
		+	>100 mg	8
antibody Fab	murine	−	1 mg (1 mg)	9
fragments		+	40 mg (150 mg)	9
single chain antibodies	murine	−	1 mg	10
interleukin-6	mammalian	+	5 mg	11

[a] In all cases shown the *cbh1* promoter was used, except for SCA production where the *Aspergillus gpd* promoter was used. Amounts obtained from a bioreactor cultivation are shown in parentheses.

References quoted: 1, Saloheimo *et al.*, 1989; 2, Saloheimo and Niku-Paavola, 1991; 3, Joutsjoki *et al.*, 1993; 4, Paloheimo *et al.*, 1993; 5, Margolles-Clark *et al.*, 1996; 6, Harkki *et al.*, 1989; 7, Uusitalo *et al.*, 1991; 8, J. Uusitalo, unpublished; 9, Nyyssönen *et al.*, 1993; 10, Nyyssönen, 1993; 11, Demolder *et al.*, 1994.

and the mature cleaved chymosin can be recovered directly from the fungal culture medium. Consequently, it is an optimal protein to be produced as a fusion since no cleavage of the product with an exogenously added protease is needed after synthesis.

In the first chymosin expression constructs, various parts of CBHI were fused to a prochymosin-coding region and these were compared with preprochymosin (with chymosin signal sequence) produced similarly from the *cbh1* promoter in the *T. reesei* strain RutC-30 (Harkki *et al.*, 1989). Since transformation occurs through integration of varying numbers of copies of the expression cassette into the fungal genome, comparisons of expression levels relied on analysis of several transformants. Higher expression levels were observed when the CBHI signal sequence was used, and the expression was improved further if 20 amino acids of the mature CBHI coding region were fused to prochymosin. It has later been shown (Jaana Uusitalo, unpublished) that on average a five-fold increase in production of secreted chymosin can be obtained if the prochymosin is fused to the CBHI core-linker region. These kinds of strains secrete well over 100 mg/l of active chymosin in shake-flask cultures.

Analysis of intracellular and extracellular chymosin indicates that the signal sequences of CBHI and chymosin are correctly processed and that some autocatalytic cleavage leading to mature chymosin already occurs intracellularly (Harkki *et al.*, 1989; Uusitalo *et al.*, 1991). Comparison of different culture conditions for chymosin production shows a correlation between the culture pH and processing of chymosin; media promoting good growth leading to lower pH in shake-flask cultures also yield more mature chymosin in a comparable cultivation period (Uusitalo *et al.*, 1991). In whey-peptone based medium, chymosin yields decreased at later growth stages, which indicated degradation by host proteases, whereas this was not observed in a cellulose-based medium (Uusitalo *et al.*, 1991).

The first results obtained with *T. reesei* demonstrated a clear improvement in heterologous chymosin production compared with other systems. Production in *E. coli* had led to intracellular accumulation of inactive chymosin in inclusion bodies (Emtage *et al.*, 1983). Secreted levels of active chymosin were significantly lower in *S. cerevisiae* (Smith *et al.*, 1985) than in *T. reesei*, and the published yields were also lower in *A. nidulans* (Cullen *et al.*, 1987). The high yields in *T. reesei* made it possible to produce several mutant forms of chymosin for structure–function studies (Dhanaraj *et al.*, 1995; Pitts *et al.*, 1991, 1993). Analysis of about 10–15 transformants (using constructs without fusions to CBHI) yielded several clones which produced 5–20 mg/l of extracellular mature chymosin that was sufficient for biochemical analysis and crystallization. Crystallization of chymosin produced in *T. reesei* demonstrated its authenticity.

It was obvious, however, that chymosin levels were much lower than the levels of CBHI naturally produced by the fungus. Analysis of the steady-state mRNA levels showed that there was much less chymosin mRNA produced from the *cbh1* promoter than mRNA resulting from the endogenous *cbh1* gene (Harkki *et al.*, 1989). It is not yet known whether production of chymosin as a CBHI fusion would result in higher mRNA levels. Another means to improve expression levels could be to raise the copy number of the integrated expression cassette. The early experiments suggested, however, that the place of integration seemed to play a more important role than copy number. The highest secreted levels (40 mg/l) were obtained when the selectable marker for transformation (*amdS*) was on the same plasmid as the chymosin expression cassette (in this case not fused to CBHI) in contrast to carrying out cotransformation. This strategy probably led to integration into favorable loci in the fungal genome, thus permitting good expression of both the transformation marker and chymosin. In one of these transformants, the chymosin cassette had actually replaced the *cbh1* locus (Harkki *et al.*, 1989). It would be interesting to see what production levels could be obtained if chymosin fused to the CBHI core-linker region were produced from the *cbh1* locus.

That chymosin behaves differently from CBHI and encounters difficulties on its way to the medium is suggested by immunofluorescence microscopy (Figure 17.2). Immuno-electronmicroscopical morphometry analysis showed that the subcellular distributions of CBHI and chymosin were significantly different (Nykänen *et al.*, manuscript in preparation). Five-fold less label for CBHI was observed in the cell wall than in the cytoplasm, whereas for chymosin there was 22-fold more label in the cell wall. Proportionally less cell-wall bound label was observed if chymosin was produced as a CBHI fusion. This indicates that the CBHI part might aid in the passage of chymosin through the cell wall.

17.3.2 Studies on antibody molecules

There is substantial interest in production of engineered antibody molecules carrying the antigen binding regions but lacking varying amounts of the other parts of

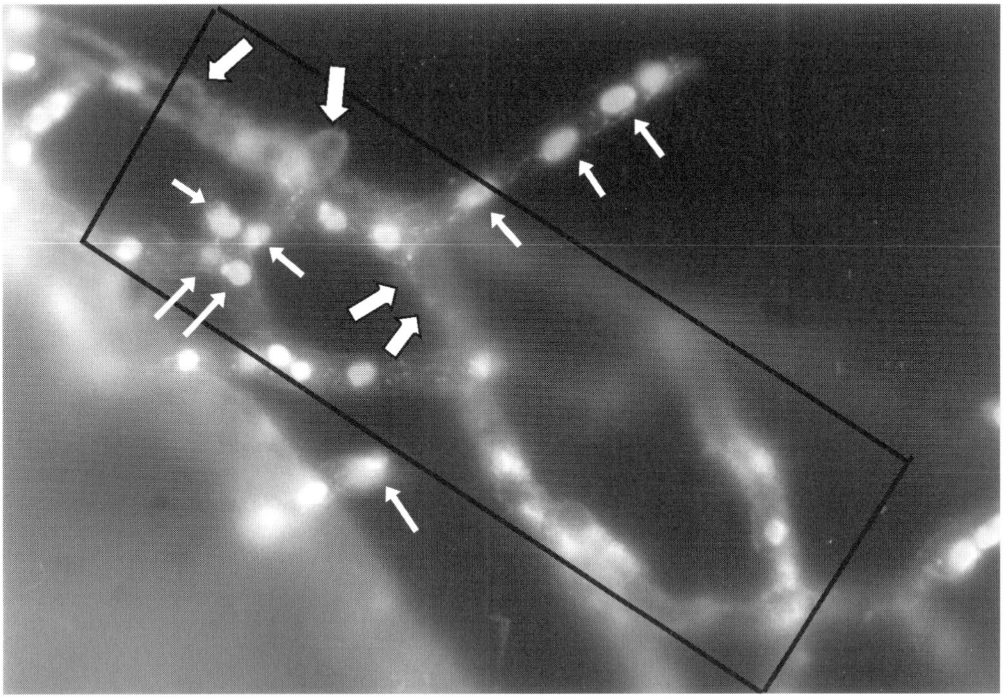

Figure 17.2 Immunofluorescent picture of *T. reesei* hyphae producing calf chymosin. Chymosin is marked by wide arrows, and nuclei by narrow arrows. Courtesy of Marko Nykänen.

the antibody chains. Fab molecules consist of the complete light chain and the Fd part of the heavy chain, assembled together with an interchain disulfide bridge. Of the different antibody forms produced in *Trichoderma*, Fab molecules have been the ones most thoroughly studied (Nyyssönen, 1993; Nyyssönen *et al.*, 1993; Nyyssönen and Keränen, 1995).

To produce Fab molecules in *T. reesei*, separate expression vectors were constructed for the light chain and the heavy Fd chain (Figure 17.3). Both chains were expressed under the *cbh1* promoter, and the *cbh1* signal sequence was used to direct secretion. The light chain was first introduced into the strain RutC-30 by using phleomycin resistance for selection of transformants. Only two light chain producers were found among over one hundred tested, and in both strains the light chain expression cassette had been integrated into the *cbh1* locus and had inactivated the endogenous *cbh1* gene. This result indicates either that the expression level was high enough to detect the secreted light chain only when expressed from the *cbh1* locus or that the lack of CBHI, produced naturally in high amounts, somehow facilitated production or detection of the light chains. These transformants secreted about 0.2 mg/l of light chain in shake-flask cultures.

A light chain-producing strain was transformed with the heavy Fd chain expression vector using the *amdS* transformation marker. A number of transformants secreted functional antigen-binding Fab molecules, and the levels produced by the best ones tested reached 1 mg/l in shake-flask cultivations. The immunological functionality of the molecules suggested that the two chains had folded and assembled together correctly. It was interesting to notice that the light chain

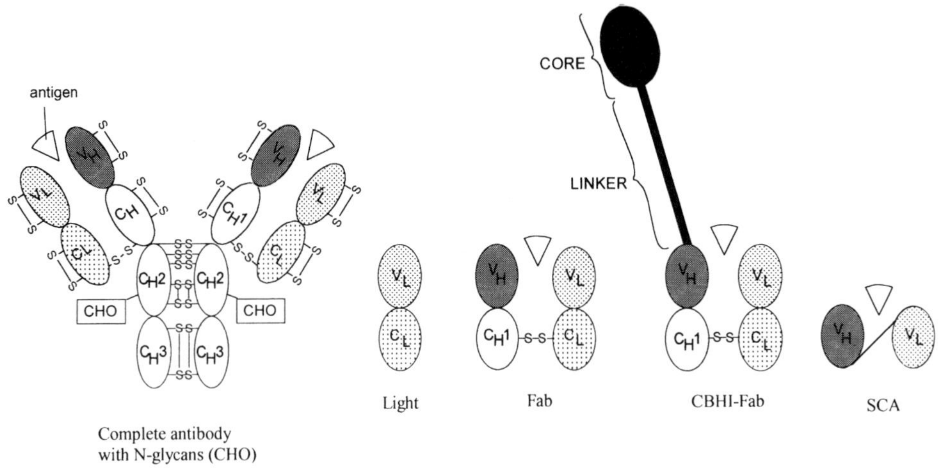

Figure 17.3 Schematic presentation of a complete antibody molecule and the shortened antibody forms produced in *T. reesei.* The linker of CBHI was used in the SCA molecule.

secretion level was increased after the introduction of the heavy chain into the same original strain first expressing only the light chain.

In order to study how the CBHI fusion affects antibody production, the CBHI cellulose binding domain (CBD) was replaced with the heavy Fd chain and the construct was transformed into the same light chain-expressing strain as used before. A significant increase in production was observed, the best transformants secreting 40 mg/l of immunologically active CBHI–Fab molecules into the culture medium. In a bioreactor cultivation the levels increased to 150 mg/l, whereas the levels obtained with the strain producing Fab not fused to CBHI remained low, about 1 mg/l.

Analysis of purified antibody molecules produced by *T. reesei* (Nyyssönen *et al.*, 1993) showed that the signal sequences were correctly cleaved from the light chain and the heavy Fd chain. In the case of the CBHI–Fab fusion, an interesting observation was made. The Fab part was released from a fraction of the fusion molecules by a yet unidentified fungal protease, which cleaved very specifically two amino acids before the heavy chain N-terminus. This cleavage could be mimicked *in vitro* by α-chymotrypsin. The uncleaved CBHI–Fab molecules were 5–12 times less immunoreactive than the idiotypic Fab molecules prepared from complete murine IgG antibodies. After cleavage from the CBHI part, the released Fab molecules attained idiotypic immunoreactivity and affinity towards the antigen.

Quantification of the separate chains using non-reducing SDS-gels indicated that all the secreted light chains were assembled with heavy chains. However, analysis of the culture supernatant of the CBHI–Fab producer revealed significant amounts (800 mg/l) of molecules consisting of the core-linker region of CBHI, suggesting that these were originating from the CBHI–heavy-Fd fusion molecule after the cleavage with the fungal protease. This cleavage occurred at least partly inside the cell, as was later confirmed using *in vivo* labelling techniques (T. Pakula, unpublished). The large amount of CBHI core raises a question: would prevention of the intracellular cleavage increase the secreted levels of CBHI–Fab? The results obtained with the different antibody-producing strains also indicate that the production of the light chains might be limiting and that their production fused to CBHI could be beneficial. Preliminary results indicate that it is possible to produce Fab molecules using constructs in which both the light and the heavy Fd chain are fused to CBHI (J. Demolder, unpublished).

Single-chain antibodies (SCA) are engineered forms in which the antigen binding sites of the heavy chain and the light chain are produced as a single molecule fused together by a linker. Circumventing the need for assembly of separate molecules, SCAs would be ideal in testing the beneficial role of CBHI because only one molecule needs to be fused to CBHI. It has already been demonstrated that *Trichoderma* is able to secrete immunologically active SCA molecules, but at present the SCAs have not been produced as CBHI fusions. The secreted level of functional SCAs reached 1 mg/l when produced without fusion (Nyyssönen, 1993).

Analysis of factors leading to differences in antibody production

The difference in yields of Fab fragments was not due to different stabilities of the molecules in the fungal culture medium. Intracellular events such as the levels of gene expression or secretion efficiency could differ between the different types of antibody molecules produced. For this reason a detailed analysis of the steady-state levels of mRNAs and intra- and extracellular proteins was carried out, and these also were correlated with expression levels of some endogenous genes and at protein level with the cellulase EGI in order to avoid discrepancies caused by possible differences in the growth stage and rate (Nyyssönen, 1993). CBHI was produced with the same efficiency as EGI in the non-transformed host strain. The production efficiencies of the light chains were calculated. It became clear that there was a definite hindrance in production after transcription, which was greatest in the strain producing the light chain only. In this case, 1000-fold less protein was secreted in relation to mRNA than in the case of cellulases. As noted previously, the introduction of the heavy Fd chain into the same cell increased the light chain production efficiency, and this was 10 times higher in the Fab producer than in the strain producing the light chain only. Fusion of the heavy Fd chain to CBHI had a significant effect on the light chain production efficiency, i.e., the ratio between MRNA and secreted protein, which in this case approached the production efficiencies of the endogenous cellulases.

Chain assembly is needed for Fab production, and the results showed that modifications in the production of the heavy chain clearly affected production efficiency of the light chain as well. This suggests that folding and chain assembly may be important for production and that fusion to CBHI may have a beneficial role in this respect, in addition to the possible role of CBHI as a general carrier for secretion. Entry into the endoplasmic reticulum can also be a bottleneck in antibody production as was suggested by the observation that light chains with a slightly higher molecular weight, and thus probably still containing the signal sequence, were detected in intracellular samples in the strain producing light chain only. These larger molecules were less abundant in the Fab producer, and the mature form was prominent in the CBHI–Fab-producing strain (Nyyssönen, 1993; Nyyssönen and Keränen, 1995).

In vivo labelling of proteins with ^{35}S-methionine combined with immunoprecipitation and 2-D gel analysis enables studies on the kinetics of translation and secretion and allows one to distinguish the various intracellular maturation forms of proteins (Pakula *et al.*, unpublished). Based on this analysis, the average synthesis time for full length CBHI molecules has been determined to be approximately four minutes and the average secretion time into the culture medium approximately ten minutes. When the synthesis and secretion of the CBHI–heavy Fd molecules and the light chain were followed in the *T. reesei* strains producing CBHI–Fab in a similar manner, differences relative to CBHI were observed. The preliminary results indicate that the intracellular appearance of full length CBHI–heavy Fd and light chains is slower compared with CBHI when normalized to the amount of amino acids in the protein but the secretion times of these fully synthesized molecules are not much different. Since it is expected that translocation of the molecules is cotranslational, this observed difference does not necessarily indicate differences in translation rates but could alternatively reflect differences in translocation and/or folding. In accordance with the possible slow folding of antibody molecules is the observation that expression of antibody molecules in *T. reesei* seems to induce the intracellular signalling pathway, the so-called "unfolded protein response", as judged by elevated levels of mRNA coding for protein disulfide isomerase (PDI) which is involved in folding of proteins in the ER (M. Saloheimo, unpublished).

17.3.3 *Other examples of heterologous proteins produced in* T. reesei

In addition to chymosin and Fab molecules, another foreign protein was produced: interleukin-6 (Demolder *et al.*, 1994). Preliminary experiments suggested that the secreted levels are about 5 mg/l. In this work a cleavage site for a Kex2-type activity was introduced into the CBHI linker. Kex2 is a Golgi-resident protease that has been shown in *S. cerevisiae* and in *Aspergillus* to cleave *in vivo* fusion proteins that carry a specific dibasic (preferentially Lys–Arg) recognition site. This type of sequence occurs in pro-region cleavage sites of proteins in filamentous fungi (Calmels *et al.*, 1991a) such as in *T. reesei* XYNII and CBHII. Cleaved interleukin-6 of apparent correct size was secreted by *T. reesei*, which indicated that *T. reesei* possesses Kex2-type activity. Cleavage was also obtained from a fusion of interleukin to the *Aspergillus* glucoamylase which similarly carried a recognition site for Kex2 activity N-terminally to interleukin. These results show that this type of an *in vivo* cleavage strategy could be used in protein production with *Trichoderma*, which would be beneficial since it circumvents the often expensive and inefficient *in vitro* cleavage methods with commercial proteases. The cloning of the gene coding for Kex2 activity has been attempted in order to overexpress it for efficient cleavage of highly expressed fusion proteins. However, the trials to clone the *T. reesei* equivalent by primers based on homologous sequences (S. P. Goller and C. P. Kubicek; M. Parriche; M. Saloheimo; all unpublished) or by complementation of a yeast *kex2* mutant (M. Saloheimo, unpublished) have been unsuccessful, although several clones showing subtilisin-like protease similarity were obtained (S. P. Goller and C. P. Kubicek, unpublished).

Fab molecules do not carry N-glycosylation sites and were not glycosylated in *T. reesei*. The major part of chymosin is also not N-glycosylated in the natural host, and no glycosylation was observed in *T. reesei* although this was not extensively analysed. The barley cysteine endopeptidase B, on the other hand, seems to become N-glycosylated when produced in *T. reesei* RutC-30, unlike in barley, as judged by

SDS-PAGE analysis (Nykänen *et al.*, 1997). The glucoamylase from another fungal species, *Hormoconis resinae*, appears differently glycosylated in the *T. reesei* strain ALKO 2221, and is migrating on a gel as a more dispersed band than the native enzyme (Joutsjoki *et al.*, 1993). This is caused by N-glycosylation as judged from endoglycosidase H treatment. The protein seems, however, not to be overglycosylated since the apparent molecular size of the protein is not larger than that of the native purified enzyme. Since glycosylation patterns of proteins might naturally vary depending on the strain or culture conditions it is difficult, however, to judge the possible real differences in glycosylation between different species based on only a few samples and gel analyses.

When analysing expression of the barley cysteine endopeptidase B (EPB), Nykänen *et al.* (1997) also studied the localization of the protein in different parts of the fungal hyphae in solid agar plate cultivations. EPB-encoding mRNA and protein were localized in young apical and subapical regions of the hyphae, whereas it seemed that CBHI was additionally localized in older parts of the mycelium. No mRNA for EPB or CBHI was seen in the fungal tips, indicating that after synthesis the proteins need to be transported to the tip regions for secretion. It would be interesting to know whether the results indicate differences in the localization of synthesis and secretion between native and heterologous proteins. On the other hand, in a similar analysis, calf chymosin was colocalized with CBHI along the fungal hyphae when produced as such or fused with CBHI (M. Nykänen, unpublished), although there was more accumulation of chymosin in the cell walls as mentioned earlier.

As indicated in Table 17.1, good yields of heterologous proteins of fungal origin can usually be obtained using the *cbh1* promoter, and there have been no attempts to produce the proteins as CBHI fusions. The production levels have been increased relative to the native production 20-fold in the case of *H. resinae* glucoamylase (Joutsjoki *et al.*, 1993) and the *T. harzianum* endochitinase (Margolles-Clark *et al.*, 1996), or even over 1000-fold as in the case of *A. niger* phytase (Paloheimo *et al.*, 1993). The yields of phytase are comparable to those produced from multiple copies from the glucoamylase promoter by recombinant *A. niger*, but the *T. reesei* enzyme preparation has an advantage for feed applications since it has high β-glucanase and low glucoamylase activities. Better yields of phytase were obtained in the *T. reesei* strains VTT-D-79125 and ALKO 2221 than in RutC-30. The exceptions to these good yields of fungal proteins so far encountered are ligninolytic enzymes, e.g., the *Plebia radiata* lignin peroxidase, which could not be detected when expressed in *T. reesei* RutC-30 (Saloheimo *et al.*, 1989), and laccase, which was expressed at levels of 20 mg/l from the *cbh1* promoter (Saloheimo and Niku-Paavola, 1991). Ligninolytic enzymes have, in general, been difficult to produce in heterologous hosts, and it would be of interest to know whether a fusion strategy would improve the yields.

The *T. harzianum* endochitinase can be produced by *T. reesei* at grams per liter level in a bioreactor, which is significant since the *T. harzianum* endochitinase has been reported to be strongly toxic to *Trichoderma* strains other than the native host and to other Fungi (see Chapter 4). In addition to the possible toxicity which may impair maximal production levels, the protein seems to encounter also other problems in this closely related heterologous host (Margolles-Clark *et al.*, 1996). An acidic extracellular protease degrades the endochitinase at later stages of the cultivation although there is still considerable potential to produce the protein based on high mRNA levels and continuing secretion of CBHI. When compared with CBHI, the ratio of secreted protein to mRNA level is also clearly lower in the case of endochitinase, indicating some post-transcriptional inefficiencies. In this same study it also became evident that transcription from the *cbh1* promoter, of endogenous *cbh1* or the foreign gene, reaches its maximum at these high expression levels and that transcription factors might become limiting to allow further increases in expression which would correlate with the copy number. It seems that already three copies of the *cbh1* promoter give maximal production levels, as suggested by the experiments carried out to increase homologous protein production in *T. reesei* (Karhunen *et al.*, 1993; Paloheimo *et al.*, 1993). Since regulation of expression of the many cellulase genes is coordinate (see Chapter 3) and deletion of some of the cellulase genes including their promoters has been shown to increase expression of some of the remaining cellulases (Suominen *et al.*, 1993), it would be of interest to examine whether increased expression levels of heterologous genes from the *cbh1* promoter could be obtained in strains deleted for the major cellulases.

In addition to these examples, Cayla Ltd has obtained good yields when producing human lysozyme in *T. reesei* as fusions to CBHI, *A. niger* glucoamylase or the bacterial phleomycin resistance protein.

17.4 Conclusions and future aspects

There is no reason to believe that the amounts of active secreted calf chymosin produced by *T. reesei* could not be raised to commercially viable levels if further optimization of culture conditions and mutagenesis of the recombinant strain were to be carried out, as was done with *A. niger* var. *awamori* in order to reach the commercial levels of several g/l (Dunn-Coleman *et al.*, 1991; Ward *et al.*, 1990). In addition, the reports on Fab production were the first ones to demonstrate that multichain molecules can be assembled into active forms and secreted in filamentous fungi at promising levels. These levels are comparable to the best yields obtained with hybridoma technology, the advantage of *Trichoderma* being the more robust and inexpensive cultivation. Because of the importance of antibodies in biotechnology and medicine, it would be important to test also the capacity of *T. reesei* to produce larger antibody molecules with several S–S bridges which might be not so easily produced in *E. coli*.

Despite the promising first results, there is still a considerable lack of information in order to establish *Trichoderma* as a heterologous production host for other than fungal proteins. After the initial results, no further optimization has been carried out either at a genetic or a cultivation level to demonstrate the limits of the production capacity of the fungus. Furthermore, the existing examples are only a few, on account of the small research community and the hesitancy of those not familiar with the system to study and develop it further. The essential tools and potential are available but the investment in filamentous fungal systems is magnitudes lower than that put into systems based on *E. coli*, insect and animal cell cultures and even the yeasts *Saccharomyces*, *Pichia* and *Hansenula*.

Since the protein fusion strategies have proven useful in heterologous protein production there is reason to optimize and develop the fusion and cleavage methods for *Trichoderma* systems further. Although the *in vivo* cleavage is designed to provide directly the authentic product, *in vitro* cleavage systems might be beneficial in cases where the fusion partner is also needed for purification of the product from the enzyme mixture produced by *Trichoderma*. For instance, the CBHI–Fab molecules could be purified based on the methods for CBHI (Nyyssönen *et al.*, 1993). Commercial proteases are being developed for more efficient *in vitro* cleavage, and it might be of interest to test also the yet unknown endogenous protease which cleaved the antibody–CBHI fusion. Using cellulase-negative strains in foreign protein production might increase the yields as discussed earlier and might also make purification of the product easier. It is important to remember, however, that if cellulose – or another plant material-based medium – is used, inactivation of cellulase genes can lead to reduced growth and also to retarded and possibly lower induction of the *cbh1* promoter (see Chapter 3). Understanding more about regulation of expression of the cellulase promoters, the inducer molecules involved and the intracellular signalling pathways would thus also contribute to heterologous protein production.

T. reesei naturally produces only low levels of extracellular proteins on glucose-based media since expression of many of the genes is under glucose repression (see Chapter 3). Using promoters functioning on glucose could, in principle, give products free of most of the endogenous proteins and could also be beneficial in production of a desired homologous protein free of unwanted side activities. Unfortunately, the production levels are not yet comparable to those obtained on cellulase-inducing conditions and are currently not economical for bulk products. Whether this is due to the weakness of the promoters currently available or due to physiological reasons demands further studies. The expression level of *cbh1* is among the highest in all organisms, and it would be interesting to see whether it is possible to construct a modified promoter or a strain with equally high expression levels on glucose-based medium. One of the factors most probably affecting expression levels is chromatin structure, which might play a role also in *T. reesei* in position- or carbon source-dependent gene expression. In order to create expression cassettes with position-independent high levels of expression, matrix attachment regions (MARs) have been isolated from *T. reesei* (Belshaw *et al.*, 1997). Their presence in vectors increases the transformation frequency, probably due to higher expression of the marker gene, but their effect on protein production has yet to be tested.

In cellulase-inducing conditions, the extracellular proteins of *T. reesei* constitute half of all protein synthesized by the fungus, and CBHI forms the major part. This raises a question as to whether the protein synthesis and secretion capacity is already at its maximum. If this is the case, overexpression of activator proteins (putative genes cloned at VTT) in order to increase the now observed limitations in transcription from the *cbh1* promoter may not increase protein yields. Physiological studies on protein production have been neglected for some years but should provide useful information. These studies can now be combined with molecular data which provides also the means to modify the production properties of the strains. *In vivo* labeling experiments can be used to measure kinetics of cellular events involved in heterologous protein production and carried out in combination with physiological studies performed in well-controlled bioreactor cultivations. These studies are expected to give insight into physiological states of the fungus both on glucose-based and other media and to provide information on genetic modification possibilities to improve the productivities.

One of the most interesting and special features of filamentous fungi is their filamentous mode of growth. Very little is known of which parts of the hyphae are productive in terms of transcription, translation and secretion, and it would be important to obtain this type of information from bioreactor cultivations. It is believed that secretion occurs in hyphal tips, which necessitates directed cellular trafficking of the secreted molecules, but it is still possible that larger parts of the mycelium take part in this process. Ultrastructural analysis would be useful in determining possible bottlenecks such as intracellular accumulation of heterologous molecules in certain cellular compartments. Although analysis of cell extracts and various intracellular maturation forms with *in vivo* labeling would give useful information concerning the intracellular fate of the molecules, these studies fail to give any idea about the possible differences of the hyphal parts in this respect. These questions are important when considering the effects of age or branching patterns of the mycelium in protein production. Image analysis can be used to study mycelial samples from bioreactor cultivations (Lejeune *et al.*, 1995) and to correlate the hyphal growth rates and branching patterns with protein productivity.

The major bottlenecks in heterologous protein production are generally believed to occur in the ER, where folding and the early steps of glycosylation are interconnected and can occur concomitant with translocation. Accumulation of incorrectly folded proteins in the ER leads to induction of the "unfolded protein response" and/or degradation of the proteins. Improvement of folding would thus be of interest in heterologous protein production. There are also reports that suggest that the amount of glycan precursors in the ER would limit the secretion capacity of *T. reesei* (Kruszewska *et al.*, 1990; J. Kruszewska and G. Palamarczyk, personal communication). One interesting future research topic would be to study the interactions between secretion, glycosylation and folding, how the possible accumulation of foreign proteins in the ER affects these phenomena, and whether there is a feedback to translation or translocation. The first attempts on the way to understanding the molecular biology of these events has been taken by cloning the *T. reesei mpg1* gene involved in the early steps of glycosylation (Kruszewska *et al.*, submitted), the *pdi1* gene involved in protein folding (Saloheimo *et al.*, 1997), genes such as *sar1* (Veldhuisen *et al.*, 1997), *ypt1* and *sec18* (M. Saloheimo, unpublished) involved in different steps of the secretory pathway and genes involved in translation initiation (M. Saloheimo and E. Alatalo, unpublished).

Glycosylation also is an important parameter when estimating the suitability of an organism for production of proteins, especially those aimed at pharmaceutical applications. Information on glycan structures has been almost completely lacking for *Trichoderma* and other filamentous fungi. Consequently, the recent analysis of glycan structures of CBHI, including NMR analysis (Maras *et al.*, 1997; see Volume 1, Chapter 6) is a significant contribution to the field. There are also other studies that show that glycan structures can vary remarkably depending on the strain and possibly also the culture conditions (H. Nevalainen, J. Ståhlberg, R. Contreras, M. Maras, personal communication). This makes it important to choose the right *T. reesei* strain as a host. Above all, the possibility of modifying *in vivo* or *in vitro* the glycan structures to resemble mammalian types (R. Contreras and M. Maras, personal communication) provides interesting future prospects for heterologous protein production by *T. reesei.*

References

Belshaw, N. J., Hakola, S., Nevalainen, H., Penttilä, M., Suominen, P., and Archer, D. 1997. *Trichoderma reesei* sequences that enchance transformation frequency bind to the nuclear matrix. Mol. Gen. Genet. **256**: 18–27.

Calmels, T., Martin, F., Durand, H., and Tiraby, G. 1991a. Proteolytic events in the processing of secreted proteins in fungi. *J. Biotechnol.* **17**: 51–66.

Calmels, T., Parriche, M., Durand, H., and Tiraby, G. 1991b. High efficiency transformation of *Tolypocladium geodes* conidiospores to phleomycin resistance. *Curr. Genet.* **20**: 309–314.

Cullen, D., Gray, G. L., Wilson, L. J., Hayenga, K. J., Lamsa, M. H., Rey, M. W., Norton, S., and Berka, R. M. 1987. Controlled expression and secretion of bovine chymosin in *Aspergillus nidulans. Bio/Technology* **5**: 369–376.

Demolder, J., Saelens, X., Penttilä, M., Fiers, W., and Contreras, R. 1994. KEX2-like processing of glucoamylase-interleukin 6 and cellobiohydrolase-interleukin 6 fusion proteins by *Trichoderma reesei.* 2nd European Conference on Fungal Genetics, Lunteren, The Netherlands. Abstract B38.

Dhanaraj, R. R., Pitts, J. E., Nugent, P., Orprayoon, P., Cooper, J. B., Blundell, T. L., Uusitalo, J., and Penttilä, M. 1995. Protein engineering of surface loops: preliminary X-ray analysis of the CHY 155-165RHI mutant. *Adv. Exp. Med. Biol.* **362**: 95–99.

Dunn-Coleman, N. S., Bloebaum, P., Berka, R. M., Bodie, E., Robinson, N., Armstrong, G., Ward, M., Przetak, M., Carter, G. L., LaCost, R., Wilson, L. J., Kodama, K. H., Baliu, E. F., Bower, B., Lamsa, M., and Heinsohn, H. 1991. Commercial levels of chymosin by *Aspergillus. Bio/Technology* **9**: 976–981.

Durand, H., Clanet, M., and Tiraby, G. 1988. Genetic improvement of *Trichoderma reesei* for large scale cellulase production. *Enzyme Microb. Technol.* **10**: 341–345.

Emtage, J. S., Angal, S., Doel, M. T., Harris, T. J. R., Jenkins, B., Lilley, G., and Lowe, P. A. 1983. Synthesis of calf prochymosin (prorennin) in *Escherichia coli. Proc. Natl. Acad. Sci. USA* **80**: 3671–3675.

Harkki, A., Uusitalo, J., Bailey, M., Penttilä, M., and Knowles, J. K. C. 1989. A novel fungal expression system: secretion of active calf chymosin from the filamentous fungus *Trichoderma reesei. Bio/Technology* **7**: 596–603.

Harkki, A., Mäntylä, A., Penttilä, M., Muttilainen, S., Bühler, R., Suominen, P., Knowles, J., and Nevalainen, H. 1991. Genetic engineering of *Trichoderma* to produce strains with novel cellulase profiles. *Enzyme Microb. Technol.* **13**: 227–233.

Ilmén, M., Thrane, C., and Penttilä, M. 1996a. The glucose repressor gene *cre1* of *Trichoderma*: isolation and expression of a full-length and a truncated mutant form. *Mol. Gen. Genet.* **251**: 451–460.

Ilmén, M., Onnela, M.-L., Klemsdal, S., Keränen, S., and Penttilä, M. 1996b. Functional analysis of the cellobiohydrolase I promoter of the filamentous fungus *Trichoderma reesei. Mol. Gen. Genet.* **253**: 303–314.

Joutsjoki, V., Torkkeli, T., and Nevalainen, H. 1993. Transformation of *Trichoderma reesei* with the *Hormoconis resinae* glucoamylase P (*gamP*) gene: production of a heterologous glucoamylase by *Trichoderma reesei. Curr. Genet.* **24**: 223–229.

Karhunen, T., Mäntylä, A., Nevalainen, K. M. H., and Suominen, P. 1993. High frequency one-step gene replacement in *Trichoderma reesei.* I: Endoglucanase overproduction. *Mol. Gen. Genet.* **241**: 515–522.

Keränen, S. and Penttilä, M. 1995. Production of recombinant proteins in the filamentous fungus *Trichoderma reesei. Curr. Opin. Biotechnol.* **6**: 534–537.

Kruszewska, J. S., Palamarczyk, G., and Kubicek, C. P. 1990. Stimulation of exoprotein secretion by choline and Tween 80 in *Trichoderma reesei* QM9414 correlates with increased activities of dolichol phosphate mannose synthase. *J. Gen. Microbiol.* **136**: 1293–1298.

KRUSZEWSKA, J. S., SALOHEIMO, M., PENTTILÄ, M., and PALAMARCZYK, G. Isolation of a *Trichoderma reesei* cDNA encoding GTP:α-D-mannose-1-phosphate guanyltransferase involved in early steps of protein glycosylation. Manuscript submitted.

KURZATKOWSKI, W., TÖRRÖNEN, A., FILIPEK, J., MACH, R., HERZOG, P., SOWKA, S., and KUBICEK, C. P. 1996. Glucose-induced secretion of *Trichoderma reesei* xylanases. *Appl. Environ. Microbiol.* **62**: 2859–2865.

LEJEUNE, R., NIELSEN, J., and BARON, V. B. 1995. Morphology of *Trichoderma reesei* QM9414 in submerged cultures. *Biotechnol. Bioeng.* **47**: 609–615.

MARAS, M., DE BRYUN, A., SCHRAML, A., HERDEWIJN, P., CLAEYSSENS, M., FIERS, W., and CONTRERAS, R. 1997. Structural characterization of N-linked oligosaccharides from cellobiohydrolase I secreted by the filamentous fungus *Trichoderma reesei* RUTC30. Eur. J. Biochem., **275**: 617–625.

MARGOLLES-CLARK, E., HAYES, C. K., HARMAN, G. E., and PENTTILÄ, M. 1996. Improved production of *Trichoderma harzianum* endochitinase by expression in *Trichoderma reesei. Appl. Environ. Microbiol.* **62**: 2145–2151.

NAKARI-SETÄLÄ, T. and PENTTILÄ, M. 1995. Production of *Trichoderma reesei* cellulases on glucose-containing media. *Appl. Environ. Microbiol.* **61**: 3650–3655.

NAKARI, T., ALATALO, E., and PENTTILÄ, M. 1993a. Isolation of *Trichoderma reesei* genes highly expressed on glucose containing media: characterization of the *tef1* gene encoding translation elongation factor 1α. *Gene* **136**: 313–318.

NAKARI, T., ONNELA, M.-L., ILMÉN, M., and PENTTILÄ, M. 1993b. New *Trichoderma* promoters for production of hydrolytic enzymes on glucose medium. In P. Suominen and T. Reinikainen (eds), Proceedings of the 2nd Tricel Symposium, Majvik, Finland. Foundation for Biotechnical and Industrial Fermentation Research, Helsinki, Vol. 8, pp. 239–246.

NAKARI-SETÄLÄ, T., ARO, N., KALKKINEN, N., ALATALO, E., and PENTTILÄ, M. 1996. Genetic and biochemical characterization of the *Trichoderma reesei* hydrophobin HFBI. *Eur. J. Biochem.* **235**: 248–255.

NEVALAINEN, H. and PENTTILÄ, M. 1995. Molecular biology of cellulolytic fungi. In U. Kück (ed.), *The Mycota II: Genetics and Biotechnology.* Springer-Verlag, Berlin/Heidelberg, pp. 303–319.

NEVALAINEN, H., HARKKI, A., PENTTILÄ, M., SALOHEIMO, M., TEERI, T., and KNOWLES, J. 1990. *Trichoderma reesei* as a production organism for enzymes for the pulp and paper industry. In T. K. Kirk and H.-M. Chang (eds), *Biotechnology of Pulp and Paper Manufacture.* Butterworth-Heinemann, Stoneham, MA, pp. 593–599.

NYKÄNEN, M., SAARELAINEN, R., RAUDASKOSKI, M., NEVALAINEN, K. M. H., and MIKKONEN, A. 1997. Expression and secretion of barley cysteine endopeptidase B and cellobiohydrolase I in *Trichoderma reesei. Appl. Environ Microbiol.* **63**: 4929–4937.

NYYSSÖNEN, E. 1993. Monoclonal antibodies: production and use in studies of the rate-limiting steps in heterologous protein production by the filamentous fungus *Trichoderma reesei.* PhD Thesis. University of Helsinki. 92 pp.

NYYSSÖNEN, E. and KERÄNEN, S. 1995. Multiple roles of the cellulase CBHI in enhancing production of fusion antibodies by the filamentous fungus *Trichoderma reesei. Curr. Genet.* **28**: 71–79.

NYYSSÖNEN, E., PENTTILÄ, M., HARKKI, A., SALOHEIMO, A., KNOWLES, J. K. C., and KERÄNEN, S. 1993. Efficient production of antibody fragments by the filamentous fungus *Trichoderma reesei. Bio/Technology* **11**: 591–595.

Paloheimo, M., Miettinen-Oinonen, A., Torkkeli, T., Nevalainen, H., and Suominen, P. 1993. Enzyme production in *T. reesei* using the *cbh1* promoter. In P. Suominen and T. Reinikainen (eds), Proceedings of the 2nd Tricel Symposium, Majvik, Finland. Foundation for Biotechnical and Industrial Fermentation Research, Helsinki, Vol. 8, pp. 229–237.

Parriche, M., Bousson, J.-C., Baron, M., and Tiraby, G. 1996. Development of heterologous protein secretion systems in filamentous fungi. 3rd European Conference on Fungal Genetics, Münster, Germany, 1996. Abstracts, p. 52.

Penttilä, M. E., André, L., Saloheimo, M., Lehtovaara, P., and Knowles, J. K. C. 1987. Expression of two *Trichoderma reesei* endoglucanases in the yeast *Saccharomyces cerevisiae*. *Yeast* **3**: 175–185.

Penttilä, M. E., André, L., Lehtovaara, P., Bailey, M., Teeri, T. T., and Knowles, J. K. 1988. Efficient secretion of two fungal cellobiohydrolases by *Saccharomyces cerevisiae*. *Gene* **63**: 103–112.

Pitts, J. E., Quinn, D., Uusitalo, J., and Penttilä, M. 1991. Protein engineering of chymosin and expression in *Trichoderma reesei*. *Biochem. Soc. Trans.* **19**: 663–665.

Pitts, J. E., Uusitalo, J. M., Mantafounis, D., Nugent, P. G., Quinn, D. D., Orprayoon, P., and Penttilä, M. E. 1993. Expression and characterization of chymosin pH optima mutants produced in *Trichoderma reesei*. *J. Biotechnol.* **28**: 69–83.

Saloheimo, M. and Niku-Paavola, M.-L. 1991. Heterologous production of a ligninolytic enzyme: expression of the *Phlebia radiata* laccase gene in *Trichoderma reesei*. *Bio/Technology* **9**: 987–990.

Saloheimo, M., Lund, M., and Penttilä, M. 1997. Characterization of the *pdi1* gene of the filamentous fungus *Trichoderma reesei*. Meeting abstracts, Protein Folding, Modification and Transport in the Early Secretory Pathway, Taos, USA, 3–9 March.

Saloheimo, M., Bajaras, V., Niku-Paavola, M.-L., and Knowles, J. K. C. 1989. A lignin peroxidase-encoding cDNA from the white-rot fungus *Phlebia radiata*: characterization and expression in *Trichoderma reesei*. *Gene* **85**: 343–351.

Salovuori, I., Makarow, M., Rauvala, H., Knowles, J., and Kääriäinen, L. 1987. Low molecular weight high-mannose type glycans in a secreted protein of the filamentous fungus *Trichoderma reesei*. *Bio/Technology* **5**: 152–156.

Schindler, M., Mach, R., Vollenhofer, S., Hodits, R., Gruber, F., de Graaff, L., and Kubicek, C. 1993. Characterization of the pyruvate kinase gene (*pki1*) of *Trichoderma reesei*. *Gene* **130**: 271–275.

Smith, R. A., Duncan, M. J., and Moir, D. T. 1985. Heterologous protein secretion from yeast. *Science* **229**: 1219–1224.

Stålbrand, H., Saloheimo, A., Vehmaanperä, J., Henrissat, B., and Penttilä, M. 1995. Cloning and expression in *Saccharomyces cerevisiae* of a *Trichoderma reesei* β-mannanase gene containing a cellulose binding domain. *Appl. Environ. Microbiol.* **61**: 1090–1097.

Suominen, P. L., Mäntylä, A. L., Karhunen, T., Hakola, S., and Nevalainen, H. 1993. High-frequency one-step gene replacement in *Trichoderma reesei*. II: Effects of deletions of individual cellulase genes. *Mol. Gen. Genet.* **241**: 523–530.

Uusitalo, J. M., Nevalainen, K. M. H., Harkki, A. M., Knowles, J. K. C., and Penttilä, M. E. 1991. Enzyme production with recombinant *Trichoderma reesei* strains. *J. Biotechnol.* **17**: 35–50.

Vanhanen, S. 1991. Isolation and characterization of genes involved in basic metabolism of the filamentous fungus *Trichoderma reesei*. Ph. D. Thesis, VTT Publications 75, Espoo, Finland, 119 pp.

Vanhanen, S., Penttilä, M., Lehtovaara, P., and Knowles, J. 1989. Isolation and characterization of the 3-phosphoglycerate kinase gene (*pgk*) from the filamentous fungus *Trichoderma reesei*. *Curr. Genet.* **15**: 181–186.

VANHANEN, S., SALOHEIMO, A., ILMÉN, M., KNOWLES, J. K. C., and PENTTILÄ, M. 1991. Promoter structure and expression of the 3-phosphoglycerate kinase-encoding gene (*pgk1*) of *Trichoderma reesei*. *Gene* **106**: 129–133.

VELDHUISEN, G., SALOHEIMO, M., FIERS, M. A., PUNT, P. J., CONTRERAS, R., PENTTILÄ, M., and VAN DEN HONDEL, C. A. M. J. J. 1997. Isolation and analysis of functional homologues of the secretion related *SAR1* gene of *S. cerevisiae* from *Aspergillus niger* and *Trichoderma reesei*. Mol. Gen. Genet. **256**: 446–455.

WARD, M., WILSON, L. J., KODAMA, K. H., REY, M. W., and BERKA, R. M. 1990. Improved production of chymosin in *Aspergillus* by expression as a glucoamylase-chymosin fusion. *Bio/Technology* **8**: 435–440.

Index